# HANDBOOK OF ANIMAL MODELS IN ALZHEIMER'S DISEASE

# Advances in Alzheimer's Disease

*Advances in Alzheimer's Disease* brings together the latest insights in Alzheimer's disease research in specific areas in which major advances have been made. This book series assembles and builds on work recently published in the *Journal of Alzheimer's Disease* (JAD) and also includes further contributions to ensure comprehensive coverage of the topic. The emphasis is on the development of novel approaches to understanding and treating Alzheimer's and related diseases.

Series Editors:
George Perry, Ph.D. and Mark A. Smith, Ph.D.†

Volume 1

ISSN 2210-5727 (print)
ISSN 2210-5735 (online)

# Handbook of Animal Models in Alzheimer's Disease

Edited by

**Gemma Casadesus, Ph.D.**
*Assistant Professor of Neurosciences, Director CWRU Rodent Behavior Core,
Case Western Reserve University, Cleveland, OH 44106, USA*

Amsterdam • Berlin • Tokyo • Washington, DC

© 2011 IOS Press and the authors.

All rights reserved. No part of this book may be reproduced, stored in a retrieval system, or transmitted, in any form or by any means, without prior written permission from the publisher.

ISBN 978-1-60750-732-1 (print)
ISBN 978-1-60750-733-8 (online)
Library of Congress Control Number: 2011928925

*Publisher*
IOS Press BV
Nieuwe Hemweg 6B
1013 BG Amsterdam
The Netherlands
fax: +31 20 687 0019
e-mail: order@iospress.nl

*Distributor in the USA and Canada*
IOS Press, Inc.
4502 Rachael Manor Drive
Fairfax, VA 22032
USA
fax: +1 703 323 3668
e-mail: iosbooks@iospress.com

LEGAL NOTICE
The publisher is not responsible for the use which might be made of the following information.

PRINTED IN THE NETHERLANDS

# Dedication

This handbook is in honor and memory of my husband and colleague, Mark A Smith, who tragically passed away on December 19, 2010. With his death, Mark left a huge void in the field of neurodegeneration but also left an incredible legacy to look up to and carry forward.

Mark was an incredibly prolific investigator and a significant contributor to the field of Alzheimer's disease. In just 20 years, he rose to become one of the most highly cited scientists in the field, authored over 800 publications, was honored with a multitude of prizes, and served on a number of editorial boards, meetings, and organizations. Scientifically, he gave the field of Alzheimer's disease and neurodegeneration seminal insights on the role of oxidative stress and cell cycle re-entry in AD pathogenesis. However, beyond these remarkable achievements, Mark will also be remembered for his tireless fight to challenge those around him to examine data from all possible angles, to not fall in line with the popular dogma but to try to disprove it. In other words, to do what science stands for at its core. He will be remembered as a creative and generous scientist who thought outside of the box, and through his zest for life and passion for his work, for his ability to deliver to audiences around the world what no one else could. His fearless approach to science has taught the field to keep our eyes on the future and to not sit too long on what is known, to question incessantly, and to keep pushing ahead. To those close to him, he emphasized the importance of generosity, collaboration, and teaching to promote the careers of the next generation. He infected everyone with the same fun, curiosity, and excitement for doing science that he carried inside him and the courage and freedom to stand true to one's convictions. To me, he has left the amazing love, strength, and knowledge that he gave me every day and SO many laughs to remember him by to continue on this journey. To his sons, he has left freedom and fortitude in their spirits to follow their hearts as they grow and a ferocious love for soccer.

It is so cruel to see such a bright light burn out so fast, but he has left a ***mark*** in all of us and with that the duty to be true to ourselves and to give our best to cure this devastating disease.

Gemma

# Preface

References of experimental use of animals to model diseases, novel experimental procedures, or test novel therapeutics date all the way back to 304–258 BCE. It is undisputable that our ability to model disease in animals has provided major breakthroughs in all fields of biomedical research and has been vastly accelerated by the development of transgenic animals.

The study of neurodegenerative diseases is highly reliant on animal models due to their complexity and plurality of pathology and symptomatology. Today we have amassed a multitude of animal models, developed through genetic, chemical, and/or lesions in multiple species with the goal of faithfully mimicking these diseases and uncover the complex nature of disease-associated mechanisms. Ultimately, the goal is to test promising therapies and manage, prevent, or cure neurodegenerative disease.

The field of neurodegenerative diseases faces unique challenges in this application. First, most animal models in this area, unlike in linear diseases, do not reproduce the full phenotypical disease spectrum. Second, for a given neurodegenerative disease, the etiology and the clinical presentation differ from one patient to the next. As such, while the current models are well suited for the study of specific pathology-driven mechanisms, more notably amyloid-β, tau, or alpha-synuclein, pharmacological testing in animal models of neurodegenerative disease often translates into poorer indices of efficacy when applied to the clinical population. With these advances and challenges in mind, this handbook, written by experts in the field of neurodegeneration, provides a rich and updated overview of a wide range of animal models that are being developed and used to study complex disease dynamics, including but also beyond pathology-associated mechanisms, with the ultimate goal to discover the neuroprotective therapeutics of the future through more accurate translation of basic to clinical outputs.

The first section of this handbook presents an overview of animal models of various species, ranging from higher mammals such as primates or dogs, to knowledge gathered for more prevalent rodent genetically-based models, as well as promising models developed in the rabbit to study metabolic endpoints and therapeutic strategies for AD. Last but not least, this first section includes the review of newer invertebrate animal models, such as *Drosophila* to study neurodegeneration. Invertebrate models provide high-throughput potential, with highly manipulable genetics and functional output that places these models in promising standing within the field.

The second section of this handbook presents the use of animal models to pinpoint disease mechanisms. Pathology driven mechanisms are well represented but not limiting. As we are learning that "bottom-up", overexpression based transgenic models do not provide an accurate representation of therapeutic effectiveness, we have also focused on more "top-down" models, for example those based on metabolic pathological endpoints that exclude pathology as the primary driver of neurodegenerative disease.

Lastly, this handbook concludes with a representation of various therapeutic interventions that are being used in models of neurodegenerative disease. Critical insight on effectiveness of clinically tested therapies in addition to novel, untested ones are aimed at providing both, the necessary critical due diligence when treatments fail – "where have we gone wrong" – "How can we do better" and a glimpse of hope for the future.

The Editor

The contributions in this book are based on articles previously published by IOS Press in the *Journal of Alzheimer's Disease*, and have in most cases been revised and updated.

# Contents

Preface ... vii

## Section 1. Overview of Animal Models of Alzheimer's Disease

Estimation of Working Memory in Macaques for Studying Drugs for the Treatment of Cognitive Disorders ... 3
*Jerry J. Buccafusco*

The Canine Model of Human Aging and Disease ... 15
*Carl W. Cotman and Elizabeth Head*

Unveiling "The Switch" from Aging to Alzheimer's Disease with the Senescence-Accelerated Mouse Model (SAMP8) ... 39
*Jaewon Chang, Merce Pallas, Xiongwei Zhu, Hyun-Jin Kim, Antoni Camins, Hyoung-gon Lee, George Perry, Mark A. Smith and Gemma Casadesus*

Alzheimer's Disease Selective Vulnerability and Modelling in Transgenic Mice ... 49
*Jürgen Götz, Nicole Schonrock, Bryce Vissel and Lars M. Ittner*

The Cholesterol-Fed Rabbit as a Model of AD: The Old, the New and the Pilot ... 59
*D. Larry Sparks*

A Rabbit Model of Alzheimer's Disease: Valid at Neuropathological, Cognitive, and Therapeutic Levels ... 77
*Diana S. Woodruff-Pak, Alexis Agelan and Luis Del Valle*

Transgenic *Drosophila* Models of Alzheimer's Amyloid-β 42 Toxicity ... 89
*Koichi Iijima and Kanae Iijima-Ando*

## Section 2. Using Animal Models to Understand Mechanisms of Disease

Estrogen, Progesterone and Hippocampal Plasticity in Rodent Models ... 109
*Michael R. Foy, Michel Baudry, Roberta Diaz Brinton and Richard F. Thompson*

Apparent Behavioral Benefits of Tau Overexpression in P301L Tau Transgenic Mice ... 129
*Dave Morgan, Sanjay Munireddy, Jennifer Alamed, Jason DeLeon, David M. Diamond, Paula Bickford, Michael Hutton, Jada Lewis, Eileen McGowan and Marcia N. Gordon*

Activation of Cell Cycle Proteins in Transgenic Mice in Response to Neuronal Loss But Not Aβ and Tau Pathology ... 139
*Joao P. Lopes, Mathew Blurton-Jones, Tritia R. Yamasaki, Paula Agostinho and Frank M. LaFerla*

Electron Microscopic 3D Reconstruction Analysis of Amyloid Deposits in 3xTg-AD Mice and Aged Canines ... 149
*Paworn Nuntagij, Naiphinich Kotchabhakdi and Reidun Torp*

The Cholesterol-Fed Rabbit as a Model System for Cholesterol-Alzheimer's Disease Studies 163
*R.P. Jaya Prasanthi, Gurdeep Marwarha and Othman Ghribi*

Hepatic Ceramide May Mediate Brain Insulin Resistance and Neurodegeneration in Type 2 Diabetes and Non-alcoholic Steatohepatitis 179
*Suzanne M. de la Monte, Lascelles E. Lyn-Cook Jr., Margot Lawton, Ming Tong, Elizabeth Silbermann, Lisa Longato, Ping Jiao, Princess Mark, Haiyan Xu and Jack R. Wands*

Animal Model of Insulin-Resistant Brain State: Intracerebroventricular Streptozotocin Injection Deteriorates Alzheimer-Like Changes in Tg2576 APP-Overexpressing Mice 201
*Konstanze Plaschke, Jürgen Kopitz, Markus Siegelin, Reinhard Schliebs, Melita Salkovic-Petrisic, Peter Riedererf and Siegfried Hoyer*

Hippocampal Alterations in Rats Submitted to Streptozotocin-Induced Dementia Model: Neuroprotection with Aminoguanidine 215
*Letícia Rodrigues, Regina Biasibetti, Alessandra Swarowsky, Marina C. Leite, André Quincozes-Santos, Matilde Achaval and Carlos-Alberto Gonçalve*

**Section 3. Alzheimer's Disease Therapies Using Animal Models**

Developing Immunotherapies for Alzheimer's Disease Using a Cholesterol-Fed Rabbit Model in the Context of Th cell Differentiation 231
*Richard Coico and Diana Woodruff-Pak*

Anti-Amyloid-β Immunotherapy in Alzheimer's Disease: Relevance of Transgenic Mouse Studies to Clinical Trials 247
*Donna M. Wilcock and Carol A. Colton*

Heterogeneity in Red Wine Polyphenolic Contents Differentially Influences Alzheimer's Disease-Type Neuropathology and Cognitive Deterioration 263
*Lap Ho, Ling Hong Chen, Jun Wang, Wei Zhao, Stephen T. Talcott, Kenjiro Ono, David Teplow, Nelson Humala, Alice Cheng, Susan S. Percival, Mario Ferruzzi, Elsa Janle, Dara L. Dickstein and Giulio Maria Pasinetti*

Delivery of NGF to the Brain: Intranasal Versus Ocular Administration in Anti-NGF Transgenic Mice 277
*Simona Capsoni, Sonia Covaceuszach, Gabriele Ugolini, Francesca Spirito, Domenico Vignone, Barbara Stefanini, Gianluca Amato and Antonino Cattaneo*

Cholinomimetic Actions of Memantine on Learning and Hippocampal Plasticity 297
*Benjamin Drever, William Anderson, Helena Johnson, Matthew O'Callaghan, Sangwan Seo, Deog-Young Choi, Gernot Riedel and Bettina Platt*

Cognitive Performances of Cholinergically Depleted Rats Following Chronic Donepezil Administration 317
*Debora Cutuli, Francesca Foti, Laura Mandolesi, Paola De Bartolo, Francesca Gelfo, Daniela Laricchiuta and Laura Petrosini*

Subject Index 337

Author Index 339

# Section 1
# Overview of Animal Models of Alzheimer's Disease

# Estimation of Working Memory in Macaques for Studying Drugs for the Treatment of Cognitive Disorders

Jerry J. Buccafusco[a,b,*]
[a] Alzheimer's Research Center, Medical College of Georgia, Augusta, GA, USA
[b] Charlie Norwood VA Medical Center, Augusta, GA, USA

**Abstract.** Non-human primates have served as subjects for studies of the cognition-enhancing potential of novel pharmacological agents for over 25 years. Only recently has a greater appreciation of the translational applicability of this model been realized. Though most Old-World monkeys do not appear to acquire an Alzheimer's-like syndrome in old age, their value resides in the brain physiology they have in common with humans. Paradigms like the delayed matching-to-sample task engender behavior that models aspects of working memory that are substrates for the actions of cognition-enhancing drugs. Our studies have provided information relevant to factors that limit the effectiveness of clinical trial design for compounds that potentially improve cognition. For example, cognition-enhancing compounds from different pharmacological classes, when administered to monkeys, can exhibit remarkable pharmacodynamic effects that outlast the presence of the drug in the body. Studies with non-human primates also can provide information regarding dose ranges and individual subject sensitivity experienced in the clinic. Components of working memory are differentially sensitive to drug effects and may be characterized by different dose ranges for certain compounds, even within the same task. Examples are provided that underscore the possible idiosyncrasies of drug action in the pharmacology of cognition –which could be of critical importance in the design of clinical trials.

Keywords: Acetylcholinesterase inhibition, aging, Alzheimer's disease, attention deficit disorder, delayed matching, drug development, nicotinic acetylcholine receptors, non-human primates, working memory

## INTRODUCTION

The ability to maintain optimal cognitive function over the entire life span has become one of the primary concerns of humans. Any number of disease entities and physical or chemical traumas can lead to a failure to maintain cognitive ability, but by far the greatest contributor is the aging process. Because of great medical achievements, and with the ever increasing life expectancy, the loss of brain function with age and disease stands to mitigate other progress made in human health and longevity. Important components of cognition include the ability to attend, to learn, and to remember. A wide variety of animal models and behavioral techniques have been applied to the study of drugs that affect memory. Animals of advanced age, usually rodents and non-human primates, have provided a good level of predictability for the clinical efficacy of proposed therapeutic agents. In fact, many drug discovery programs continue to use rodents in general screening procedures for identifying potential cognitive enhancing agents, electing to continue with subsequent testing of potential lead compounds in non-human primates. It has been our experience that evaluation of such compounds in non-human primates allows for a greater level of predictability in terms of clinical potency and efficacy as compared with lower species. Various operant tasks, usually food-motivated, allow for the measurement of

---

*Address for correspondence: Jerry J. Buccafusco, Ph.D., Director, Alzheimer's Research Center, Medical College of Georgia, Augusta, GA 30912-2300, USA. Tel.: +1 706 721 6355; Fax: +1 706 721 9861; E-mail: jbuccafu@mcg.edu.

abilities which are relevant to human aging such as attention, strategy formation, reaction time in complex situations and memory for recent events (e.g., [13,18]). Aged monkeys generally are impaired in their ability to attain efficient performance of these tasks, and they often exhibit a reduced level of task efficiency relative to their younger cohorts [9]. Rhesus monkeys over 19 years old show both the behavioral and neuropathological signs of aging. In our lab, we (as other have) identified significant amyloid-plaque-like lesions in areas of the brain important for cognitive function and in areas that are equivalent to those reported for humans [23, 24]. Whereas plaque density does not always correlate with pre-morbid degrees of behavioral impairment (as it often fails to correlate in the human condition), it is more likely that age-related defects in cognitive function are associated with more subtle changes in brain neurochemistry and pathology [32]. Aging in rhesus monkeys is not usually associated with the rapid cognitive decline commonly observed in Alzheimer's disease patients, and as such, these animals are perhaps even more useful when modeling the slower cognitive decline associated with "normal" aging in humans.

The complex functions involved in cognition and memory are known to occur over a widely varying anatomical architecture and over numerous neural pathways. A variety of neurotransmitter molecules and secondary messenger systems also appear to play myriad roles. Over the past 20 years, this laboratory has been fortunate in having an almost unique opportunity to study the potential cognitive-enhancing actions of a large number of standard and novel pharmacological agents in young and aged Old-World monkeys. These agents represent drugs from different chemical, and thus pharmacological, classes. From these studies (and from others in the field) six statements may be made with reasonable certainty: 1) there exist a large number of potential neural targets for cognitive/memory enhancement; 2) operant tasks in non-human primates provide a high degree of predictability for the clinical effectiveness of cognition-enhancing drugs in humans; 3) there appears to be a dose limit (usually very close to the maximal therapeutic effect) for each compound; 4) certain drugs that possess memory-enhancing actions can induce protracted improvements lasting long after the disappearance of the compound from the body; 5) the dose-response relationship for certain cognition-enhancing drugs can be different for the components of memory; and 6) cognition-enhancing drugs are effective independent of the age of the subject.

## NEURAL TARGETS FOR COGNITION/MEMORY ENHANCEMENT

There exist a surprising number of potential molecular targets for drug development in the field of cognition pharmacology [6,10]. The participation of numerous neurotransmitter substances involved in cognition and the large number of molecular therapeutic targets for neurodegenerative disorders such as Alzheimer's disease is perhaps not too surprising, since memory is dependent upon several distinct processes, and different types of memory are relegated to different (but sometimes overlapping) brain regions. For example, components of the hippocampal formation have been implicated in processing spatial, declarative, and episodic types of memory in humans, primates, and rodents [12, 21,25,30,33]. A reasonable argument has been made for the possibility that the hippocampus does not play as important a role in semantic memory [11,31], with habit learning more dependent upon the striatum [26]. Also, emotional or conditioning learning processes appear to reside within the amygdala [3]. Even within what has been termed working memory, there are separable and interacting components that may include discrimination/attention, encoding, retention, and retrieval [18].

Much early memory research was focused on the cholinergic system; not too surprising as the rather selective loss and reduced function of basal forebrain cholinergic neurons were shown to play a role in the cognitive deficits associated with Alzheimer's disease. Accordingly, the acetylcholinesterase inhibitors were the first, at least partly, successful drug class to be used for the treatment of Alzheimer's disease. Indeed, the "cholinergic hypothesis" of Alzheimer's disease was successfully tested by using such compounds in animal models and in normal humans, and eventually individuals with Alzheimer's disease [28]. Studies by Bartus and his colleagues (for review [2]) in the 1980s clearly demonstrated the utility of aged monkeys in predicting the human pharmacology of cholinesterase inhibitors and other cholinergic compounds. For example, his research team was first to demonstrate the individual sensitivity to the therapeutic effects of physostigmine in aged Cebus monkeys in their performance of an automated delayed response task. They coined the term Best Dose; defined as that dose of compound producing the most effective response for the individual subject. These findings preceded clinical studies that demonstrated not only the effectiveness but also the limitations of cholinesterase inhibitors for cognition enhancement, including individual sensitivity to the beneficial effects. The Bartus laboratory went on to study the potential cognitive enhancing actions of other cholinergic and non-cholinergic compounds [2]. In their model, only the cholinergic drugs (cholinesterase inhibitors and muscarinic receptor agonists) produced improvements in task accuracy. However, subtype selective compounds that activate or inhibit a wide variety of neurotransmitter receptors were not available at the time.

The field has significantly advanced since the first centrally-acting cholinergic drugs were tested in animals and in the clinic. Table 1 presents a list of potential cognition-enhancing compounds from several pharmacological classes that have been evaluated in essentially the same delayed matching-to-sample (DMTS) task that we routinely use. These compounds represent only a small number of the compounds evaluated at our Primate Center since 1990. The compounds that could not be listed represent proprietary molecules still held or under evaluation by pharmaceutical companies and other concerns. However, even this partial listing provides support for the concept that there exist many potential drug targets for improving cognition and memory. After years of study, there is still no drug target or pathway that appears to stand alone as the most important in this regard. Much of this target uncertainty is compound-dependent, and off-target actions often limit compound effectiveness (see below). Targets that have staying power include both muscarinic and nicotinic cholinergic receptors. The discovery of subtypes of each major class of receptor has greatly increased the number of targets for potential for drug development. Disadvantages with ligand gated ion channel receptors, e.g., nicotinic acetylcholine receptors, include the uncertainty regarding their individual stoichiometry (channel receptors are composed of different subunits which can come together to form a variety of active ion channels with differing affinities for ligands, and differing biophysical properties such as ion selectivity), their anatomical distribution, and their relative expression patterns in the human brain [17]. Accordingly primate models have demonstrated greater levels of clinical predictability than rodent models. Indeed, we had personal experience with proprietary compounds initially found effective in rodent models that subsequently failed to have similar effects in monkeys. In these few cases where compounds exhibited effectiveness in rodent models, but not in monkeys, but still in any case progressed to clinical trials, the clinical results have been less than satisfactory. One compound listed in Table 1 that exemplifies this scenario is the cholinesterase inhibitor velnacrine (HP-029). This compound showed very good promise in rodent models, but was not particularly effective in monkeys. The compound was modestly effective in humans and did not progress very far in clinical trials.

## THE DELAYED MATCHING-TO-SAMPLE (DMTS) TASK

The subjects are macaques (either *Macaca mulatta* or *Macaca nemestrina*) well trained (>100 individual sessions) in the DMTS task. The animals are maintained on tap water (unlimited) and standard laboratory monkey chow (Harlan Teklad Laboratory monkey diet, Madison, WI) supplemented with fruits and vegetables. Food is removed from cages at about 0630 hours, and replaced after the completion of testing of all subjects for the day (at about 1630 hours). Additional nourishment is derived from 300 mg reinforcement food pellets (commercial composition of standard monkey chow and banana flakes, Noyes Precision food pellets, P.J. Noyes Co., Lancaster, NH) obtained during experimental sessions. On weekends, animals are fed without time restrictions. Room temperature and humidity are maintained at $22 \pm 0.6°C$ and $52 \pm 2\%$, respectively. Test panels attached to each animal's home cage present the task by using a computer-automated system. The test systems include touch-sensitive screen (15 inch AccuTouch LCD Panelmount TouchMonitor)/pellet dispenser units (Med Associates) mounted in light-weight aluminum chasses. The stimuli include red, blue, and yellow rectangles presented against a black background. A trial is initiated by presentation of a sample stimulus composed of one of the three colors. The sample stimulus (located above and centered between the two choice stimuli) remains in view until the monkey touches the screen within the borders of the sample rectangle to initiate a pre-programmed delay interval. Touching a stimulus gives the illusion that the figure was actually depressed. Following the delay interval, the two choice stimuli are presented. One of the two choice colors is presented so that the color of one stimulus matches the color of the sample stimulus. A correct (matching) choice is reinforced. Non-matching choices are neither reinforced nor punished. The inter-trial interval is 5 sec and each session consists of 96 trials. The presentation of stimulus color, choice colors, and choice position are fully counterbalanced so as to relegate non-matching (mediating) strategies to chance

Table 1
Comparison of outcomes of drug trials in non-human primates and humans

| | Observations in Non-human primates | Observations in humans |
|---|---|---|
| *Nicotinic cholinergic* | | |
| Nicotine | ↑ DMTS accuracy | ↑ Cognition in AD |
| ABT-418 | ↑ DMTS accuracy | ↑ Cognition in AD |
| ABT-089 | ↑ Distractor DMTS accuracy | ↑ Attention in ADHD |
| ABT-594 ($\alpha 4\beta 2$) | ↑ DMTS accuracy | Analgesic |
| A-582941 ($\alpha 7$) | ↑ DMTS accuracy | |
| GTS-21 (partial $\alpha 7$) | ↑ DMTS accuracy | ↑ Cognition in healthy vols; ↑ Cognition in SCHIZ |
| Cotinine | ↑ Distractor DMTS accuracy/↑ DMTS accuracy | Tested as anti-smoking aid – no side effects |
| Isoarecolone | ↑ DMTS accuracy | |
| SIB-1553A | ↑ Distractor DMTS accuracy/↑ DMTS accuracy | ↑ Cognition in healthy vols; ↑ Attention |
| Varenicline | ↑ DMTS accuracy | Anti-smoking aid |
| MEM 3454 (partial $\alpha 7$) | ↑ DMTS accuracy | Phase 2A POC in cognition for schizophrenia |
| *Muscarinic Cholinergic* | | |
| WAY-132983 (M1 preferring agonist) | ↑ DMTS accuracy – GI side effects | Withdrawn from clinical trials |
| Talsaclidine (M1 preferring agonist) | ↑ DMTS accuracy – GI side effects | Withdrawn from clinical trials |
| *Cholinesterase Inhibitor* | | |
| Physostigmine | ↑ DMTS accuracy | ↑ Cognition in AD |
| Velnacrine (HP-029) | Modest effect at best dose | Modest ↑ cognition in AD |
| Ladostigil (TV3326) + MAO inhibitory activity | ↑ DMTS accuracy | |
| Donepezil | Modest ↑ DMTS accuracy | Approved for AD |
| *Adrenergic agonist* | | |
| Clonidine | ↑ DMTS accuracy | ↑ Working memory in PD |
| Methylphenidate | ↑ Distractor DMTS accuracy | Effective treatment for ADHD |
| *Serotonin receptor* | | |
| RS-56812 (5-HT$_3$ receptor antagonist) | ↑ DMTS accuracy | |
| RS 17017 (5-HT$_4$ receptor agonist) | ↑ DMTS accuracy | |
| EMD 281014 (5HT$_{2A}$ antagonist) | ↑ DMTS accuracy | |
| Lecozotan (SRA-333) (5-HT$_{1A}$ antagonist) | ↑ DMTS accuracy | Phase I |
| *Glutamate receptor* | | |
| IDRA 21 (positive modulator AMPA receptor) | ↑ DMTS accuracy | |
| Memantine | Modest ↑ DMTS accuracy | Approved for AD |
| *Aiminopyridine* | | |
| Besipiridine (HP-749) | No effect on DMTS in NHPs | Withdrawn from clinical trials |
| *Amnestic agents* | | |
| Scopolamine | Amnestic | Amnestic |
| Mecamylamine | Amnestic | Amnestic |
| Ketamine | Amnestic | Amnestic |

Abbreviations: DMTS – delayed matching-to-sample; AD – Alzheimer's disease; Vols – volunteers; SCHIZ – schizophrenia; Phase 2A POC – initial efficacy clinical trial - proof of concept; GI – gastro-intestinal; ↑ – increase; PD – Parkinson's disease; ADHD – attention deficit hyperactivity disorder; NHPs – non-human primates.

levels of accuracy. Three to five different presentation sequences are rotated through each daily session to prevent the subjects from memorizing the first several trials. Delay intervals are established during several non-drug or vehicle sessions prior to initiating the study. The duration for each delay interval is adjusted for each subject until three levels of group performance accuracy are approximated: zero delay (85–100% of trials answered correctly); short delay interval (75–84% correct); medium delay interval (65–74% correct); and long delay interval (55–64% correct). The assignment of delay intervals is necessary to avoid ceiling effects in the most proficient animals during drug studies, and to insure that each animal begins testing at relatively the same level of task difficulty. Failure to respond during a trial initiates the next trial in the sequence. The % trials correct is determined only from the total number of trials actually completed.

Distractor stimuli (interference trials) are presented on 24 of the 96 trials completed during distractor DMTS sessions. The stimuli are presented simultaneously on the sample and choice keys for 3 sec and they consist of a random pattern of the three colored rectangles flashing in an alternating manner. The distractor rectangles are comprised of the same 3 colors used for sample and choice stimuli presentation. The total duration of presentation for a given colored light is 0.33 sec. Distractor stimuli are presented an equal number of times on trials with Short, Medium, and Long delay intervals. The distractor sequence is initiated 1 second into the delay interval.

## TRANSLATIONAL EFFECTIVENESS OF DMTS STUDIES IN MONKEYS

Table 1 also compares the overall effects of compounds in DMTS studies in monkeys with their outcome in clinical trials. The clear take-away message is that compounds that are effective in improving cognitive performance in monkeys as assessed in the DMTS task are often also effective in humans. This relationship also extends to drugs used for the treatment of attention deficit disorders. For the purpose of modeling Attention Deficit Hyperactivity Disorder, we use a distractor DMTS task that includes task relevant distractors [19]. Aged monkeys are more significantly impaired during distractor trials than younger animals and their performance deficit is more difficult to reverse by pharmacological means [19]. Therefore we mainly use young adult animals as test subjects for the distractor DMTS task. As indicated in Table 1, compounds that effectively improve attention in attention deficit disorders can ameliorate distractor-impaired accuracy in the monkeys. As above, treatment-induced changes in the memory retention curves for the standard DMTS task can offer some clues to the components of memory most affected by the drug. Indeed many compounds that we found to reverse distractibility had earlier been shown to improve Short delay trial accuracy during the standard DMTS task. An example is the nicotinic receptor agonist ABT-089 which improved Short delay accuracies [16] and was later demonstrated to reverse distractor-impaired accuracy in the distractor-DMTS task [20].

Part of the translational effectiveness of our primate model goes beyond its predictability for cognition enhancement. Animals often stop or reduce responding after they sense discomfort caused by the treatment. This type of behavior occurs well prior to any serious toxic effect of the treatment encountered during dosing. As an example, animals stopped or reduced testing after specific doses of either talsaclidine or WAY-132983. Both compounds are partially selective M1 muscarinic acetylcholine receptor agonists and both were effective cognition-enhancing agents in the animals at the lower doses tested. However as the dose was increased in an attempt to further increase task performance, the animals exhibited side effects, possibly related to overlapping effects at M3 receptors [1,27]. The side effects observed in monkeys predicted the appearance of similar side effects reported in clinical trials which proved too severe to permit further human study. Thus the translational benefit of pre-clinical studies with non-human primates extends to the prediction of clinically-encountered side effects, some of which might not be obvious in rodent studies.

## DOSE LIMITATIONS FOR COGNITION-ENHANCING AGENTS

One of the characteristic features of the pharmacological profiles for cognition or memory enhancing compounds is the inverted U-shaped relationship. After the maximum in task accuracy is attained the efficacy returns towards baseline as the dose is increased. Thus, cognition enhancing agents are often limited by a narrow dose range for therapeutic efficacy [7]. Assessment of the maximal therapeutic potential also is complicated by individual subject sensitivity to the drug. Together, these factors – narrow therapeutic window and individual sensitivity – cause difficulties in the evaluation of drug efficacy and in determining a proper dosing strategy for patients. It has been suggested that narrow dose windows might derive from the onset of side effects as the dose is increased. This is certainly possible considering the somatic and parasympathomimetic side effects produced by acetylcholinesterase inhibition. But a variety of potential cognition-enhancing compounds from many other pharmacological classes exhibit similarly narrow effective dose windows. Thus, even if it is not possible to improve on the efficacy of the compounds evaluated to date, new drugs that exhibit wider dose windows for their positive mnemonic actions would have a clear therapeutic advantage. The preclinical evaluation of new cognition-enhancing agents in the primate DMTS task has the advantage of employing doses ranges closest to those used ultimately in humans, and exhibit wider effective dose windows

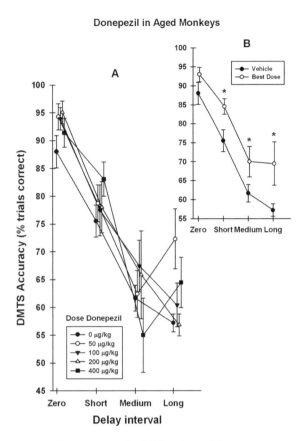

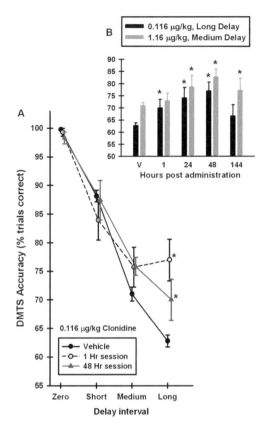

Fig. 1. Effect of donepezil administration to 7 Rhesus macaques (10–28 years of age) on their performance of a delayed matching-to-sample (DMTS) task. Donepezil was administered by the oral route (compound dissolved in orange juice) 30 minutes prior to testing. A. Comparison of all four doses of donepezil with vehicle (0 μg/kg). Each value indicates the mean ± S.E.M., and symbols were slightly displaced so that the error bars could be visualized. There were no statistical differences among the treatments, either alone, or across delay intervals ($P > 0.10$). B. The effect of the individual Best Dose of donepezil on DMTS accuracies. The Best Dose was that dose of compound producing the most effective response for the individual subject. There was a statistically significant effect of the Best Dose relative to vehicle ($F_{1,3} = 17.4$, $P < 0.0001$). *Significantly different ($P < 0.05$) from respective mean vehicle accuracy.

Fig. 2. Effect of clonidine administration to 6 pigtail monkeys (Macaca nemestrina) (6–20 years of age) on their performance of a delayed matching-to-sample (DMTS) task. Clonidine (0.116 and 1.16 μg/kg) was administered 1 hour prior to testing. Data were also obtained 24 hours, 48 hours, and 6 days after drug administration. A. Comparison of sessions run 1 and 48 hours after the 0.116 μg/kg dose of clonidine. There was a statistically significant effect of clonidine treatment ($F_{2,5} = 4.80$, $P = 0.0093$) and a significant effect of treatment across delay intervals ($F_{6,171} = 5.46$, $P < 0.0001$). *Significantly different ($P < 0.05$) from respective mean vehicle accuracy. Also there was a nearly statistically significant difference between the mean accuracies during Long delay trials between the 1 hour and the 48 hour time points after clonidine administration ($t = 1.97$, $P = 0.051$) B. Task accuracies after the 0.116 μg/kg dose during Long delay trials; and after the 1.16 μg/kg dose during Medium delay trials (Medium delay trials were improved to a greater extent by clonidine than were the Long delay trials after the 1.16 mg/kg dose – data not shown) plotted as a function of the time after clonidine was administered. *Significantly different from respective (0 μg/kg) vehicle (V) mean ($P < 0.05$). Each value indicates the mean ± S.E.M., and symbols were slightly displaced so that the error bars could be visualized.

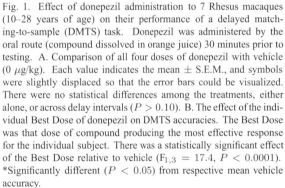

that would be encountered in clinical trials. But there is more to defining the broadness of the therapeutic window than encountering side effects at higher doses. Though it is possible to envision some mnemonic mechanism that could become saturated as the dose is increased, the downturn in efficacy at high doses is more difficult to explain.

Because of the often individual differences in sensitivity to cognition-enhancing agents, it is sometimes difficult to evaluate the true efficacy of a drug, particularly when making comparisons across pharmacological classes. In evaluating new compounds in our model, some of the most promising drugs are those for which determination of a statistically significant effect of drug treatment versus baseline performance is obtained during dose-response studies. This outcome usually suggests a combination of reasonable efficacy and low response variability. Perhaps less promising are those compounds that initially fail to show a significant improvement in task accuracy, but for which selection of a Best Dose results in a statistically significant improvement. This is the case for most of the acetylcholinesterase inhibitors that we have evaluated. We define the Best Dose for each subject as that which evokes the greatest improvement in accuracy for the average of all four delay intervals. There are other calculations that could be applied to dose-response data sets, e.g., the most effective dose associated with the group's most improved delay interval, or the most effective dose associated with the individual's most improved delay interval. However defined, the Best Dose removes the intra-subject variability from consideration. This concept is illustrated in Fig. 1 which summarizes the effects of four oral doses of donepezil on the mean accuracies achieved by seven aged macaques in their performance of the DMTS task. The initial dose-response data (Fig. 1A) resulted in no statistically significant differences from vehicle (0 $\mu$g/kg) even though there were some suggestive trends towards improvement. In contrast, selection of a single Best Dose for each subject resulted in highly significant improvements in accuracies relative to vehicle ($F_{1,6} = 17.4$, $P < 0.0001$). Indeed, the entire retention curve (accuracy vs. delay interval) was shifted to the right of vehicle, with perhaps the greatest effect for the Long delay trial accuracy (Fig. 1B). If anything, these results indicate that careful titration of dosage along with careful estimation of some indices of cognition could maximize effectiveness of this class of agents in patients. As importantly, the average Best Dose of donepezil in this study was 129 $\mu$g/kg – a very good estimate of the clinical dose range for the orally-administered drug: $\sim$70–140 $\mu$g/kg.

## PROTRACTED IMPROVEMENT IN TASK ACCURACY BY CERTAIN COGNITION-ENHANCING AGENTS

Most often, the magnitude of the pro-mnemonic action produced by cognition-enhancing drugs is greatest at the time of testing just after administration. But on the day or days following drug administration, accuracy is again at baseline. However we have encountered compounds for which the opposite was the case. One of these compounds was the partial $\alpha 7$ nicotinic acetylcholine receptor agonist GTS-21 [4]. It has not been possible to predict, even knowing the pharmacological or chemical class, or knowing the profile of responsiveness in sessions run shortly after drug administration, the potential for a compound to produce a protracted improvement in cognition [8]. Thus, we routinely run sessions on the day or days after drug administration to directly assess this potential. The pharmacokinetic-pharmacodynamic mismatch for cognition-enhancing drugs was first noted for nicotine [5]. Nicotine's short plasma half-life did not predict the protracted improvement in task accuracies in Rhesus monkeys. Levin and colleagues [15] reported similar findings for nicotine in rats. In fact, they demonstrated that the protracted beneficial effect of nicotine on radial arm maze accuracy was not dependent upon the presence of the drug at the time of behavioral training. But all compounds that act on nicotine acetylcholine receptors do not share nicotine's prolonged actions. The nicotinic receptor agonists – nicotine, ABT-089, and isoarecolone – maintained the same relative order in terms of mnemonic effectiveness during the first test sessions as they did when animals were tested on the day after drug administration. In contrast, SIB-1553A and ABT-418, which were very effective when tested on the day of treatment, evoked no significant task improvement when tested 24 hours later. Perhaps even more surprisingly, for the analogs GTS-21 and 4OH-GTS-21, the degree of task improvement measured 24 hours after treatment were, on average, greater than those seen on the first day of treatment [8]. The mechanism underlying the protracted positive mnemonic actions produced by nicotine has not been determined, though we have speculated that the mechanism could include the induction of long-term potentiation (LTP), a strengthening of synaptic connections that is associated with learning and memory formation [8]. The compound most exemplifying this protracted improvement in cognitive enhancing action is the $\alpha_2$-adrenergic receptor agonist clonidine. In our standard paradigm, a single injection of clonidine produced a time-dependent improvement in task accuracies, with the maximal improvement occurring during sessions run three days after administration. Some residual improvement continued to be evident on the sixth day after administration (Fig. 2). Though clonidine perhaps represents an extreme example of the complexities of the pharmacodynamic effects of cognition-enhancing drugs, consideration of the pharmacodynamic actions of potential cognition enhancing agents could improve the outcome of clinical trials for many compounds developed for the treatment of Alzheimer's disease and other disorders of cognition. Here again the value of preclinical evaluations of these types of novel compounds in non-human primates is underscored.

## THE DOSE-RESPONSE RELATIONSHIP CAN BE DIFFERENT FOR EACH OF THE COMPONENTS OF MEMORY

As noted above, one of the advantages of using operant tasks that include delay intervals is that some information can be gained regarding the component of working memory affected by drug treatment: discrimination/attention, encoding, and retention. In some tasks, like our distractor-DMTS task, both attentional and working memory components can be assessed within the same session. One of the more interesting observations made in assessing these type of data is that dose-response relationships can differ for different components of working memory, even within the same task. This is exemplified in Fig. 3 which summarizes the dose-response data for a novel cognition-enhancing compound JWB1-84-1 [22]. Mean improvements in task accuracy during Short and Medium delay intervals were maximal in the higher dose range – about 100 μg/kg. Mean improvement in task accuracy during Long delay intervals were maximal in the lower dose range – about 10 μg/kg. JWB1-84-1 improved working memory in the standard DTMS task, and the compound also reversed distractor-impaired accuracies in the distractor-version of the task. These results suggest that the doses appropriate for treating general cognitive impairment and attention deficits could be quite different.

Another, somewhat different example of this phenomenon is exemplified by our studies with nicotine [20]. Figure 4 compares the dose-response relationship for nicotine between non-distractor trials and distractor trials in a distractor-DMTS task performed by adult Rhesus monkeys. Like the previous example, the data are derived from within the same task. In this case, however, the comparison is between more attentional (distractor trials) versus the more working memory (non-distractor trials) components of the task. Again there was a clear difference between dose sensitivity for the two conditions. Compound reversibility of the distractor-impaired accuracy during Short delay intervals was maximal at the highest (20 μg/kg) dose. Mean improvement in non-distractor task accuracy during Long delay intervals was maximal at the 5 μg/kg dose. Thus, not only was there a clear differential in terms of dose-sensitivity within the distractor-DMTS task, but the results were similar to those in the previous example with JWB1-84-1. It would be difficult, if not impossible, to predict these dose sensitivity issues in advance of actually performing the animal studies.

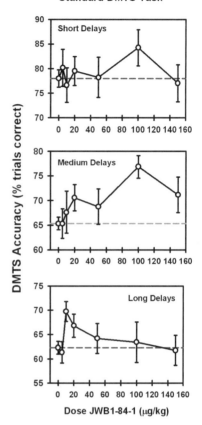

Fig. 3. Effect of JWB1-84-1 administration to 11 Rhesus monkeys (18–31 years of age) on their performance of a delayed matching-to-sample (DMTS) task. JWB1-84-1 was administered by the intramuscular route (thigh muscle) 10 minutes prior to testing. Changes in task accuracy are plotted as a function of dose. Mean improvements in task accuracy during Short and Medium delay intervals were maximal in the higher dose range – about 100 μg/kg. Mean improvement in task accuracy during Long delay intervals were maximal in the lower dose range – about 10 μg/kg. Each value indicates the mean ± S.E.M. These results suggest that the dose-range for treating general cognitive impairment and attention deficits could be quite different.

The data are of significant importance for the design of clinical studies which are most critical from a safety and cost consideration.

## COGNITION-ENHANCING DRUGS CAN BE EFFECTIVE INDEPENDENT OF THE AGE OF THE SUBJECT

Since our first studies with nicotine in the late 1980's, we noted that nicotine could improve DMTS accuracies in both young and aged monkeys. These early results were not congruent with those from rodent mod-

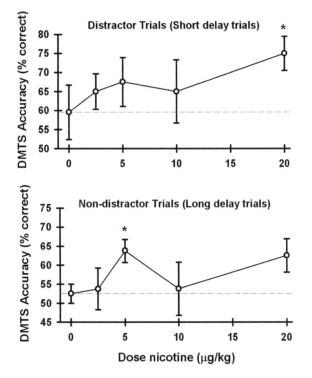

Fig. 4. Effect of nicotine administration to 6 Rhesus monkeys on their performance of the distractor version of the delayed matching-to-sample (DMTS) task. The task is divided into 24 randomly presented distractor trials and 72 non-distractor trials. Nicotine 5–20 μg/kg was administered 10 minutes before testing. Mean accuracy values are presented only for the delay interval associated with the respective maximal improvement in accuracy – Short delay trials for distractor trials, and Long delay intervals for non-distractor trials. Each value indicates the mean ± S.E.M. *Significantly different ($P < 0.05$) from respective mean vehicle (0 μg/kg) accuracy.

els wherein it was difficult to show positive mnemonic actions of cognition-enhancing agents in un-impaired rodents. In order to show improved memory some memory-impairing process was often used prior to dosing with the test compound. Some of these approaches included treatment with scopolamine, chemical or electrolytic lesions of specific brain regions such as the nucleus basalis or medial septum, or more recently the use of a transgenic murine model of high cerebral amyloid deposition. Also aged rodents have been shown to be impaired in memory-dependent behavioral tasks (for review [29]). The problem with using un-impaired young rodents was that they often performed the tasks too efficiently such that a "ceiling" effect was evident. For many rodent tasks, it is difficult to increase task difficulty sufficiently to avoid this ceiling effect. This is particularly problematic when longitudinal studies are required and for which baseline creep over time and repetition is inevitable for young healthy subjects.

In our primate DMTS task, all subjects' performance efficiencies are normalized to a common memory retention algorithm so that irrespective of baseline proficiency, all animals experience the same memory load prior to drug studies (see Fig. 1). This is accomplished by adjusting each subject's delay intervals (time intervals between sample presentation and choice presentation) so that baseline accuracies fit the algorithm prior to drug study [14]. Delay intervals can be further adjusted between studies should the animals' performance become more efficient over time. Using this approach we have noted that, in general, young monkeys perform more efficiently on the DMTS task than do aged subjects, i.e., aged animals generally require shorter delay intervals to achieve the same accuracies as younger animals [9]. However, there are individuals that do not conform to this general view, such that there are young animals that exhibit poor task efficiency and aged animals that exhibit very good task efficiency. This situation makes aging studies in monkeys difficult, since it is often difficult to know the intelligence of an aged subject when it was young – as most aged monkeys are acquired when they are adults. Thus it is not often clear whether an aged subject is truly age-impaired in task performance or whether the subject would have been poor at task performance even when young.

By pushing the mnemonic limit through the use of progressively longer delay intervals, ample room for improvement for the effects of cognition-enhancing compounds both for young and aged subjects can be established. Returning to the donepezil example, this compound was studied in 24 subjects over several years. The data were divided into those derived from 10 younger subjects (10.8 ± 1.03 years of age) and 14 older subjects (28.4 ± 0.78 years of age). Donepezil (10, 25, 50, and 100 μg/kg) was administered intramuscularly 15 minutes prior to testing. As with the data derived from the oral administration study used to generate Fig. 1, there were no significant drug treatment effects for either age group with respect to their control baselines (data not shown). Selection of a Best Dose for each cohort resulted in the data presented in Fig. 5. The Best Dose of donepezil produced significant improvements in task accuracies for the young ($F_{2,6} = 3.37$, $P = 0.037$) and the old ($F_{2,6} = 11.8$, $P < 0.0001$) groups and there were no significant differences between the two groups. Thus donepezil was equally effective in the two age groups supporting the original premise. The mean value for the Best Dose obtained for the young subjects was 44 μg/kg. The mean value for the Best Dose obtained for the old sub-

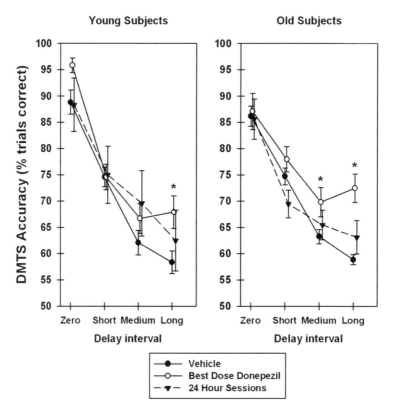

Fig. 5. Effect of donepezil administration to 24 macaques (10 young and 14 old) on their performance of a delayed matching-to-sample (DMTS) task. Young subjects were 5–17 years of age; old subjects were 19–31 years of age. Donepezil was administered by the intramuscular route (thigh muscle) 15 minutes prior to testing. Symbols were slightly displaced so that the error bars could be visualized. *Significantly different ($P < 0.05$) from respective mean vehicle accuracy.

jects was 74 μg/kg; but the difference was not statistically significant ($P = 0.19$). Also of note is the lack of carryover effect produced by donepezil to the sessions run 24 hours after administration. Though acetylcholinesterase inhibitors indirectly activate nicotinic receptors (via the release of acetylcholine), unlike nicotine, donepezil is not able to activate the downstream pathways that mediate the protracted mnemonic response to nicotine. Therefore, the pharmacodynamic effects of new cognition-enhancing compounds cannot be predicted. They can only be discovered by testing in relevant animal models.

## CONCLUSIONS

Despite almost two decades of research the predominant treatment of the cognitive deficits associated with Alzheimer's disease (currently the only approved indication for cognition-enhancing drugs in the US) is based on acetylcholinesterase inhibition. There is almost an embarrassment of riches in terms of potential drug targets for improving cognition and memory. Yet translation of this information to the clinic has been dismal. Whereas some of the clinical failures have been based on side effect issues – many central nervous system targets are not dissimilar from off-target molecules in the peripheral nervous system – it is also highly likely that clinical failures could be due to species differences between rodents and humans in terms of target distribution and expression. Perhaps even more importantly, inattention to the pharmacological idiosyncrasies of cognition enhancing agents described above has contributed to the formulation of non-optimal regimens for clinical studies. Though intensive pre-clinical study in non-human primate models will not mitigate all risk in down-stream drug development, the outcome of primate studies should provide for more rational decision-making in selecting lead compounds, in predicting potential side effects in human subjects, and in developing optimal dose regimens for clinical trials. Reliance simply on pharmacokinetic profiles of new compounds to establish dose regimens for clinical trials is problematic when the therapeutic agent is designed to improve cognition and memory in humans. It is clear that drugs developed to enhance cognition will work maximally in humans exhibiting mild forms of impairment. By late stage Alzheimer's disease, it is too much to ask of the currently available Alzheimer's medications. Cognition enhancing drugs work equally well to improve accuracy in both young and aged monkeys; occasionally the effects are superior in younger animals. Thus it is important that earlier stages of cognitive impairment such as "mild cognitive impairment" or "benign senescent forgetfulness" should be approved as disease entities deserving of treatment. Finally, there should be no hesitation about using cognition enhancing drugs as early as possible in approved disease states as patients will derive the most benefit early on when the quality of life may be most important.

## ACKNOWLEDGMENTS

The author would like to thank all who have participated in our primate studies as colleagues, fellows, graduate students, and especially the excellent animal study technicians – currently Nancy Kille and Donna Blessing. Throughout the years this work has been funded by many pharmaceutical companies, federal, and private funding agencies. A portion of the work described for the compound JWB1-84-1 was funded through a grant from the Institute for the Study of Aging. Partial salary support for the author was provided by a Merit Review Award from the Department of Veterans Affairs. The work described above was carried out without significant financial interest that could present an actual or potential conflict of interest for the author.

## References

[1] A.C. Bartolomeo, H. Morris, J.J. Buccafusco, N. Kille, S. Rosenzweig-Lipson, M.G. Husbands, A.L. Sabb, M. Abou-Gharbia, J.A. Moyer and C.A Boast, The preclinical pharmacological profile of WAY-132983, a potent M1 preferring agonist, *J Pharmacol Exp Ther* **292** (2000), 584-596.

[2] R.T. Bartus, On neurodegenerative diseases, models, and treatment strategies: lessons learned and lessons forgotten a generation following the cholinergic hypothesis, *Exp Neurol* **163** (2000), 495-529.

[3] A. Bechara, D. Tranei, H. Damasio, R. Adolphs, C. Rockland and A.R. Damasio, Double dissociation of conditioning and declarative knowledge relative to the amygdala and hippocampus in humans, *Science* **269** (1995), 1115-1118.

[4] C.A. Briggs, D.J. Anderson, J.D. Brioni, J.J. Buccafusco, M.J. Buckley, J.E. Campbell, M.W. Decker, D. Donnelly-Roberts, R.L. Elliott, M.W. Holladay, Y-H.Hui, W.J. Jackson, D.J.B. Kim, K.C. Marsh, A. O'Neill, M.A. Prendergast, K.B. Ryther, J.P. Sullivan and S.P. Arneric, Functional characterization of the novel neuronal nicotinic acetylcholine receptor ligand GTS-21. *In vitro* and *in vivo*, *Pharmacol Biochem Behav* **57** (1997), 231-241.

[5] J.J. Buccafusco and W.J. Jackson, Beneficial effects of nicotine administered prior to a delayed matching-to-sample task in young and aged monkeys, *Neurobiol Aging* **12** (1991), 233-238.

[6] J.J. Buccafusco and A.V. Terry Jr., Multiple CNS targets for eliciting beneficial effects on memory and cognition, *J Pharmacol Exp Ther* **295** (2000), 438-446.

[7] J.J. Buccafusco and A.V. Terry Jr., Nicotine and cognition in young and aged non-human primates, in: *Nicotine and the Nervous System*, E. Levin, ed., CRC Press, New York, 2002, pp. 179-197.

[8] J.J. Buccafusco, S.R. Letchworth, M. Bencherif and P.M. Lippillo, Long-lasting cognitive improvement with nicotinic receptor agonists: mechanisms of pharmacokinetic-pharmacodynamic discordance, *Trends Pharmacological Sci* **26** (2005), 352-360.

[9] J.J. Buccafusco, Cognitive pharmacology in aging macaques, in: *Animal Models of Cognitive Impairment*, E. Levin and J.J. Buccafusco, eds., CRC Press, New York, 2006, pp. 285-300.

[10] J.J. Buccafusco, Dementia and pharmacotherapy: memory drugs, in: *Handbook of Contemporary Neuropharmacology, Vol. 3*, D.R. Shibley, I. Hanin, M. Kuhar and P. Skolnick, eds, John Wiley & Sons, Inc., Hoboken, New Jersey, 2007, pp. 461-478.

[11] H. Eichenbaum, How does the brain organize memories, *Science* **277** (1997), 330-332.

[12] P.E. Gilbert, R.P. Kesner, and W.E. DeCoteau, Memory for spatial location: Role of the hippocampus in mediating spatial pattern separation. *J Neurosci* **18** (1998), 804-810.

[13] E. Irle, J. Kessler, H.J. Markowitz and W. Hofmann, Primate learning tasks reveal strong impairments in patients with presenile or senile dementia of the Alzheimer's type, *Brain Cogn* **6** (1987), 429-449.

[14] W.J. Jackson, K. Elrod and J.J. Buccafusco, Delayed matching-to-sample in monkeys as a model for learning and memory deficits: Role of brain nicotinic receptors, in: *Novel Approaches to the Treatment of Alzheimer's Disease*, E.M. Meyer, J. Simpkins and J. Yamamoto, eds, Plenum Press, New York, 1989, pp. 39-52.

[15] E.D. Levin, S.J. Briggs, N.C. Christopher and J.E. Rose, Persistence of chronic nicotine-induced cognitive facilitation, *Behav Neural Biol* **58** (1992), 152-158.

[16] N-H. Lin, D.E. Gunn, K.B. Ryther, D.S. Garvey, D.L. Donnelly-Roberts, M.W. Decker, J.D. Brioni, M.J. Buckley, A.D. Rodrigues, K.G. Marsh, D.J. Anderson, J.J. Buccafusco, M.A. Prendergast, J.P. Sullivan, M. Williams, S.P. Arneric and M.W. Holladay, Structure-activitiy studies on 2-methyl-3-(2(S)-pyrrolidinylmethoxy)pyridine (ABT-089): An orally bioavailable 3-pyridyl ether nicotinic acetylcholine receptor ligand with cognitive-enhancing properties, *J Med Chem* **40** (1997), 385-390.

[17] N.S. Millar and P.C. Harkness, Assembly and trafficking of nicotinic acetylcholine receptors, *Mol Membrane Biol* **25** (2008), 279-292.

[18] M.G. Paule, P.J. Bushnell, J.P.J. Maurissen, G.R. Wenger, J.J. Buccafusco, J.J. Chelonis and R. Elliott, Symposium

[18] overview: the use of delayed matching-to-sample procedures in studies of short-term memory in animals and humans, *Neurotoxicol Teratol* **20** (1998), 493-502.
[19] M.A. Prendergast, W.J. Jackson, A.V. Terry Jr., N.J. Kille, S.P. Arneric and J.J. Buccafusco, Age-related: differences in distractibility and response to methylphenidate in monkeys, *Cerebral Cortex* **8** (1998), 164-172.
[20] M.A. Prendergast, W.J. Jackson, A.V. Terry Jr., M.W. Decker, S.A. Arneric and J.J. Buccafusco, Central nicotinic receptor agonists ABT-418, ABT-089, and (-)-Nicotine reduce distractibility in young-adult monkeys, *Psychopharmacology* **136** (1998), 50 58.
[21] G. Riedel, J. Micheau, A.G.M. Lam, E.V.L. Roloff, S.J. Martin, H. Bridge, L. de Hoz, B. Poeschel, J. McCulloch and R.G.M. Morris, Reversible neural inactivation reveals hippocampal participation in several memory processes, *Nat Neurosci* **2** (1999), 898-905.
[22] A. Sood, J.W. Beach, S.J. Webster, A.V. Terry, Jr. and J.J. Buccafusco, The effects of JWB1-84-1 on memory-related task performance by amyloid A$\beta$ transgenic mice and by young and aged monkeys, *Neuropharmacology* **53** (2007), 588-600.
[23] J.B. Summers, M.A. Prendergast, W.D. Hill and J.J.Buccafusco, Localization of ubiquitin in the plaques of five aged primates by dual-label fluroescent immunhistochemistry, *Alzheimers Res* **3** (1997), 11-21.
[24] J.B. Summers, W.D. Hill, M.A. Prendergast and J.J.Buccafusco, Co-localization of apolipoprotein E and beta-amyloid in plaques and cerebral blood vessels of aged non-human primates, *Alzheimers Rep* **1** (1998), 119-128.
[25] E. Sybirska, L. Davachi and P.S. Goldman-Rakic, Prominance of direct entorhinal-CA1 pathway activation in sensorimotor and cognitive tasks revealed by 2-DG functional mapping in nonhuman primates, *J Neurosci* **20** (2000), 5827-5834.
[26] E. Teng, L. Stefanacei, L.R. Squire and S.M. Zola, Contrasting effects on discrimination learning after hippocampal lesions and conjoint hippocampal-caudate lesions in monkeys, *J Neurosci* **20** (2000), 3853-3863.
[27] A.V. Terry Jr., J.J. Buccafusco, F. Borsini and A. Leusch, Improvements in memory-related task performance by aged rhesus monkeys administered the muscarinic M1-preferring agonist talsaclidine, *Psychopharmacology* **162** (2002), 292-300.
[28] A.V. Terry Jr. and J.J. Buccafusco, The cholinergic hypothesis of Alzheimer's disease: recent challenges and their implications for novel drug development, *J Pharmacol Exp Ther* **306** (2003), 821-827.
[29] D. Van Dam and P.P. De Deyn, Drug discovery in dementia: the role of rodent models, *Nat Rev Drug Discovery* **5** (2006), 956-970.
[30] S.D. Vann, M.W. Brown, J.T. Erichsen and J.P. Aggleton, Fos imaging reveals differential patterns of hippocampal and parahippocampal subfield activation in rats in response to different spatial memory tests, *J Neurosci* **20** (2000), 2711-2718
[31] F. Vargha-Khadem, D.G. Gadian, K.E. Watkins, A. Connelly, W. Van Paesschen and M. Mishkin, Differential effects of early hippocampal pathology on episodic and semantic memory, *Science* **277** (1997), 376-380.
[32] M.L. Voytko, Nonhuman primates as models for aging and Alzheimer's disease, *Lab Animal Sci* **48** (1998), 611-617.
[33] S.M. Zola, L.R. Squire, E. Teng, L. Stefanacei, E.A. Buffalo and R.E. Clark, Impaired recognition memory in monkeys after damage limited to the hippocampal region, *J Neurosci* **20** (2000), 451-463.

# The Canine Model of Human Aging and Disease

Carl W. Cotman[a] and Elizabeth Head[a,b,1]
[a]*Institute for Brain Aging & Dementia, Department of Neurology, University of California, Irvine, CA, 92697, USA*
[b]*Sanders-Brown Center on Aging, University of Kentucky, Lexington, KY, 40536, USA*

**Abstract.** Aged dogs (beagles) develop losses in executive function, learning and memory. The severity of decline in these cognitive domains represents a spectrum that captures normal aging, mild cognitive impairment and early/mild Alzheimer disease (AD) in humans. In parallel, dogs naturally accumulate several types of neuropathology (although not all) consistent with human brain aging and AD including cortical atrophy, neuron loss, loss of neurogenesis, beta-amyloid (Aβ) plaques, cerebral amyloid angiopathy, mitochondrial dysfunction and oxidative damage. Many of these neuropathological features correlate with the extent of cognitive decline in a brain region-dependent manner. Dogs are ideally suited for longitudinal studies and we provide a summary of the beneficial effects of an antioxidant diet, of behavioral enrichment and of Aβ immunotherapy. In addition, combinatorial treatment approaches can be a powerful strategy for improving brain function through enhancement of multiple molecular pathways.

Keywords: antioxidants, beagle, behavioral enrichment, beta-amyloid, immunotherapy, learning, memory, neuron loss

## INTRODUCTION

The preservation of cognitive function throughout life is becoming more critical as our population ages and more individuals are reaching extreme old age. The single biggest risk factor for Alzheimer disease (AD) and other neurodegenerative diseases is advanced age. This may suggest that aging itself and associated changes in neuron function may be "permissive" for the development of disease. For example, oxidative damage progressively accumulates in the aging brain in virtually all animal models studies to date, similar to humans, and damaged molecules can lead to impaired neuronal function and vulnerability to the development of AD [1-3]. Alzheimer disease (AD) accounts for the largest proportion of patients with dementia and the number of individuals that will develop the disease is rapidly rising [4]. It is critical to continue to identify risk factors and develop new therapeutics to prevent or treat AD. These therapeutics may target the reduction of specific types of AD pathology (e.g. Aβ immunotherapy) or may modify environmental factors to promote successful aging (e.g. antioxidants/physical exercise). Animal models allow controlled intervention experiments and minimize the time between cognitive testing and neurobiological studies. Despite successful modeling of numerous aspects of aging, no single animal model has fully replicated *all* aspects of human aging and AD. Thus, it is necessary to continue to develop additional models, to take advantage of unique aspects of each model, and to combine information from convergent studies.

Our aging research with dogs suggests that this model complements existing models but with distinctive characteristics that support unique research opportunities. In the early 90's, we initiated behavioral studies in dogs [5-7] with parallel examination of brain tissues from aged dogs [8] with the long-term goal of relating neuropathology to cognitive dysfunction [9-11]. Dogs exhibit age-dependent losses in cognitive function with progressive accumulation of neuropathology, and are uniquely well-suited to studies of dietary, environmental and AD-pathology targeted interventions.

---

[1]Corresponding author.

## THE AGING DOG MODEL

The dog, or *canis lupus familiaris* is a domesticated subspecies of the wolf and includes over 400 different breeds [12]. There can be significant variation in longevity in dogs depending on their breed, their body weight and their environment. Typically, larger breeds of dogs have shorter lifespans than smaller breeds [13-15]. For the majority of the studies we will describe in this review and in many aging studies in other laboratories, beagles are the most commonly used breed of dog. Given the large differences in lifespan across different breeds of dogs, and breed-specific vulnerabilities to age-associated diseases, some caution in extending our results in beagles to other breeds is required. This could be a critical and unique feature to studying dog aging as there is a suggestion that brain pathology, including the age of onset and extent may vary across breeds [16].

In laboratory beagles, those maintained in a kennel setting, the median life-span (i.e. 50% have died) can be as high as 12-14 years depending on the colony evaluated [17,18] (Fig. 1A) and sexual maturity is reached between the ages of 1 to 6 months [19]. Of note, there are examples of "exceptional longevity" in pet/companion dogs [20] consistent with increasing numbers of "oldest-old" human in our aging population [21]. There are several ways to estimate the relationship between human and beagle age. Using linear models, it is estimated that, 5.5-7 human years is approximately equivalent to 1 year of a beagle life [17]. Alternative polynomial modeling suggests that beagles considered to be "aged" are over 9 years of age (in our studies we use a cut-off of 10 years), which represents humans between the ages of 66-96 years [15]. Using this same model, middle aged beagles are between 5-9 years (~40-60 years in humans) and young beagles are under 5 years (<40 years) (Fig. 1B). The most common cause of death in aging beagles is cancer (22-26%), with rates that are comparable to reports in humans [17]. Interestingly, the age of onset and vulnerabilities to specific types of cancer can be quite similar in beagles and humans [17,22]. Additional significant causes of death in aged beagles include convulsive seizures, infection, chronic kidney disease, congestive heart failure, intervertebral disc disease, vascular thromboses and intestinal disorders [17]. However, in studies of beagles outside of a laboratory setting (i.e. pet or companion beagles drawn from veterinary clinics), the median lifespan is lower (e.g. 10 years) as additional events including accidents and behavioral problems can be leading causes of death [23]. It is interesting to consider pet/ companion dogs (and these may represent many different breeds including mixed breeds) in aging studies as environmental factors may be virtually identical in dogs and humans in this setting. As such, clinical populations of companion dogs are a very interesting sample to study and several reports from these types of cohorts are described in this review.

### Does the dog model normal aging and/or Alzheimer disease in humans?

It is interesting to address this question in two ways and that is to distinguish between the cognitive and the neuropathological aspects to both normal human aging and AD. As will be described in the following sections, when we consider cognition, aged dogs (between 10-15 years of age) show a spectrum of intact and impaired function similar to reports in aging humans and those individuals with early/mild AD. In human aging, individual variability in cognitive function is a consistent feature and can include individuals that have intact cognition, those with mild or selective cognitive impairments (MCI) or those with severe cognitive decline (dementia)[24-26]. Dementia in humans can have multiple etiologies with the most common being AD [27]. Thus, individual dogs can be categorized as normally aged whereas another subset is severely impaired relative to their age-matched peers [28].

In terms of neuropathology, our severely impaired dogs exhibit some but not all features of AD in humans [29]. For example, aged impaired dogs develop more severe beta-amyloid (Aβ) pathology than non impaired animals, larger and selective losses in brain tissue volume, cell loss in the dentate gryus and an impaired ability to generate new neurons (see following sections). However, the aged beagle brain does not naturally develop neurofibrillary tangles. Early tangles have been suggested in some dog brains based upon overlapping tau phosphorylation sites with humans [30-34]. One possible reason for the lack of tau pathology in aged dogs relative to humans is significant differences in the tau protein sequence (http://www.ensembl.org/Canis_familiaris/).

Thus, we suggest that the aging beagle represents a spectrum of both normal aging (cognitively intact), clinically similar features of MCI (i.e. selective

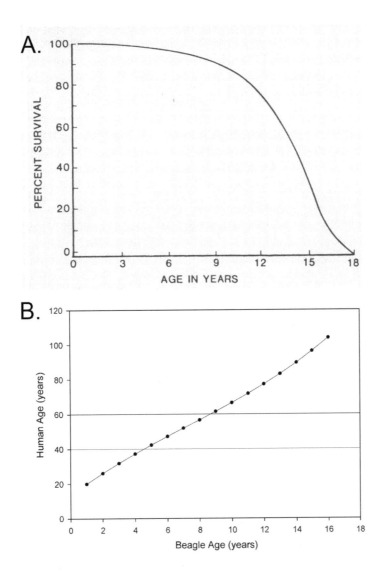

Fig. 1. Dog survival and human aging equivalents. (A). The percent survival of beagles at the Lovelace Respiratory Research Institute as a function of age shows that the median lifespan is approximately 13.2 years (reproduced with permission from [18] with permission from Blackwell Publishing). (B). Estimates of chronological age in beagles and equivalents in human age (reproduced with permission from Gerontological Society of America [15]).

losses in specific cognitive domains)[26] and more global dysfunction (i.e multiple cognitive domains and functional decline) similar to early/mild AD in measures of cognition. Neuropathologically, aged dogs may develop features consistent with normal aging in humans and prodromal AD but given reports of full-blown AD pathology in humans with MCI [35], this appears to be the point in which the dog model diverges from human disease. However, aging dogs might represent very early disease similar to the type of AD neuropathology progression observed in Down syndrome. In this condition, virtually all adults with Down syndrome develop AD pathology by the age of 40 years [36]. However, Aβ plaques appear first with a strikingly similar distribution and appearance to those observed in the aged beagle brain ([36] and see Fig 2D&F). After Aβ plaques appear in the DS brain, neurofibrillary tan-

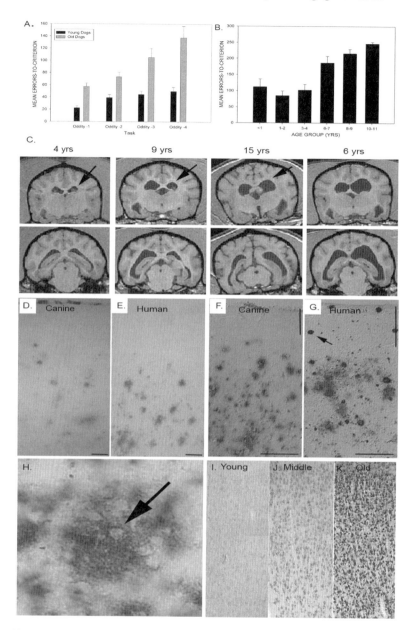

Fig. 2. Features of cognitive and neurobiological aging in beagles. (A). On a complex learning task, the oddity discrimination problem, aged dogs commit more errors when reaching criterion levels of responding relative to young animals. In addition, progressive increases in error scores for each oddity problem suggest that aged dogs solve each successive problem individually whereas young dogs learn the concept of the task. (B). Acquisition of a spatial memory task is age-dependent in dogs and deficits first appear in middle age (6-7 years). (C). Magnetic resonance images (GE Signa 1.5 T) from a 4-year-old, 9-year-old, and 15-year-old dog taken from locations through thalamus and hippocampus. The old dogs show marked increases in ventricular volume and cortical atrophy (with deep gyri and widened sulci). One middle-aged dog (6-year-old) who showed unusually large ventricles also had blood brain barrier dysfunction and early onset β-amyloid accumulation (reproduced with permission [58]). Prefrontal Aβ immunohistochemistry in a cognitive unimpaired 13-year old beagle (D), an unimpaired 90 year old human (E), impaired 12-year old beagle (F) and an 86 year old patient with AD (G) illustrating similar extent and distribution of plaques. The arrow in G shows that dense plaques are observed in human brain in superficial cortical layers but are absent in the dog. H. A higher magnification of Aβ immunostaining within a diffuse plaque characterized by the presence of intact neurons (arrow). Oxidized DNA/RNA (anti-8oxodG immunohistochemistry), representing oxidative damage, increases progressively with age in the prefrontal cortex of a young (I), middle aged (J) and old beagle (K) (unpublished data). Bars represent means and error bars represent standard errors of the mean. Figures are reproduced or modified with permission from Elsevier Limited [44,54,73,197,221].

gles subsequently develop almost a decade later [37,38]. If we extrapolate this to dogs, aged animals with significant Aβ pathology without neurofibrillary tangles may represent the earliest form of AD pathogenesis. More recently, we have described the presence of oligomers in the aged canine brain[39], thought to represent an assembly state of beta-amyloid that impairs synaptic function [40] To our knowledge, there have not been reports of pathology in dogs representing other neurodegenerative disease in humans. Vascular dementia, if present in the aged beagle, has not been described nor systematically evaluated despite the presence of significant cerebral amyloid angiopathy (to be described later).

## Age associated cognitive decline in aging dogs

### Cognitive domain-specific vulnerabilities

Cognitive aging in dogs has several key features including domain specific vulnerabilities and individual variability in the extent of decline. We first reported age-dependent cognitive deficits in dogs in 1994 [7], based on performance on tasks modified from those used in nonhuman primates. Table 1 provides a list of cognitive tasks that we have used to assess function in various cognitive domains in the dog model that will be described, analogous tasks used to measure function in these same cognitive domains in nonhuman primates and humans, and functional localization (when available in dogs). Subsequently we developed additional tasks to detect cognitive deterioration with age, leading to the discovery that aged dogs show deficits in complex learning tasks, including size concept learning [41,42], oddity discrimination learning [43,44] (Fig. 2A), size discrimination learning [11,45] and spatial learning [46] (Fig. 1B). Tasks sensitive to prefrontal cortex function, including reversal learning and visuospatial working memory, also deteriorate with age [11,45,47]. Further, egocentric spatial learning and reversal, measuring the ability of animals to select a correct object based on their own body orientation is also age-sensitive [46]. We have optimized for dogs a measure of spatial attention, landmark discrimination, originally developed in nonhuman primates, and have shown that this cognitive domain is also vulnerable to aging [48,49]. Interestingly, on simple learning tasks and procedural learning measures, aged dogs perform as well as younger animals [7] suggesting that a subset of cognitive functions

remains intact with age and that it is unlikely that sensory deficits are a significant contributor to increased error scores.

Memory also declines with age in dogs. To test memory we use the object recognition task based on a procedure similar to that described in nonhuman primates [50], which employs a delayed non-matching to sample procedure (DNMS) and also reveals age deficits in acquisition [7]. These age-dependent cognitive deficits are not linked to obvious sensory deficits or locomotor impairment [5]. Perhaps the most useful age-sensitive task we have used is a spatial memory task, in which dogs are required to recognize the location of a sample stimulus and then respond to a different location during the test trial. We refer to this as a delayed non-match to position task (DNMP). Results published in 1995 [6] suggested that the task was age-sensitive. We subsequently developed a 3-choice visuospatial working memory task that allows determination of the differential age-dependent strategies (e.g. cognitive or stimulus-dependent strategies) dogs use in solving the problem [51]. Further, this task shows minimal practice effects in longitudinal studies [52]. We have now identified the time course of the development of cognitive decline, and have found that deterioration in spatial ability occurs early in the aging process, between 6 and 7 years of age in dogs [47] (Fig. 2B). Thus, cognitive decline in aged dogs is domain-specific and involves memory and executive function cortical systems as observed in other animal models, in aging humans and in individuals with mild cognitive impairment and AD.

### With age, individual variability in cognitive ability increases in dogs

Cognitive aging in dogs has several key features including domain specific vulnerabilities and individual variability in the extent of decline. We first reported age-dependent cognitive deficits in dogs in 1994 [7], based on performance on tasks modified from those used in nonhuman primates. Table 1 provides a list of cognitive tasks that we have used to assess function in various cognitive domains in the dog model that will be described, analogous tasks used to measure function in these same cognitive domains in nonhuman primates and humans, and functional localization (when available in dogs). Subsequently we developed additional tasks to detect cognitive deterioration with age, leading to the

Table 1
Cognitive domains assessed in dog aging and comparison with nonhuman primate tasks and analogous tasks used in human neuropsychological testing. *Proposed localization – not confirmed in lesions studies in dogs; ** Neuropsychological tasks for humans that assess function in similar cognitive domains

| Cognitive Domain | Dog Task | Localization in Dog Brain | Nonhuman Primate Tasks | Examples of Human Neuropsychological Tasks** |
|---|---|---|---|---|
| Learning | Visual discrimination learning | Medial temporal lobe/parietal lobe* | Visual discrimination learning [223,224] | digit copy, rotary pursuit, face discrimination [225], object discrimination [226,227] |
|  | Reward and object approach learning | Nigrostriatal and motor cortex* | Food pickup task, fine motor learning [228,229] |  |
| Memory | delayed nonmatching to sample acquisition | Rhinal cortex [230] | Object recognition memory task [50] | delayed recognition and recall, digit span [231] |
|  | delayed nonmatching to sample memory | Rhinal cortex [230] | Object recognition memory task [50] |  |
|  | spatial delayed nonmatch to sample acquisition | Dorsolateral prefrontal cortex [230] | Delayed Response Task [232,233] |  |
|  | spatial delayed nonmatch to sample memory | Hippocampus [234] | Delayed Response Task [232,233] |  |
| Executive Function | Visual reversal learning | Prefrontal cortex/ medial temporal lobe [235] | Visual reversal learning [223,224] | card or object sorting tasks, set shifting, response inhibition [236] |
|  | Oddity discrimination | Prefrontal cortex/ medial temporal lobe* | N/A |  |
|  | Egocentric spatial reversal learning | Hippocampal/prefrontal cortex* | Spatial reversal [223] |  |
|  | Size concept learning | Prefrontal cortex/ medial temporal lobe* | Hierarchical/Relational learning [237] |  |
| Visuospatial Function | Landmark discrimination | Prefrontal cortex/ parietal cortex* | Landmark discrimination [238] | Visual construction, block design, spatial learning [226,227] |
|  | Egocentric spatial learning | Hippocampus/medial temporal lobe* | Spatial learning [223] |  |

discovery that aged dogs show deficits in complex learning tasks, including size concept learning [41,42], oddity discrimination learning [43,44] (Fig. 2A), size discrimination learning [11,45] and spatial learning [46] (Fig. 1B). Tasks sensitive to prefrontal cortex function, including reversal learning and visuospatial working memory, also deteriorate with age [11,45,47]. Further, egocentric spatial learning and reversal, measuring the ability of animals to select a correct object based on their own body orientation is also age-sensitive [46]. We have optimized for dogs a measure of spatial attention, landmark discrimination, originally developed in nonhuman primates, and have shown that this cognitive domain is also vulnerable to aging [48,49]. Interestingly, on simple learning tasks and procedural learning measures, aged dogs perform as well as younger animals [7] suggesting that a subset of cognitive functions remains intact with age and that it is unlikely that sensory deficits are a significant contributor to increased error scores.

Memory also declines with age in dogs. To test memory we use the object recognition task based on a procedure similar to that described in nonhuman primates [50], which employs a delayed nonmatching to sample procedure (DNMS) and also reveals age deficits in acquisition [7]. These age-dependent cognitive deficits are not linked to obvious sensory deficits or locomotor impairment [5]. Perhaps the most useful age-sensitive task we have used is a spatial memory task, in which dogs are required to recognize the location of a sample stimulus and then respond to a different location during the test trial. We refer to this as a delayed non-match to position task (DNMP). Results published in 1995 [6] suggested that the task was age-sensitive. We subsequently developed a 3-choice visuospatial working memory task that allows determination of the differential age-dependent strategies (e.g. cognitive or stimulus-dependent strategies) dogs use in solving the problem [51]. Further, this task shows minimal practice effects in longitudinal studies [52]. We have now identified the time course of the development of cognitive decline, and have found that deterioration in spatial ability occurs early in the aging process, between 6 and 7 years of age in dogs [47] (Fig. 2B). Thus, cognitive decline in aged dogs is domain-specific and involves memory and executive function cortical systems as observed in other animal models, in aging humans and in individuals with mild cognitive impairment and AD.

*Summary of neurobiological changes underlying cognitive aging*

There are a number of morphological features of aging in the dog brain that are similar to that observed in both normal aging and in early/mild AD [53-57] and that in descriptive studies correlate with cognitive decline. *In vivo* imaging shows significant cortical atrophy, microscopically there is neuron loss within the hippocampus, reduced neurogenesis, a progressive accumulation of Aβ pathology, cerebrovascular pathology and progressive oxidative damage.

*Cortical atrophy and neuron loss/neurogeneis*
*In vivo* imaging studies show cortical atrophy [58] and ventricular widening [58-60] occurs with age in dogs (Fig. 2C). More recent magnetic resonance imaging studies suggest differential vulnerabilities of specific areas of cortex to aging. For example, the prefrontal cortex loses tissue volume at an earlier age than the hippocampus in aging dogs [61]. Further, more extensive prefrontal cortical volume loss is associated with poorer cognition [61]. However, the hippocampus shows progressive atrophy reaching significantly lower volumes when animals are over 11 years when compared to young adult dogs.

There is a significant association between the extent of cortical atrophy and cognition; animals with more extensive atrophy perform more poorly on measures of learning and memory. These results were confirmed in neurobiological experiments in a study of 30 dogs demonstrating a correlation between cortical atrophy (measured in coronal sections) and cognitive dysfunction [62] similar to that seen in humans [63,64].

Interestingly, cortical and subcortical volume variation in measures of atrophy occur as a function of sex in dogs [65] suggesting differential vulnerabilities to the aging process in both grey and white matter in males and females. For these experiments, voxel based morphometry analyses of magnetic resonance imaging scans was used in a study of 62 beagles (31 males, 31 females) from 6 months to 15 years of age [65]. Although several regions show overlap in the extent of cortical atrophy as a function of age in males and females such as in the parietal cortex, the prefrontal cortex shows larger losses in males relative to females. In contrast, females show larger losses of volume in the temporal cortex. Additional differences were noted in the white matter of aged males and females. Although both males and females showed equivalent atrophy of the internal capsule, females showed greater atrophy of the alveus of the hippocampus relative to males. Males, on the other hand, showed a reduction in the white matter tract volume of the optic nerve bundle. Although we have not observed any significant differences in cognition nor in neuropathology between males and females, these results suggest that there may be differential structural changes with age as a function of sex in dogs. There have been similar patterns of gender-associated structural aging also reported in the human literature (e.g. [66]). Female beagles undergo a progressive rather than abrupt (i.e. menopause) loss of reproductive ability with age but to what extent changes in hormonal status affect brain structures has not been explored.

Cortical atrophy, particularly in the hippocampus may result as a consequence of neuron loss, as reported in normal human brain aging [67,68] with more extensive losses occurring in AD [69,70]. Specifically, in human aging a pattern of neuron loss is observed that can be distinguished from AD. In AD, neurons are lost in area CA1 [69-72] and in the subiculum with more severely affected cases showing losses in CA2 and CA4 of the hippocampus but not in dentate granule cells [70](but see [67] for a different pattern of neuron loss in AD). In contrast, with aging, neurons are lost primarily in the hilus of the dentate gyrus [68,71]. Thus, neurons were counted using unbiased stereological methods within individual subfields of the hippocampus and in the entorhinal cortex of young and aged dogs. The hilus of the dentate gryus showed a significant loss of neurons (~30%) in the aged brain compared to young dogs [73]. Although the sample size was relatively small (n=5 young, n=5 old), there was individual variability in numbers of hilar neurons in aged animals but only one aged dog (14.2 years) had hilar neuron counts within the range of the young dogs. To some extent, this variability related to cognitive function, dogs with higher numbers of neurons performed a size discrimination task, thought to be dependent on the medial temporal lobe, with fewer errors but these results are preliminary. Whether or not old dogs with larger neuron losses might represent severely impaired animals (much like AD) and whether neuron losses might first appear when aged dogs are impaired relative to unimpaired aged dogs, requires a much larger study. Differences in neuron number were not detected in the remaining regions sampled including area CA1, CA3, dentate granule cells, subiculum and entorhinal cortex. Additional work is warranted as neuron loss has been reported in other dog hippocampal aging studies [74,75]. Reduced neurogenesis may also contribute to age-associated cognitive decline. In counts of new neurons in the subgranular layer of the dentate gyrus labeled by bromodeoxyuridine given to animals prior to euthanasia, there was a significant loss in neurogenesis (90-95%) with age in the dogs. Further, the number of new neurons was correlated with cognitive function; animals with lower new neuron numbers had higher error scores of measures of learning and memory sensitive to medial temporal lobe function [76]. Similar losses in neurogenesis have been reported in other laboratories [75,77]. Given that environmental enrichment and physical exercise can increase neurogenesis in rodents [78,79], aged dogs may also be a good model in which to evaluate therapeutics that can enhance neurogenesis and potentially improve cognition.

*Aβ pathology*

Neuron loss and cortical atrophy in vulnerable brain regions of the aged dog may be due to the accumulation of pathological proteins. Dogs deposit endogenous levels of Aβ that has an identical amino acid sequence to humans [80,81] and with similar posttranslational vulnerabilities (e.g. oxidation, racemization, isomerization) [82,83] as they age. The dog β-amyloid precursor protein (APP) is virtually identical to human APP (~98% homology), as established from the sequence published online for the dog genome (http://www.ensembl.org/Canis_familiaris/). Most of the deposits in the dog brain are of the diffuse subtype (Fig. 2H), but are fibrillar at the ultrastructural level and at an advanced stage, which models early plaque formation in humans [84-86]. We also observe intracellular Aβ using immunohistochemistry [53]. Our work and the work of others demonstrate that specific brain regions show differential accumulation of Aβ, paralleling some reports in the aged human brain [81,87-94]. When cortical regions are sampled for Aβ deposition, each region shows a different age of Aβ onset [91]. Aβ deposition occurs earliest in the prefrontal cortex of the dog and later in temporal and occipital cortex, providing us with a rationale for expecting that (1) prefrontal-dependent tasks would deteriorate with age earlier than medial temporal lobe tasks and (2) that development of neuronal dysfunction proceeds in a brain-region specific pattern. Dog Aβ deposition follows a similar although not identical pattern of accumulation as reported in human brain. In human brain, Aβ also appears early in neocortical regions, including the frontal cortex, with entorhinal and hippocampal Aβ appearing later [94]. Braak and colleagues also observe the earliest deposits of Aβ in the neocortex and particularly within basal portions of the frontal, temporal and occipital lobes while the hippocampus remains devoid of pathology until later stages of AD [92]). It is critical to note that initial Aβ deposits occur in a 3- to 4-year time window in dogs that starts between the ages of 8 and 9 years, suggesting that longitudinal studies for evaluating interventions to slow or halt Aβ are feasible.

The extent of Aβ plaque deposition in the dog brain is linked to the severity of cognitive deficits [11,62,95,96]. Age and cognitive status can predict Aβ pathology in discrete brain structures. For example, dogs with prefrontal cortex-dependent reversal learning deficits show significantly higher amounts of Aβ in this brain region [10,11] (Fig. 2D-G). On the other hand, dogs deficient on a size discrimination learning task, thought to be sensitive to temporal lobe function, show large amounts of Aβ deposition in the entorhinal cortex [11]. As in laboratory beagles, the extent of Aβ plaques varies as a function of age in companion dogs (including a wide variety of breeds and mixed breeds) [62,97,98]. Further, the extent of Aβ plaques correlates with behavior changes and this association remains significant even if age is removed as a covariate [62,95]. Aβ plaques in the parietal lobe correlates with behavioral changes in aged companion animals related to appetite, drinking, incontinence, day and night rhythm, social behavior (interaction with owners and other dogs; personality), orientation, perception and memory [62]. Thus, cognitive decline in aged dogs correlates with the age-related accumulation of Aβ as it is with humans [11,99-106]. However, not all reports in studies of human autopsy samples observe a correlation between Aβ pathology and cognition [107-110] and a similar lack of association has been reported in nonhuman primates [111,112]. There may be many different reasons for discrepancies in the human literature including the method used to visualize Aβ, the species of Aβ examined and the measure of cognition [113]. The lack of association between Aβ and cognition in nonhuman primates is interesting and there is a report that nonhuman primate Aβ is primarily the shorter less toxic species Aβ1-40 [114] but these differences in dogs and monkeys is interesting and could be explored further. In addition, there are many other types of neuropathology (described previously and in the upcoming sections) that may also explain some of the individual animal variability in cognitive function.

Interestingly, the accumulation of Aβ plaques in the brains of dogs does not begin until approximately 8 years of age. Thus, cognition declines prior to Aβ plaque accumulation. However in a more recent study, we observed that biochemically extracted Aβ can be detected in middle age [39]. In addition, the ratio of Aβ1-42:Aβ1-40 in the cerebrospinal fluid (CSF) declines linearly with age and is predictive of brain Aβ. Thus, we hypothesize that cognitive impairment may be more tightly coupled to the production of toxic soluble assembly states of Aβ as reported in transgenic mice [115-117]. Indeed, oligomers, a soluble and highly diffusible assembly state of Aβ can be detected in canine brain and cerebrospinal fluid [39]. The link of the extent of oligomers to level of cognitive decline has yet to be established. Alternatively, other pathologies may contribute to cognitive decline in middle age including cerebrovascular dysfunction, progressive oxidative damage and synaptic dysfunction.

*Cerebrovascular pathology*

A common type of pathology observed in both normal human brain aging and particularly in AD is the accumulation of cerebrovascular amyloid angiopathy (CAA) [118-120]. CAA, involving the deposition of Aβ in association with blood vessels, may compromise the blood brain barrier, impair vascular function (constriction and dilation) [121] and cause microhemorrhages [122]. CAA in the dog brain was first observed by Braunmuhl [123] as early as 1956 and was subsequently confirmed by Wisniewski and colleagues [90]. Vascular and perivascular abnormalities and cerebrovascular Aβ pathology are frequently found in aged dogs [87,88,124-131]. Cultured vascular smooth muscle cells from dog brain can mimic the pathological process (e.g. Aβ production and accumulation) that occurs in humans with AD and Down syndrome [121,132-134]. Vascular Aβ is primarily the shorter 1-40 species, which is identical in dogs and humans [135]. Dog CAA can be associated with cerebral hemorrhage [129,130] and the distribution of CAA in dog brain also appears similar to humans with the occipital cortex being particularly vulnerable [119]. The extent of CAA in aged dog brains can correlate with clinical signs of cognitive dysfunction in companion dogs [95]. Overall, dogs are thought to be a good natural model for examining CAA and treatments for CAA [136]. CAA similarities in dogs and in humans suggest they may be a useful model in which to test Aβ immunotherapy approaches as will be described later.

*Oxidative damage*

Aging and the production of free radicals can lead to oxidative damage to proteins, lipids and nucleotides that, in turn, may cause neuronal dysfunction and ultimately neuronal death. Normally, several mechanisms are in place that balances the production of free radicals such as endogenous antioxidants. However with age, a number of these protective mechanisms begin to fail. In AD, oxidative damage is particularly pronounced and significant increases in protein oxidation, lipid peroxidation and DNA/RNA oxidation have all been reported [1,137-150]. Further, in humans with MCI, which is thought to represent early AD ([26,35], already show either intermediate or similar levels of oxidative damage as observed in AD [1,2,137,139,141,146,151-153].

There are a number of downstream consequences of oxidative modifications to proteins, lipids and DNA/RNA including a reduction in protein synthesis [3], altered proteasome function [154], and impaired protein/enzyme function [155,156]. Further, selective oxidative modifications to key proteins identified using proteomics approaches may lead to neuronal dysfunction through abnormalities in pathways associated with energy metabolism, excitotoxicity, proteasomal dysfunction, lipid, synaptic dysfunction and pH buffering [1].

In dog brain, the accumulation of carbonyl groups, which is a measure of oxidative damage to proteins, increases with age [157,158] and is associated with reduced endogenous antioxidant enzyme activity or protein levels such as in glutamine synthetase and superoxide dismutase (SOD) [157,159-161]. In several studies, a relation between age and increased oxidative damage has been inferred by measuring the amount of end products of lipid peroxidation (oxidative damage to lipids) including the extent of 4-hydroxynonenal (4HNE) [62,98,161,162], lipofuscin (LF)[62], lipofuscin-like pigments (LFP) [98,162], or malondialdehyde [157]. Increased oxidative damage to DNA or RNA (8OHdG) in aged dog brain has been reported [62], a feature we have also subsequently observed (Fig. 2I-K). It is likely that one source of damaging free radicals is mitochondrial dysfunction and mitochondria isolated from aged dogs show higher ROS production [163].

Oxidative damage may also be associated with behavioral decline in dogs. Rofina and collaborators found that increased oxidative end products (lipofuscin-like pigment and protein carbonyls) in aged companion dog brain (including several breeds and mixed breeds) [62,98,164] correlates with severity of behavior changes due to cognitive dysfunction. Similarly, in our own studies of aging beagles, higher protein oxidative damage (3-nitrotyrosine) and lower endogenous antioxidant capacity (superoxide dismutase and glutathione-S-transferase) are all associated with poorer prefrontal-dependent and spatial learning [160]. These correlative studies suggest a link between cognition and progressive oxidative damage in the dog suggesting this is a useful model in which to test antioxidant treatment strategies. Indeed, as will be described next, manipulating oxidative status in the brains of aged dogs leads to significant improvements in cognition strongly supporting a role of oxidative damage in neurodegeneration.

*Summary of interventions altering cognitive and neurobiological aging in dogs*

Given the age-sensitivity of our cognitive testing protocols and the multiple neuropathology markers that are influenced by aging in dogs and that are correlated with cognitive decline, we have developed sensitive outcome measures that are useful for intervention studies. It is important now to manipulate neurobiological features of the aging dog brain and determine links to cognition by systematic and hypothesis-driven experimental studies. Significant cognitive decline and progressive neuropathology occurs within a relatively narrow period of time allowing for longitudinal studies. Following animals over time, indeed over years, provides a powerful approach in which to detect signs of decline but also evidence of cognitive maintenance in response to treatments. Much of our focus for over 9 years now has been to identify therapeutics that may improve or maintain cognition in aged dogs and reduce neuropathology thought to be critical for human brain aging and disease. However, we have approached this problem from a unique perspective in that we have been using combinatorial treatment approaches that target different molecular cascades to detect additive benefits.

*Antioxidants and behavioral enrichment: effects on cognition and neuropathology*

In our first series of studies, we targeted oxidative damage and possible neuron loss and dysfunction as being critical components causing cognitive decline by treating aged dogs with an antioxidant enriched diet, behavioral enrichment or a combination of both. The rationale for studying the effects of behavioral enrichment on cognition and brain pathology stems from previous experiments in rodents and humans. In the rodent literature, an enriched environment can increase synaptic growth, stimulate angiogenesis, and drive neurogenesis [165,166]. In turn, exercise, usually a component of behavioral enrichment, increases brain-derived neurotrophic factor (BDNF) and cerebral blood flow, induces molecules mediating synaptic plasticity, generates new neurons and improves learning [167,168] (also see review by [167,169]). In human studies, aerobically fit elderly individuals show less age-related brain atrophy [170] and improved attention and executive function [171]. Several studies have shown that increased mental activity can reduce the risk of dementia [172-174]. Older adults (>75 yrs) with a high frequency of participation in cognitive and physical activities had a reduced risk of dementia [173]. Consistent with these observational studies, in systematic and controlled experiments using cognitive training interventions, significant cognitive improvements can be observed in older adults [175-177]. The challenge of human studies on behavioral enrichment is the difficulty in controlling the exact enrichment and identifying an appropriate control group that is inactive/non-enriched and randomized with respect to all other variables.

Antioxidants may also be beneficial for brain aging. Work in animal models consistently supports the hypothesis that antioxidants can improve function and behavior. In aged rodents, antioxidants in the form of supplements and fruits and vegetables can improve learning, memory, and motor function and reduce oxidative damage (e.g.[178-181]). However, epidemiological or intervention studies in aged humans or patients with AD have less clear outcomes. While some epidemiological studies have shown a positive effect of antioxidant supplementation on cognition and reduction of risk for developing AD [182,183], other studies have failed to report significant effects [184-186]. However, in one epidemiological study, combinations of antioxidants were superior to single supplementation [187] and further reduced the risk of developing AD. Moreover, dietary intake of antioxidants has been shown to be superior to supplements in human studies (i.e., the use of vitamins) [183,188]. Recent reviews of the human epidemiological literature have emphasized that antioxidants may be one of the most promising treatments for preventing AD [189] as the AD brain accumulates extensive oxidative damage [137].

There have been few systematic studies evaluating antioxidants in clinical trials. Vitamin E has been shown to delay institutionalization in AD patients, suggesting some beneficial effects [190]. However, vitamin E alone did not improve cognition in patients with MCI, which is thought to be a precursor to AD [191]. In normally aging individuals, vitamin E alone has little benefit in elderly women [192], however, one study showed that supplementation with a combination of vitamins E and C led to improved cognition [193]. Difficulties in interpreting human studies (either epidemiological or clinical trials) stem from differences in the amount of supplements taken, the form and source, duration and regularity of use, and the challenges of determining the exact dietary intake of antioxidants. For example, in clinical trials, the amount of antioxidants derived from diet may exceed that provided in the treatment supplement. In addition, it is difficult to determine what dose should be used to raise antioxidant capacity sufficiently to reduce oxidative damage. Too high of a level of a single antioxidant may be detrimental when not balanced with additional antioxidants – for example, vitamin C helps to recycle vitamin E. Thus, this is an area that has yet to be fully explored.

We hypothesized that single treatments with antioxidants or behavioral enrichment would be beneficial for improving cognition and reducing pathology but that the combination treatment would lead to larger improvements in signs of aging compared to the individual treatments. Studies are needed in higher animal models, in which the enrichment conditions can be precisely controlled over an extended period of time. We used 48 aged beagles (between ~8-12 years) that were divided into four groups – balanced with respect to baseline cognitive ability, sex, and age – and provided either or both an antioxidant diet or behavioral enrichment. In a subset of experiments, we included an additional 17 young beagles (<5 years of age) for comparison to aged dogs. The groups were as follows: (1) no behavioral enrichment/control diet group; (2) behavioral enrichment/control diet; (3) no behavioral enrichment/antioxidant diet; and (4) combined behavioral enrichment and antioxidant diet. Young dogs were all placed in the behavioral enrichment condition with half provided with the antioxidant enriched diet (i.e. similar to groups 2 and 4).

An antioxidant-enriched dog diet was formulated to include a broad spectrum of antioxidants and two mitochondrial co-factors. Both the control and antioxidant diets were formulated to meet the nutrient profile for the American Association of Feed Control Officials recommendations for adult dogs [194] and, with the exception of the supplements in the antioxidant diet, were identical in composition [48]. The control and test diet had the following differences in formulation on an as fed basis, respectively: dl-alpha-tocopherol acetate (120 vs 1050 ppm), ascorbic acid as Stay-C (30 vs 80 ppm), l-carnitine (20 vs 260 ppm), dl-alpha-lipoic acid (20 vs 128 ppm). Based on an average weight of 10 kg per animal, the daily doses for each compound were 800IU or 210 mg/day (21 mg/kg/day) of vitamin E, 16 mg/day (1.6 mg/kg/day) of vitamin C, 52 mg/day (5.2 mg/kg/day) of carnitine, 26 mg/day (2.6 mg/kg/day) of lipoic acid. Fruits and vegetables were also incorporated at a 1 to 1 exchange ratio for corn, resulting in 1% inclusions of each of the following: spinach flakes, tomato pomace, grape pomace, carrot granules, and citrus pulp. This was equivalent to raising fruits and vegetable servings from 3 to 5-6/day. Vitamin E was increased ~75% by the antioxidant diet in treated dogs [48]. The behavioral enrichment condition consisted of additional cognitive experience (20-30 min/day, 5 days/week), an enriched sensory environment (housing with a kennel-mate, rotation of play toys in kennel once/week), and physical exercise (2x20min walks/week outdoors) [195].

To evaluate longitudinal age changes and the effects of behavioral enrichment, all dogs were tested annually for total treatment duration (including both the antioxidant diet and behavioral enrichment interventions) of 2.8 years. Treatment with the antioxidant diet lead to cognitive improvements in learning that were rapid and within two weeks of beginning the diet, aged animals began showing significant improvements in spatial attention (landmark task) [48]. Subsequent testing of animals with a more difficult complex learning task, oddity discrimination, also revealed benefits of the diet [44] (Fig. 3A). Improved learning ability was maintained over time with the antioxidant treatment while untreated animals showed a progressive decline [195]. We observed similar improvements in cognition in animals provided with behavioral enrichment. In addition, for each cognitive task that allowed a comparison between all 4 treatment conditions (control, antioxidant diet alone, behavioral enrichment alone or a combination), the combination treatment group showed consistently more benefits than either treatment alone [195,196](Fig. 3B). For example, spatial memory showed a trend towards improvement in singly treated animals but only after long term treatment (>2 years) with a combination of the two interventions were these differences statistically significant [197] (Fig. 3C). Interestingly, cognitive improvements were limited to aged animals as young dogs treated with the antioxidant diet were not different from control diet fed dogs [198]. These results suggest a selective improvement in aged dogs fed an antioxidant diet indicating that dietary antioxidants repair age-associated oxidative damage.

We next hypothesized that oxidative damage would be reduced in brain tissue from aged treated dogs. Thus, at the end of the longitudinal study, after 2.8 years of treatment the brains of a subset of 24 beagles were examined. First, parietal cortex biopsies were taken to isolate live mitochondria. The antioxidant diet treated dogs but not those receiving behavioral enrichment showed a significant reduction in mitochondrial ROS production [163]. A biochemical study of the parietal cortex from treated aged animals revealed novel and interesting outcomes [160]. First, oxidative damage to proteins, measured biochemically by derivitization with 2,4-dinitrophenylhydrazine (DNPH) was reduced in the parietal cortex of antioxidant treated animals but interestingly, this was most pronounced in the aged animals that received both the antioxidant diet and behavioral enrichment [160]. These results suggest that oxidative damage to proteins overall was reduced in the parietal cortex, thought to be involved with landmark discrimination learning and other visual learning tasks that were improved in the cognitive portion of the treatment study. Second, in parallel, increased endogenous antioxidant activity measured biochemically by superoxide dismutase and glutathione-S-transferase activity was found in animals receiving the combination treatment [160]. Thus, the ability of neurons in the parietal cortex to buffer oxidative damage by neutralizing radical oxygen species is improved and may lead to improved function. Third, using redox proteomics, which involves the coupling of 2-D gel electrophoresis separation of proteins with mass spectrometric techniques, the total carbonyl level/protein (i.e extent of protein oxidation) was estimated [1]. Using this approach with dog parietal cortex, several specific proteins were identified that showed reduced oxidation and are thought to be involved with energy metabolism (glyceraldehyde-3-phosphate dehydrogenase, fructose-biphosphate aldolase C, creatine kinase), maintenance and stabilization of cellular integrity (neurofilament triplet L protein, fascin) and endogenous antioxidant systems (Cu/Zn superoxide dismutase, glutathione-S-transferase, glutamate dehydrogenase) [160]. In combination, reducing protein oxidation of key molecules involved with metabolism and neuronal integrity could be instrumental for cognitive improvements.

Importantly, reduced oxidative damage to proteins and increased endogenous antioxidant activity were associated with improved cognition in the aging dog study. Higher levels of one measure of protein oxidation (3-nitrotyrosine) and lower levels of antioxidant activity (glutathione-S-transferase) was correlated with higher error scores on a reversal learning problem and on a visuospatial task (i.e .impaired function)[160]. These results strongly suggest that

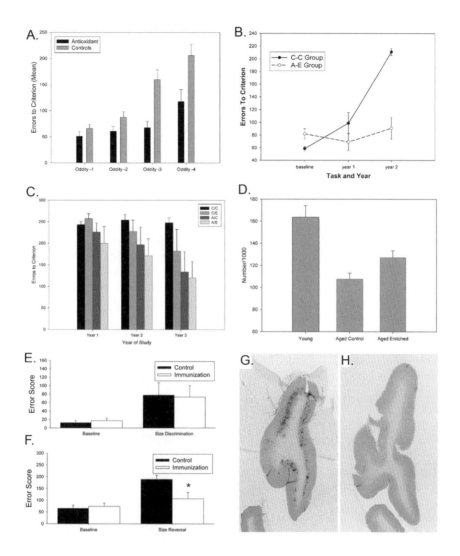

Fig. 3. Changes in cognition and neuropathology in treated aged beagles. (A). Aged dogs provided with an antioxidant-enriched diet show significantly lower error scores on an oddity task and a possible shift in the use of strategies that more closely mimics young animals. (B) A combined treatment with both an antioxidant diet and behavioral enrichment leads to maintenance of cognitive function as measured by repeated reversal measures taken at annual intervals. (C). Acquisition of a spatial memory task improves in aged dogs after long term treatment with either or both an antioxidant diet and behavioral enrichment. (D) Substantial neuron loss occurs in the hilus of the hippocampus of aged dogs that can be partially reversed with behavioral enrichment (2.8 years of treatment). Immunization with fibrillar Aβ1-42 does not improve performance on a visual (size) discrimination learning task (E) but maintains reversal (size) learning (F) over a 2 year period of time in aged beagles. In the same group of aged beagles, Aβ immunization leads to a significant decrease in Aβ plaques (anti-Aβ immunohistochemistry) in the prefrontal cortex (G) relative to adjuvant-only treated controls (H). Bars represent means and error bars represent standard errors of the mean. C-C – aged dogs in control food diet and control behavioral enrichment condition, C/E – control diet condition with behavioral enrichment, A/C – antioxidant diet condition and no behavioral enrichment, A/E – combined treatment with both antioxidant diet and behavioral enrichment. *p<.05. Figures are reproduced or modified with permission from Elsevier Limited [44,73,197,221] or Society for Neuroscience [222].

oxidative damage, particularly to vulnerable proteins involved with energy metabolism, neuronal integrity and antioxidant systems are key contributors to cognitive decline associated with aging in the dog.

Oxidation of critical proteins and enzymes with age in beagles identified using proteomics also overlap with observations reported in MCI or AD brain (e.g. enolase, glyceraldehyde-3-phosphate-dehydrogenase, creatine kinase, and glutathione-S-transferase) [1,146,160]. Thus, when comparing aged dog to aged humans and AD, proteomics approaches indicate that similar families of proteins are vulnerable to aging and AD pathology and specific molecular cascades might be sensitive to oxidative modification and in turn, compromise neuron function.

Cognitive benefits of the behavioral enrichment treatment that was also provided for 2.8 years may be mediated through a separate molecular cascade in the brain. Animals provided with behavioral enrichment treatment showed less neuronal loss in the hilus of the dentate gyrus in unbiased stereological counts of neurons (immunohistochemistry with NeuN, a protein found only in neurons), a region of the brain important for visuospatial learning and memory, compared to untreated dogs and those treated with the antioxidant diet alone[73](Fig. 3D). Interestingly, this is the same area of the brain that showed age-associated losses and further, higher hilar neuron number was correlated with lower error scores on a medial temporal lobe-dependent task, visual discrimination learning. Preservation of neuron number may occur through upregulation of growth factors in the brain such as brain-derived neurotrophic factor [167] as described in rodent studies.

Oxidative damage may also influence Aβ production and accumulation [199] and thus we measured the extent of Aβ plaque accumulation in several cortical regions in treated animals. Aβ deposition (plaque 'loads") detected by immunohistochemistry using free-floating paraformaldehyde fixed sections in the antioxidant treated animals was reduced by 27-84% in the parietal, entorhinal, and occipital cortices, but not in the prefrontal cortex [200]. Aβ-lowering effects were not observed in animals receiving the behavioral enrichment intervention alone. Given that dogs began treatment between 8 and 12 years of age, which is a time when Aβ pathology should already be present within the prefrontal cortex [91], these results suggest that the antioxidant treatment could reduce new Aβ accumulation but could not reverse existing pathology.

When neuropathology outcomes measures are considered together there is evidence to suggest that combination treatment approaches can target independent molecular cascades that either alone or together can improve neuron function. However, the results of this study raises the question of to what extent does Aβ reduction explain cognitive benefits observed in response to the antioxidant diet? To address this question, a new longitudinal study in aged dogs with pre-existing Aβ pathology and cognitive decline were included in a experiment that was intended to selectively target the reduction of Aβ.

*Aβ targeted therapeutics*

Aged dogs may be very useful models for assessing Aβ targeted therapies. Given correlative studies both in our laboratory and in others, there is a link between the extent of Aβ and cognition in aged dogs. However, to test this hypothesis directly it is important to manipulate Aβ level and measure cognitive outcomes. These experiments are relevant for human clinical trials; in particular, given that Aβ is identical in dogs and humans and that APP is very similar, pharmacological approaches targeting APP processing could be tested in dogs. For example, compounds that may affect Aβ production such as non-steroidal anti-inflammatories [201] or statins [202,203] could reduce Aβ and improve cognition in dogs. Due to similarities between the dog and human in terms of responsiveness and in drug handling and metabolism, the dog can be considered to be a useful model for chronic statin treatment in humans [204,205]. Indeed, dogs are unique in that they were used to establish efficacy and safety in the majority of statins currently on the market and have been used in chronic studies over 2 years in length at doses relevant for humans.

In transgenic mouse models of AD, deposition of Aβ may be prevented or significantly reduced after immunization with fibrillar Aβ1-42 [206-209]. Further, learning and memory is improved by either active [207,208,210] or passive immunization [211-214]. We recently completed a study evaluating the cognitive effects of Aβ immunization on dog cognition and neuropathology in aged animals [52]. Using an active immunization approach, we vaccinated aged dogs with pre-existing Aβ pathology and cognitive dysfunction with fibrillar Aβ1-42 formulated with aluminum hydroxide (an adjuvant safe for use in humans).

Over a 2 year treatment period including 25 vaccinations, we were surprised to find little cognitive improvement on multiple measures of learning and memory using the same tasks that were established as being both age and intervention sensitive in our previous studies (e.g. landmark discrimination, oddity discrimination, visual discrimination learning (Fig. 3E), reversal learning and spatial learning and memory). Interestingly, after 22 months of treatment we observed a significant improvement on reversal learning in treated animals. Overall there was an increase in error scores in control dogs over time reflecting both an increase in task difficulty and an aging process but this was differentially affected by treatment. A significant time by treatment effect indicated that fibrillar Aβ immunized animals showed a maintenance of reversal learning ability over time (Fig. 3F). We further confirmed that frontal function, specifically, was affected by treatment by analyzing the types of errors individuals animals made. Fibrillar Aβ immunized animals made significantly fewer perseverative responses relative to the other two treatment groups. These results suggest that immunization with fibrillar Aβ1-42 leads to improved and maintained executive function in aged dogs with both pre-existing Aβ and cognitive deficits.

At the end of the study, we detected significant reductions in Aβ plaque accumulation detected by immunohistochemistry and soluble and insoluble Aβ1-40 and Aβ1-42 measured biochemically using a sandwich enzyme-linked immunosorbant assay. Reductions in Aβ were observed in almost all brain regions sampled (prefrontal, parietal, occipital, entorhinal cortex) that are essential for performance of the learning and memory tasks used in the study (Fig. 3G,H).

As described previously, dogs may be a good model in which to examine CAA changes in response to treatment. We examined the extent of CAA using sections that were immunostained for Aβ in immunized animals relative to controls using a semiquantitative ranking system. The same regions as used for the Aβ plaque load analysis were used (prefrontal, parietal, occipital and entorhinal cortex). There was a tendency for less CAA in immunized animals, which contrasts with reports in human autopsy cases and may reflect a chronic response to Aβ immunization as opposed to perhaps more transient effects with short term treatment in AD patients.

Outcomes from the longitudinal dog vaccination study are similar to reports of the clinical trial in patients with AD using a similar formulation of immunizing with fibrillar Aβ1-42 (AN1792 study). In the Swiss cohort of the trial there was a maintenance of function on a global test of cognition (Mini mental state examination) and on a hippocampal-dependent task (visual paired associated test) in treated individuals who developed antibodies capable of binding to plaques [215]. In a second, larger study no differences between antibody responders and placebo groups were observed on several cognitive and disability scales [216]. However, a composite score of a neuropsychological test battery indicated "less worsening" of decline in antibody responders after 12 months and an improvement in the memory domain [216]. A small number of patients enrolled in the AN1792 study have come to autopsy and show Aβ plaque reduction without any effect on the extent of neurofibrillary tangles or CAA [217-219]. Interestingly, in the case report by Masliah and colleagues (2005), the frontal cortex showed the largest response to immunotherapy [219], which is similar to our observations in the dog. The most recent autopsy study of 8 patients that were in the AN1792 study further confirm reduced Aβ pathology in response to treatment, 5 years after the last injection [220]. However, reduction of brain Aβ did not slow disease progression and 7 of 8 patients had severe end stage dementia prior to death.

Although there were several important differences between the dog and human vaccination studies (discussed in [52]), the dog study suggests that reducing plaque accumulation or total Aβ may be insufficient to restore neuronal function without directly targeting neuron health. Thus, antioxidant-associated Aβ plaque reduction described in the previous sections may not be the only mechanism underlying cognitive improvements in aged treated beagles. However, this does not rule out that the alternative approach of Aβ prevention may be more strongly associated with cognitive maintenance. Additionally, we propose that combining Aβ immunotherapy with a second treatment, such as behavioral enrichment or possibly an antioxidant diet that may independently restore neuron health after Aβ removal, may lead to significant improvements in cognition and enhance cognitive maintenance.

*Conclusion and implications for human cognitive aging and Alzheimer's disease*

The aged dog shows decline in learning and memory across many different types of cognitive domains, which clinically simulates MCI and early/mild AD. Accordingly, the dog model provides opportunities to examine potential preventative and therapeutic treatments that may impact the development of disease in patients with AD. We are able to measure function in multiple cognitive domains with learning and memory tasks established to be both age and treatment sensitive. Many of these tasks assess cognitive domains that are also vulnerable to aging and AD in humans. Longitudinal studies, in line with human clinical trials, can be completed in aged dogs and also provide measures of maintenance of function. Further, the development of some of the features of normal human aging and neuropathology observed in AD (e.g.A$\beta$) is sporadic, age-dependent and at human levels. We show that dogs may be particularly useful for studies of diet and environmental manipulation. The impact of environmental/lifestyle factors on the aging process and potential for enhancing vulnerability to AD may be explored in clinical cohorts of aging dogs (companion animals living in human environments). Further, the result of our A$\beta$ vaccination study also suggests that dogs may be useful for A$\beta$-targeted intervention studies. For example, the newest autopsy studies of patients with AD that were vaccinated with fibrillar beta-amyloid1-42 is consistent our results in aged dogs; patients showed a significant reduction of brain A$\beta$ without evidence of cognitive improvement [220]. These differences can provide new insights into treatments to promote successful aging in humans and emphasize the need to test therapeutics in higher mammalian species such as dogs or nonhuman primates that naturally develop similar signs of aging and AD as humans. The results of several different types of interventions using the dog model can be considered together and suggests new intervention approaches.

## ACKNOWLEDGEMENTS

Funding provided by NIH/NIA grants AG12694 and AG20242. The authors appreciate the helpful discussions with Dr. Julene Johnson.

## References

[1] Butterfield DA, Sultana R.(2007) Redox proteomics identification of oxidatively modified brain proteins in Alzheimer's disease and mild cognitive impairment: insights into the progression of this dementing disorder. *J Alzheimers Dis*;**12**(1):61-72.

[2] Lovell MA, Markesbery WR.(2007) Oxidative DNA damage in mild cognitive impairment and late-stage Alzheimer's disease. *Nucleic Acids Res*;**35**(22):7497-504.

[3] Ding Q, Dimayuga E, Keller JN.(2007) Oxidative damage, protein synthesis, and protein degradation in Alzheimer's disease. *Curr Alzheimer Res*;**4**(1):73-9.

[4] Hebert LE, Scherr PA, Bienias JL, Bennett DA, Evans DA.(2003) Alzheimer disease in the US population: prevalence estimates using the 2000 census. *Arch Neurol*;**60**(8):1119-22.

[5] Head E, Callahan, H., Cummings, B.J., Cotman, C.W., Ruehl, W.W., Muggenberg, B.A. and Milgram, N.W.(1997) Open field activity and human interaction as a function of age and breed in dogs. *Physiology & Behavior*;**62**(5):963-71.

[6] Head E, Mehta R, Hartley J, Kameka AM, Cummings BJ, Cotman CW, Ruehl WW, Milgram NW.(1995) Spatial learning and memory as a function of age in the dog. *Behav Neurosci*;**109**:851-8.

[7] Milgram NW, Head E, Weiner E, Thomas E.(1994) Cognitive functions and aging in the dog: Acquisition of nonspatial visual tasks. *Behav Neurosci*;**108**:57-68.

[8] Cummings BJ, Su, J.H., Cotman, C.W., White, R. and Russell, M.J.(1993) Beta-amyloid accumulation in aged canine brain: a model of plaque formation in Alzheimer's disease. *Neurobiology of Aging*;**14**:547-60.

[9] Cummings BJ, Head, E., Ruehl, W.W., Milgram, N.W., and Cotman, C.W.(1996) The canine as an animal model of human aging and dementia. *Neurobiol Aging*;**17**:259-68.

[10] Cummings BJ, Head, E., Ruehl, W.W.,Milgram, N.W., and Cotman, C.W.(1996) Beta-amyloid accumulation correlates with cognitive dysfunction in the aged canine. *Neurobiology of Learning & Memory*;**66**:11-23.

[11] Head E, Callahan H, Muggenburg BA, Cotman CW, Milgram NW.(1998) Visual-discrimination learning ability and beta-amyloid accumulation in the dog. *Neurobiol Aging*;**19**(5):415-25.

[12] Parker HG, Kim LV, Sutter NB, Carlson S, Lorentzen TD, Malek TB, Johnson GS, DeFrance HB, Ostrander EA, Kruglyak L.(2004) Genetic structure of the purebred domestic dog. *Science*;**304**(5674):1160-4.

[13] Greer KA, Canterberry SC, Murphy KE.(2007) Statistical analysis regarding the effects of height and weight on life span of the domestic dog. *Res Vet Sci*;**82**(2):208-14.

[14] Galis F, Van der Sluijs I, Van Dooren TJ, Metz JA, Nussbaumer M.(2007) Do large dogs die young? *J Exp Zoolog B Mol Dev Evol*;**308**(2):119-26.

[15] Patronek GJ, Waters DJ, Glickman LT.(1997) Comparative longevity of pet dogs and humans: implications for gerontology research. *J Gerontol A Biol Sci Med Sci*;**52**(3):B171-8.

[16] Bobik M, Thompson, T. and Russell, M.J.(1994) Amyloid deposition in various breeds of dogs. *Society for Neuroscience Abstracts*;**20**:172.

[17] Albert RE, Benjamin SA, Shukla R.(1994) Life span and cancer mortality in the beagle dog and humans. *Mech Ageing Dev*;**74**(3):149-59.

[18] Lowseth LA, Gillett NA, Gerlach RF, Muggenburg BA.(1990) The effects of aging on hematology and serum chemistry values in the beagle dog. *Vet Clin Path*;**19**(1):13-9.

[19] Feldman EC, Nelson RW. Breeding, pregnancy, and parturition. In: Feldman EC, Nelson RW, editors. Canine and Feline Endocrinology and Reproduction, 3rd Edition. St. Louis, MO: Saunders; 2004.

[20] Cooley DM, Schlittler DL, Glickman LT, Hayek M, Waters DJ.(2003) Exceptional longevity in pet dogs is accompanied by cancer resistance and delayed onset of major diseases. *J Gerontol A Biol Sci Med Sci*;**58**(12):B1078-84.

[21] Head E, Corrada MM, Kahle-Wrobleski K, Kim RC, Sarsoza F, Goodus M, Kawas CH.(2007) Synaptic

proteins, neuropathology and cognitive status in the oldest-old. *Neurobiol Aging*.

[22] Waters DJ, Shen S, Glickman LT.(2000) Life expectancy, antagonistic pleiotropy, and the testis of dogs and men. *Prostate*;**43**(4):272-7.

[23] Proschowsky HF, Rugbjerg H, Ersboll AK.(2003) Mortality of purebred and mixed-breed dogs in Denmark. *Prev Vet Med*;**58**(1-2):63-74.

[24] Albert MS, Funkenstein HH. The effects of age: Normal variation and its relation to disease. In: Asburg AK, McKhanney GM, McDonald WI, editors. Disorders of the Nervous System: Clinical Neurology, 2nd edition. Philadelphia: Saunders Inc; 1992. p 598-611.

[25] Petersen RC, Smith, G. E., Waring, S. C., Ivnik, R. J., Kokmen, E., Tangelos, E. G.(1997) Aging, memory, and mild cognitive impairment. *International Psychogeriatrics*;**9 Suppl 1**:65-9.

[26] Petersen RC, Smith, G. E., Waring, S. C., Ivnik, R. J., Tangelos, E. G., Kokmen, E.(1999) Mild cognitive impairment: clinical characterization and outcome. *Archives of Neurology*;**56**(3):303-8.

[27] McKhann G, Drachman, D., Folstein, M., Katzman, R., Price, D., and Stadlan, E.M.(1984) Clinical Diagnosis of Alzheimer's disease: Report of the NINCDS-ADRDA work group under the auspices of Department of Health and Human Services task force on Alzheimer's disease. *Neurology*;**34**:939-44.

[28] Adams B, Chan, A., Callahan, H. and Milgram, N.W.(2000) The Canine as a Model of Human Cognitive Aging: Recent Developments. *Progress in Neuro-Psychopharmacology & Biological Psychiatry*;**24**(5):675-92.

[29] Mirra SS, Heyman A, McKeel D, Sumi SM, Crain BJ, Brownlee LM, Vogel FS, Hughes JP, van Belle G, Berg L.(1991) The Consortium to Establish a Registry for Alzheimer's Disease (CERAD). Part II. Standardization of the neuropathologic assessment of Alzheimer's disease. *Neurology*;**41**(4):479-86.

[30] Head E, Moffat, K., Das, P., Sarsoza, F., Poon, W.W., Landsberg, G., Cotman, C.W., Murphy, M.P.(2005) b-Amyloid Deposition and Tau Phosphorylation in Clinically Characterized Aged Cats. *Neurobiol Aging*;**26**:749-63.

[31] Kuroki K, Uchida, K., Kiatipattanasakul, W., Nakamura, S., Yamaguchi, R., Nakayama, H., Doi, K., and Tateyama, S.(1997) Immunohistochemical detection of tau proteins in various non-human animal brains. *Neuropathology*;**17**:174-80.

[32] Papaioannou N, Tooten PC, van Ederen AM, Bohl JR, Rofina J, Tsangaris T, Gruys E.(2001) Immunohistochemical investigation of the brain of aged dogs. I. Detection of neurofibrillary tangles and of 4-hydroxynonenal protein, an oxidative damage product, in senile plaques. *Amyloid*;**8**(1):11-21.

[33] Pugliese M, Mascort J, Mahy N, Ferrer I.(2006) Diffuse beta-amyloid plaques and hyperphosphorylated tau are unrelated processes in aged dogs with behavioral deficits. *Acta Neuropathol*;**112**(2):175-83.

[34] Wegiel J, Wisniewski, H.M., and Soltysiak, Z.(1998) Region- and cell-type-specific pattern of tau phosphorylation in dog brain. *Brain Research*;**802**:259-66.

[35] Morris JC, Storandt M, Miller JP, McKeel DW, Price JL, Rubin EH, Berg L.(2001) Mild cognitive impairment represents early-stage Alzheimer disease. *Arch Neurol*;**58**(3):397-405.

[36] Mann DMA, Esiri MM.(1989) The pattern of acquisition of plaques and tangles in the brains of patients under 50 years of age with Down's syndrome. *J Neurol Sci*;**89**:169-79.

[37] Head E, Lott IT.(2004) Down syndrome and beta-amyloid deposition. *Curr Opin Neurol*;**17**(2):95-100.

[38] Wisniewski K, Wisniewski H, Wen G.(1985) Occurrence of neuropathological changes and dementia of Alzheimer's disease in Down's syndrome. *Ann Neurol*;**17**:278-82.

[39] Head E, Pop V, Sarsoza F, Kayed R, Beckett TL, Studzinski CM, Tomic JL, Glabe CG, Murphy MP.(Amyloid-beta Peptide and Oligomers in the Brain and Cerebrospinal Fluid of Aged Canines. *J Alzheimers Dis*.

[40] Walsh DM, Klyubin I, Fadeeva JV, Cullen WK, Anwyl R, Wolfe MS, Rowan MJ, Selkoe DJ.(2002) Naturally secreted oligomers of amyloid beta protein potently inhibit hippocampal long-term potentiation in vivo. *Nature*;**416**(6880):535-9.

[41] Tapp PD, Siwak,C., Head, E., Cotman, C.W., Murphey, H., Muggenburg, B.A., Ikeda-Douglas, C., Milgram, N.W.(2004) Concept abstraction in the aging dog: development of a protocol using successive discrimination and size concept tasks. *Behav Brain Res*;**153**:199-210.

[42] Tapp D, Siwak CT, Zicker SC, Head E, Muggenburg BA, Cotman CW, Murphey HL, Ikeda-Douglas CJ, Milgram NW.(2003) An Antioxidant Enriched Diet Improves Concept Learning in Aged Dogs. *Society for Neuroscience Abstracts*;**Abstract 836.12**.

[43] Milgram NW, Zicker SC, Head E, Muggenburg BA, Murphey H, Ikeda-Douglas C, Cotman CW.(2002) Dietary enrichment counteracts age-associated cognitive dysfunction in canines. *Neurobiology of Aging*;**23**:737-45.

[44] Cotman CW, Head E, Muggenburg BA, Zicker S, Milgram NW.(2002) Brain Aging in the Canine: A Diet Enriched in Antioxidants Reduces Cognitive Dysfunction. *Neurobiology of Aging*;**23**(5):809-18.

[45] Tapp PD, Siwak CT, Estrada J, Muggenburg BA, Head E, Cotman CW, Milgram NW.(2003) Size and Reversal Learning in the Beagle Dog as a Measure of Executive Function and Inhibitory Control in Aging. *Learning and Memory*;**10**(1):64-73.

[46] Christie LA, Studzinski CM, Araujo JA, Leung CS, Ikeda-Douglas CJ, Head E, Cotman CW, Milgram NW.(2005) A comparison of egocentric and allocentric age-dependent spatial learning in the beagle dog. *Prog Neuropsychopharmacol Biol Psychiatry*;**29**(3):361-9.

[47] Studzinski CM, Christie LA, Araujo JA, Burnham WM, Head E, Cotman CW, Milgram NW.(2006) Visuospatial function in the beagle dog: an early marker of cognitive decline in a model of human aging and dementia. *Neurobiol Learn Mem*;**86**(2):197-204.

[48] Milgram NW, Head E, Muggenburg BA, Holowachuk D, Murphey H, Estrada J, Ikeda-Douglas CJ, Zicker SC, Cotman CW.(2002) Landmark discrimination learning in the dog: effects of age, an antioxidant fortified diet, and cognitive strategy. *Neuroscience and Biobehavioral Reviews*;**26**(6):679-95.

[49] Milgram NW, Adams, B., Callahan, H., Head, E., Mackay, W., Thirlwell, C., and Cotman, C.W.(1999)

[50] Mishkin M, Delacour J.(1975) An analysis of short-term visual memory in the monkey. *Journal of Experimental Psychology: Animal Behavior Proceedings*;**1**:326-34.
 Landmark discrimination learning in the dog. *Learning & Memory*;**6**(1):54-61.
[51] Chan AD, Nippak PM, Murphey H, Ikeda-Douglas CJ, Muggenburg B, Head E, Cotman CW, Milgram NW.(2002) Visuospatial impairments in aged canines (Canis familiaris): the role of cognitive-behavioral flexibility. *Behav Neurosci*;**116**(3):443-54.
[52] Head E, Pop V, Vasilevko V, Hill M, Saing T, Sarsoza F, Nistor M, Christie L, Milton S, Glabe C, Barrett E, Cribbs D.(2008) A two-year study with fibrillar beta-amyloid (Abeta) immunization in aged canines: effects on cognitive function and brain Abeta. *J Neurosci*;**28**(14):3555-66.
[53] Cummings BJ, Head E, Ruehl WW, Milgram NW, Cotman CW.(1996) The canine as an animal model of human aging and dementia. *Neurobiology of Aging*;**17**:259-68.
[54] Head E, Milgram, N.W. and Cotman, C.W. Neurobiological Models of Aging in the Dog and Other Vertebrate Species. In: In. P. Hof and Mobbs C, editor. Functional Neurobiology of Aging. San Diego: Academic Press; 2001. p 457-68.
[55] Wisniewski HM, Wegiel J, Morys J, Bancher C, Soltysiak Z, Kim KS. Aged dogs: an animal model to study beta-protein amyloidogenesis. In: K. Maurer PR, and H. Beckman, editor. Alzheimer's disease Epidemiology, Neuropathology, Neurochemistry and Clinics. New York: Springer-Verlag; 1990. p 151-67.
[56] Borras D, Ferrer I, Pumarola M.(1999) Age-related changes in the brain of the dog. *Vet Pathol*;**36**(3):202-11.
[57] Tapp PD, Siwak C. The canine model of human brain aging: cognition, behavior and neuropathology. In: Conn PM, editor. Handbook of Models of Human Aging. Burlington, MA: Elsevier Inc.; 2006. p 415-34.
[58] Su M-Y, Head, E., Brooks, W.M., Wang, Z., Muggenberg, B.A., Adam, G.E., Sutherland, R.J., Cotman, C.W. and Nalcioglu, O.(1998) MR Imaging of anatomic and vascular characteristics in a canine model of human aging. *Neurobiology of Aging*;**19**(5):479-85.
[59] Kimotsuki T, Nagaoka T, Yasuda M, Tamahara S, Matsuki N, Ono K.(2005) Changes of magnetic resonance imaging on the brain in beagle dogs with aging. *J Vet Med Sci*;**67**(10):961-7.
[60] Gonzalez-Soriano J, Marin Garcia P, Contreras-Rodriguez J, Martinez-Sainz P, Rodriguez-Veiga E.(2001) Age-related changes in the ventricular system of the dog brain. *Ann Anat*;**183**(3):283-91.
[61] Tapp PD, Siwak CT, Gao FQ, Chiou JY, Black SE, Head E, Muggenburg BA, Cotman CW, Milgram NW, Su MY.(2004) Frontal lobe volume, function, and beta-amyloid pathology in a canine model of aging. *J Neurosci*;**24**(38):8205-13.
[62] Rofina JE, van Ederen AM, Toussaint MJ, Secreve M, van der Spek A, van der Meer I, Van Eerdenburg FJ, Gruys E.(2006) Cognitive disturbances in old dogs suffering from the canine counterpart of Alzheimer's disease. *Brain Res*;**1069**(1):216-26.
[63] Du AT, Schuff N, Chao LL, Kornak J, Ezekiel F, Jagust WJ, Kramer JH, Reed BR, Miller BL, Norman D, Chui HC, Weiner MW.(2005) White matter lesions are associated with cortical atrophy more than entorhinal and hippocampal atrophy. *Neurobiol Aging*;**26**(4):553-9.
[64] Ezekiel F, Chao L, Kornak J, Du AT, Cardenas V, Truran D, Jagust W, Chui H, Miller B, Yaffe K, Schuff N, Weiner M.(2004) Comparisons between global and focal brain atrophy rates in normal aging and Alzheimer disease: Boundary Shift Integral versus tracing of the entorhinal cortex and hippocampus. *Alzheimer Dis Assoc Disord*;**18**(4):196-201.
[65] Tapp PD, Head K, Head E, Milgram NW, Muggenburg BA, Su MY.(2006) Application of an automated voxel-based morphometry technique to assess regional gray and white matter brain atrophy in a canine model of aging. *Neuroimage*;**29**(1):234-44.
[66] Coffey CE, Lucke JF, Saxton JA, Ratcliff G, Unitas LJ, Billig B, Bryan RN.(1998) Sex differences in brain aging: a quantitative magnetic resonance imaging study. *Arch Neurol*;**55**(2):169-79.
[67] Simic G, Kostovic I, Winblad B, Bogdanovic N.(1997) Volume and number of neurons of the human hippocampal formation in normal aging and Alzheimer's disease. *J Comp Neurol*;**379**(4):482-94.
[68] West MJ.(1993) Regionally specific loss of neurons in the aging human hippocampus. *Neurobiology of Aging*;**14**:287-93.
[69] West MJ, Kawas CH, Martin LJ, Troncoso JC.(2000) The CA1 region of the human hippocampus is a hot spot in Alzheimer's disease. *Ann N Y Acad Sci*;**908**:255-9.
[70] Bobinski M, Wegiel J, Tarnawski M, Bobinski M, Reisberg B, de Leon MJ, Miller DC, Wisniewski HM.(1997) Relationships between regional neuronal loss and neurofibrillary changes in the hippocampal formation and duration and severity of Alzheimer disease. *J Neuropathol Exp Neurol*;**56**(4):414-20.
[71] West MJ, Coleman, P.D., Flood, D.G. and Troncoso, J.C.(1994) Differences in the pattern of hippocampal neuronal loss in normal ageing and Alzheimer's disease. *Lancet*;**344**:769-72.
[72] Price JL, Ko AI, Wade MJ, Tsou SK, McKeel DW, Morris JC.(2001) Neuron number in the entorhinal cortex and CA1 in preclinical Alzheimer disease. *Arch Neurol*;**58**(9):1395-402.
[73] Siwak-Tapp CT, Head E, Muggenburg BA, Milgram NW, Cotman CW.(2008) Region specific neuron loss in the aged canine hippocampus is reduced by enrichment. *Neurobiol Aging*;**29**(1):521-8.
[74] Hwang IK, Li H, Yoo KY, Choi JH, Lee CH, Chung DW, Kim DW, Seong JK, Yoon YS, Lee IS, Won MH.(2008) Comparison of glutamic acid decarboxylase 67 immunoreactive neurons in the hippocampal CA1 region at various age stages in dogs. *Neurosci Lett*;**431**(3):251-5.
[75] Hwang IK, Yoo KY, Li H, Choi JH, Kwon YG, Ahn Y, Lee IS, Won MH.(2007) Differences in doublecortin immunoreactivity and protein levels in the hippocampal dentate gyrus between adult and aged dogs. *Neurochem Res*;**32**(9):1604-9.
[76] Siwak-Tapp CT, Head E, Muggenburg BA, Milgram NW, Cotman CW.(2007) Neurogenesis decreases with age in the canine hippocampus and correlates with cognitive function. *Neurobiol Learn Mem*;**88**(2):249-59.
[77] Pekcec A, Baumgartner W, Bankstahl JP, Stein VM, Potschka H.(2008) Effect of aging on neurogenesis in the canine brain. *Aging Cell*;**7**(3):368-74.
[78] van Praag H, Kempermann G, Gage FH.(2000) Neural consequences of environmental enrichment. *Nat Rev Neurosci*;**1**(3):191-8.

[79] van Praag H, Shubert T, Zhao C, Gage FH.(2005) Exercise enhances learning and hippocampal neurogenesis in aged mice. *J Neurosci*;**25**(38):8680-5.

[80] Johnstone EM, Chaney MO, Norris FH, Pascual R, Little SP.(1991) Conservation of the sequence of the Alzheimer's disease amyloid peptide in dog, polar bear and five other mammals by cross-species polymerase chain reaction analysis. *Brain Res Mol Brain Res*;**10**(4):299-305.

[81] Selkoe DJ, Bell, D.S., Podlisny, M.B., Price, D.L., and Cork, L.C.(1987) Conservation of brain amyloid proteins in aged mammals and humans with Alzheimer's disease. *Science*;**235**:873-7.

[82] Azizeh BY, Head E, Ibrahim MA, Torp R, Tenner AJ, Kim RC, Lott IT, Cotman CW.(2000) Molecular Dating of Senile Plaques in Aged Down's Syndrome and Canine Brains. *Experimental Neurology*;**163**(1):111-22.

[83] Satou T, Cummings, B.J., Head, E., Nielson, K.A., Hahn, F.F., Milgram, N.W., Velazquez, P., Cribbs, D.H., Tenner, A.J. and Cotman, C.W.(1997) The progression of beta-amyloid deposition in the frontal cortex of the aged canine. *Brain Research*;**774**:35-43.

[84] Torp R, Head, E., and Cotman, C.W.(2000) Ultrastructural analyses of beta-amyloid in the aged dog brain: Neuronal beta-amyloid is localized to the plasma membrane. *Progress in Neuro-Psychopharmacology & Biological Psychiatry*;**24**:801-10.

[85] Torp R, Head, E., Milgram, N.W., Hahn, F., Ottersen, O.P. and Cotman, C.W.(2000) Ultrastructural evidence of fibrillar β–amyloid associated with neuronal membranes in behaviorally characterized aged dog brains. *Neuroscience*;**93**(3):495-506.

[86] Torp R, Ottersen OP, Cotman CW, Head E.(2003) Identification of neuronal plasma membrane microdomains that colocalize beta-amyloid and presenilin: implications for beta-amyloid precursor protein processing. *Neuroscience*;**120**(2):291-300.

[87] Giaccone G, Verga, L., Finazzi, M., Pollo, B., Tagliavini, F., Frangione, B. and Bugiani, O.(1990) Cerebral preamyloid deposits and congophilic angiopathy in aged dogs. *Neuroscience Letters*;**114**:178-83.

[88] Ishihara T, Gondo, T., Takahashi, M., Uchino, F., Ikeda, S., Allsop, D. and Imai, K.(1991) Immunohistochemical and immunoelectron microscopial characterization of cerebrovascular and senile plaque amyloid in aged dogs' brains. *Brain Research*;**548**:196-205.

[89] Wisniewski HM, Johnson, A.B., Raine, C.S., Kay, W.J. and Terry, R.D.(1970) Senile plaques and cerebral amyloidosis in aged dogs. *Laboratory Investigations*;**23**:287-96.

[90] Wisniewski HM, Wegiel, J., Morys, J., Bancher, C., Soltysiak, Z. and Kim, K.S. Aged dogs: an animal model to study beta-protein amyloidogenesis. In: K. Maurer PR, and H. Beckman, editor. Alzheimer's disease Epidemiology, Neuropathology, Neurochemistry and Clinics. New York: Springer-Verlag; 1990. p 151-67.

[91] Head E, McCleary R, Hahn FF, Milgram NW, Cotman CW.(2000) Region-specific age at onset of beta-amyloid in dogs. *Neurobiol Aging*;**21**(1):89-96.

[92] Braak H, Braak E.(1991) Neuropathological stageing of Alzheimer-related changes. *Acta Neuropathol*;**82**(4):239-59.

[93] Braak H, Braak E, Bohl J.(1993) Staging of Alzheimer-related cortical destruction. *Review in Clin Neurosci*; **33**:403-8.

[94] Thal DR, Rub U, Orantes M, Braak H.(2002) Phases of A beta-deposition in the human brain and its relevance for the development of AD. *Neurology*;**58**(12):1791-800.

[95] Colle M-A, Hauw, J.-J., Crespeau, F., Uchiara, T., Akiyama, H., Checler, F., Pageat, P., and Duykaerts, C.(2000) Vascular and parenchymal Aβ deposition in the aging dog: correlation with behavior. *Neurobiology of Aging*;**21**:695-704.

[96] Cummings BJ, Head E, Afagh AJ, Milgram NW, Cotman CW.(1996) Beta-amyloid accumulation correlates with cognitive dysfunction in the aged canine. *Neurobiol Learn Mem*;**66**(1):11-23.

[97] Rofina J, van Andel I, van Ederen AM, Papaioannou N, Yamaguchi H, Gruys E.(2003) Canine counterpart of senile dementia of the Alzheimer type: amyloid plaques near capillaries but lack of spatial relationship with activated microglia and macrophages. *Amyloid*;**10**(2):86-96.

[98] Rofina JE, Singh K, Skoumalova-Vesela A, van Ederen AM, van Asten AJ, Wilhelm J, Gruys E.(2004) Histochemical accumulation of oxidative damage products is associated with Alzheimer-like pathology in the canine. *Amyloid*;**11**(2):90-100.

[99] Alafuzoff L, Iqbal, K., Friden, H., Adolfsson, R. and Winblad, B.(1987) Histopathological criteria for progressive dementia disorders: clinical-pathological correlation and classification by multivariate analysis. *Acta Neuropathologica (Berlin)*;**74**:209-25.

[100] Delaere P, Duyckaerts, C., Masters, C., Beyreuther, K., Piette, F. and Hauw, J-J.(1990) Large amounts of neocortical beta A4 deposits without neuritic plaques nor tangles in a psychometrically assessed, non-demented person. *Neuroscience Letters*;**116**:87-93.

[101] Dayan AD.(1970) Quantitative histological studies on the aged human brain. I. Senile plaques and neurofibrillary tangles in "normal" patients. *Acta Neuropathologica (Berlin)*;**16**:85-94.

[102] Dickson DW, Crystal, H.A.,Bevona, C., Honer, W., Vincent, I. and Davies, P.(1995) Correlations of synaptic and pathological markers with cognition of the elderly. *Neurobiology of Aging*;**16**(3):285-304.

[103] Langui D, Probst, A. and Ulrich, J.(1995) Alzheimer's changes in non-demented and demented patients: a statistical approach to their relationships. *Acta Neuropathol*;**89**:57-62.

[104] Tomlinson BE, Blessed, G. and Roth, M.(1968) Observations on the brains of non-demented old people. *Journal of Neurological Science*;**7**:331-56.

[105] Wisniewski HM. The aging brain. In: Andrews EJ, Ward, B.C., and Altman, N.H., editor. Spontaneous Animal Models of Human Disease. New York: Academic Press; 1979. p 148-52.

[106] Cummings BJ, and Cotman, C.W.(1995) Image analysis of beta-amyloid "load" in Alzheimer's disease and relation to dementia severity. *Lancet*;**346**:1524-8.

[107] Forman MS, Mufson EJ, Leurgans S, Pratico D, Joyce S, Leight S, Lee VM, Trojanowski JQ.(2007) Cortical biochemistry in MCI and Alzheimer disease: lack of correlation with clinical diagnosis. *Neurology*;**68** (10):757-63.

[108] Giannakopoulos P, Herrmann FR, Bussiere T, Bouras C, Kovari E, Perl DP, Morrison JH, Gold G, Hof PR.(2003) Tangle and neuron numbers, but not amyloid load, predict cognitive status in Alzheimer's disease. *Neurology*;**60**(9):1495-500.

[109] Arriagada PV, Growdon, J.H., Hedley-White, E.T., Hyman, B.T.(1992) Neurofibrillary tangles but not senile plaques parallel duration and severity of disease. *Neurology*;**42**:631-9.

[110] Terry RD, Masliah E, Salmon DP, Butters N, DeTeresa R, Hill R, Hansen LA, Katzman R.(1991) Physical basis of cognitive alterations in Alzheimer's disease: synapse loss is the major correlate of cognitive impairment. *Ann Neurol*;**30**(4):572-80.

[111] Cork LC.(1993) Plaques in prefrontal cortex of aged, behaviorally-tested Rhesus monkeys: incidence, distribution and relationship to task performance. *Neurobiology of Aging*;**1993**:675-6.

[112] Sloane JA, Pietropaolo MF, Rosene DL, Moss MB, Peters A, Kemper T, Abraham CR.(1997) Lack of correlation between plaque burden and cognition in the aged monkey. *Acta Neuropathol*;**94**(5):471-8.

[113] Cummings BJ.(1997) Plaques and tangles: searching for primary events in a forest of data. *Neurobiology of Aging*;**18**(4):358-62.

[114] Gearing M, Tigges, J., Mori, H., and Mirra, S.S.(1996) Aβ40 is a major form of β-amyloid in nonhuman primates. *Neurobiol Aging*;**17**:903-8.

[115] Oddo S, Caccamo A, Tran L, Lambert MP, Glabe CG, Klein WL, LaFerla FM.(2006) Temporal profile of amyloid-beta (Abeta) oligomerization in an in vivo model of Alzheimer disease. A link between Abeta and tau pathology. *J Biol Chem*;**281**(3):1599-604.

[116] Lacor PN, Buniel MC, Chang L, Fernandez SJ, Gong Y, Viola KL, Lambert MP, Velasco PT, Bigio EH, Finch CE, Krafft GA, Klein WL.(2004) Synaptic targeting by Alzheimer's-related amyloid beta oligomers. *J Neurosci*;**24**(45):10191-200.

[117] Westerman MA, Cooper-Blacketer, D., Mariash, A., Kotilinek, L., Kawarabayashi, T., Younkin, L.H., Carlson, G.A., Younkin, S.G., Ashe, K.H.(2002) The relationship between Aβ and memory in the Tg2576 mouse model of Alzheimer's disease. *J Neurosci*;**22**(5):1858-67.

[118] Attems J.(2005) Sporadic cerebral amyloid angiopathy: pathology, clinical implications, and possible pathomechanisms. *Acta Neuropathol*;**110**(4):345-59.

[119] Attems J, Jellinger KA, Lintner F.(2005) Alzheimer's disease pathology influences severity and topographical distribution of cerebral amyloid angiopathy. *Acta Neuropathol*;**110**(3):222-31.

[120] Herzig MC, Van Nostrand WE, Jucker M.(2006) Mechanism of cerebral beta-amyloid angiopathy: murine and cellular models. *Brain Pathol*;**16**(1):40-54.

[121] Prior R, D'Urso, D., Frank,R., Prikulis, I., Pavlakovic, G.(1996) Loss of vessel wall viability in cerebral amyloid angiopathy. *NeuroReport*;**7**:562.

[122] Deane R, Zlokovic BV.(2007) Role of the blood-brain barrier in the pathogenesis of Alzheimer's disease. *Curr Alzheimer Res*;**4**(2):191-7.

[123] Braunmuhl A.(1956) Kongophile angiopathie und senile plaques bei greisen hunden. *Arch Psychiatr Nervenkr*;**194**:395-414.

[124] Shimada A, Kuwamura, M., Akawkura, T., Umemura, T. , Takada, K., Ohama, E. and Itakura, C.(1992) Topographic relationship between senile plaques and cerebrovascular amyloidosis in the brain of aged dogs. *J Vet Med Sci*;**54**(1):137-44.

[125] Uchida K, Tani, Y., Uetsuka, K., Nakayama, H. and Goto, N.(1992) Immunohistochemical studies on canine cerebral amyloid angiopathy and senile plaques. *J Vet Med Sci*;**54**(4):659-67.

[126] Uchida K, Nakayama, H., Tateyama, S. and Goto, N.(1992) Immunohistochemical analysis of constituents of senile plaques and cerebro-vascular amyloid in aged dogs. *J Vet Med Sci*;**54**(5):1023-9.

[127] Uchida K, Okuda, R., Yamaguchi, R., Tateyama, S., Nakayama, H., and Goto, N.(1993) Double-labeling immunohistochemical studies on canine senile plaques and cerebral amyloid angiopathy. *J Vet Med Sci*;**55**(4):637-42.

[128] Uchida K, Kuroki, K., Yoshino, T., Yamaguchi, R., and Tateyama, S.(1997) Immunohistochemical study of constituents other than beta-protein in canine senile plaques and cerebral amyloid angiopathy. *93* (277-284).

[129] Uchida K, Miyauchi Y, Nakayama H, Goto N.(1990) Amyloid angiopathy with cerebral hemorrhage and senile plaque in aged dogs. *Nippon Juigaku Zasshi*;**52**(3):605-11.

[130] Uchida K, Nakayama H, Goto N.(1991) Pathological studies on cerebral amyloid angiopathy, senile plaques and amyloid deposition in visceral organs in aged dogs. *J Vet Med Sci*;**53**(6):1037-42.

[131] Yoshino T, Uchida, K., Tateyama, S., Yamaguchi, R., Nakayama, H. and Goto, N.(1996) A retrospective study of canine senile plaques and cerebral amyloid angiopathy. *Vet Pathol*;**33**:230-4.

[132] Frackowiak J, Mazur-Kolecka, B., Wisniewski, H.M., Potempska, A., Carroll, R.T., Emmerling, M.R., and Kim, K.S.(1995) Secretion and accumulation of Alzheimer's beta-protein by cultured vascular smooth muscle cells from old and young dogs. *Brain Research*;**676**:225-30.

[133] Prior R, D'Urso, D., Frank, R., Prikulis, I., Pavlakovic, G.(1995) Experimental deposition of Alzheimer amyloid beta-protein in canine leptomeningeal vessels. *NeuroReport*;**6**:1747-51.

[134] Prior R, D'Urso, D., Frank, R., Prikulis, I., Wihl, G. and Pavlakovic, G.(1996) Canine leptomeningeal organ culture: a new experimental model for cerebrovascular beta-amyloidosis. *J Neurosci Meth*;**68**:143-8.

[135] Wisniewski T, Lalowski, M., Bobik, M., Russell, M., Strosznajder, J. and Frangione, B.(1996) Amyloid Beta 1-42 deposits do not lead to Alzheimer's neuritic plaques in aged dogs. *Biochem J*;**313**:575-80.

[136] Walker LC.(1997) Animal models of cerebral beta-amyloid angiopathy. *Brain Research Reviews*;**25**:70-84.

[137] Lovell MA, Markesbery WR.(2007) Oxidative damage in mild cognitive impairment and early Alzheimer's disease. *J Neurosci Res*;**85**(14):3036-40.

[138] Lovell MA, Gabbita, S.P., and Markesbery, W.R.(1999) Increased DNA oxidation and decreased levels of repair products in Alzheimer's disease ventricular CSF. *Journal of Neurochemistry*;**72**:771-6.

[139] Lovell MA, Markesbery WR.(2008) Oxidatively modified RNA in mild cognitive impairment. *Neurobiol Dis*;**29**(2):169-75.

[140] Pratico D, Lee, M.Y., Trojanowski, J.Q., Rokach, J., and Fitzgerald, G. A.(1998) Increased F2-isoprostanes in Alzheimer's disease: evidence for enhanced lipid peroxidation in vivo. *Faseb Journal*;**12**(15):1777-83.

[141] Pratico D, Clark, C.M., Liun, F., Lee, V. Y.-M., Trojanowski, J.Q.(2002) Increase of brain oxidative stress in mild cognitive impairment. *Arch Neurol*;**59**:972-6.

[142] Pratico D, Delanty N.(2000) Oxidative injury in diseases of the central nervous system: focus on Alzheimer's disease. *Am J Med*;**109**(7):577-85.

[143] Montine TJ, Beal MF, Cudkowicz ME, O'Donnell H, Margolin RA, McFarland L, Bachrach AF, Zackert WE, Roberts LJ, Morrow JD.(1999) Increased CSF F2-isoprostane concentration in probable AD. *Neurology*;**52**(3):562-5.

[144] Montine TJ, Neely, M.D., Quinn, J.F., Beal, M.F., Markesbery, W.R., Roberts, L.J., and Morrow, J.D.(2002) Lipid peroxidation in aging brain and Alzheimer's disease. *Free Radical Biology & Medicine*;**33**(5):620-6.

[145] Butterfield DA.(2004) Proteomics: a new approach to investigate oxidative stress in Alzheimer's disease brain. *Brain Res*;**1000**(1-2):1-7.

[146] Butterfield DA, Reed T, Newman SF, Sultana R.(2007) Roles of amyloid beta-peptide-associated oxidative stress and brain protein modifications in the pathogenesis of Alzheimer's disease and mild cognitive impairment. *Free Radic Biol Med*;**43**(5):658-77.

[147] Smith CD, Carney JM, Starke-Reed PE, Oliver CN, Stadtman ER, Floyd RA, Markesbery WR.(1991) Excess brain protein oxidation and enzyme dysfunction in normal aging and in Alzheimer disease. *Proc Natl Acad Sci U S A*;**88**(23):10540-3.

[148] Smith MA, Sayre, L.M., Monnier, V.M., and Perry, G.(1996) Oxidative posttranslational modifications in Alzheimer's Disease. *Molecular and Chemical Neuropathology*;**28**:41-8.

[149] Smith MA, Rottkamp, C.A., Nunomura, A., Raina, A.K., and Perry, G.(2000) Oxidative stress in Alzheimer's disease. *Biochimica et Biophysica Acta*;**1502**:139-44.

[150] Ames BN, Shigenaga, M.K., and Hagen, T.M.(1993) Oxidants, antioxidants, and the degenerative diseases of aging. *Proc Natl Acad Sci USA*;**90**:7915-22.

[151] Rinaldi P, Polidori, M.C., Metastasio, A., Mariani, E., Mattioli, P., Cherubini, A., Catani, M., Cecchetti, R., Senin, U., MEcocci, P.(2003) Plasma antioxdiants are similarly depleted in mild cognitive impairment and in Alzheimer's disease. *Neurbiol Aging*;**24**:915-9.

[152] Markesbery WR, Lovell MA.(2007) Damage to lipids, proteins, DNA, and RNA in mild cognitive impairment. *Arch Neurol*;**64**(7):954-6.

[153] Keller JN, Schmitt FA, Scheff SW, Ding Q, Chen Q, Butterfield DA, Markesbery WR.(2005) Evidence of increased oxidative damage in subjects with mild cognitive impairment. *Neurology*;**64**(7):1152-6.

[154] Ding Q, Keller, J.N.(2001) Proteosomes and proteosome inhibition in the central nervous system. *Free Radical Biology & Medicine*;**31**(5):574-84.

[155] Stadtman ER.(1992) Protein oxidation and aging. *Science*;**257**:1220-4.

[156] Stadtman ER, Berlett BS.(1997) Reactive oxygen-mediated protein oxidation in aging and disease. *Chem Res Toxicol*;**10**(5):485-94.

[157] Head E, Liu J, Hagen TM, Muggenburg BA, Milgram NW, Ames BN, Cotman CW.(2002) Oxidative Damage Increases with Age in a Canine Model of Human Brain Aging. *Journal of Neurochemistry*;**82**:375-81.

[158] Skoumalova A, Rofina J, Schwippelova Z, Gruys E, Wilhelm J.(2003) The role of free radicals in canine counterpart of senile dementia of the Alzheimer type. *Exp Gerontol*;**38**:711-9.

[159] Kiatipattanasakul W, Nakamura S, Kuroki K, Nakayama H, Doi K.(1997) Immunohistochemical detection of anti-oxidative stress enzymes in the dog brain. *Neuropathology*;**17**:307-12.

[160] Opii WO, Joshi G, Head E, Milgram NW, Muggenburg BA, Klein JB, Pierce WM, Cotman CW, Butterfield DA.(2008) Proteomic identification of brain proteins in the canine model of human aging following a long-term treatment with antioxidants and a program of behavioral enrichment: relevance to Alzheimer's disease. *Neurobiol Aging*;**29**(1):51-70.

[161] Hwang IK, Yoon YS, Yoo KY, Li H, Choi JH, Kim DW, Yi SS, Seong JK, Lee IS, Won MH.(2008) Differences in Lipid Peroxidation and Cu,Zn-Superoxide Dismutase in the Hippocampal CA1 Region Between Adult and Aged Dogs. *J Vet Med Sci*;**70**(3):273-7.

[162] Papaioannou N, Tooten, P.C.J., van Ederen, A.M., Bohl, J.R.E., Rofina, J., Tsangaris, T., Gruys, E.(2001) Immunohistochemical investigation of the brain of aged dogs. I. Detection of neurofibrillary tangles and of 4-hydroxynonenal protein, an oxidative damage product, in senile plaques. *Amyloid: J Protein Folding Disord*;**8**:11-21.

[163] Head E, Nukala VN, Fenoglio KA, Muggenburg BA, Cotman CW, Sullivan PG.(2009) Effects of age, dietary, and behavioral enrichment on brain mitochondria in a canine model of human aging. *Exp Neurol*.

[164] Skoumalova A, Rofina J, Schwippelova Z, Gruys E, Wilhelm J.(2003) The role of free radicals in canine counterpart of senile dementia of the Alzheimer type. *Exp Gerontol*;**38**(6):711-9.

[165] Kempermann G, Kuhn, H.G. and Gage, F.H.(1997) More hippocampal neurons in adult mice living in an enriched evironment. *Nature*;**386**:493-5.

[166] Kempermann G, Kuhn, H.G., and Gage, F.H.(1998) Experience-induced neurogenesis in the senescent dentate gyrus. *J Neurosci*;**18**(9):3206-12.

[167] Cotman CW, Berchtold NC.(2002) Exercise: a behavioral intervention to enhance brain health and plasticity. *TINS*;**25**(6):295-30.

[168] Adlard PA, Perreau VM, Cotman CW.(2005) The exercise-induced expression of BDNF within the hippocampus varies across life-span. *Neurobiol Aging*;**26**(4):511-20.

[169] Poo MM.(2001) Neurotrophins as synaptic modulators. *Nat Rev Neurosci*;**2**(1):24-32.

[170] Colcombe SJ, Erickson, K.I., Raz, N., Webb, A.G., Cohen, N.J., McAuley, E., Kramer, A.F.(2003) Aerobic fitness reduces brain tissue loss in aging humans. *J Gerontol A Biol Sci Med Sci*;**58**(2):176-80.

[171] Colcombe SJ, Kramer, A.F., Erickson, K.I., Scalf, P., McAuley, E., Cohen, N.J., Webb, A., Jerome, G.J., Marquez, D.X., Elavsky, S.(2004) Cardiovascular fitness, cortical plasticity, and aging. *Proc Natl Acad Sci U S A*;**101**(9):3316-21.

[172] Wilson RS, Mendes de Leon, C.F., Barnes, L.L., Schneider, J.A., Bienias, J.L., Evans, D.A., Bennett, D.A.(2002) Participation in cognitively stimulating activities and risk of incident Alzheimer disease. *JAMA*;**287**:742-8.

[173] Verghese J, Lipton, R.B., Katz, M.J., Hall, C.B., Derby, C.A., Kuslansky, G., Ambrose, A.F.,Sliwinski, M., Buschke, H.(2003) Leisure activities and the risk of dementia in the elderly. *N Engl J Med*;**348**(25):2508-16.

[174] Friedland RP, Fritsch,T., Smyth, K.A., Koss, E., Lerner, A.J., Chen, C.H., Petot, G.J., Debanne, S.M.(2001) Patients with Alzheimer's disease have reduced activities in midlife compared with healthy control-group members. *PNAS*;**98**(6):3440-5.

[175] Ball K, Berch, D.B., Helmers, K.F., Jobe, J.B., Leveck, M.D., Marsiske, M., Morris, J.N., Rebok, G.W., Smith, D.M., Tennstedt, S.L., Unverzagt, F.W., Willis, S.L. for the ACTIVE Study Group.(2002) Effects of cognitive training interventions with older adults: A randomized controlled trial. *JAMA*;**288**:2271-81.

[176] Yesavage JA.(1985) Nonpharmacologic treatments for memory losses with normal aging. *Am J Psychiatry*;**142**(5):600-5.

[177] Willis SL, Tennstedt SL, Marsiske M, Ball K, Elias J, Koepke KM, Morris JN, Rebok GW, Unverzagt FW, Stoddard AM, Wright E.(2006) Long-term effects of cognitive training on everyday functional outcomes in older adults. *Jama*;**296**(23):2805-14.

[178] Bickford PC, Gould T, Briederick L, Chadman K, Pollock A, Young D, Shukitt-Hale B, Joseph J.(2000) Antioxidant-rich diets improve cerebellar physiology and motor learning in aged rats. *Brain Res*;**866**(1-2):211-7.

[179] Socci DJ, Crandall BM, Arendash GW.(1995) Chronic antioxidant treatment improves the cognitive performance of aged rats. *Brain Research*;**693**(1-2):88-94.

[180] Joseph JA, Shukkitt-Hale, B., Denisova, N.A., Prior, R.L., Cao, G., Martin, A., Taglialatela, G., and Bickford, P.C.(1998) Long-term dietary strawberry, spinach or vitamin E supplementation retards the onset of age-related neuronal signal-transduction and cognitive behavioral deficits. *J Neurosci*;**18**(19):8047-55.

[181] Joseph JA, Shukkitt-Hale, B., Denisova, N.A., Bielinski, D., Martin, A., McEwen, J.J., Bickford, P.C.(1999) Reversals of age-related declines in neuronal signal transduction, cognitive and motor behavioral deficits with blueberry, spinach, or strawberry dietary supplementation. *The Journal of Neuroscience*;**19**(18):8114-21.

[182] Engelhart MJ, , Geerlings, M.I., Ruitenberg, A., van Swieten, J.C., Hofman, A., Witteman, J.C., Breteler, M.M.(2002) Dietary intake of antioxidants and risk of Alzheimer disease. *JAMA*;**287**(24):3223-9.

[183] Morris MC, Evans, D.A., Bienias, J.L., Tangney, C.C., Wilson, R.S.(2002) Vitamin E and cognitive decline in older persons. *Arch Neurol*;**59**:1125-32.

[184] Luchsinger JA, Tang, M.X., Shea, S., Mayeux, R.(2003) Antioxidant vitamin intake and risk of Alzheimer disease. *Arch Neurol*;**60**(2):203-8.

[185] Masaki KH, Losonczy, K.G., Izmirlian, G., Foley, D.J., Ross, G.W., Petrovitch, H., Havlik, R., and White, L.R.(2000) Association of vitamin E and C supplement use with cogntive function and dementia in elderly men. *Neurology*;**54**:1265-72.

[186] Luchsinger JA, Tang MX, Shea S, Mayeux R.(2003) Antioxidant vitamin intake and risk of Alzheimer disease. *Arch Neurol*;**60**(2):203-8.

[187] Zandi PP, Anthony, J.C., Khachaturian, A.S., Stone, S.V., Gustafson, D., Tschanz, J.T., Norton,C., Welsh-Bohmer, K.A., Breitner, J.C., Group. CCS.(2004) Reduced risk of Alzheimer disease in users of antioxidant vitamin supplements: the Cache County Study. *Arch Neurol*;**61**(1):82-8.

[188] Morris MC, Evans DA, Bienias JL, Tangney CC, Bennett DA, Aggarwal N, Wilson RS, Scherr PA.(2002) Dietary intake of antioxidant nutrients and the risk of incident Alzheimer disease in a biracial community study. *Jama*;**287**(24):3230-7.

[189] Rutten BP, Steinbusch, H.W., Korr, H., Schmitz, C.(2002) Antioxidants and Alzheimer's disease: from bench to bedside (and back again). *Curr Opin Clin Nutr Metab Care*;**5**(6):645-51.

[190] Sano M, Ernesto, C., Thomas, R.G., Klauber, M.R., Schafer, K., Grundman, M., Woodbury, P., Growdon, J., Cotman, C.W., Pfeiffer, E., Schneider, L.S. and Thal, L.J.(1997) A controlled trial of selegiline, alpha-tocopherol, or both as treatment for Alzheimer's disease. *The New England Journal of Medicine*;**336**:1216-22.

[191] Petersen RC, Thomas RG, Grundman M, Bennett D, Doody R, Ferris S, Galasko D, Jin S, Kaye J, Levey A, Pfeiffer E, Sano M, van Dyck CH, Thal LJ.(2005) Vitamin E and donepezil for the treatment of mild cognitive impairment. *N Engl J Med*;**352**(23):2379-88.

[192] Kang JH, Cook N, Manson J, Buring JE, Grodstein F.(2006) A randomized trial of vitamin E supplementation and cognitive function in women. *Arch Intern Med*;**166**(22):2462-8.

[193] Chandra RK.(2001) Effect of vitamin and trace-element supplementation on cognitive function in elderly subjects. *Nutrition*;**17**:709-12.

[194] AAFCO. Association of American Feed Control Officials, AAFCO dog and cat food nutrient profiles. . Incorporated PoAFCO, editor. West Lafayette, IN: AAFCO; 1999. 134-44 p.

[195] Milgram NW, Head E, Zicker SC, Ikeda-Douglas CJ, Murphey H, Muggenburg B, Siwak C, Tapp D, Cotman CW.(2005) Learning ability in aged beagle dogs is preserved by behavioral enrichment and dietary fortification: a two-year longitudinal study. *Neurobiol Aging*;**26**(1):77-90.

[196] Milgram NW, Head, E., Zicker, S.C., Ikeda-Douglas, C., Murphey, H., Muggenberg, B.A., Siwak, C.T., Dwight, Tapp. P., Lowry, S.R., Cotman. C,W.(2004) Long-term treatment with antioxidants and a program of behavioral enrichment reduces age-dependent impairment in discrimination and reversal learning in beagle dogs. *Exp Gerontol*;**39**(5):753-65.

[197] Nippak PM, Mendelson J, Muggenburg B, Milgram NW.(2007) Enhanced spatial ability in aged dogs following dietary and behavioural enrichment. *Neurobiol Learn Mem*;**87**(4):610-23.

[198] Siwak CT, Tapp PD, Head E, Zicker SC, Murphey HL, Muggenburg BA, Ikeda-Douglas CJ, Cotman CW, Milgram NW.(2005) Chronic antioxidant and mitochondrial cofactor administration improves discrimination learning in aged but not young dogs. *Prog Neuropsychopharmacol Biol Psychiatry*;**29**(3):461-9.

[199] Butterfield DA.(1997) beta-Amyloid-associated free radical oxidative stress and neurotoxicity: implications for Alzheimer's disease. *Chem Res Toxicol*;**10**(5):495-506.

[200] Pop V, Head E, Nistor M, Milgram NW, Muggenburg BA, Cotman CW.(2003) Reduced Aβ deposition with long-term antioxidant diet treatment in aged canines. *Society for Neuroscience, Abstract Viewer/Itinerary Planner* **525.4**.

[201] Weggen S, Eriksen JL, Das P, Sagi SA, Wang R, Pietrzik CU, Findlay KA, Smith TE, Murphy MP, Bulter T, Kang DE, Marquez-Sterling N, Golde TE, Koo EH.(2001) A subset of NSAIDs lower amyloidogenic Abeta42 independently of cyclooxygenase activity. *Nature*;**414**(6860):212-6.

[202] Refolo LM, Pappolla MA, LaFrancois J, Malester B, Schmidt SD, Thomas-Bryant T, Tint GS, Wang R, Mercken M, Petanceska SS, Duff KE.(2001) A cholesterol-lowering drug reduces beta-amyloid pathology in a transgenic mouse model of Alzheimer's disease. *Neurobiol Dis*;**8**(5):890-9.

[203] Sparks DL, Sabbagh M, Connor D, Soares H, Lopez J, Stankovic G, Johnson-Traver S, Ziolkowski C, Browne P.(2006) Statin therapy in Alzheimer's disease. *Acta Neurol Scand Suppl*;**185**:78-86.

[204] Alberts AW.(1990) Lovastatin and simvastatin - inhibitors of HMG CoA reductase and cholesterol biosynthesis. *Cardiology*;**77**(4):14-21.

[205] Gerson RJ, MacDonald JS, Alberts AW, Kornbrust DJ, Majka JA, Stubbs RJ, Bokelman DL.(1989) Animal safety and toxicology of simvastatin and related hydroxy-methylglutaryl-coenzyme A reductase inhibitors. *Am J Med*;**87**(4A):28S-38S.

[206] Das P, Howard V, Loosbrock N, Dickson D, Murphy MP, Golde TE.(2003) Amyloid-beta immunization effectively reduces amyloid deposition in FcRgamma-/- knock-out mice. *J Neurosci*;**23**(24):8532-8.

[207] Janus C, Pearson J, McLaurin J, Mathews PM, Jiang Y, Schmidt SD, Chishti MA, Horne P, Heslin J, French J, Mount HT, Nixon RA, Mercken M, Bergeron C, Fraser PE, St George-Hyslop P, Westaway D.(2000) A beta peptide immunization reduces behavioural impairment and plaques in a model of Alzheimer's disease. *Nature*;**408**(6815):979-82.

[208] Morgan D, Diamond DM, Gottschall PE, Ugen KE, Dickey C, Hardy J, Duff K, Jantzen P, DiCarlo G, Wilcock D, Connor K, Hatcher J, Hope C, Gordon M, Arendash GW.(2000) A beta peptide vaccination prevents memory loss in an animal model of Alzheimer's disease. *Nature*;**408**(6815):982-5.

[209] Schenk D, Barbour, R., Dunn, W., Gordon, G., Grajeda, H., Guido, T., Hu, K., Huang, J., Johnson-Wood, K., Khan, K., Kholodenko, D., Lee, M., Liao, Z., Lieberburg, I., Motter, R., Mutter, L., Soriano, F., Shopp, G., Vasquez, N., Vandervert, C., Walker, S., Wogulis, M., Yednock, T., Games, D., and Seubert, P.(1999) Immunization with amyloid-β attentuates Alzheimer-disease-like pathology in the PDAPP mouse. *Nature*;**400**:173-7.

[210] Sigurdsson EM, Knudsen E, Asuni A, Fitzer-Attas C, Sage D, Quartermain D, Goni F, Frangione B, Wisniewski T.(2004) An attenuated immune response is sufficient to enhance cognition in an Alzheimer's disease mouse model immunized with amyloid-beta derivatives. *J Neurosci*;**24**(28):6277-82.

[211] Dodart JC, Bales KR, Gannon KS, Greene SJ, DeMattos RB, Mathis C, DeLong CA, Wu S, Wu X, Holtzman DM, Paul SM.(2002) Immunization reverses memory deficits without reducing brain Abeta burden in Alzheimer's disease model. *Nat Neurosci*;**5**(5):452-7.

[212] Kotilinek LA, Bacskai B, Westerman M, Kawarabayashi T, Younkin L, Hyman BT, Younkin S, Ashe KH.(2002) Reversible memory loss in a mouse transgenic model of Alzheimer's disease. *J Neurosci*;**22**(15):6331-5.

[213] Morley JE, Farr SA, Flood JF.(2002) Antibody to amyloid beta protein alleviates impaired acquisition, retention, and memory processing in SAMP8 mice. *Neurobiol Learn Mem*;**78**(1):125-38.

[214] Wilcock DM, Rojiani A, Rosenthal A, Levkowitz G, Subbarao S, Alamed J, Wilson D, Wilson N, Freeman MJ, Gordon MN, Morgan D.(2004) Passive amyloid immunotherapy clears amyloid and transiently activates microglia in a transgenic mouse model of amyloid deposition. *J Neurosci*;**24**(27):6144-51.

[215] Hock C, Konietzko U, Streffer JR, Tracy J, Signorell A, Muller-Tillmanns B, Lemke U, Henke K, Moritz E, Garcia E, Wollmer MA, Umbricht D, de Quervain DJ, Hofmann M, Maddalena A, Papassotiropoulos A, Nitsch RM.(2003) Antibodies against beta-amyloid slow cognitive decline in Alzheimer's disease. *Neuron*;**38**(4):547-54.

[216] Gilman S, Koller M, Black RS, Jenkins L, Griffith SG, Fox NC, Eisner L, Kirby L, Boada Rovira M, Forette F, Orgogozo JM.(2005) Clinical effects of A{beta} immunization (AN1792) in patients with AD in an interrupted trial. *Neurology*.

[217] Ferrer I, Boada Rovira M, Sanchez Guerra ML, Rey MJ, Costa-Jussa F.(2004) Neuropathology and pathogenesis of encephalitis following amyloid-beta immunization in Alzheimer's disease. *Brain Pathol*;**14**(1):11-20.

[218] Nicoll JA, Wilkinson D, Holmes C, Steart P, Markham H, Weller RO.(2003) Neuropathology of human Alzheimer disease after immunization with amyloid-beta peptide: a case report. *Nat Med*;**9**(4):448-52.

[219] Masliah E, Hansen L, Adame A, Crews L, Bard F, Lee C, Seubert P, Games D, Kirby L, Schenk D.(2005) Abeta vaccination effects on plaque pathology in the absence of encephalitis in Alzheimer disease. *Neurology*;**64**(1):129-31.

[220] Holmes C, Boche D, Wilkinson D, Yadegarfar G, Hopkins V, Bayer A, Jones RW, Bullock R, Love S, Neal JW, Zotova E, Nicoll JA.(2008) Long-term effects of Abeta42 immunisation in Alzheimer's disease: follow-up of a randomised, placebo-controlled phase I trial. *Lancet*;**372**(9634):216-23.

[221] Milgram NW, Head, E., Zicker, SC., Ikeda-Douglas, C.J., Murphey, H. Muggenburg, B., Siwak, C., Tapp, D, Cotman, C.W.(2005) Learning ability in aged beagle dogs is preserved by behavioral enrichment and dietary fortification: A two-year longitudinal study. *Neurobiol Aging*;**26**:77-90.

[222] Head E, Pop V, Vasilevko V, Hill M, Saing T, Sarsoza F, Nistor M, Christie LA, Milton S, Glabe C, Barrett E, Cribbs D.(2008) A two-year study with fibrillar beta-amyloid (Abeta) immunization in aged canines: effects on cognitive function and brain Abeta. *J Neurosci*; **28**(14):3555-66.

[223] Lai ZC, Moss MB, Killiany RJ, Rosene DL, Herndon JG.(1995) Executive system dysfunction in the aged monkey: spatial and object reversal learning. *Neurobiol Aging*;**16**:947-54.

[224] Rapp PR.(1990) Visual discrimination and reversal learning in the aged monkey (*Macaca mulatta*). *Behav Neurosci*;**6**:876-84.

[225] Cronin-Golomb A. Color vision, object recognition, and spatial localization in aging and Alzheimer's disease. In: Hof PR, Mobbs CV, editors. Functional Neurobiology of Aging. San Diego: Academic Press; 2001. p 517-29.

[226] Boutet I, Milgram NW, Freedman M.(2007) Cognitive decline and human (Homo sapiens) aging: an investigation using a comparative neuropsychological approach. *J Comp Psychol*;**121**(3):270-81.

[227] Freedman M, Oscar-Berman M.(1989) Spatial and visual learning deficits in Alzheimer's disease and Parkinson's disease. *Brain and Cognit*;**11**:114-26.

[228] Emborg ME, Ma SY, Mufson EJ, Levey AI, Taylor MD, Brown WD, Holden JE, Kordower JH.(1998) Age-related declines in nigral neuronal function correlate with motor impairments in rhesus monkeys. *J Comp Neurol*; **401**(2):253-65.

[229] Kordower JH, Liu YT, Winn S, Emerich DF.(1995) Encapsulated PC12 cell transplants into hemiparkinsonian monkeys: a behavioral, neuroanatomical, and neurochemical analysis. *Cell Transplant*;**4**(2):155-71.

[230] Christie L, Saunders RC, Kowalska D, MacKay WA, Head E, Cotman CW, Milgram NW.(In press) Rhinal and Dorsolateral Prefrontal Cortex Lesions Produce Selective Impairments in Object and Spatial Learning and Memory in the Canine. *Journal of Comparative Anatomy*.

[231] Lezak MD, Howieson DB, Loring DW. Neuropsychological Assessment. 4th ed. New York: Oxford University Press; 2004.

[232] Arnsten AFT, and Goldman-Rakic, P.S.(1985) Alpha 2-adrenergic mechanisms in prefrontal cortex associated with cognitive decline in aged nonhuman primates. *Science*;**230**:1273-6.

[233] Walker LC, Kitt CA, Struble RJ, Wagster MV, Price DL, Cork LC.(1988) The neural basis of memory decline in aged monkeys. *Neurobiol Aging*;**9**:657-66.

[234] Kowalska DM.(1995) Effects of hippocampal lesions on spatial delayed responses in dog. *Hippocampus*;**5**:363-70.

[235] Warren JM. The behavior of carnivores and primates with lesions in the prefrontal cortex. In: Warren JM, Akert K, editors. The Frontal Granular Cortex and Behavior. New York: McGraw-Hill Book Company; 1964. p 168-91.

[236] Kramer JH, Quitania L. Bedside Frontal Lobe Testing. In: Miller BL, Cummings, J.L., editor. The Human Frontal Lobes, 3rd Edition. New York: The Guilford Press; 2007. p 279-91.

[237] Rapp PR, Kansky MT, Eichenbaum H.(1996) Learning and memory for hierarchical relationships in the monkey: effects of aging. *Behav Neurosci*;**110**(5):887-97.

[238] Pohl W.(1973) Dissociation of spatial discrimination deficits following frontal and parietal lesions in monkeys. *J Comp Physiol Psychol*;**82**:227-39.

# Unveiling "The Switch" from Aging to Alzheimer's Disease with the Senescence-Accelerated Mouse Model (SAMP8)

Jaewon Chang[d], Merce Pallas[a], Xiongwei Zhu[b], Hyun-Jin Kim[d], Antoni Camins[a], Hyoung-gon Lee[b], George Perry[b,c], Mark A. Smith[b], and Gemma Casadesus[d,*]

[a]Unitat de Farmacologia, Facultat de Farmàcia Institut de Biomedicina Universitat de Barcelona (IBUB).
Nucli Universitari de Pedralbes E-08028 Barcelona, Spain.
[b]Department of Pathology, Case Western Reserve University, Cleveland, Ohio, USA
[c]College of Sciences, University of Texas at San Antonio, San Antonio, Texas, USA
[d]Department of Neurosciences, Case Western Reserve University, Cleveland, Ohio, USA

**Abstract.** Current mouse models of Alzheimer's disease (AD) are restricted to the expression of AD-related pathology associated with specific mutations present in early-onset familial AD and thus represent <5% of AD cases. To date there are no mouse lines that model late-onset/age-related AD, the feature which accounts for the vast majority of cases. As such, based on current mutation-associated models, the chronology of events that lead to the disease in the aged population is difficult to establish. However, published data show that senescence-accelerated mouse (SAMP8), as a model of aging, display many features that are known to occur early in the pathogenesis of AD such as increased oxidative stress, amyloid-β alterations, and tau phosphorylation. Therefore, SAMP8 mice may be an excellent model for studying the earliest neurodegenerative changes associated with AD and provide a more encompassing picture of human disease, a syndrome triggered by a combination of age-related events. Here, the neurochemical, neuropathological, and behavioral alterations, characterized in SAMP8 mice are critically reviewed and discussed in relation to the potential use of this mouse model in the study of AD pathogenesis.

Keywords: aging, Alzheimer's disease, amyloid-β, animal models, cognition, gliosis, neurodegeneration, oxidative stress, tau phosphorylation, SAMP8

## INTRODUCTION

The necessity for a deeper understanding of neurological diseases such as Alzheimer's disease (AD) that increase in frequency as a function of age has become of paramount importance with the coming of age of the baby boom generation and the increasing social demands for individuals to perform better and longer. By the year 2050, 30% of the total population will be over 65 years of age and a large number of those individuals will show signs of neurodegenerative disease which will present significant economical and emotional burdens to not only the health system but, more importantly, to families of those affected and to the individuals themselves.

Several signaling cascades such as IGF-1 and PIP3 have been intimately associated with the aging process, and progress is being made towards mechanistically linking them to AD pathology [23]. Another promising target is ceramide, a lipid precursor in the production sphingomyelin, a crucial component of membranes that is increased during aging. Ceramide is involved in many crucial events associated with the development of AD, including cell cycle regulation, inflammatory activity and cholesterol metabolism to name a few. The fact that it increases during aging and that it may have a pivotal role in mediating the relationship between lipid composi-

---

*Address for correspondence: Gemma Casadesus, Ph.D., Case Western Reserve University, 2109 Adelbert Road, Cleveland, Ohio 44106 USA.Tel.: 216-368-8903; Email: gxc40@case.edu.

tion and the regulation and production of the amyloid-β (Aβ) peptide, makes this molecule a potential mechanism associated with the switch from aging to AD [22,23]. However, as discussed and contrasted by Casadesus and colleagues [16], the pleuripotent activity of ceramide indicates an intrinsic function in rescue machinery to protect cells subject to degeneration [55]. In relation to AD, one of the earliest events that occur in AD is oxidative stress [65] and increased ceramide also occurs early in AD [36]. Therefore, one possibility, in accord with the stress-response activation of the spyngomyalin pathway, would be that increased oxidative stress would lead to activation of this pathway that could then signal the biochemical messages associated with cell protection [72,98]. Increasing our mechanistic understanding of brain aging and identifying the chronology of appearance of pathological markers and how they interact to activate the "molecular switch" that turns benign aging into neurodegeneration seem a crucial step in the development of future therapeutic strategies to prevent or hinder the progression of neurodegenerative disease such as AD.

This task is currently difficult for several reasons. One in particular is the fact that mouse models of AD are mainly restricted to the overexpression of genes with specific mutations associated with early-onset familial AD, which only account for <5% of total AD cases. To date there are no mouse lines that model late-onset/age-related AD. However, the senescence-accelerated mouse strains, in particular the SAMP8 strain, may provide an excellent model to study the earliest neurodegenerative changes associated with AD and provide a more encompassing picture of the disease; a syndrome brought about by a combination of age-related events [99,101].

The senescence-accelerated mouse (SAM), was originally generated from AKR/J mice in the University of Kyoto by Dr. Takeda [91]. After conventional inbred mating, the observation that several litters showed features indicative of accelerated aging, including loss of hair, lordokyphosis, periophthalmic disorders, loss of activity and shortened life expectancy, among other characteristics, led to selective breeding for these phenotypes using sister-brother mating, which through subsequent generations became the senescence-prone, short-lived inbred strain (SAMP) (mean lifespan of 9.7 months). Littermates of the age-accelerated mice which did not show senescence-related phenotypes were bred in the same fashion to obtain the senescence-resistant, longer-lived mouse strain (SAMR) (mean life expectancy of 16.3 months); this is longer than the SAMP8 but shorter than a normal inbred mouse which has a mean life expectancy of 28 months. Nine SAMP and three SAMR sub-strains exist, each exhibiting slightly different phenotypic abnormalities. However, the main characteristic of all SAMP strains is normal development and maturity of reproductive function followed by an early manifestation of senescence-related phenotypes. While no animal model mimics exactly all of the features that are present in physiological disease or aging, the SAMP8 strain is particularly well-suited to study the "transitional switch" between aging and AD as it shares many neuropathological, neurochemical, and importantly cognitive abnormalities found in aged individuals and, to a greater extent, in patients with AD.

## AD-LIKE COGNITIVE ABNORMALITIES IN THE SAMP8

In normal aging humans and, individuals with AD to a more severe degree, episodic memories [56,68] as well as working and spatial memories [80,93] show progressive decline. Unfortunately, the mechanisms responsible for this type of behavioral decline and associated neuronal changes are yet to be elucidated. However, it is known that the hippocampus, a highly plastic area of the brain, is crucial in the modulation of episodic and spatial memory but not working memory. Importantly, the hippocampus is also one of the most age-sensitive areas in the brain and it is thought that the aging process greatly diminishes the plastic capabilities of this region and these declines lead to the age-related impairments in cognitive output (for review, see [25]s).

As such, a valid AD-transgenic model needs to faithfully reflect the behavioral changes observed in human AD patients. To accurately interpret behavioral results from transgenic mouse models of AD, it is important to intimately understand the behavioral tasks that are most often used to test cognitive changes in mice as well as what each cognitive test is actually measuring [10]. The SAMP8 mouse has been extensively phenotyped for behavioral abnormalities and results indicate that cognitive abnormalities in this model are abundant and mimic behavioral/cognitive deficits observed in AD patients and other transgenic models.

*Spatial learning and memory tasks*

The Morris water maze (MWM), Radial arm maze (RAM) and Radial arm water maze (RAWM) are three cognitive tasks that measure spatial memory, the part of memory responsible for recording information about one's environment and its spatial orientation, and is dependent on hippocampal function [54]. These tasks therefore are particularly sensitive to examine age-related/AD-like deficits. The goal of these tasks often requires finding an escape or a reward by remembering their location in relation to cues positioned in the environment. Tasks like the MWM can be used to measure spatial learning and memory as well as the capacity of the animal to retrieve and retain learned information. Additionally, the flexibility to purge and re-learn new strategies can be determined using a probe trial and reversal trial [63,82]. On the other hand, one of the limitations of the MWM is that different components of memory, i.e., reference and working memories, cannot be tested simultaneously. As such, one task that can accommodate simultaneous measurement of memory components and one that has also been widely used to study spatial memory performance in rodents is the RAM [43,83]. In this task, the rodent has to remember the configuration of rewarded arms using environmental cues. Specifically, the animal must remember which arms are baited (reference memory) as well as which arms have already entered (working memory). However, while this task permits the examination of both reference and working memory, major limitations are the use of food or water deprivation in this task as well as the presence of odor confounds.

One task that encompasses the best of both worlds is the relatively new spatial memory task, the RAWM. The RAWM differs from the MWM and RAM in that performance in the RAWM requires finding a platform that is submerged in water located in one of several arms. This makes the task slightly more difficult, but forces the animal to use spatial cues and working memory (keeping track of the arms it has already visited) to remember where the platform is located [2,15].

Learning and memory changes in the SAMP8 mice have been extensively studied using these cognitive measures [30]. Spatial memory impairments are detected using spatial tasks such as the MWM and the RAM beginning at approximately 4 months of age [21,29,42]. However, other studies suggest that spatial impairments appear earlier when using more sensitive tests such as the RAWM. In this regard, it has been shown in the RAWM, rather than in the MWM, that impairments in spatial learning can be detected in SAMP8 mice as early as 3 months old, and the impairment in spatial memory in SAMP8 mice aged 5 months. These results indicate that spatial learning and memory is affected in the SAMP8 mice early on, and that RAWM and MWM each measures different aspects of spatial learning and memory [20].

*Associative and non-spatial learning tasks*

Fear conditioning is a widely used test to measure hippocampal-dependent and associative memory. This test is thought to be sensitive to emotion-associated learning and therefore is a useful measure of amygdalar-hippocampal communication. This test involves measuring a fear response in the animal (expressed as freezing posture) after it has been trained to associate a tone with a mild shock. Hippocampal function is measured by placing the animal in the context in which it previously received the aversive stimulus (Contextual fear conditioning). An animal that has associated the spatial configuration (hippocampal) with the shock will freeze in the absence of the aversive stimulus. The associative conditioning component is measured by presenting the animals with the tone in a context different than that in which it previously received the aversive stimulus (cued fear conditioning). In this portion of the test, the animal will freeze in response to the tone regardless of the context. The SAMP8 mouse, as well as transgenic mouse models of AD, display impairments in this task. As such, after conditioning, SAMP8 mice at 4 and 8 months show weaker contextual fear (hippocampal) compared to SAMR1 mice. However, these differences are not evident at earlier ages indicating that SAMP8 mice show age-related deficits in contextual (context – no tone) but not associative (tone only) memory [67]. This is particularly relevant since age-related hippocampal contextual memory function rather than an associative memory dysfunction is an earlier deficit in AD patients.

A simpler associative-type task is the passive-avoidance learning test in which the animal must learn to avoid a mild aversive stimulus presented in a preferred environment (dark), by remaining in a non-preferred environment (bright) [50,57,79]. Interestingly, while associative learning in the fear conditioning task is not affected in the SAMP8 mouse, both passive and active avoidance (i.e., learning to escape by exiting the chamber in which it previously received the aversive stimulus) show significant declines starting as early as 2 months of age [59,60]. Also, the SAMP8 mouse exhibits reduced anxiety-like behavior [58]. Since it is well known that anxiety/fear potentiates the encoding of memory by the hippocampus, deficits in this task may therefore be, at least partially, due to the lack of efficient processing of the amygdala (emotional center) which may then lead to poorer encoding of emotional-related memory. Such a notion is supported by the lack of differences between SAMP8 and controls in the associative-memory fear conditioning task (tone – no shock) which uses a more robust form of conditioning (loud tone rather than environmental preference) presented repeatedly and not in a single trial. Furthermore, Flood and Morley [27] also failed to detect any impairment of learning or retention on step-down passive avoidance as a function of age in the SAMP8 mice from 4 to 12 months of age.

Taken together, and again similar to AD patients and other AD-mouse models, these findings indicate that tasks associated with hippocampal involvement are more consistently affected in the SAMP8 mouse than tasks that are not associated. Furthermore, as the difficulty of the task increases, the SAMP8 strains demonstrate progressive age-related deficits at earlier ages. Likewise, while tasks such as passive avoidance are sensitive at early ages in the SAMP8 mouse, it is unclear whether these deficits are related to either memory function or differences in how emotional-related stimuli are processed.

## AD-RELATED NEUROPATHOLOGICAL ABNORMALITIES IN THE SAMP8 MOUSE

### Oxidative stress

The free radical theory of aging postulates that oxidative modifications by reactive oxygen species (ROS) on proteins, DNA, lipid membranes and other molecules lead to cellular dysfunction and aging in humans and animals [39]. Physiologically, ROS are found in all aerobic organisms and arise from the secondary production of superoxide by mitochondrial reduction of molecular oxygen, the production of $H_2O_2$ by oxidases such as monoamine oxidase as well as the reaction of superoxide with nitric oxide to yield peroxynitrite, which is capable of both oxidation and nitration reactions. Although a host of individual ROS have been identified, the hydroxyl radical is the principal ROS implicated in biologically relevant oxidative stress and is responsible, either directly or indirectly, for most of the free radical damage seen in AD [18,66,71,84-88]. Furthermore, mitochondrial manganese superoxide dismutase (MnSOD) and the constitutive cytoplasmic CuZn-SOD enzymes convert reactive superoxide to harmless $O_2$ and $H_2O_2$, the latter is then removed by catalase and peroxidases, which are ubiquitous in tissues. Both these enzymes have been investigated for their role in oxidative stress and AD and shown to be altered in this disease [6,24,32,70].

In this light, a valid animal model of accelerated aging/AD needs to show alterations in oxidative stress similar to humans and that modulation of oxidative stress consequently leads to improvements of age-related symptoms such as cognitive dysfunction. Indeed, there are ample data to support this in the SAMP8 mice. In this regard, MnSOD [49], catalase [77] and glutathione peroxidase [69] are all decreased in the SAMP8 compared to its control, the SAMR1. In parallel, SAMP8 mice as early as 5 months of age exhibit increased lipid peroxidation and carbonyl damage as well as decreased levels of superoxide dismutase activity compared with age-matched SAMR1 [1,73,75]. Early increases in lipid peroxidation are also supported by earlier studies demonstrating that SAMP8 show strain specific increases in this ROS as early as 2 months of age [97]. However, the literature on oxidative stress in the SAMP8 mouse is not always unequivocal. For example, while Alvarez-Garcia [1] found no differences in activity of catalase and glutathione reductase, another study in animals of the same age shows the opposite, demonstrating that while glutathione-S-transferase was not affected, glutathione reductase and catalase were significantly inhibited compared to the SAMR1 [89]. Interestingly some of the changes in these enzymes were only significantly altered in males, therefore, at least partially explaining some of the discrepancies associated with the oxidative stress findings in the SAMP8. This latter issue underscores the importance of sex-specific analysis and analyzing large enough sample numbers when carrying out such studies.

### Tau phosphorylation

Tau proteins constitute a group of 6 isoforms that are involved in tubulin polymerization, microtubule stabilization, and thus axonal transport as well as cytoskeleton signaling. Moreover, tau phosphorylation is involved in various neurodegenerative disorders, since hyperphosphorylation inhibits microtubule assembly, promoting microtubule instability and, ultimately, degeneration of the affected neurons [9,17,45]. In fact, hyperphosphorylated tau is the main component of paired helical filament of neurofibrillary lesions in neurodegenerative diseases like AD, which are found in neural formations where neurons degenerate: in cellular bodies forming neurofibrillary tangles, in neuritic extensions and dystrophic neurites, surrounding amyloid plaques and in senile plaques [41]. A number of serine/threonine kinases that are responsible for this phosphorylation have been identified, including GSK3β, CDK-5, PKC and SAPK [3,26,44]. Importantly, alterations of tau processing affect entorhinal cortex and the hippocampus [7,8], the first area that is deteriorated in AD [12], and regions highly associated with cognitive function [90]. In humans, it has been demonstrated that alterations in tau are highly associated with the aging process [3]. The fact that age-related tau hyperphosphorylation is associated to degeneration of neurons that parallel the earliest pattern of degeneration observed in AD, makes this hallmark an important one to have in a valid model of aging/early AD regardless of their etiological significance [19,52].

Indeed, earlier studies have demonstrated that abnormal levels of phosphorylated tau are present in aged (11 month old) SAMP8 compared to SAMR1 mice [94]. However, recent studies provide a more in depth understanding of tau phosphorylation dynamics in the SAMP8 mouse. As such, data indicates that various forms of hyperphosphorylated tau are present in the SAMP8 compared to SAMR1 [13,14]. Importantly, these comprehensive studies have also demonstrated that the hyperphosphorylation of tau in the SAMP8 mouse is driven by AD-related mechanisms such as increases in Cdk5 expression [14]. Treatments, such as melatonin [34] and lithium chloride, that reduce Cdk5 and GSK3β activation and activation of the cdk5/p35 pathway at its cleavage to cdk5/p25, all known key players in AD-related hyperphosphorylation of tau during aging and neurodegenerative diseases [35,100], lead to reduction in tau hyperphosphorylation when administered to SAMP8 mice. Importantly, tau hyperphosphorylation increases occur as early as 5 months of age [1], suggesting that this process is an integral part of aging and, like oxidative stress, an early event in AD.

## Amyloid-β

Senile plaques are another pathological hallmark of AD and have fostered the leading hypothesis for the development of this disease [37,38]. The Aβ cascade hypothesis, while not unrefuted [53], is driven by the fact that genetically-linked AD is associated with mutations in either the amyloid-β protein precursor (AβPP) or presenilins-1/2 (PS1/2) that affect the processing of Aβ and contribute to the accumulation of Aβ in neurons and consequent formation of senile plaques [78]. Additionally, the accumulation of senile plaques in some, but not all, models of AD is associated with cognitive impairment [11]. In the SAMP8 mice, early studies showed an age-related increase of Aβ peptide using a polyclonal antibody to $Aβ_{1-42}$ [92] and $Aβ_{1-40}$ [31] compared to control SAMR1 that was localized in AD-affected regions including the hippocampus. Similarly, SAMP8 mice also show age-related increases in AβPP and mRNA [47,61,64]. Importantly, cognitive deficits observed in the SAMP8 mouse were significantly improved by downregulating the expression of AβPP using an antisense oligonucleotide specific to AβPP mRNA in aged SAMP8 mice [47,48] and oxidative stress [74]. Similarly, other markers associated with Aβ dynamics and AD such as apolipoprotein E (ApoE) and PS2 mRNA are also altered in the SAMP8 mouse [94,95]. Taken together these findings suggest that abnormal expression of AD-associated genes may play a key role in the AD-like cognitive deterioration observed in the SAMP8 model. More recent studies support earlier findings that antibodies against Aβ in the SAMP8 mouse administered intracerebrally [62], or intravenously using antibodies that cross the blood brain barrier [4], improve cognitive function in the SAMP8 mouse. Cognitive decline and improvement by Aβ antibodies in these mice is likely to be related to the overproduction of this protein and consequent alteration in membrane fatty acids. This altered membrane mobility has been hypothesized to lead to a decline in neurotransmitter activity including acetylcholine [28,29]. Because of the tight association of acetylcholine to learning and memory [40], it is thought that it is these reductions that ultimately lead to decreased ability of the P8 mouse to learn and retain new information.

Table 1
Chronology of appearance of AD-like pathology and cognitive decline in various transgenic mouse models of AD. Studies in AD individuals suggest that the chronology of appearance of pathology [7,51] and symptoms is more closely associated to that observed in the SAMP8 mouse compared to other transgenic mouse models. Notably, the chronology of appearance of pathology and behavioral abnormalities in the exhibiting mutation-based AD models is highly variable across models thus suggesting that these models may be more relevant to study the mechanisms of the mutations rather than the disease itself.

| AD-Like Phenotypes | 2 Months | 4 Months | 6 Months | 8 Months | 10 Months | 12 Months | 12+ Months |
|---|---|---|---|---|---|---|---|
| Cognitive Deficits (Hippocampal) | | APP23 JNPL3 TauP301L 3xTg-AD | PDAPP CRND8 SAMP8 | | Tg 2576 | PSAPP | hPS1-2 Tg |
| Oxidative Stress | PSAPP SAMP8 (3M) | | | Tg 2576 | APP23 | PDAPP | |
| Amyloid-β | | CRND8 3xTg-AD PSAPP | APP23 | | Tg 2576 hPS1-2 Tg | PDAPP | SAMP8[#] |
| Tau phosphorylation | | JNPL3 TauP301L 3xTg-AD | SAMP8 (5M) | | | APP23 | PDAPP Tg 2576 |
| Gliosis | | | | SAMP8 | | | Tg 2576 |

[#]Amyloid-β plaques occur around 18 months. Increased levels of soluble amyloid-β are detected at 8 months.

*Gliosis*

Glial pathology occurs in the aging brain and is thought to be another key contributor to age-related neurodegenerative pathology as cells that are actively involved in inflammatory and neurodegenerative processes are associated with AD [76]. In this regard, rapid activation of astrocytes occurs during pathogenic events and leads to increased synthesis of the intermediate filament protein glial fibrillary acidic protein (GFAP). In turn, high levels of GFAP expression and astrocyte hypertrophy are present in the brains of AD cases [5,46]. Similarly, astrocyte-derived S100B is also overexpressed by activated astrocytes in AD [81].

Various studies demonstrate that the SAMP8 shows similar, however not identical, expression of these markers to those observed in the AD brain. As such, early studies demonstrated that the binding activity of [3H] 1-(2-chlorophenyl)-N-methyl-N-(1-methylpropyl)-3-isoquinolinecarboxam ide (PK-11195) as a neurochemical marker of gliosis is increased with aging in the cerebral cortex and hippocampus of SAMP8 as is the immunoreactivity for GFAP [64]. These findings are supported by more recent studies showing marked astrogliosis and microgliosis in SAMP8 compared to controls [89].

Furthermore, similar reports using immunohistochemistry, western blot, and RT-PCR techniques showed a significant age-related increase in both protein and mRNA levels of GFAP in the hippocampi of aged SAMP8 compared to SAMR1 mice. However, no differences were found in S100β, and astroglial-derived Ca2+-binding protein with a neurotrophic role in neurons and glial cells. This suggests that at least in this mouse model upregulation of GFAP rather than changes in S100β is a more important contributor to the functional deficits observed in the SAMP8 mouse [96].

## CONCLUSIONS

The senescence-accelerated prone mouse SAMP8 is a well-established aging mouse model. Importantly this mouse model shows most histopathological signatures of AD, namely Aβ alterations, hyperphosphorylation of tau protein, increased oxidative stress and gliosis. While the chronology of appearance of pathological events in this mouse has not yet been formally established in this strain, based on the reviewed literature above, the SAMP8 mouse seems to follow more closely an AD-like trajectory compared to other existing AD-models (Table 1). It is important to note that no model is perfect. Because the SAMP8 and R1 strains were bred independently based on phenotypical traits, after consecutive breeding through many generations, there is a question as to whether the R1 is the appropriate control for the SAMP8. As such, genetic analyses have shown that the SAMP8 and SAMR1 show genetic sequence differences to a higher degree than, for example, when the R1 is compared to the related SAMP10 strain [33]. However, while the controls for this model may not be optimal, this mouse is still extremely valuable both mechanistically and pharmacologically because it presents with spontaneous neurodegenerative changes associated with age-related AD (95% of the cases), rather than pathological changes associated with the overexpression of specific AD-related pathology driven by mutations present in early-onset AD cases (<5% of total AD cases). Therefore, rather than studying the effects of overexpression of AD-related proteins, the SAMP8 mice is an excellent model to study the "pathological switch" from aging to AD and dissect the interaction and contribution of each of these pathological entities in the development and progression of this syndrome called AD, which, after all, is brought about by a combination of age-related events [99,101].

## ACKNOWLEDGMENTS

Work in the authors' laboratories is supported by the Alzheimer's Association, NIA, NINDS, Ministerio de Educación y Ciencia (Spain), Instituto de Salud Carlos III, Generalitat de Catalunya, and la Fundació La Marató.

Dr. Smith is, or has in the past been, a paid consultant for, owns equity or stock options in and/or receives grant funding from Neurotez, Neuropharm, Edenland, Panacea Pharmaceuticals, and Voyager Pharmaceuticals. Dr. Perry is a paid consultant for and/or owns equity or stock options in Takeda Pharmaceuticals, Voyager Pharmaceuticals and Panacea Pharmaceuticals.

## References

[1] O. Alvarez-Garcia, I. Vega-Naredo, V. Sierra, B. Caballero, C. Tomas-Zapico, A. Camins, J.J. Garcia, M. Pallas and A. Coto-Montes, Elevated oxidative stress in the brain of senescence-accelerated mice at 5 months of age, *Biogerontology* **7** (2006), 43-52.

[2] G.W. Arendash, M.N. Gordon, D.M. Diamond, L.A. Austin, J.M. Hatcher, P. Jantzen, G. DiCarlo, D. Wilcock and D. Morgan, Behavioral assessment of Alzheimer's transgenic mice following long-term Abeta vaccination: task specificity and correlations between Abeta deposition and spatial memory, *DNA Cell Biol* **20** (2001), 737-744.

[3] J. Avila, The influence of aging in one tauopathy: Alzheimer 's disease, *Arch Immunol Ther Exp (Warsz)* **52** (2004), 410-413.

[4] W.A. Banks, S.A. Farr, J.E. Morley, K.M. Wolf, V. Geylis and M. Steinitz, Anti-amyloid beta protein antibody passage across the blood-brain barrier in the SAMP8 mouse model of Alzheimer's disease: an age-related selective uptake with reversal of learning impairment, *Exp Neurol* **206** (2007), 248-256.

[5] T.G. Beach and E.G. McGeer, Lamina-specific arrangement of astrocytic gliosis and senile plaques in Alzheimer's disease visual cortex, *Brain Res* **463** (1988), 357-361.

[6] C. Bergeron, S. Muntasser, M.J. Somerville, L. Weyer and M.E. Percy, Copper/zinc superoxide dismutase mRNA levels are increased in sporadic amyotrophic lateral sclerosis motorneurons, *Brain Res* **659** (1994), 272-276.

[7] H. Braak and E. Braak, Staging of Alzheimer-related cortical destruction, *Int Psychogeriatr* **9 Suppl 1** (1997), 257-261; discussion 269-272.

[8] H. Braak, E. Braak, D. Yilmazer, R.A. de Vos, E.N. Jansen and J. Bohl, Neurofibrillary tangles and neuropil threads as a cause of dementia in Parkinson's disease, *J Neural Transm Suppl* **51** (1997), 49-55.

[9] J.P. Brion, Immunological demonstration of tau protein in neurofibrillary tangles of Alzheimer's disease, *J Alzheimers Dis* **9** (2006), 177-185.

[10] K.J. Bryan, H.G. Lee, G. Perry, M.A. Smith and G. Casadesus, Transgenic mouse models of Alzheimer's disease: behavioral testing and considerations, in: *Methods of Behavioral Analysis in Neuroscience*, 2nd edition, J.J. Buccafusco, eds., Taylor & Francis Group, Boca Raton, 2008, pp. in press.

[11] L.H. Bryan KJ, Perry G, Smith MA, Casadesus G *Transgenic mouse models of Alzheimer's disease: behavioral testing and considerations.*, Boca Raton, Taylor & Francis Group, Oxford, U.K, 2008.

[12] G.F. Busatto, G.E. Garrido, O.P. Almeida, C.C. Castro, C.H. Camargo, C.G. Cid, C.A. Buchpiguel, S. Furuie and C.M. Bottino, A voxel-based morphometry study of temporal lobe gray matter reductions in Alzheimer's disease, *Neurobiol Aging* **24** (2003), 221-231.

[13] B. Caballero, I. Vega-Naredo, V. Sierra, C. Huidobro-Fernandez, C. Soria-Valles, D.D. Gonzalo-Calvo, D. Tolivia, J. Gutierrez-Cuesta, M. Pallas, A. Camins, M.J. Rodriguez-Colunga and A. Coto-Montes, Favorable effects of a prolonged treatment with melatonin on the level of oxidative damage and neurodegeneration in senescence-accelerated mice, *J Pineal Res* (2008),

[14] A.M. Canudas, J. Gutierrez-Cuesta, M.I. Rodriguez, D. Acuna-Castroviejo, F.X. Sureda, A. Camins and M. Pallas, Hyperphosphorylation of microtubule-associated protein tau in senescence-accelerated mouse (SAM), *Mech Ageing Dev* **126** (2005), 1300-1304.

[15] G. Casadesus, B. Shukitt-Hale, H.M. Stellwagen, X. Zhu, H.G. Lee, M.A. Smith and J.A. Joseph, Modulation of hippocampal plasticity and cognitive behavior by short-term blueberry supplementation in aged rats, *Nutr Neurosci* **7** (2004), 309-316.

[16] S.M. Casadesus G, Perry G. By Claudio Costantini, Rekha M.K. Kolasani, and Luigi Puglielli, Commentary:

[17] "Ceramide and cholesterol: Possible connections between normal aging of the brain and Alzheimer's disease. Just hypotheses or molecular pathways to be identified?" *Alzheimer's & Dementia* **1** (2005), 51–52.

[17] A.D. Cash, G. Aliev, S.L. Siedlak, A. Nunomura, H. Fujioka, X. Zhu, A.K. Raina, H.V. Vinters, M. Tabaton, A.B. Johnson, M. Paula-Barbosa, J. Avila, P.K. Jones, R.J. Castellani, M.A. Smith and G. Perry, Microtubule reduction in Alzheimer's disease and aging is independent of tau filament formation, *Am J Pathol* **162** (2003), 1623-1627.

[18] R.J. Castellani, P.L. Harris, L.M. Sayre, J. Fujii, N. Taniguchi, M.P. Vitek, H. Founds, C.S. Atwood, G. Perry and M.A. Smith, Active glycation in neurofibrillary pathology of Alzheimer disease: N(epsilon)-(carboxymethyl) lysine and hexitol-lysine, *Free Radic Biol Med* **31** (2001), 175-180.

[19] R.J. Castellani, H.G. Lee, X. Zhu, A. Nunomura, G. Perry and M.A. Smith, Neuropathology of Alzheimer disease: pathognomonic but not pathogenic, *Acta Neuropathol (Berl)* **111** (2006), 503-509.

[20] G.H. Chen, Y.J. Wang, X.M. Wang and J.N. Zhou, Accelerated senescence prone mouse-8 shows early onset of deficits in spatial learning and memory in the radial six-arm water maze, *Physiol Behav* **82** (2004), 883-890.

[21] H. Cheng, J. Yu, Z. Jiang, X. Zhang, C. Liu, Y. Peng, F. Chen, Y. Qu, Y. Jia, Q. Tian, C. Xiao, Q. Chu, K. Nie, B. Kan, X. Hu and J. Han, Acupuncture improves cognitive deficits and regulates the brain cell proliferation of SAMP8 mice, *Neurosci Lett* **432** (2008), 111-116.

[22] K.R. Costantini C, Puglielli L Ceramide and Cholesterol: Possible Connections Between Normal Aging of the Brain and Alzheimer's Disease. Just hypotheses or molecular pathways to be identified?, *Alzheimer's and Dementia* **1** (2005), 43 - 50.

[23] S.H. Costantini C, Puglielli L, An aging pathway controls the TrkA to p75NTR receptor switch and amyloid beta-peptide generation., *EMBO J* **25** (2006 ), 1997-2006.

[24] B.S. De Leo ME, Passantino M, Palazzotti B, Mordente A, Daniele A, Filippini V, Galeotti T , Masullo C, Oxidative stress and overexpression of manganese superoxide dismutase in patients with Alzheimer's disease., *Neurosci Lett* **250** (1998), 173-176.

[25] I. Driscoll and R.J. Sutherland, The aging hippocampus: navigating between rat and human experiments, *Rev Neurosci* **16** (2005), 87-121.

[26] I. Ferrer, T. Gomez-Isla, B. Puig, M. Freixes, E. Ribe, E. Dalfo and J. Avila, Current advances on different kinases involved in tau phosphorylation, and implications in Alzheimer's disease and tauopathies, *Curr Alzheimer Res* **2** (2005), 3-18.

[27] J.F. Flood and J.E. Morley, Early onset of age-related impairment of aversive and appetitive learning in the SAM-P/8 mouse, *J Gerontol* **47** (1992), B52-59.

[28] J.F. Flood, F.J. Harris and J.E. Morley, Age-related changes in hippocampal drug facilitation of memory processing in SAMP8 mice, *Neurobiol Aging* **17** (1996), 15-24.

[29] J.F. Flood, S.A. Farr, K. Uezu and J.E. Morley, Age-related changes in septal serotonergic, GABAergic and glutamatergic facilitation of retention in SAMP8 mice, *Mech Ageing Dev* **105** (1998), 173-188.

[30] J.F. Flood and J.E. Morley, Learning and memory in the SAMP8 mouse, *Neurosci Biobehav Rev* **22** (1998), 1-20.

[31] A. Fukunari, A. Kato, Y. Sakai, T. Yoshimoto, S. Ishiura, K. Suzuki and T. Nakajima, Colocalization of prolyl endopeptidase and amyloid beta-peptide in brains of senescence-accelerated mouse, *Neurosci Lett* **176** (1994), 201-204.

[32] A. Furuta, D.L. Price, C.A. Pardo, J.C. Troncoso, Z.S. Xu, N. Taniguchi and L.J. Martin, Localization of superoxide dismutases in Alzheimer's disease and Down's syndrome neocortex and hippocampus, *Am J Pathol* **146** (1995), 357-367.

[33] J.A. Greenhall, M.A. Zapala, M. Caceres, O. Libiger, C. Barlow, N.J. Schork and D.J. Lockhart, Detecting genetic variation in microarray expression data, *Genome Res* **17** (2007), 1228-1235.

[34] J. Gutierrez-Cuesta, F.X. Sureda, M. Romeu, A.M. Canudas, B. Caballero, A. Coto-Montes, A. Camins and M. Pallas, Chronic administration of melatonin reduces cerebral injury biomarkers in SAMP8, *J Pineal Res* **42** (2007), 394-402.

[35] M. Hamdane, A.V. Sambo, P. Delobel, S. Begard, A. Violleau, A. Delacourte, P. Bertrand, J. Benavides and L. Buee, Mitotic-like tau phosphorylation by p25-Cdk5 kinase complex, *J Biol Chem* **278** (2003), 34026-34034.

[36] G. Han, K. Gable, S.D. Kohlwein, F. Beaudoin, J.A. Napier and T.M. Dunn, The Saccharomyces cerevisiae YBR159w gene encodes the 3-ketoreductase of the microsomal fatty acid elongase, *J Biol Chem* **277** (2002), 35440-35449.

[37] J. Hardy, Alzheimer's disease: the amyloid cascade hypothesis: an update and reappraisal, *J Alzheimers Dis* **9** (2006), 151-153.

[38] J.A. Hardy and G.A. Higgins, Alzheimer's disease: the amyloid cascade hypothesis, *Science* **256** (1992), 184-185.

[39] D. Harman, Aging: a theory based on free radical and radiation chemistry, *J Gerontol* **11** (1956), 298-300.

[40] M.E. Hasselmo and L.M. Giocomo, Cholinergic modulation of cortical function, *J Mol Neurosci* **30** (2006), 133-135.

[41] Y. Ihara, Neurofibrillary tangles/paired helical filaments (1981-83), *J Alzheimers Dis* **9** (2006), 209-217.

[42] S. Ikegami, S. Shumiya and H. Kawamura, Age-related changes in radial-arm maze learning and basal forebrain cholinergic systems in senescence accelerated mice (SAM), *Behav Brain Res* **51** (1992), 15-22.

[43] S. Ikegami, Behavioral impairment in radial-arm maze learning and acetylcholine content of the hippocampus and cerebral cortex in aged mice, *Behav Brain Res* **65** (1994), 103-111.

[44] K. Iqbal, C. Alonso Adel, S. Chen, M.O. Chohan, E. El-Akkad, C.X. Gong, S. Khatoon, B. Li, F. Liu, A. Rahman, H. Tanimukai and I. Grundke-Iqbal, Tau pathology in Alzheimer disease and other tauopathies, *Biochim Biophys Acta* **1739** (2005), 198-210.

[45] K. Iqbal and I. Grundke-Iqbal, Discoveries of tau, abnormally hyperphosphorylated tau and others of neurofibrillary degeneration: a personal historical perspective, *J Alzheimers Dis* **9** (2006), 219-242.

[46] M.L. Kashon, G.W. Ross, J.P. O'Callaghan, D.B. Miller, H. Petrovitch, C.M. Burchfiel, D.S. Sharp, W.R. Markesbery, D.G. Davis, J. Hardman, J. Nelson and L.R. White, Associations of cortical astrogliosis with cognitive performance and dementia status, *J Alzheimers Dis* **6** (2004), 595-604; discussion 673-581.

[47] V.B. Kumar, S.A. Farr, J.F. Flood, V. Kamlesh, M. Franko, W.A. Banks and J.E. Morley, Site-directed antisense oligonucleotide decreases the expression of amyloid precursor protein and reverses deficits in learning and memory in aged SAMP8 mice, *Peptides* **21** (2000), 1769-1775.

[48] V.B. Kumar, K. Vyas, M. Franko, V. Choudhary, C. Buddhiraju, J. Alvarez and J.E. Morley, Molecular cloning, expression, and regulation of hippocampal amyloid precursor protein of senescence accelerated mouse (SAMP8), *Biochem Cell Biol* **79** (2001), 57-67.

[49] T. Kurokawa, S. Asada, S. Nishitani and O. Hazeki, Age-related changes in manganese superoxide dismutase activity in the cerebral cortex of senescence-accelerated prone and resistant mouse, *Neurosci Lett* **298** (2001), 135-138.

[50] P.A. Lawlor, R.J. Bland, P. Das, R.W. Price, V. Holloway, L. Smithson, B.L. Dicker, M.J. During, D. Young and T.E. Golde, Novel rat Alzheimer's disease models based on AAV-mediated gene transfer to selectively increase hippocampal Abeta levels, *Mol Neurodegener* **2** (2007), 11.

[51] H.G. Lee, G. Casadesus, X. Zhu, A. Takeda, G. Perry and M.A. Smith, Challenging the amyloid cascade hypothesis: senile plaques and amyloid-beta as protective adaptations to Alzheimer disease, *Ann N Y Acad Sci* **1019** (2004), 1-4.

[52] H.G. Lee, G. Perry, P.I. Moreira, M.R. Garrett, Q. Liu, X. Zhu, A. Takeda, A. Nunomura and M.A. Smith, Tau phosphorylation in Alzheimer's disease: pathogen or protector?, *Trends Mol Med* **11** (2005), 164-169.

[53] H.G. Lee, X. Zhu, R.J. Castellani, A. Nunomura, G. Perry and M.A. Smith, Amyloid-beta in Alzheimer disease: the null versus the alternate hypotheses, *J Pharmacol Exp Ther* **321** (2007), 823-829.

[54] E.A. Maguire, E.R. Valentine, J.M. Wilding and N. Kapur, Routes to remembering: the brains behind superior memory, *Nat Neurosci* **6** (2003), 90-95.

[55] M.P. Mattson, Y. Goodman, H. Luo, W. Fu and K. Furukawa, Activation of NF-kappaB protects hippocampal neurons against oxidative stress-induced apoptosis: evidence for induction of manganese superoxide dismutase and suppression of peroxynitrite production and protein tyrosine nitration, *J Neurosci Res* **49** (1997), 681-697.

[56] J.M. McDowd and F.I. Craik, Effects of aging and task difficulty on divided attention performance, *J Exp Psychol Hum Percept Perform* **14** (1988), 267-280.

[57] J.L. McGaugh, Time-dependent processes in memory storage, *Science* **153** (1966), 1351-1358.

[58] M. Miyamoto, Y. Kiyota, M. Nishiyama and A. Nagaoka, Senescence-accelerated mouse (SAM): age-related reduced anxiety-like behavior in the SAM-P/8 strain, *Physiol Behav* **51** (1992), 979-985.

[59] M. Miyamoto, [Experimental techniques for developing new drugs acting on dementia (8)--Characteristics of behavioral disorders in senescence-accelerated mouse (SAMP8): possible animal model for dementia], *Nihon Shinkei Seishin Yakurigaku Zasshi* **14** (1994), 323-335.

[60] M. Miyamoto, Characteristics of age-related behavioral changes in senescence-accelerated mouse SAMP8 and SAMP10, *Exp Gerontol* **32** (1997), 139-148.

[61] J.E. Morley, V.B. Kumar, A.E. Bernardo, S.A. Farr, K. Uezu, N. Tumosa and J.F. Flood, Beta-amyloid precursor polypeptide in SAMP8 mice affects learning and memory, *Peptides* **21** (2000), 1761-1767.

[62] J.E. Morley, S.A. Farr and J.F. Flood, Antibody to amyloid beta protein alleviates impaired acquisition, retention, and memory processing in SAMP8 mice, *Neurobiol Learn Mem* **78** (2002), 125-138.

[63] R. Morris, Developments of a water-maze procedure for studying spatial learning in the rat, *J Neurosci Methods* **11** (1984), 47-60.

[64] Y. Nomura, Y. Yamanaka, Y. Kitamura, T. Arima, T. Ohnuki, Y. Oomura, K. Sasaki, K. Nagashima and Y. Ihara, Senescence-accelerated mouse. Neurochemical studies on aging, *Ann N Y Acad Sci* **786** (1996), 410-418.

[65] A. Nunomura, G. Perry, G. Aliev, K. Hirai, A. Takeda, E.K. Balraj, P.K. Jones, H. Ghanbari, T. Wataya, S. Shimohama, S. Chiba, C.S. Atwood, R.B. Petersen and M.A. Smith, Oxidative damage is the earliest event in Alzheimer disease, *Journal of Neuropathology and Experimental Neurology* **60** (2001), 759-767.

[66] A. Nunomura, G. Perry, G. Aliev, K. Hirai, A. Takeda, E.K. Balraj, P.K. Jones, H. Ghanbari, T. Wataya, S. Shimohama, S. Chiba, C.S. Atwood, R.B. Petersen and M.A. Smith, Oxidative damage is the earliest event in Alzheimer disease, *J Neuropathol Exp Neurol* **60** (2001), 759-767.

[67] A. Ohta, I. Akiguchi, N. Seriu, K. Ohnishi, H. Yagi, K. Higuchi and M. Hosokawa, Deterioration in learning and memory of fear conditioning in response to context in aged SAMP8 mice, *Neurobiol Aging* **22** (2001), 479-484.

[68] R.J. Ohta, D.A. Walsh and I.K. Krauss, Spatial perspective-taking ability in young and elderly adults, *Exp Aging Res* **7** (1981), 45-63.

[69] Y. Okatani, A. Wakatsuki, R.J. Reiter and Y. Miyahara, Melatonin reduces oxidative damage of neural lipids and proteins in senescence-accelerated mouse, *Neurobiol Aging* **23** (2002), 639-644.

[70] M.A. Pappolla, R.A. Omar, K.S. Kim and N.K. Robakis, Immunohistochemical evidence of oxidative [corrected] stress in Alzheimer's disease, *Am J Pathol* **140** (1992), 621-628.

[71] G. Perry, R.J. Castellani, K. Hirai and M.A. Smith, Reactive oxygen species mediate cellular damage in Alzheimer disease, *J Alzheimers Dis* **1** (1998), 45-55.

[72] G. Perry, H. Roder, A. Nunomura, A. Takeda, A.L. Friedlich, X. Zhu, A.K. Raina, N. Holbrook, S.L. Siedlak, P.L. Harris and M.A. Smith, Activation of neuronal extracellular receptor kinase (ERK) in Alzheimer disease links oxidative stress to abnormal phosphorylation, *Neuroreport* **10** (1999), 2411-2415.

[73] A.L. Petursdottir, S.A. Farr, J.E. Morley, W.A. Banks and G.V. Skuladottir, Lipid peroxidation in brain during aging in the senescence-accelerated mouse (SAM), *Neurobiol Aging* **28** (2007), 1170-1178.

[74] H.F. Poon, J. Joshi, R. Sultana, S.A. Farr, W.A. Banks, J.E. Morley, V. Calabrese and D.A. Butterfield, Antisense directed at the Abeta region of APP decreases brain oxidative markers in aged senescence accelerated mice, *Brain Res* **1018** (2004), 86-96.

[75] M.I. Rodriguez, G. Escames, L.C. Lopez, A. Lopez, J.A. Garcia, F. Ortiz, V. Sanchez, M. Romeu and D. Acuna-Castroviejo, Improved mitochondrial function and increased life span after chronic melatonin treatment in senescent prone mice, *Exp Gerontol* **43** (2008), 749-756.

[76] M. Sastre, T. Klockgether and M.T. Heneka, Contribution of inflammatory processes to Alzheimer's disease:

molecular mechanisms, *Int J Dev Neurosci* **24** (2006), 167-176.
[77] E. Sato, N. Oda, N. Ozaki, S. Hashimoto, T. Kurokawa and S. Ishibashi, Early and transient increase in oxidative stress in the cerebral cortex of senescence-accelerated mouse, *Mech Ageing Dev* **86** (1996), 105-114.
[78] D.J. Selkoe, Amyloid beta-protein and the genetics of Alzheimer's disease, *J Biol Chem* **271** (1996), 18295-18298.
[79] Y. Senechal, P.H. Kelly and K.K. Dev, Amyloid precursor protein knockout mice show age-dependent deficits in passive avoidance learning, *Behav Brain Res* (2007),
[80] M.J. Sharps, Spatial memory in young and elderly adults: category structure of stimulus sets, *Psychol Aging* **6** (1991), 309-312.
[81] J.G. Sheng, R.E. Mrak and W.S. Griffin, S100 beta protein expression in Alzheimer disease: potential role in the pathogenesis of neuritic plaques, *J Neurosci Res* **39** (1994), 398-404.
[82] B. Shukitt-Hale, G. Mouzakis and J.A. Joseph, Psychomotor and spatial memory performance in aging male Fischer 344 rats, *Exp Gerontol* **33** (1998), 615-624.
[83] B. Shukitt-Hale, J.J. McEwen, A. Szprengiel and J.A. Joseph, Effect of age on the radial arm water maze-a test of spatial learning and memory, *Neurobiol Aging* **25** (2004), 223-229.
[84] M.A. Smith, R.K. Kutty, P.L. Richey, S.D. Yan, D. Stern, G.J. Chader, B. Wiggert, R.B. Petersen and G. Perry, Heme oxygenase-1 is associated with the neurofibrillary pathology of Alzheimer's disease, *Am J Pathol* **145** (1994), 42-47.
[85] M.A. Smith, S. Taneda, P.L. Richey, S. Miyata, S.D. Yan, D. Stern, L.M. Sayre, V.M. Monnier and G. Perry, Advanced Maillard reaction end products are associated with Alzheimer disease pathology, *Proc Natl Acad Sci U S A* **91** (1994), 5710-5714.
[86] M.A. Smith, M. Rudnicka-Nawrot, P.L. Richey, D. Praprotnik, P. Mulvihill, C.A. Miller, L.M. Sayre and G. Perry, Carbonyl-related posttranslational modification of neurofilament protein in the neurofibrillary pathology of Alzheimer's disease, *J Neurochem* **64** (1995), 2660-2666.
[87] M.A. Smith, G. Perry, P.L. Richey, L.M. Sayre, V.E. Anderson, M.F. Beal and N. Kowall, Oxidative damage in Alzheimer's, *Nature* **382** (1996), 120-121.
[88] M.A. Smith, P.L. Richey Harris, L.M. Sayre, J.S. Beckman and G. Perry, Widespread peroxynitrite-mediated damage in Alzheimer's disease, *J Neurosci* **17** (1997), 2653-2657.
[89] F.X. Sureda, J. Gutierrez-Cuesta, M. Romeu, M. Mulero, A.M. Canudas, A. Camins, J. Mallol and M. Pallas, Changes in oxidative stress parameters and neurodegeneration markers in the brain of the senescence-accelerated mice SAMP-8, *Exp Gerontol* **41** (2006), 360-367.
[90] D.F. Swaab, E.J. Dubelaar, M.A. Hofman, E.J. Scherder, E.J. van Someren and R.W. Verwer, Brain aging and Alzheimer's disease; use it or lose it, *Prog Brain Res* **138** (2002), 343-373.
[91] T. Takeda, Senescence-accelerated mouse (SAM): a biogerontological resource in aging research, *Neurobiol Aging* **20** (1999), 105-110.
[92] M. Takemura, S. Nakamura, I. Akiguchi, M. Ueno, N. Oka, S. Ishikawa, A. Shimada, J. Kimura and T. Takeda, Beta/A4 proteinlike immunoreactive granular structures in the brain of senescence-accelerated mouse, *Am J Pathol* **142** (1993), 1887-1897.
[93] R.J. Weber, L.T. Brown and J.K. Weldon, Cognitive maps of environmental knowledge and preference in nursing home patients, *Exp Aging Res* **4** (1978), 157-174.
[94] X. Wei, Y. Zhang and J. Zhou, Alzheimer's disease-related gene expression in the brain of senescence accelerated mouse, *Neurosci Lett* **268** (1999), 139-142.
[95] X. Wei, Y. Zhang and J. Zhou, Differential display and cloning of the hippocampal gene mRNas in senescence accelerated mouse, *Neurosci Lett* **275** (1999), 17-20.
[96] Y. Wu, A.Q. Zhang and D.T. Yew, Age related changes of various markers of astrocytes in senescence-accelerated mice hippocampus, *Neurochem Int* **46** (2005), 565-574.
[97] F. Yasui, M. Ishibashi, S. Matsugo, Y. Kojo, Y. Oomura and K. Sasaki, Brain lipid hydroperoxide level increases in senescence-accelerated mice at an early age, *Neurosci Lett* **350** (2003), 66-68.
[98] X. Zhu, R.J. Castellani, A. Takeda, A. Nunomura, C.S. Atwood, G. Perry and M.A. Smith, Differential activation of neuronal ERK, JNK/SAPK and p38 in Alzheimer disease: the 'two hit' hypothesis, *Mech Ageing Dev* **123** (2001), 39-46.
[99] X. Zhu, A.K. Raina, G. Perry and M.A. Smith, Alzheimer's disease: the two-hit hypothesis, *Lancet Neurol* **3** (2004), 219-226.
[100] X. Zhu, A.K. Raina, G. Perry and M.A. Smith, Alzheimer's disease: the two-hit hypothesis, *Lancet Neurology* **3** (2004), 219-226.
[101] X. Zhu, H.G. Lee, G. Perry and M.A. Smith, Alzheimer disease, the two-hit hypothesis: an update, *Biochim Biophys Acta* **1772** (2007), 494-502.

# Alzheimer's Disease Selective Vulnerability and Modelling in Transgenic Mice

Jürgen Götz[a,*], Nicole Schonrock[a], Bryce Vissel[b] and Lars M. Ittner[a]
[a]*Alzheimer's and Parkinson's Disease Laboratory, Brain and Mind Research Institute, University of Sydney, NSW, Australia*
[b]*Neural Plasticity and Regeneration Group, Neuroscience Program, Garvan Institute of Medical Research, Darlinghurst, NSW, Australia*

**Abstract.** Neurodegenerative diseases are characterized by 'hot spots' of degeneration. The regions of primary vulnerability vary between different neurodegenerative diseases. Within these regions, some neurons are lost whereas others that are morphologically indiscriminable survive. The enigma of this selective vulnerability is tightly linked to two fundamental problems in the neurosciences. Firstly, it is not understood how many neuronal cell types make up the mammalian brain; estimates are in the order of more than thousand. Secondly, the mechanisms by which some nerve cells undergo functional impairment followed by degeneration while others do not, remain elusive. Understanding the basis for this selective vulnerability has significant implications for understanding the pathogenesis of disease and for developing treatments. Here, we review what is known about selective vulnerability in Alzheimer's disease, frontotemporal dementia and Parkinson's disease. We suggest, since transgenic animal models of disease reproduce aspects of selective vulnerability, that these models offer a valuable system for future investigations into the physiological basis of selective vulnerability.

Keywords: Aβ amyloid, Alzheimer's disease, amygdala, frontotemporal dementia, hippocampus, neurofibrillary tangles, Parkinson's disease, tau

## THE ENIGMA OF SELECTIVE VULNERABILITY

The factors that determine which neurons degenerate in neurodegenerative diseases are largely unknown. At issue are brain areas that may be referred to as 'hot spots' as they undergo selective degeneration. Within these hotspots some individual neurons are spared and others are lost as the disease progresses. This phenomenon has been termed 'selective vulnerability' [1]. It is tightly linked to two fundamental problems in the neurosciences. Firstly, it is not understood how many cell types there are in the mammalian brain partly due to a lack of definitive criteria as we outline further below [2].

Secondly, the mechanisms by which nerve cells first undergo a functional impairment and eventual degeneration have remained elusive.

It could be speculated that different molecular mechanisms exist in different neural subtypes that underpin their survival or conversely, vulnerability to death. It may even be possible that subtle differences between neurons have quite significant effects. In this review, we suggest that identifying the molecular signatures of different neural subtypes in the central nervous system will be essential to elucidate the mechanisms that underlie their selective vulnerability. This will be important, both for an understanding of the disease process and ultimately, for developing treatments.

## HOW ARE NEURONAL CELL-TYPES DEFINED?

Brain function is orchestrated by a highly diverse array of neuronal cell types, many of which are not

---

*Corresponding Author: Dr. J. Götz, Alzheimer's and Parkinson's Disease Laboratory, Brain and Mind Research Institute, University of Sydney, 100 Mallett St, Camperdown, NSW 2050, Australia; Tel. +61-2-9351 0789, Fax. +61-2-9351 0731; E-mail: jgoetz@med.usyd.edu.au or juergen.goetz@sydney.edu.au

well defined. There may be several hundred or tens of thousands of types of neuronal cells. In addition to neuronal cell types there are microglia and astrocytes. The complexity is enormous when considered in the context of the vast number of processes and dendritic spines that establish an intricate network of connectivity effectively linking the informational and operational domains of these different cell types. Brain function is impaired when the number of cells of particular types in a brain area falls below a critical threshold or when they fail to communicate properly.

But are the current criteria sufficient to define neuronal cell types, even in the context of neurodegeneration? In defining cell types, the ultimate goal has always been to single out a group of neurons that carry out a distinct function, although the strategic path has traditionally been first to identify cell types and then to discover their function. Morphology has been regarded as the prime criterion by which neuronal cell types are distinguished, in which neuronal shape is the main criterion because it directly reflects synaptic connectivity. The criterion of shape allows not only one cell to be distinguished from another, but it is a first step towards understanding the underlying wiring. Cell types have been further defined on the basis of location, electrophysiological properties, synaptic physiology, and marker gene expression. Clearly, however, a simple well defined neuronal taxonomy remains elusive.

For the mammalian brain, extrapolating numbers in the experimentally more readily accessible retina provides some idea of the scale we can expect. In mammals, the retina contains five major classes of neurons, together represented by around 60 individual cell types; for example, there are projection neurons (around a dozen in most mammalian species), and intrinsic neurons, which have been subdivided into horizontal, bipolar and amacrine cells, using a simple letter code such as A1, A2 or A3 to define cell types based on morphological criteria and anatomical location [3,4]. The different shapes can often be associated with distinct, defining combinations of proteins, of which those that are involved in synaptic transmission have been particularly helpful. For example, bipolar cells that depolarize in response to light have axons that form arborizations deep in the inner plexiform layer; these cells express the metabotropic glutamate receptor isoform mGluR6. Bipolar cells that hyperpolarize to light form arborizations high in the inner plexiform layer, and express ionotropic (AMPA/kainite) receptors; this helps in the identification of functional cell types. By extrapolation, it is possible to speculate that with around 60 cell types in the retina, there may be 1,000 neuronal cell types in the cortex alone, integrating knowledge of neuronal spacing, cell numbers and the diameter of the dendritic field [5]. However, most of these cell types are not defined.

To address this issue, more recently, a number of highly sophisticated tools have become available. Ways to mark different neurons include the 'Cre' recombination technology and 'Green Fluorescent Protein (GFP)' staining, to simultaneously stain neighboring neurons so that they are spectacularly isolated, exhibiting up to 166 different 'color shades' based on four types of fluorescent proteins [6]. While the technology requires refining, the hope exists that it will be possible to use such technology to assist in identifying neuronal subtypes. It seems reasonable to suggest that combining such approaches with gene-expression profiling could be a useful alternative strategy in dividing neurons into functional subtypes [7,8]. We suggest this approach offers great hope for creating a molecularly defined taxonomy of neuronal subtypes that will, in turn, establish the molecular basis of selective vulnerability.

## SELECTIVE VULNERABILITY IN ALZHEIMER'S DISEASE AND FRONTOTEMPORAL DEMENTIA

Neurodegenerative diseases of the human brain comprise a variety of disorders that, for demographic reasons, affect an increasing percentage of the aging population [9]. Alzheimer's (AD) and Parkinson's disease (PD) are examples of such late-onset diseases [10]. AD is the most common form of dementia, whereas PD is the most common movement disorder. The pathological changes in the AD and PD brain precede the onset of clinical symptoms by decades [11]. Histopathologically, the AD brain is characterized by abundant amyloid plaques, neurofibrillary lesions and the concomitant loss of nerve cells and synapses [12,13]. This neurodegeneration spreads in a predictable, non-random manner across the brain [14]. The neurofibrillary tangles (NFTs), which contain massively phosphorylated, aggregated forms of the microtubule-associated protein tau and represent a hallmark lesion of the disease, in addition to Aβ peptide-containing amyloid plaques, develop in specific predilection sites [15-17]. The fact that there is selective vulnerability to NFTs formation is beyond question. In the basal forebrain all neurons that die appear to contain NFTs [18]. In contrast, up to 20% of the neuronal loss in the CA1 region cannot be explained by NFT formation, which is a slow process [19-21].

The spreading of the tau pathology is subject to little inter-individual variation and provides a basis for distinguishing six stages: the transentorhinal stages I and II representing silent cases; the limbic stages III and IV; and the neocortical stages V and VI [17]. The cellular and molecular foundation of this staging is not at all understood. Several hypotheses have been put forward. For example, a correlation has been proposed with the pattern of consecutive myelination in the course of the development of the nervous system. More specifically, it has been claimed that neurons in association areas with minimal myelination are more vulnerable than those in primary cortices that are characterized by a more extensive and developmentally delayed myelination [22]. For the hippocampus, neuronal loss has been correlated with the distribution of glucocorticoid receptors [23].

There is a second level of vulnerability: remarkably, within brain areas susceptible to neurodegeneration, some neurons that appear morphologically indistinguishable from neighboring neurons that die can be spared for decades. Wherever neurons are lost in affected brain areas, protected neurons are found in their immediate vicinity [19,24,25]. Similar findings have been reported for other dementias, such as frontotemporal dementia (FTD) that is characterized by a tau pathology, in the absence of amyloid plaques [26]. Together this raises the major question as to whether the differences in selective vulnerability reflect molecular differences between cells that are not obvious based on current definitions of 'similar neurons'. If true then these different cells may potentially have slightly different intracellular signaling mechanisms that either lead to, or resist, cell death.

## SELECTIVE VULNERABILITY IN PARKINSON'S DISEASE

Parkinson's disease (PD) is the most frequent neurodegenerative disorder with impaired motor functions. The PD brain is characterized by a selective loss of a subset of dopaminergic neurons. Tyrosine hydroxylase (TH) is the rate-limiting enzyme in the synthesis of dopamine, and TH-reactivity is used as marker for dopaminergic neurons. The clinical features of parkinsonism such as tremor, bradykinesia and rigidity become evident only after around 80% of the TH-expressing neurons in the Substantia Nigra pars compacta (SNpc) have died [27-30]. It has been shown that upon the loss of TH-positive SNpc neurons in the rodent 6-hydroxy-DA (6-OHDA) model of PD, TH-positive neurons can partially recover [31]. There is evidence that this recovery occurs via a phenotype 'shift' from TH-negative to TH-positive cells [32]. Given that 80% of dopaminergic neurons must be lost from the SN before profound symptoms occur, just some preservation or restoration of dopaminergic neurons will have a dramatic therapeutic impact. There is therefore every reason to hope that innovative therapeutic strategies such as gene or cell replacement therapy may work for PD.

The PD brain is marked by fibrillar cytoplasmic inclusions that are abundant in degenerating dopaminergic neurons of the SN [33]. The lesions are known as Lewy bodies (LBs) and neurites (LN); they are ubiquitin-positive and mainly contain α-synuclein. In PD, neuronal vulnerability is known to characterize the SNpc, but abnormal protein deposition extends far beyond. It includes additional, specific neurons in autonomic ganglia, the spinal cord, brainstem, basal forebrain, limbic lobe, and even the neocortex [34].

While there is a selective neuronal loss of the TH-positive A9 dopaminergic neuron group in the SNpc, with a survival rate of 10%, the TH-positive A10 group in the medial and ventral tegmentum is largely spared, with a survival rate of 60%, even in severe cases [28,35]. A9 neurons mainly project to the dorsolateral striatum involved in motor control, whereas A10 neurons connect to the ventromedial striatum, thalamus and cortex, and are involved in reward and emotional behavior.

Again, there is a second level of vulnerability: within the A9 group, the caudally and laterally located ventral TH-positive neurons are the most vulnerable [28,36]. This is unlikely a random process but may reflect distinct characteristics of TH-producing neurons that are not picked up using conventional criteria. Vulnerability of SN neurons may be caused by the greater susceptibility of dopamine and its metabolites in their production of reactive oxygen species (ROS) that eventually kill neurons [37]. Differential protein expression has been implicated in selective vulnerability, such as the presence of the K$^+$ channel GIRK2, which is exclusively expressed in vulnerable A9 neurons [38,39]. A role for GIRK2 is further supported by the loss of dopaminergic neurons in the *weaver* mouse model, which carries a spontaneous *GIRK2* mutation [40]. This indicates that differential gene expression patters may determine selective vulnerability not only in the PD, but also AD brain.

It is therefore likely that the vulnerability of different subtypes of dopaminergic neurons depends on their specific molecular signatures, as this will in turn determine the intracellular signaling pathways that define their vulnerability to cell death. The process of therapeutic development for PD will require understanding the basis for this selective vulnerability for two reasons. Firstly, understanding the reasons for lack of vulnerability of some neurons may allow the development of approaches to protect the neurons that are vulnerable. Secondly, with the current suggestions that cell replacement strategies may eventually work, it likely will be important to replace the correct subtypes of dopaminergic neurons if these are to generate therapeutic benefits. It is therefore critical to develop an effective and robust taxonomy of dopaminergic neurons and to subsequently elucidate the basis of their selective vulnerability.

## SELECTIVE VULNERABILITY IN OTHER NEURODEGENERATIVE CONDITIONS

Pathways regulating neuronal vulnerability are generally poorly understood. Within the hippocampus some sectors are susceptible to ischemia, whereas others are resistant. More specifically, the CA4 region was found to be the most vulnerable, the CA2 and the medial CA1 the second and third most vulnerable regions, whereas the lateral part of the CA1, the CA3 and the dentate gyrus were found to be resistant [41]. By incubating organotypic slice cultures with NMDA this was found to induce much larger Ca$^{2+}$ elevations in CA1 than CA3 neurons. Consequently, CA1 mitochondria were found to exhibit stronger Ca$^{2+}$ accumulation, more extensive swelling and damage, stronger depolarization of their membrane potential, and a significant increase in ROS generation [42]. A related report suggests that the selective sensitivity of the CA1 region may be attributable to the density of mature neurons expressing polyamine-sensitive NMDA receptor (NR2B) subunits [43]. Likewise, a role for AMPA signaling and metabotropic excitotoxicity have been proposed to explain the selective neuronal vulnerability of subplate as compared to cortical neurons [44].

In Niemann-Pick type C (NPC) disease, a lysosomal storage disorder characterized by impaired cholesterol trafficking and neurodegeneration, the relative contribution of neuronal and glial defects to neuron loss is unclear. When mutant mice were generated which lack Npc1 expression in mature cerebellar Purkinje cells, this caused an age-dependent impairment in motor tasks. Surprisingly however these mice did not show the early death or weight loss that characterizes mice with a complete loss of Npc1. In the context of this review article it is interesting that there was a progressive loss of Purkinje cells in an anterior-to-posterior gradient, with a subset of neurons being resistant. As this also occurred in global knockout mice this would argue that this neurodegeneration is cell autonomous. Why a subpopulation of Purkinje cells in the posterior cerebellum exhibited marked resistance to cell death despite Npc1 deletion is not understood [45].

## SELECTIVE VULNERABILITY IN TRANSGENIC MOUSE MODELS

Patterns of selective vulnerability comparable to the human AD brain occur also in rodents [46,47]; nerve cell loss, the end-point, however has been reproduced only in a small subset of transgenic mouse models [48-50]. We will mention just a few of these models here, to elaborate on our point that selective vulnerability of neurons to cell loss is observed in mouse models of neurodegeneration.

One of these mouse models is the K3 strain [51], which expresses human tau together with the pathogenic K369I mutation found in Pick's disease [52]. Pick's disease belongs to the FTD complex, that is often characterized by parkinsonism [53]. K3 mice express the K369I tau transgene in neurons within the hippocampus (CA1 pyramidal), cortex (pyramidal), amygdala, striatum and SNpc, among other brain areas. Transgene expression affects the dopaminergic system, causing early-onset parkinsonism (rigidity, tremor, bradykinesia and postural instability) in these mice. We identified an underlying molecular mechanism: the phenotype is caused by an impaired anterograde transport of distinct cargos, as shown for the nigrostriatal system and sciatic nerve [51,54]. Importantly, functional impairment has an early onset and occurs in the absence of overt nerve cell loss. However, in the absence of better cell-type-specific markers, it is unclear whether at this age one or more neuronal cell types have already been lost that would account for the phenotype. Eventually, as the mice age, up to 60% of TH-positive neurons in the SNpc are lost – but again, whether there is a selectivity for specific neuronal cell types is still in question [51].

In the K3 mice, two questions arise: (a) K369I tau expression of TH-positive neurons is moderate compared with other brain areas, but not all SNpc neurons degenerate, and CA1 neurons, for example, do not. What makes SNpc neurons particularly vulnerable compared with neurons in the CA1 region? (b) The loss of TH-positive, K369I tau-expressing neurons in the SNpc is only partial (60% loss by 24 months of age). What protects a subset of morphologically indiscriminable neurons within this brain area while others degenerate? Clearly the K3 mice we have generated offer a valuable tool to address these questions.

Meanwhile, different mice carrying similar mutations can exhibit different vulnerabilities. For example, to understand tau hyperphosphorylation and aggregation in AD [55], we established P301L mutant tau transgenic pR5 mice that are characterized by tau-containing NFT formation in the hippocampus and amygdala, as well as memory impairment [56-62]. In the absence of an unbiased stereological analysis, we found no obvious cell loss. In contrast, a different line of mice that expresses P301L mutant tau under inducible control (the rTg(tauP301L)$_{4510}$ line) shows an NFT pathology similar to that of the pR5 mice, but is characterized by massive brain weight loss and gross atrophy of the forebrain [63]. In fact unlike our pR5 mice, in the rTg(tauP301L)$_{4510}$ mice, 60% of CA1 hippocampal pyramidal neurons have already been lost by 5.5 months and, only 23% remain by 8.5 months [63]. Reducing transgene expression in the latter mice using doxycycline in the drinking water partly rescues brain atrophy and nerve cell loss (as shown for CA1 neurons), and improves the phenotype. The explanation for the different vulnerability to cell loss observed between the two strains of mice is most likely that the two strains of mice express different levels of the P301L mutant tau transgene. Additionally, genetic background issues could be important as different strains of mice show differential vulnerability to cell loss. This points to the fact that a range of factors in any given cell, such as expression levels of a 'toxic' gene and the genetic context in which a 'toxic' gene is expressed, will determine the vulnerability of that neuron to cell loss.

When the Aβ pathology is combined with the tau pathology, this causes an increased NFT formation [64]. Interestingly, this enhanced pathology is restricted to specific brain areas. When P301L tau-expressing JNPL3 mice were crossed with Aβ-forming Tg2576 mice, these showed a more than sevenfold increase in NFT numbers in restricted brain areas, the olfactory bulb, entorhinal cortex and amygdala, compared to P301L single transgenic mice; plaque formation, in comparison, was unaffected by the presence of the tau lesions [65].

We used an alternative approach to reveal the phenomenon of selective vulnerability: we stereotaxically injected synthetic preparations of fibrillar Aβ42 into the somatosensory cortex and the CA1 region of the P301L tau transgenic pR5 mice described above [56], wild-type human tau transgenic ALZ17 mice [66,67] and nontransgenic littermate controls, causing a five-fold increase of NFTs specifically in the amygdala of pR5, but not at all in ALZ17 or control mice [57]. This implies that not all brain areas are similarly susceptible to Aβ-mediated NFT induction. In both studies, the amygdala turns out to be a 'hot spot' of NFT induction. Unless in both mouse models tau levels in the amygdala are particularly high compared to other brain areas, a different transcriptional profile may account for the observed differences [68]. Support for the latter is provided by data obtained with long-term neuronal cultures from wild-type mice, as cortical neurons are less susceptible to Aβ- and staurosporine-induced toxicity than hippocampal neurons [69].

Other AD mouse models with selective neuronal loss, to name several prominent examples, include P301S tau transgenic mice [70,71], the N279K tau transgenic strain T-279 [72], ΔK280 tau transgenic TauRD/ΔK280 mice with an inducible tau expression [73], Pin1 knockout mice [74,75], APP transgenic APP23 mice [76], APP(SL)PS1 knock-in mice [77], and apoE4Δ(272–299) mice [78], as reviewed by us recently [49]. Among the neuronal populations that are lost are motor neurons, pyramidal CA1 or dopaminergic neurons. In all these model mice selective vulnerability is clearly evident, as only a subset of all neurons are lost in any of the affected brain regions. The identity of these selectively vulnerable neurons is unknown.

In conclusion, aspects of selective vulnerability known to characterize the AD brain, in particular its susceptibility to specific toxic insults such as the Aβ peptide and selective neuronal cell loss, have been reproduced in transgenic animal models. These model systems are therefore excellent tools for longitudinal studies that aim to define neuronal cell types and to obtain a transcriptomic and proteomic profile of vulnerable compared to protected brain areas.

## TRANSCRIPTOMICS IN DEFINING CELL TYPES AND DISSECTING SELECTIVE VULNERABILITY

The advent of increasingly sophisticated transcriptomic and proteomic techniques made it possible to identify differentially expressed genes and proteins in the AD and PD brain, and to pinpoint pathogenic mechanism, such as mitochondrial dysfunction or an impairment of the unfolded protein response (UPR) [79-82]. We used the tools of functional genomics to characterize our AD and FTD mouse and tissue culture systems and to dissect pathogenic mechanisms that not only operate in the transgenic mouse but also in human diseased brain [83-89].

Functional genomics not only assisted in dissecting disease mechanisms, but also emerged as a powerful tool in identifying cell-type-specific gene expression (i.e., obtaining a transcriptomic profile under physiological conditions) that may ultimately assist in identifying neuronal cell types [90]. However, thus far only a few instances of truly cell-type-specific gene expression profiles have been reported [91]. For example, serial analysis of gene expression (SAGE) was used to identify gene markers in the developing retina [92].

In an impressive, laborious study, Sugino et al. obtained 11 fluorescently labelled neuronal populations from different brain areas using four GFP-expressing transgenic mouse lines [2]. They triturated the neurons and by a panning process manually isolated 30–120 neurons from each brain area that they had before characterized via current-clamp recordings. The subsequent analysis of the transcriptomic profile using the *Gene Ontology* software allowed them to construct a taxonomic tree that showed clear distinctions between neuronal cell types such as cortical interneurons and projection neurons [2]. As the authors point out this dataset should be useful for the classification of unknown neuronal subtypes, the investigation of specifically expressed genes and the genetic manipulation of specific neuronal circuit elements [2].

Transgenic mice such as the K3 model of parkinsonism described earlier in this review lose a significant subset of TH-positive neurons in the SNpc as they age. These mice are therefore an excellent system to apply an integrative functional genomics approach; it will link differential gene expression in the SNpc to the identification of novel cell types in this brain area. We believe that studies such as these will shed light on the selective vulnerability that characterizes diseases such as AD and PD and that this new knowledge can be used to develop treatment strategies [12,93]. Ultimately, it may be able to equip neurons that are prone to degenerate with genes or combinations of genes that will protect them from neurodegeneration.

Of course the complexity of this approach will be profound. The gene sets expressed in different neurons will differ depending on the brain area that needs to be protected, the patient's burden of disease-associated mutations and risk alleles, the disease variant and the type of toxic insult.

## SUMMARY AND OUTLOOK

In summary, the hypothesis we put forward is that subtle differences in molecular signaling pathways between different neurons define the neuronal subtype and this, in turn will define their selective vulnerability to death [94]. It is not hard to imagine that different cell types may show a predisposition for, or a differential protection from, alternative death mechanisms such as apoptosis, which is relatively fast, disruption of the Golgi apparatus, oxidative damage or other mechanisms of death [95-97]. Defining the molecular signatures of different neurons and relating it to their vulnerability will be a major undertaking but an important one. It will not be easy, however: unraveling all of this is likely to be confounded by effects of the micro-environment of neurons. While a major focus of research into selective vulnerability is on neuronal dysfunction, there is also a contribution of the glial compartment in neuronal cell loss [98]. Furthermore this complexity will be further layered on a complexity defined by genetic background and environmental influences.

# ACKNOWLEDGEMENTS

JG is a Medical Foundation Fellow. This work has been supported by the University of Sydney, the National Health & Medical Research Council (NHMRC), the Australian Research Council (ARC), the New South Wales Government through the Ministry for Science and Medical Research (BioFirst Program), the Nerve Research Foundation, the Medical Foundation (University of Sydney) and the Judith Jane Mason & Harold Stannett Williams Memorial Foundation. LI has been supported by the NHMRC and ARC. NS is a Human Frontiers Science Program (HFSP) Fellow. BV was supported by the Baxter Foundation, Ronald Geoffrey Arnott Foundation and by the NSW State Government's BioFirst Award and Spinal Cord and Related Neurological Conditions Grant. JG thanks Dr Tony Souter for editing.

# References

[1] Ball MJ (1978) Topographic distribution of neurofibrillary tangles and granulovacuolar degeneration in hippocampal cortex of aging and demented patients. A quantitative study. *Acta Neuropathol* 42, 73-80.
[2] Sugino K, Hempel CM, Miller MN, Hattox AM, Shapiro P, Wu C, Huang ZJ, Nelson SB (2006) Molecular taxonomy of major neuronal classes in the adult mouse forebrain. *Nat Neurosci* 9, 99-107.
[3] Kolb H, Nelson R, Mariani A (1981) Amacrine cells, bipolar cells and ganglion cells of the cat retina: a Golgi study. *Vision Res* 21, 1081-1114.
[4] Pourcho RG, Goebel DJ (1987) A combined Golgi and autoradiographic study of 3H-glycine-accumulating cone bipolar cells in the cat retina. *J Neurosci* 7, 1178-1188.
[5] Dacey DM, Petersen MR (1992) Dendritic field size and morphology of midget and parasol ganglion cells of the human retina. *Proc Natl Acad Sci U S A* 89, 9666-9670.
[6] Livet J, Weissman TA, Kang H, Draft RW, Lu J, Bennis RA, Sanes JR, Lichtman JW (2007) Transgenic strategies for combinatorial expression of fluorescent proteins in the nervous system. *Nature* 450, 56-62.
[7] Mott DD, Dingledine R (2003) Interneuron Diversity series: Interneuron research--challenges and strategies. *Trends Neurosci* 26, 484-488.
[8] Markram H, Toledo-Rodriguez M, Wang Y, Gupta A, Silberberg G, Wu C (2004) Interneurons of the neocortical inhibitory system. *Nat Rev Neurosci* 5, 793-807.
[9] Ferri CP, Prince M, Brayne C, Brodaty H, Fratiglioni L, Ganguli M, Hall K, Hasegawa K, Hendrie H, Huang Y, Jorm A, Mathers C, Menezes PR, Rimmer E, Scazufca M (2005) Global prevalence of dementia: a Delphi consensus study. *Lancet* 366, 2112-2117.
[10] Kurosinski P, Guggisberg M, Gotz J (2002) Alzheimer's and Parkinson's disease - overlapping or synergistic pathologies? *Trends Mol Med* 8, 3-5.
[11] Braak H, Braak E (1998) Evolution of neuronal changes in the course of Alzheimer's disease. *J Neural Transm Suppl* 53, 127-140.
[12] Gotz J, Deters N, Doldissen A, Bokhari L, Ke Y, Wiesner A, Schonrock N, Ittner LM (2007) A Decade of Tau Transgenic Animal Models and Beyond *Brain Pathol* 17, 91-103.
[13] Gotz J, Ittner LM, Schonrock N, Cappai R (2008) An update on the toxicity of Abeta in Alzheimer's disease. *Neuropsychiatr Dis Treat* 4, 1033-1042.
[14] Goedert M, Jakes R, Spillantini MG, Crowther RA, Cohen P, Vanmechelen E, Probst A, Gotz J, Burki K (1995) Tau protein in Alzheimer's disease. *Biochem Soc Trans* 23, 80-85.
[15] Braak H, Braak E (1995) Staging of Alzheimer's disease-related neurofibrillary changes. *Neurobiol Aging* 16, 271-278; discussion 278-284.
[16] Thal DR, Rub U, Orantes M, Braak H (2002) Phases of A beta-deposition in the human brain and its relevance for the development of AD. *Neurology* 58, 1791-1800.
[17] Braak H, Braak E (1991) Neuropathological stageing of Alzheimer-related changes. *Acta Neuropathol (Berl)* 82, 239-259.
[18] Cullen KM, Halliday GM (1998) Neurofibrillary degeneration and cell loss in the nucleus basalis in comparison to cortical Alzheimer pathology. *Neurobiol Aging* 19, 297-306.
[19] Gomez-Isla T, Hollister R, West H, Mui S, Growdon JH, Petersen RC, Parisi JE, Hyman BT (1997) Neuronal loss correlates with but exceeds neurofibrillary tangles in Alzheimer's disease. *Ann Neurol* 41, 17-24.
[20] Giannakopoulos P, Herrmann FR, Bussiere T, Bouras C, Kovari E, Perl DP, Morrison JH, Gold G, Hof PR (2003) Tangle and neuron numbers, but not amyloid load, predict cognitive status in Alzheimer's disease. *Neurology* 60, 1495-1500.
[21] Kril JJ, Patel S, Harding AJ, Halliday GM (2002) Neuron loss from the hippocampus of Alzheimer's disease exceeds extracellular neurofibrillary tangle formation. *Acta Neuropathol (Berl)* 103, 370-376.
[22] Braak H, Del Tredici K, Schultz C, Braak E (2000) Vulnerability of select neuronal types to Alzheimer's disease. *Ann N Y Acad Sci* 924, 53-61.
[23] McEwen BS, Sapolsky RM (1995) Stress and cognitive function. *Curr Opin Neurobiol* 5, 205-216.
[24] Vickers JC, Riederer BM, Marugg RA, Buee-Scherrer V, Buee L, Delacourte A, Morrison JH (1994) Alterations in neurofilament protein immunoreactivity in human hippocampal neurons related to normal aging and Alzheimer's disease. *Neuroscience* 62, 1-13.
[25] Sampson VL, Morrison JH, Vickers JC (1997) The cellular basis for the relative resistance of parvalbumin and calretinin immunoreactive neocortical neurons to the pathology of Alzheimer's disease. *Exp Neurol* 145, 295-302.
[26] Kersaitis C, Halliday GM, Kril JJ (2004) Regional and cellular pathology in frontotemporal dementia: relationship to stage of disease in cases with and without Pick bodies. *Acta Neuropathol* 108, 515-523.
[27] McGeer PL, Itagaki S, Akiyama H, McGeer EG (1988) Rate of cell death in parkinsonism indicates active neuropathological process. *Ann Neurol* 24, 574-576.
[28] Damier P, Hirsch EC, Agid Y, Graybiel AM (1999) The substantia nigra of the human brain. II. Patterns of loss of dopamine-containing neurons in Parkinson's disease. *Brain* 122 ( Pt 8), 1437-1448.

[29] Pakkenberg B, Moller A, Gundersen HJ, Mouritzen Dam A, Pakkenberg H (1991) The absolute number of nerve cells in substantia nigra in normal subjects and in patients with Parkinson's disease estimated with an unbiased stereological method. *J Neurol Neurosurg Psychiatry* 54, 30-33.

[30] Rodriguez M, Barroso-Chinea P, Abdala P, Obeso J, Gonzalez-Hernandez T (2001) Dopamine cell degeneration induced by intraventricular administration of 6-hydroxydopamine in the rat: similarities with cell loss in parkinson's disease. *Exp Neurol* 169, 163-181.

[31] Lee J, Zhu WM, Stanic D, Finkelstein DI, Horne MH, Henderson J, Lawrence AJ, O'Connor L, Tomas D, Drago J, Horne MK (2008) Sprouting of dopamine terminals and altered dopamine release and uptake in Parkinsonian dyskinaesia. *Brain* 131, 1574-1587.

[32] Aumann TD, Gantois I, Egan K, Vais A, Tomas D, Drago J, Horne MK (2008) SK channel function regulates the dopamine phenotype of neurons in the substantia nigra pars compacta. *Exp Neurol* 213, 419-430.

[33] Goedert M (2001) Alpha-synuclein and neurodegenerative diseases. *Nat Rev Neurosci* 2, 492-501.

[34] Braak H, Del Tredici K, Rub U, de Vos RA, Jansen Steur EN, Braak E (2003) Staging of brain pathology related to sporadic Parkinson's disease. *Neurobiol Aging* 24, 197-211.

[35] McRitchie DA, Cartwright HR, Halliday GM (1997) Specific A10 dopaminergic nuclei in the midbrain degenerate in Parkinson's disease. *Exp Neurol* 144, 202-213.

[36] German DC, Manaye K, Smith WK, Woodward DJ, Saper CB (1989) Midbrain dopaminergic cell loss in Parkinson's disease: computer visualization. *Ann Neurol* 26, 507-514.

[37] Olney JW, Zorumski CF, Stewart GR, Price MT, Wang GJ, Labruyere J (1990) Excitotoxicity of L-dopa and 6-OH-dopa: implications for Parkinson's and Huntington's diseases. *Exp Neurol* 108, 269-272.

[38] Schein JC, Hunter DD, Roffler-Tarlov S (1998) Girk2 expression in the ventral midbrain, cerebellum, and olfactory bulb and its relationship to the murine mutation weaver. *Dev Biol* 204, 432-450.

[39] Karschin C, Dissmann E, Stuhmer W, Karschin A (1996) IRK(1-3) and GIRK(1-4) inwardly rectifying K+ channel mRNAs are differentially expressed in the adult rat brain. *J Neurosci* 16, 3559-3570.

[40] Roffler-Tarlov S, Graybiel AM (1984) Weaver mutation has differential effects on the dopamine-containing innervation of the limbic and nonlimbic striatum. *Nature* 307, 62-66.

[41] Yang G, Kitagawa K, Ohtsuki T, Kuwabara K, Mabuchi T, Yagita Y, Takazawa K, Tanaka S, Yanagihara T, Hori M, Matsumoto M (2000) Regional difference of neuronal vulnerability in the murine hippocampus after transient forebrain ischemia. *Brain Res* 870, 195-198.

[42] Stanika RI, Winters CA, Pivovarova NB, Andrews SB (2010) Differential NMDA receptor-dependent calcium loading and mitochondrial dysfunction in CA1 vs. CA3 hippocampal neurons. *Neurobiol Dis* 37, 403-411.

[43] Butler TR, Self RL, Smith KJ, Sharrett-Field LJ, Berry JN, Littleton JM, Pauly JR, Mulholland PJ, Prendergast MA (2010) Selective vulnerability of hippocampal cornu ammonis 1 pyramidal cells to excitotoxic insult is associated with the expression of polyamine-sensitive N-methyl-D-asparate-type glutamate receptors. *Neuroscience* 165, 525-534.

[44] Nguyen V, McQuillen PS (2010) AMPA and metabotropic excitotoxicity explain subplate neuron vulnerability. *Neurobiol Dis* 37, 195-207.

[45] Elrick MJ, Pacheco CD, Yu T, Dadgar N, Shakkottai VG, Ware C, Paulson HL, Lieberman AP (2010) Conditional Niemann-Pick C mice demonstrate cell autonomous Purkinje cell neurodegeneration. *Hum Mol Genet* 19, 837-847.

[46] Berger B, Gaspar P, Verney C (1991) Dopaminergic innervation of the cerebral cortex: unexpected differences between rodents and primates. *Trends Neurosci* 14, 21-27.

[47] Joel D, Weiner I (2000) The connections of the dopaminergic system with the striatum in rats and primates: an analysis with respect to the functional and compartmental organization of the striatum. *Neuroscience* 96, 451-474.

[48] Gotz J, Streffer JR, David D, Schild A, Hoerndli F, Pennanen L, Kurosinski P, Chen F (2004) Transgenic animal models of Alzheimer's disease and related disorders: Histopathology, behavior and therapy. *Mol Psychiatry* 9, 664-683.

[49] Gotz J, Ittner LM (2008) Animal models of Alzheimer's disease and frontotemporal dementia. *Nat Rev Neurosci* 9, 532-544.

[50] Gotz J, Gotz NN (2009) Animal models for Alzheimer's disease and frontotemporal dementia: a perspective. *ASN Neuro* 1, pii: e00019.

[51] Ittner LM, Fath T, Ke YD, Bi M, van Eersel J, Li KM, Gunning P, Gotz J (2008) Parkinsonism and impaired axonal transport in a mouse model of frontotemporal dementia. *Proc Natl Acad Sci U S A* 105, 15997-16002.

[52] Neumann M, Schulz-Schaeffer W, Crowther RA, Smith MJ, Spillantini MG, Goedert M, Kretzschmar HA (2001) Pick's disease associated with the novel Tau gene mutation K369I. *Ann Neurol* 50, 503-513.

[53] Lee VM, Goedert M, Trojanowski JQ (2001) Neurodegenerative tauopathies. *Annu Rev Neurosci* 24, 1121-1159.

[54] Ittner LM, Ke YD, Gotz J (2009) Phosphorylated Tau Interacts with c-Jun N-terminal Kinase-interacting Protein 1 (JIP1) in Alzheimer Disease. *J Biol Chem* 284, 20909-20916.

[55] Chen F, David D, Ferrari A, Gotz J (2004) Posttranslational modifications of tau - Role in human tauopathies and modeling in transgenic animals. *Curr Drug Targets* 5, 503-515.

[56] Gotz J, Chen F, Barmettler R, Nitsch RM (2001) Tau Filament Formation in Transgenic Mice Expressing P301L Tau. *J Biol Chem* 276, 529-534.

[57] Gotz J, Chen F, van Dorpe J, Nitsch RM (2001) Formation of neurofibrillary tangles in P301L tau transgenic mice induced by Abeta 42 fibrils. *Science* 293, 1491-1495.

[58] Gotz J (2001) Tau and transgenic animal models. *Brain Res Brain Res Rev* 35, 266-286.

[59] Pennanen L, Welzl H, D'Adamo P, Nitsch RM, Gotz J (2004) Accelerated extinction of conditioned taste aversion in P301L tau transgenic mice. *Neurobiol Dis* 15, 500-509.

[60] Pennanen L, Wolfer DP, Nitsch RM, Gotz J (2006) Impaired spatial reference memory and increased exploratory behavior in P301L tau transgenic mice. *Genes Brain Behav* 5, 369-379.

[61] Deters N, Ittner LM, Gotz J (2008) Divergent phosphorylation pattern of tau in P301L tau transgenic mice. *Eur J Neurosci.* 28, 137-147.

[62] Deters N, Ittner LM, Gotz J (2009) Substrate-specific reduction of PP2A activity exaggerates tau pathology. *Biochem Biophys Res Commun* 379, 400-405.
[63] Santacruz K, Lewis J, Spires T, Paulson L, Kotilinek L, Ingelsson M, Guimaraes A, DeTure M, Ramsden M, McGowan E, Forster C, Yue M, Orne J, Janus C, Mariash A, Kuskowski M, Hyman B, Hutton M, Ashe KH (2005) Tau suppression in a neurodegenerative mouse model improves memory function. *Science* 309, 476-481.
[64] Gotz J, Schild A, Hoerndli F, Pennanen L (2004) Amyloid-induced neurofibrillary tangle formation in Alzheimer's disease: insight from transgenic mouse and tissue-culture models. *Int J Dev Neurosci* 22, 453-465.
[65] Lewis J, Dickson DW, Lin W-L, Chisholm L, Corral A, Jones G, Yen S-H, Sahara N, Skipper L, Yager D, Eckman C, Hardy J, Hutton M, McGowan E (2001) Enhanced neurofibrillary degeneration in transgenic mice expressing mutant Tau and APP. *Science* 293, 1487-1491.
[66] Probst A, Gotz J, Wiederhold KH, Tolnay M, Mistl C, Jaton AL, Hong M, Ishihara T, Lee VM, Trojanowski JQ, Jakes R, Crowther RA, Spillantini MG, Burki K, Goedert M (2000) Axonopathy and amyotrophy in mice transgenic for human four-repeat tau protein. *Acta Neuropathol (Berl)* 99, 469-481.
[67] Gotz J, Nitsch RM (2001) Compartmentalized tau hyperphosphorylation and increased levels of kinases in transgenic mice. *Neuroreport* 12, 2007-2016.
[68] Zirlinger M, Kreiman G, Anderson DJ (2001) Amygdala-enriched genes identified by microarray technology are restricted to specific amygdaloid subnuclei. *Proc Natl Acad Sci U S A* 98, 5270-5275.
[69] Lim YA, Ittner LM, Lim YL, Gotz J (2008) Human but not rat amylin shares neurotoxic properties with Abeta42 in long-term hippocampal and cortical cultures. *FEBS Lett* 582, 2188-2194.
[70] Allen B, Ingram E, Takao M, Smith MJ, Jakes R, Virdee K, Yoshida H, Holzer M, Craxton M, Emson PC, Atzori C, Migheli A, Crowther RA, Ghetti B, Spillantini MG, Goedert M (2002) Abundant tau filaments and nonapoptotic neurodegeneration in transgenic mice expressing human P301S tau protein. *J Neurosci* 22, 9340-9351.
[71] Yoshiyama Y, Higuchi M, Zhang B, Huang SM, Iwata N, Saido TC, Maeda J, Suhara T, Trojanowski JQ, Lee VM (2007) Synapse Loss and Microglial Activation Precede Tangles in a P301S Tauopathy Mouse Model. *Neuron* 53, 337-351.
[72] Dawson HN, Cantillana V, Chen L, Vitek MP (2007) The tau N279K exon 10 splicing mutation recapitulates frontotemporal dementia and parkinsonism linked to chromosome 17 tauopathy in a mouse model. *J Neurosci* 27, 9155-9168.
[73] Mocanu MM, Nissen A, Eckermann K, Khlistunova I, Biernat J, Drexler D, Petrova O, Schonig K, Bujard H, Mandelkow E, Zhou L, Rune G, Mandelkow EM (2008) The potential for beta-structure in the repeat domain of tau protein determines aggregation, synaptic decay, neuronal loss, and coassembly with endogenous Tau in inducible mouse models of tauopathy. *J Neurosci* 28, 737-748.
[74] Liou YC, Sun A, Ryo A, Zhou XZ, Yu ZX, Huang HK, Uchida T, Bronson R, Bing G, Li X, Hunter T, Lu KP (2003) Role of the prolyl isomerase Pin1 in protecting against age-dependent neurodegeneration. *Nature* 424, 556-561.

[75] Pastorino L, Sun A, Lu PJ, Zhou XZ, Balastik M, Finn G, Wulf G, Lim J, Li SH, Li X, Xia W, Nicholson LK, Lu KP (2006) The prolyl isomerase Pin1 regulates amyloid precursor protein processing and amyloid-beta production. *Nature* 440, 528-534.
[76] Sturchler-Pierrat C, Abramowski D, Duke M, Wiederhold KH, Mistl C, Rothacher S, Ledermann B, Burki K, Frey P, Paganetti PA, Waridel C, Calhoun ME, Jucker M, Probst A, Staufenbiel M, Sommer B (1997) Two amyloid precursor protein transgenic mouse models with Alzheimer disease-like pathology. *Proc Natl Acad Sci U S A* 94, 13287-13292.
[77] Casas C, Sergeant N, Itier JM, Blanchard V, Wirths O, van der Kolk N, Vingtdeux V, van de Steeg E, Ret G, Canton T, Drobecq H, Clark A, Bonici B, Delacourte A, Benavides J, Schmitz C, Tremp G, Bayer TA, Benoit P, Pradier L (2004) Massive CA1/2 neuronal loss with intraneuronal and N-terminal truncated Abeta42 accumulation in a novel Alzheimer transgenic model. *Am J Pathol* 165, 1289-1300.
[78] Harris FM, Brecht WJ, Xu Q, Tesseur I, Kekonius L, Wyss-Coray T, Fish JD, Masliah E, Hopkins PC, Scearce-Levie K, Weisgraber KH, Mucke L, Mahley RW, Huang Y (2003) Carboxyl-terminal-truncated apolipoprotein E4 causes Alzheimer's disease-like neurodegeneration and behavioral deficits in transgenic mice. *Proc Natl Acad Sci U S A* 100, 10966-10971.
[79] Butterfield DA, Boyd-Kimball D, Castegna A (2003) Proteomics in Alzheimer's disease: insights into potential mechanisms of neurodegeneration. *J Neurochem* 86, 1313-1327.
[80] Pienaar IS, Daniels WM, Gotz J (2008) Neuroproteomics as a promising tool in Parkinson's disease research. *J Neural Transm* 115, 1413-1430.
[81] Eckert A, Hauptmann S, Scherping I, Meinhardt J, Rhein V, Drose S, Brandt U, Fandrich M, Muller WE, Gotz J (2008) Oligomeric and fibrillar species of β-amyloid (Aβ42) both impair mitochondrial function in P301L tau transgenic mice. *J Mol Med* 86, 1255-1267.
[82] Rhein V, Song X, Wiesner A, Ittner LM, Baysang G, Meier F, Ozmen L, Bluethmann H, Drose S, Brandt U, Savaskan E, Czech C, Gotz J, Eckert A (2009) Amyloid-beta and tau synergistically impair the oxidative phosphorylation system in triple transgenic Alzheimer's disease mice. *Proc Natl Acad Sci U S A* 106, 20057-20062.
[83] Chen F, Wollmer MA, Hoerndli F, Münch G, Kuhla B, Rogaev EI, Tsolaki M, Papassotiropoulos A, Gotz J (2004) Role for glyoxalase I in Alzheimer's disease. *Proc Natl Acad Sci U S A* 101, 7687-7692.
[84] David D, Hoerndli F, Gotz J (2005) Functional Genomics meets neurodegenerative disorders Part I: Transcriptomic and proteomic technology. *Prog Neurobiol* 76, 153-168.
[85] David DC, Hauptmann S, Scherping I, Schuessel K, Keil U, Rizzu P, Ravid R, Dröse S, Brandt U, Müller WE, Eckert E, Gotz J (2005) Proteomic and functional analysis reveal a mitochondrial dysfunction in P301L tau transgenic mice. *J Biol Chem* 280, 23802-23814.
[86] David DC, Ittner LM, Gehrig P, Nergenau D, Shepherd C, Halliday G, Gotz J (2006) ß-Amyloid treatment of two complementary P301L tau-expressing Alzheimer's disease models reveals similar deregulated cellular processes. *Proteomics* 6, 6566-6577.
[87] Hoerndli F, David D, Gotz J (2005) Functional genomics meets neurodegenerative disorders. Part II: Application and data integration. *Prog Neurobiol* 76, 169-188.

[88] Hoerndli FJ, Pelech S, Papassotiropoulos A, Götz J (2007) Abeta treatment and P301L tau expression in an Alzheimer's disease tissue culture model act synergistically to promote aberrant cell cycle re-entry. *Eur J Neurosci.* 26, 60-72.

[89] Hoerndli FJ, Toigo M, Schild A, Gotz J, Day PJ (2004) Reference genes identified in SH-SY5Y cells using custom-made gene arrays with validation by quantitative polymerase chain reaction. *Anal Biochem* 335, 30-41.

[90] Blackshaw S, Livesey R (2002) Applying genomics technologies to neural development. *Curr Opin Neurobiol* 12, 110-114.

[91] Masland RH (2004) Neuronal cell types. *Curr Biol* 14, R497-500.

[92] Blackshaw S, Harpavat S, Trimarchi J, Cai L, Huang H, Kuo WP, Weber G, Lee K, Fraioli RE, Cho SH, Yung R, Asch E, Ohno-Machado L, Wong WH, Cepko CL (2004) Genomic analysis of mouse retinal development. *PLoS Biol* 2, E247.

[93] Gotz J, Ittner LM, Schonrock N (2006) Alzheimer's disease and frontotemporal dementia: prospects of a tailored therapy? *Med J Aust* 185, 381-384.

[94] Gotz J, Ittner LM, Fandrich M, Schonrock N (2008) Is tau aggregation toxic or protective: a sensible question in the absence of sensitive methods? *J Alzheimers Dis* 14, 423-429.

[95] Salehi A, Ravid R, Gonatas NK, Swaab DF (1995) Decreased activity of hippocampal neurons in Alzheimer's disease is not related to the presence of neurofibrillary tangles. *J Neuropathol Exp Neurol* 54, 704-709.

[96] Broe M, Shepherd CE, Milward EA, Halliday GM (2001) Relationship between DNA fragmentation, morphological changes and neuronal loss in Alzheimer's disease and dementia with Lewy bodies. *Acta Neuropathol* 101, 616-624.

[97] Love S, Barber R, Wilcock GK (1999) Increased poly(ADP-ribosyl)ation of nuclear proteins in Alzheimer's disease. *Brain* 122 ( Pt 2), 247-253.

[98] Kurosinski P, Gotz J (2002) Glial cells under physiologic and pathological conditions. *Arch Neurol* 59, 1524-1528.

# The Cholesterol-Fed Rabbit as a Model of AD: The Old, the New and the Pilot

D. Larry Sparks*
*Roberts Laboratory for Neurodegenerative Disease Research, Sun Health Research Institute, Sun City, AZ, USA*

**Abstract**. Pioneering autopsy studies revealed a possible link between coronary artery disease, cholesterol and Alzheimer's disease (AD). In the cholesterol-fed rabbit model of human coronary artery disease, we identified numerous neuropathologic features of AD including central accumulation of amyloid-$\beta$ (A$\beta$) and cognitive deficits compared to rabbits fed unaltered diet. Removing cholesterol from the diet or treatment with cholesterol-lowering medications reversed the severity of AD-like alterations. This fostered the rationale for testing a cholesterol-lowering statin medication for benefit in the treatment of AD. Further studies suggested that the cholesterol-fed rabbit was a viable model for AD, but the severity of the neuropathology produced exhibited gender-related differences. Furthermore the induction of AD-like neuropathology by dietary cholesterol was found to depend on the quality of water the animal was drinking. Cholesterol-fed rabbits drinking distilled water showed minimal central changes, whereas animals drinking distilled water supplemented with low levels of copper were severely affected. It was clear that cholesterol caused the over-production of A$\beta$ in the brain and copper influenced its clearance to the blood. Emerging data suggest that low-density lipoprotein receptor-related protein-1 (LRP) on brain capillaries clears A$\beta$ from brain and that excess circulating copper negatively influences this process.

Keywords: Alzheimer's disease, cholesterol, coronary artery disease, rabbit

## INTRODUCTION

In the early 1980s, it was commonly held that amyloid-$\beta$ (A$\beta$) containing senile plaques (SP) occurred exclusively in the brains of individuals with Alzheimer's disease (AD). As a fledgling AD researcher working in the forensic arena, I observed abundant SP in ostensibly cognitively normal individuals, the youngest being 28 years of age. Simultaneously, we found that our methods of determining postmortem interval from brain neurochemical content [66,67] followed predictably different patterns in critical coronary artery disease (cCAD) compared to non-heart disease [68]. Based on these data, silver stained sections of hippocampus from our normal control population were grouped according to cause of death, and essentially all of the SP showed up in the cCAD [65,70] and/or hypertension [73] populations. These findings were not well received as it took over four years from discovery in 1985 and initial publication of the data in 1990.

The finding of SP in cognitively normal individuals with cCAD was instrumental in curtailing the practice of neuropathologically affixing the diagnosis of AD in the absence of clinical data. Heretofore, non-demented individuals with hypertension or significant coronary artery disease could be neuropathologically diagnosed as AD in cases where it was not clear that the subject

---

*Address for correspondence: D. Larry Sparks, Sun Health Research Institute, 10515 W. Santa Fe Drive, Sun City, AZ 85351, USA. E-mail: Larry.Sparks@sunhealth.org.

was cognitively normal. It is most likely that these individuals died of their heart disease prior to manifesting cognitive symptoms, and had they been sufficiently resilient to not succumb to their heart disease – eventually their cognitive performance would have caught up with mounting neuropathology. This is supported by observations in Down's syndrome (DS), where neuronal alterations, microgliosis and SP [69,98] are present, often decades before neurofibrillary tangles (NFT) and dementia [32].

About the same time we observed SP in coronary artery disease (CAD), investigators capitalized on identified amyloid-$\beta$ protein precursor (A$\beta$PP) mutations, associated with production and accumulation of A$\beta$ as SP in the brains of individuals with familial AD, by inserting human genetic material containing human A$\beta$PP mutations into the mouse genome. Most consider the A$\beta$ peptide to be an aberrant metabolic by-product of the larger precursor protein (A$\beta$PP). Memory deficits and accumulation of SP-like deposits of A$\beta$ occur in these transgenic mouse models of AD. Based on finding SP in CAD, we were presented with the choice of looking for heart disease in transgenic mouse models of AD, or looking for AD-like pathology in the brain of a small-animal model of human heart disease. We chose the latter.

## CHOLESTEROL-FED RABBIT MODEL OF AD

### A$\beta$ accumulation in brain

We first reported studies performed in cholesterol-fed New Zealand white rabbits as the best small animal model of human coronary artery disease (CAD) – based on the finding of SP in the brains of non-demented individuals with CAD – indicating that dietary cholesterol induces a pronounced accumulation of neuronal A$\beta$ compared to animals fed normal chow diet [71]. The cholesterol-fed rabbit was initially used as a model for experimental atherosclerosis studies following findings by Nikolaj Nikolajewitsch Anitschkow in 1913 that cholesterol causes atherosclerotic changes in the rabbit arterial intima [2,41]. Although the cholesterol-fed rabbit model of human CAD clearly pre-dated any transgenic animal model of AD, careful assessment of previous studies in the model indicated that no one had ever investigated the effect of increased circulating cholesterol on rabbit brain function or pathology. In the inaugural investigations pointed toward that end, we explored the time-on-diet relationship between cholesterol and the time-course of severity and regional locations of A$\beta$ immunoreactivity accumulations produced during the early 90s; again it took a number of years to publish these findings [71]. Experimental animals were fed high cholesterol diet for 4, 6 or 8 weeks and compared to animals fed control diet 6 weeks. Rare or no A$\beta$ immunoreactive neurons were observed in the brains of control rabbits. With relatively prolonged dietary exposure to cholesterol, A$\beta$ deposition became more pronounced. A$\beta$ immunoreactive neurons were observed in the hippocampus and adjacent cortex by 4 weeks of the experimental diet and progressively increased in staining intensity and number of affected cells with longer time-on-diet. A similar relationship occurred in frontal cortex but lagged behind the hippocampus and adjacent cortex by at least 2 additional weeks of cholesterol diet.

Such A$\beta$ deposits were predominantly intraneuronal, although some animals exhibited diffuse extracellular A$\beta$ plaque-like deposits, which were not argyrophilic (silver positive). A$\beta$ was transiently apparent in neuronal processes of regio superior of the hippocampus during the time-course of the diet. Increased numbers of pyramidal neurons, granule cell neurons, and large neurons of the cortex and hilus were also affected with increased time on the diet.

A$\beta$ staining in AD, DS and cCAD subjects also show intraneuronal staining in hippocampus (hilus, granule cell layer and pyramidal cells) and neocortex, in a pattern strikingly similar to that found in rabbits fed cholesterol [74]. Laminar distribution of intraneuronal A$\beta$ in cholesterol-fed rabbit cortex [83] was found to be similar to the laminar distribution of SP in CAD and AD [78]. Twice as many SP were found in the upper 3 layers of the cortex compared to the lower 3 layers in both cCAD and AD; Layers 2 and 3 of the upper half of the cortex and layer 5 of the lower half were prominently affected in both groups [78].

These observations suggested a possible relationship between cardiac function and deposition of A$\beta$ as part of a neuropathologic final common pathway in both heart disease and AD, while establishing relevance of the observed changes in cholesterol-fed rabbits with those identifiable in AD.

### Cholesterol levels

We established conditions enabling baseline resolution of cholesterol by HPLC, and found 9 to 14% increases in cholesterol in various regions of rabbit brain after 8 weeks of dietary cholesterol. The increase of hippocampal and frontal cortex cholesterol in the experimental animals was nearly significant ($0.05 < p < 0.1$), while the 14.6% in hippocampal cortex did achieve significance ($p < 0.02$) [75,76]. The severity and level of significance of increased brain cholesterol levels induced by dietary administration was consistently greater in the hippocampal cortex compared to frontal cortex. It was interesting that the number of A$\beta$ immunolabeled neurons and the intensity of immunoreactivity were significantly greater in hippocampal cortex compared to frontal cortex among 8 week cholesterol-fed rabbits. The mild increase in rabbit brain cholesterol content after dietary administration could be highly significant in relation to maintaining normal metabolism for continued function and possibly survival.

*Apolipoprotein-E (ApoE) immunoreactivity*

Apolipoprotein E (ApoE) is known to facilitate cholesterol transport in the brain [46], and it is well accepted that an individual's ApoE genotype and metabolism are closely linked to their risk of developing AD. Identical to AD, SP in cCAD are readily immunostained by ApoE antibody in an almost one-to-one distribution with A$\beta$ immunoreactive SP in the neuropil. Occasionally neurons also expressed the ApoE epitope in cCAD. These findings prompted us to determine the effect of cholesterol diet on ApoE immunoreactivity in the rabbit brain [72]. Scattered ApoE immunoreactive neurons are observable after 6 weeks of cholesterol diet. After 8 weeks of cholesterol, dramatic differences in ApoE immunoreactivity are found in hippocampus and cortex compared to controls. Although onset of ApoE immunoreactivity seems to precede A$\beta$ accumulation in the cholesterol-fed rabbit brain, demonstrable increases in ApoE immunoreactivity clearly supported the evidence that cholesterol levels are increased.

*Lysosomal activity – Cathepsin-D*

Cathepsin D is a lysosomal enzyme capable of releasing the N-terminal sequence of A$\beta$ and prior to the advent of $\beta$-secretase was thought to be involved as an early metabolic step in senile plaque formation in AD [53]. The activation of the lysosomal system in cCAD parallels changes demonstrable in AD. Senile plaques occurred in every AD and most cCAD subjects, and were immunoreactive with cathepsin D antibody in both groups [29]. The enzymatic activity of cathepsin D in fresh tissue paralleled the histological observations.

Because of the proposed relationship between metabolic production of A$\beta$ and the lysosomal enzyme cathepsin D in AD, cholesterol-fed rabbits were assessed for changes in cathepsin D immunoreactivity. Increased cathepsin D immunoreactivity was apparent by 6 weeks on the diet and became severe by 8 weeks in hippocampus; this anatomically paralleled A$\beta$ immunoreactive neurons, but was delayed [75] and thus did not directly cause the production of A$\beta$.

*Microglia in cholesterol-fed rabbit brain*

In an effort to determine the breadth of similarities between AD and this animal model of AD, sections of hippocampus from cholesterol-fed and control rabbits were stained for microglial cells by lectin impregnation. Signs of central nervous system (CNS) inflammation were observed in hypercholesterolemic rabbits. Early signs of microglia activation were seen by 6 weeks on the cholesterol diet [99]. Fully activated microglial cells were observed in rabbits fed cholesterol diet for 8 weeks. The presence of such activated microglia is an amply documented histopathologic feature of AD [14]. This microglial activation seemingly lags behind the accumulation of A$\beta$ immunoreactivity and persists after regression of the cholesterol diet and apparent clearance of the peptide(s). Clearance of A$\beta$ may be directly due to activated microglia. The absence of A$\beta$PP and the inability to detect A$\beta$PP mRNA in microglial cells [61], together with a lack of conclusive evidence showing production of A$\beta$ by microglia, suggest that presence in microglia may be a result of phagocytosis of neuronally accumulating peptide. This is supported by the observation in double-labeling studies with A$\beta$ and microglial antibodies that microglia appear to embrace neurons expressing A$\beta$ immunoreactivity (Fig. 1).

On the one hand, the foregoing data suggested that microglia are activated by the build-up of A$\beta$ peptide(s) caused by cholesterol diet and are involved in the clearance of the peptide(s) upon regression of the diet. Microglial cells are in an early activated state after A$\beta$ immunoreactivity is consistently demonstrable, and they are fully activated after immunoreactivity is waning due to the removal of cholesterol from the diet [99]. On the other hand, recent data suggests that microgliosis occurs in cholesterol-fed rabbit brain regardless of whether or not there is accumulation of A$\beta$ [106].

Careful inspection of the sections stained for microglia revealed apoptotic neurons in the granule cell

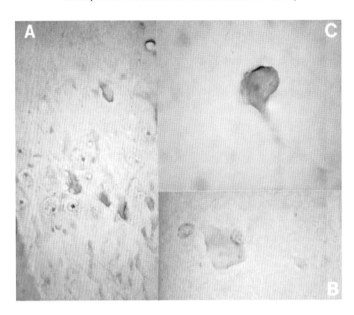

Fig. 1. Microglia associated with A$\beta$ immunoreactive neurons. Section of hippocampus and adjacent cortex from a rabbit fed cholesterol for 10 weeks were initially stained with A$\beta$ antibody (10D5; visualized with Nickel-DAB to produce a blue color) and counter-stained with *Griffionia simplicifolia* B4-isolectin for microglial cells (visualized with DAB to yield brown labeling). Numerous A$\beta$ immunoreactive neurons in the pyramidal cell layer were double-labeled for microglia (A). In the cortex, many A$\beta$ immunoreactive neurons have adherent cells expressing the lectin epitope (B). Some cortical neurons are encapsulated by microglial processes (C). (Provided by Dr. Wolfgang 'Jake' Streit).

layer of hippocampus and cortical perivascular cuffs, characteristic of vascular inflammation, are readily apparent in the cholesterol-fed rabbit brain stained for microglia by the lectin epitope [99]. This observation supported the hypothesis that vascular inflammation may be involved in the mechanism of A$\beta$ accumulation in the cholesterol-fed rabbit. In contrast to all other work where we tested for changes in markers known to be altered in AD, we sought to identify vascular inflammation in the cholesterol-fed rabbit brain and having done so, thereafter demonstrated similar findings in AD.

*Vascular inflammation – Monoclonal Endothelial Cell Antigen-32 (MECA-32)*

We reported for the first time that there was increased vessel MECA-32 immunoreactivity, and therefore vascular inflammation, in AD [80]. Only one of the six human controls showed faint, but discernable vascular MECA-32 immunoreactivity, all other age-matched human controls showed no MECA-32 immunoreactivity. All AD subjects exhibited pockets of cerebrovascular MECA-32 immunoreactivity. MECA-32 antibody interacts exclusively with endothelial cells in the CNS that are undergoing inflammation in multiple sclerosis (MS) and an accepted animal model of MS, experimental allergic encephalitis (EAE) [15,18]. Such vascular inflammation is widespread in MS and EAE. In the case of AD, increased vascular MECA-32 immunoreactivity occurs only in isolated pockets. This could suggest a micro-environmental alteration or attack of blood vessels in AD rather than a generalized disorder of the vasculature. Previous reports suggest a reactive process is occurring in the AD vasculature [36], which may be related to endothelial degeneration and localized breaches of the blood-brain barrier (BBB) in the disorder [37,38]. We suggest that localized endothelial inflammation in AD is part of the above-noted reactive process leading to proposed breaches of the BBB.

We reported that isolated pockets of cerebrovascular MECA-32 immunoreactivity are produced by dietary cholesterol in the rabbit. No control-fed rabbit exhibited vascular MECA-32 immunoreactivity and each cholesterol-fed rabbit exhibited MECA-32 immunoreactivity. Vascular MECA-32 immunoreactivity could not be identified in any control rabbit brain, whether from archival tissue or briefly stored tissue. Each of the archival cholesterol-fed rabbit brains exhibited faint vascular MECA-32 immunoreactivity. All of the cholesterol-fed rabbits investigated within months of sacrifice showed more intense cerebrovascular MECA-32 immunoreactivity.

It would seem that excess cholesterol in the blood in rabbits induces change at the surface of the endothelial cell leading to isolated patches of inflammation, degeneration and breach of the BBB. In turn, this process may set in motion a cascade of biochemical derangement in the underlying brain tissue leading to $\beta$-amyloidogenesis in the rabbit brain. We would further suggest that it might be increased circulating concentrations of free radicals caused by increased levels of cholesterol that initiates endothelial inflammation, and that a similar mechanism could be active in AD [77].

*Free radical enzyme markers in brain and blood after cholesterol diet*

As circulating free radical load is implicated in the mechanism of atheroma production in CAD and is proposed to have an influence in AD, we determined the activity of superoxide dismutase (SOD) activity in the plasma and red blood cells (RBC) of rabbits fed control or 2% cholesterol diet for 8 weeks. In the same animals, SOD activity was measured in the hippocampus, hippocampal cortex and frontal cortex [76]. Changes in the brain paralleled changes in the blood where SOD activity was increased. Evaluation of the relationship between regional cholesterol levels and SOD activity in brains of rabbits fed control or cholesterol diet revealed a significant correlation between the mean regional SOD activity and cholesterol (rank order relationship with r squared = 0.99). These data suggest that there may be a regional association between SOD activity and cholesterol levels and the possibility that free radicals could play a role in A$\beta$ production or accumulation.

*SOD immunoreactivity in cholesterol-fed rabbit brain*

Because we observed an increase in SOD enzymatic activity in cholesterol-fed rabbits and SOD activity varied on a region-to-region basis with cholesterol content in rabbit brain, we performed preliminary immunocytochemical investigations with a Cu/Zn SOD antibody. The intensity of SOD immunoreactivity increased as time on cholesterol diet increased, as did the number of immunoreactive neurons [75]. The time course of increased SOD immunoreactivity with increased time on the cholesterol diet was established by counting the number of immunodecorated neurons after 4, 6 and 8 weeks of cholesterol diet compared to control-fed rabbits. These data suggest that increased SOD activity doubles after four weeks on the diet, when increased A$\beta$ immunoreactivity is minimal. This may indicated free radicals play a role in cholesterol-induced A$\beta$ accumulation.

Re-evaluation of sections stained with a Cu/Zn SOD antibody in both human and rabbits revealed 'halos' of SOD immunoreactivity surrounding vessels in the brains of AD patients and cholesterol-fed animals [77]. Similar features were not found in age-matched human controls or in rabbits fed control diet. Perivascular 'halos' of SOD immunoreactivity were apparent in both small and large bore vessels in both AD and cholesterol-fed rabbit brain. Such vessels appear to have significantly enhanced SOD immunoreactivity themselves. This tended to support the contention that there may be flow of free radicals from the blood, through the vasculature, and into the CNS, with SOD increases in each compartment as free radicals increase.

*Breach of the blood-brain barrier in the cholesterol fed rabbit*

Two pairs of rabbits were fed cholesterol diet (2%) or control diet for 12 weeks. One hour prior to sacrifice by removal of the heart and perfusion of the brain with 2% paraformaldehyde, each animal was injected with an Evans Blue solution to determine if there might be breaches of the BBB [80]. Distinct areas of the microcirculation exhibited surrounding pockets of Evans Blue in the cholesterol-fed rabbits only. This would suggest that there are isolated breaches of the BBB which could account for seepage of blood-borne factors into the parenchyma, activation of SOD, disruption of neuronal metabolism, activation of neuronal SOD, production of A$\beta$ immunoreactive material, activation of microglia, induction of the complement cascade, and subsequent neuronal death in the cholesterol-fed rabbit.

## GENDER-RELATED DIFFERENCES IN CHOLESTEROL-INDUCED AD-LIKE NEUROPATHOLOGY

During the course of performing studies in the late 90s, we found that dietary cholesterol was less effective in producing AD-like neuropathology. It turns out that in contrast to the previous routine, female rabbits rather than males had been acquired for investigation. Once this was discovered, we actively sought to identify any gender-related differences in this animal model of AD. In a previous effort to ascertain the scope of the effect of increased circulating cholesterol on A$\beta$ accumulation across various animal models, we had performed pilot studies in 12-month old male Watanabe rabbits where A$\beta$ immunoreactive neurons were abundant in

Table 1
Effect of dietary cholesterol on levels of Aβ in rabbit frontal pole. Values of Aβ concentrations are in mg/mg protein ± SEM

| Treatment/Time | N | Total | 40 | 42 |
| --- | --- | --- | --- | --- |
| Control | 3 | 0.53 + 0.43 | 0.17 + 0.17 | 0.35 + 0.27 |
| Cholesterol 12 weeks (females) | 3 | 1.83 + 0.73* | 1.27 + 0.36** | 0.54 + 0.35 |
| Cholestero l8 weeks (males) | 3 | 1.53 + 0.58** | 0.99 + 0.39** | 0.53 + 0.21 |

*= $p < 0.1$.
**= $p < 0.05$.

the brain; to date, essentially every animal model of AD – from transgenic mice to aged beagles – accumulate Aβ earlier or in greater abundance, or both, if the animals are administered cholesterol in the diet [3, 16,22,44,58,62,89]. The Watanabe rabbit is a back-bred homogenous line of New Zealand white rabbit that spontaneously exhibit increased circulating cholesterol levels due to a genetic mutation eliminating the HDL receptor.

*Aβ immunoreactivity in spontaneously hypercholesterolemic female Watanabe rabbits*

To test the effect of gender on the relationship between circulating cholesterol and central Aβ accumulation, in the late 90's we investigated the hippocampal cortex in a pilot age-related study of female Watanabe rabbits compared to control New Zealand white female rabbits. Control rabbits of 6, 8 and 12 months of age, stained for Aβ showed little or no immunoreactivity. Watanabe rabbits, one each of 3, 6 and 12 months of age, and two animals 36 months of age were studied [82]. In the 3 month-old female Watanabe rabbit, abundant neurons exhibited very intense Aβ immunoreactivity. In the 6 month-old female Watanabe rabbit numerous neurons, but fewer than at 3 months of age, exhibited moderately intense Aβ immunoreactivity. In the 12 month-old female Watanabe rabbit brain, a very limited number of neurons showed faint Aβ immunoreactivity. In the 36 months old female Watanabe rabbits, a moderate to abundant number of neurons in the hippocampal cortex exhibit Aβ immunoreactivity. It is of note that female Watanabe rabbits begin ovulation at about 6 months of age, which subsides between 30 and 40 months of age. The data suggested that estrus protects the spontaneously hypercholesterolemic rabbit from Aβ production in the brain [82].

*Effect on levels of Aβ in cholesterol-fed rabbits*

To test the effect of gender in the cholesterol-fed New Zealand white rabbit, fresh frozen frontal cortex stored at $-70°C$ was investigated for levels of Aβ (total, 40 and 42 amino acid lengths). The frontal cortex was investigated from female rabbits fed cholesterol or control diet for 12 weeks, and male rabbits fed cholesterol diet for 8 weeks. It was clear that levels of Aβ are increased by 8 weeks of feeding cholesterol to male rabbits to a level comparable to feeding cholesterol to female rabbits for 12 weeks (Table 1).

This seems to be consistent with the observation of less robust accumulation of Aβ identified immunohistochemically in cholesterol-fed female rabbits compared to male counter-parts. This supported the use of only male rabbits in future studies.

## CHOLINERGIC MARKERS

Based on the well-recognized acetyl-cholinergic alterations in AD brain, we quantified choline acetyltransferase (ChAT) activity and visualized acetylcholinesterase (AChE) immunoreactive fibers in cholesterol-fed rabbit brain compared to rabbits fed control diet. (These two pilot experiments were undertaken using female cholesterol-fed rabbits prior to discovering the aforementioned gender-related differences.)

*Choline acetyltransferase (ChAT) activity in mid-frontal cortex (females)*

ChAT activity was assessed in the mid-frontal cortex of rabbits fed 2% cholesterol for 12 weeks [23]. ChAT activity was reduced 39%, and even with the small number of animals investigated in this pilot study the difference approached significance ($p = 0.0701$) (Table 2).

*Cholinergic fibers in rabbit brain (females)*

We investigated the hippocampus and hippocampal cortex in two rabbits fed a control diet and two rabbits fed cholesterol for 12 weeks and sacrificed by perfusion of the brain via the heart with 2% paraformalde-

Table 2
Choline acetyltransferase (ChAT) activity in mid-frontal cortex. Values presented are means ± SEM with units of pmoles/mg protein/min

| Group | N | ChAT activity |
|---|---|---|
| Control diet | 2 | 2.37 ± 0.57 |
| Cholesterol diet | 3 | 1.44 ± 0.15 |

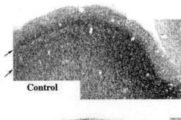
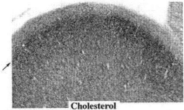

Fig. 2. AChE fibers in hippocampal cortex of rabbits fed normal diet or 2% cholesterol diet.

hyde. Coronal sections were subsequently immersion fixed in 4% paraformaldehyde for 36 hours and then placed in a sucrose solution until sectioning. Sections were stained immunohistochemically for AChE activity (Tago method [100]). Regionally specific AChE immunoreactive fibers and neurons were observable in the control rabbits.

In the hippocampal cortex of both control rabbits, there was intense fiber AChE immunoreactivity along the temporal fissure which takes on a laminar pattern away from the temporal fissure (Fig. 2, arrows). In the hippocampal cortex of both cholesterol-fed rabbits, there was a significant reduction of fiber AChE immunoreactivity along the temporal fissure and a loss of the lower fiber band in adjacent cortex (Fig. 2, arrow). In the hippocampus, there was no apparent difference between control and cholesterol-fed rabbits (Fig. 2).

These pilot studies suggest a cholinergic alteration in the cholesterol-fed rabbit consistent with findings in AD. Further assessment of cholinergic markers in this animal model of AD could be fruitful and provide yet another avenue of assessing prospective medications in the treatment of AD.

## CLEARANCE STUDIES

Human studies in the late 80s indicated possible clearance of A$\beta$, in that none of 6 individuals over 60 years of age surviving greater than 5 years after coronary by-pass grafting (CABAG) had SP in their brains compared to >60% prevalence of SP of similarly aged individuals with significant CAD. To directly test for possible clearance of A$\beta$, 6 rabbits were fed cholesterol diet for 8 weeks and then 3 were put on control diet for 2 weeks, and 3 were fed experimental diet for an additional 2 weeks [74]. This was supported by the observation that deposition of diffuse form SP may be a reversible process [34]. In animals regressed from cholesterol diet, there were fewer A$\beta$ reactive neurons in the cortex and hippocampus, and neurons were less intensely stained. After 10 weeks on the cholesterol diet, there is a significant increase in the areal density of affected neurons in the cortex compared to densities previously reported in rabbits fed the cholesterol diet for 8 weeks [71]. Upon regression of the high cholesterol (2%) diet for two weeks, the number of A$\beta$ immunoreactive neurons and intensity of staining is significantly reduced. Therefore, the deposition of A$\beta$ immunoreactive material induced by dietary cholesterol is apparently a reversible process. This suggests that, no matter the mechanism of A$\beta$ accumulation – a direct or indirect effect of cholesterol – neuronal A$\beta$ peptide(s) can be and is/are cleared from the brain if conditions permit. This suggested that if conditions never permit clearance of A$\beta$ peptide(s) from neurons that accumulated A$\beta$ may be deposited in the neuropil as an alternate means of clearing the protein from functional cells.

The foregoing studies provided a clear rationale for assessing a cholesterol-lowering medication in the treatment of AD [79,82]. Eventually we showed that statin therapy was of benefit in the treatment of mild-to-moderate AD in the Alzheimer Disease Cholesterol-Lowering Trial (ADCLT) [87,88], but as we initiated the clinical investigation, the animal model serving as the foundation stopped working. We were no longer able to identify AD-like neuropathology in the brains of cholesterol-fed rabbits. For the first 16 months of the ADCLT, we frantically tried to determine why the animal model no longer worked. Finally the animal vivarium was inspected for possible differences from previously conducted studies. The first thing noticed was the storage of large blue water cooler bottles. Upon questioning, it was noted that this was the water supply for the animals in the facility, and that it was dis-

Table 3
Aβ levels in the blood of rabbits fed cholesterol (2%) or control diet and administered tap or distilled drinking water *ad libitum*

Circulating Aβ levels (Aβ40; pg/ml)

| Diet group | 2% Cholesterol | Control |
|---|---|---|
| Tap water | 199.1 + 20.0 | 138.9 + 20.0 |
| Distilled drinking water | 290.7 + 20.0* | 137.5 + 20.0 |

*= $p < 0.05$ compared to respective group fed control diet.

Table 4
Number of animals receiving saline, nicotine or AF267b administered either normal or 2% cholesterol diet and administered tap or distilled water to test the effect on Aβ accumulation

| Food | N | Water | Treatment |
|---|---|---|---|
| Normal chow | 8 | Tap | Saline |
| Normal chow | 10 | Distilled | Saline |
| Normal chow | 5 | Tap | Nicotine |
| Normal chow | 7 | Distilled | Nicotine |
| 2% cholesterol | 11 | Tap | Saline |
| 2% cholesterol | 13 | Distilled | Saline |
| 2% cholesterol | 10 | Tap | Nicotine |
| 2% cholesterol | 10 | Distilled | Nicotine |
| 2% cholesterol | 5 | Tap | AF267B |
| 2% cholesterol | 5 | Distilled | AF267B |

tilled water. Instructions were passed on to replicate the cholesterol-fed rabbit studies that had not worked for the previous 16 months with half of the animals administered 'blue bottled water' and half to receive drinking water from a specific tap water supply. Dietary cholesterol again produced AD-like neuropathology, but only in the animals on tap water [81].

## WATER QUALITY AND THE CHOLESTEROL-FED RABBIT

Recent studies suggested the accumulation of Aβ in the brains of cholesterol-fed rabbits was dependent on the quality of the water the animals were administered. Animals given tap water accumulated considerably more Aβ in the brain than animals administered distilled water [81]. Both the intensity of the immunoreactivity observed and the number of neurons affected was greatly diminished in animals given distilled water. The difference in the intensity of the immunoreactivity is important because there must be a relationship between the intensity of antibody immunoreactivity and concentration of its antigen, in this case Aβ. Morphologically, the neurons with Aβ immunoreactivity often appeared shrunken in size among cholesterol-fed rabbits on tap water compared to animals on distilled water. Investigation of Aβ levels in the rabbit blood showed lower levels in cholesterol-fed animals suggested Aβ accumulated in the brain because of reduced clearance from the brain to the blood (Table 3) [81]. This is analogous to findings in mouse models of AD where the equilibrium of Aβ in the brain and blood may vary with the level of its deposition in the brain [12,13].

These data were taken by us as the first direct evidence of the ability to manipulate clearance of Aβ from the brain. When Aβ accumulates in the brain, only minimal increases are identified in the blood (cholesterol/tap water; Table 3) and when there is only minimal central accumulation of Aβ, there is considerably more of the toxin occurring in the blood (cholesterol/distilled water; Table 3).

## CHOLINERGIC MEDICATIONS IN THE CHOLESTEROL-FED RABBIT

AD is characterized by a dysfunction of cholinergic neurotransmission. The cholinergic system plays an important role in cognitive function and deficits in this system correlate well with the cognitive impairment seen in AD. Because of this, AChE inhibitors are the most widely used drugs to treat AD, and although we did not return our attention to expanding the pilot studies suggesting reduced acetyl-choline production (ChAT) and fibers (AChE) in the cholesterol-fed rabbit, we did however assess the effectiveness of both muscarinic and nicotinic medications for the ability to reduce the accumulation of Aβ in the cholesterol-fed rabbit the model. The muscarinic type-1 receptor agonist AF267b and nicotine were tested in a single experimental protocol.

Adolescent male New Zealand white rabbits (3000–4000 g) were housed in the rabbit facility at the Sun Health Research Institute operating under the guidelines of the USDA with a 12:12 light cycle, at $67 \pm 7°F$, and 45–50% humidity. Animals were randomly assigned to one of ten groups for the study (Table 4). Animals were injected daily (s.c.) with isotonic saline or nicotine at a dose of 1.0 mg/kg body weight, or AF267b at a dose of 1.0 mg/kg body weight. Two separate 10-week experiments, each containing all 10 groups, were performed. Food intake was limited to one cup per day (236 ml) and *ad libitum* water consumption varied between 950 and 1180 ml/day (Table 4).

*Muscarinic-1 receptor agonist (AF267b)*

Limited numbers of Aβ immunoreactive neurons were observed in cortex and hilus of rabbits fed normal chow drinking either tap water or distilled water.

Among the cholesterol-fed animals injected with saline, abundant neurons contained identifiable A$\beta$ but the intensity was much greater in animals drinking tap water. Such neurons were considerably smaller than those occasional encountered in animals fed normal chow. The number of neurons expressing A$\beta$ immunoreactivity was reduced 25–30% in the animals administered AF267b, and the intensity of the immunoreactivity was reduced approximately 50%, but only in cholesterol-fed rabbits drinking tap water [85]. This suggested that the medication influenced the clearance of A$\beta$ and not cholesterol-induced A$\beta$ production by either interacting with the agent in tap water inhibiting clearance of A$\beta$ or via the M-1 receptor to produce the same effect [85].

### Nicotine

Male New Zealand white rabbits fed control diet for 10 weeks exhibit limited numbers of faintly A$\beta$ immunoreactive neurons in the hippocampal cortex of animals on tap water (Fig. 3A) and distilled water (Fig. 3D). Those rabbits drinking tap water and fed cholesterol (2%) for 10 weeks exhibit numerous intensely stained A$\beta$ immunoreactive neurons (Fig. 3B), and as noted in previous experiments the intensity of A$\beta$ immunoreactivity was greatly diminished in animals drinking distilled water. Among rabbits fed cholesterol and administered nicotine daily for 10 weeks there were much fewer A$\beta$ immunoreactive neurons that were faintly stain in animals drinking tap water (Fig. 3C) and distilled water (not shown). This would suggest that chronic nicotine may possibly attenuate the cholesterol-induced production and accumulation of A$\beta$ in the brain.

We counted neurons exhibiting A$\beta$ immunoreactivity in five random fields of inferior temporal cortex among the treated rabbits. We found that there was a significant increase in the number of neurons labeled with A$\beta$ immunoreactivity among cholesterol-fed rabbits (Table 5). We also discovered that the number of labeled neurons was less among cholesterol-fed rabbits chronically administered nicotine (Table 5), but variability precluded a significant difference.

Because nicotine and nicotinic acetylcholine receptors are responsible for numerous biological effects, there are a number of possible mechanisms by which nicotine could potentially lower A$\beta$ accumulation. Nicotine enhances $\alpha$-secretase activity as the normal metabolic route for A$\beta$PP, causing increased release of the resultant non-amyloidogenic A$\beta$PP fragment from cells in culture [17,42]. The purported antioxidant properties of nicotine [28,52] may also play a role in the reduced neuronal accumulation of A$\beta$ because of the role that chronic oxidative stress has been shown to play in AD [7,20,26,47,56,92]. It is also possible that nicotine activating acetylcholine receptors could have anti-inflammatory properties [104], which may be an important process in the development of AD [1,24,31, 45,51].

Table 5
Number of neurons exhibiting A$\beta$ immunoreactivity in the inferior (temporal) hippocampal cortex of New Zealand white rabbits fed 2% cholesterol diet or normal chow, injected with either sterile isotonic saline or nicotine, and allowed either tap water or bottled distilled water

| Water ad libitum Diet/treatment | Tap water | Distilled water |
| --- | --- | --- |
| Normal chow/saline | 31.4 ± 4.8 | 26.3 ± 4.1 |
| Normal chow/nicotine | 34.7 ± 6.7 | 23.4 ± 3.5 |
| Cholesterol/saline | 65.7 ± 4.2*,** | 50.9 ± 4.3* |
| Cholesterol/nicotine | 44.7 ± 15.9 | 36.6 ± 6.6 |

ANOVA: $p = 0.02$.
*= $p < 0.05$ compared to normal chow saline injected.
**= $p < 0.05$ compared to cholesterol-saline on distilled water.

Our results are similar to previous studies that demonstrated the effects of nicotine [30,43,54,102] or M1 receptor agonists [6] on A$\beta$ deposition in brain and reinforce the idea that cholinergic agents may be of therapeutic value in the treatment of AD. The lack of an effect of AF267B on animals fed distilled water indicates that water quality is important when administering this drug and opens the possibility that water quality can have on different physiological effects depending on the drug administered, and more importantly that the efficacy of a medication prescribed to a patient might be dependent of the quality of water an individual may be drinking. It is notable that copper cations, which are in the tap water used, can increase the affinity of the muscarinic receptor agonists, unveiling a high affinity agonistic site [21]. Indeed, when copper cations are added to the distilled water at the same concentrations that are present in the tap water, we find A$\beta$ deposition in the brains of cholesterol-fed rabbits to be as much and often more severe than found in the cholesterol-fed rabbits given distilled water [83].

## COPPER

A role for altered copper metabolism as a cause of neurodegenerative disorders other than AD is clearly recognized [103], and a role for copper in AD is emerging. The link between copper and ceruloplasmin are

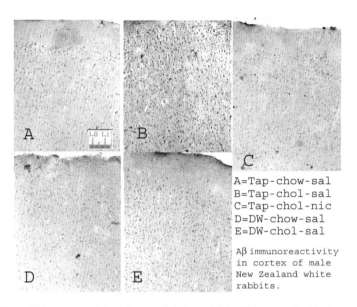

Fig. 3. Aβ immunoreactivity in brains of cholesterol-fed rabbits injected with nicotine.

clear. Ceruloplasmin, an inflammation-sensitive plasma protein [19], transports 95% of the copper in human blood [103]. Circulating levels of ceruloplasmin are positively correlated with levels of copper in the blood [35,57]. It has been shown that ceruloplasmin levels in humans in response to dietary copper supplementation is a good indicator of changing circulating copper [49].

Both copper and ceruloplasmin levels have been shown to be elevated in the blood of patients with AD compared to controls by many [55,63,93,94,97], but not all investigators [64]. Squitti and colleagues have reported a significant increase in circulating copper in AD and a trend for increased ceruloplasmin [95]. These authors also reported a significant negative correlation in AD between increased copper/ceruloplasmin and decreased scores on the Mini-Mental State Examination (MMSE), AVLT-A7 and the clock draw [95,96]. Furthermore, epidemiologic studies have shown that among AD patients eating foods containing high levels of saturated and trans fats, there was a significant relationship between increased dietary copper intake and an increased rate of cognitive decline [50].

Based on the premise that it might be copper in tap water that promoted accumulation of Aβ in the cholesterol-fed rabbit brain, we determined the effect of adding trace levels of copper to distilled water given to cholesterol-fed animals compared to cholesterol-fed animals given unaltered distilled water. As observed in cholesterol-fed animals, drinking tap water (containing copper) compared to animals administered distilled water, addition of 0.12 PPM copper ion to distilled water promoted the neuronal accumulation of Aβ [83]. Initial studies testing the ability of cholesterol-fed rabbits to acquire complex memory using the eye-blink behavioral paradigm yielded conflicting results. The cholesterol-fed rabbits, allowed local tap water (Morgantown WV), actually performed somewhat better than animals on normal chow (diet without cholesterol) [60]. These animals exhibited limited neuronal accumulation and no extracellular deposition of Aβ [60], similar to cholesterol-fed animals administered distilled drinking water. Analysis of Morgantown, WV tap water revealed the probable reason; there were negligible levels of copper compared to Sun City tap water (Table 6).

Subsequent studies indicated that introduction of copper into the drinking water of cholesterol-fed rabbits produced an 80% deficit in the ability of animals to acquire complex memory compared to animals fed the same diet and allowed unaltered distilled drinking water [83]. As with tap water, it was suggested that copper in the drinking water led to the inhibition of Aβ clearance from the brain, over-produced as a result of elevated cholesterol, thus leading to its accumulation and the subsequent memory deficits.

Further studies reveal that intensely immunoreactive neurons, extracellular deposits of Aβ and brain vessels in cholesterol-fed rabbits given copper supplemented water were stained by thioflavin-S [84]. Thioflavin-S reactive features were not observed in cholesterol-fed rabbits given unaltered distilled drinking water. The

Table 6
Trace metal levels in distilled water and tap water from Sun City, AZ and Morgantown, WV, USA. Values presented are in parts per million (PPM)

| Factor | Detection limits | Distilled water | Sun City tap water | Morgantown tap water |
|---|---|---|---|---|
| Calcium | 0.1 | 0.1 | 37.9 | 54.3 |
| Magnesium | 0.1 | ND | 19.4 | 10 |
| Sodium | 0.1 | 0.2 | 44.6 | 33.4 |
| Potassium | 0.1 | ND | 3.1 | 1.9 |
| Strontium | 0.05 | ND | 0.56 | 0.36 |
| Arsenic | 0.0002 | ND | 0.000592 | ND |
| Barium | 0.01 | ND | 0.04401 | 0.03501 |
| Iron | 0.05 | ND | ND | ND |
| Manganese | 0.02 | ND | ND | ND |
| Copper | 0.003 | ND | 0.21 | 0.006 |
| Zinc | 0.05 | ND | ND | 0.06 |
| Mercury | 0.0002 | ND | ND | ND |
| Chloride | 0.5 | ND | 90.2 | 17.1 |
| Nitrate | 0.5 | ND | 5.4 | 0.5 |
| Sulfate | 3 | ND | 31 | 153 |
| Bicarbonate | 0 | 0 | 107.8 | 43.5 |
| Fluoride | 0.1 | ND | 0.4 | 1.0 |
| Aluminum | 0.002 | ND | ND | ND |
| Silica | 0 | 0.02 | 23.8 | 4.46 |
| Organics | 0.1 | 0.16 | 0.20 | 1.57 |

data suggest that similar to findings in AD, there is an accumulation of fibrillar A$\beta$ induced in the brains of rabbits fed a cholesterol diet and administered trace levels of copper in their drinking water.

There is evidence that A$\beta$ binds ApoE and may be chaperoned in transport of cholesterol from the brain as a means of clearance of the toxin [46,107]. Deane and colleagues have shown that low-density lipoprotein receptor-related protein-1 (LRP) on brain capillaries is intimately involved in the clearance A$\beta$ from brain [10]. Recently, Deane and colleagues have shown that administering copper to mice in their drinking water causes increased copper and reduced LRP in the brains microvasculature while concurrently reducing A$\beta$ clearance and increasing central accumulation [11]. We proposed that A$\beta$ is cleared from the brain by tagging along with cholesterol/ApoE and LRP in traversing the BBB, with subsequent incorporation into HDL for delivery of the toxin to the liver [90]. We showed that in a setting of elevated cholesterol levels, overproduced A$\beta$ is cleared to the blood and can simultaneously be identified in the liver if copper is absent from the animal's drinking water, but if trace levels copper (0.12 PPM) are added to the drinking water, A$\beta$ accumulates in the brain, while the levels in the liver are greatly reduced. Consistent with these observations, previous studies have shown that A$\beta$ introduced into the blood or periphery is cleared predominantly by the liver [25, 33,39,101].

In order to assess if copper levels in the circulation of humans influenced clearance of A$\beta$, we investigated the relationship between circulating levels of copper and A$\beta$ and cognitive performance. Similar to Squitti's data, we have shown that there are significant increases in blood copper and ceruloplasmin (its chaperone) in AD compared to age-matched control [86]. The copper and ceruloplasmin increases were related to decreasing performance on the MMSE and AVLT-A7 in controls and the clock draw among individuals with mild cognitive impairment (MCI). Moreover, circulating copper/ceruloplasmin levels increased in controls with lower performance on the AVLT-A7 and MMSE and remained elevated in MCI and AD as cognitive ability progressively deteriorates. At the same time circulating A$\beta$ levels were increased in controls exhibiting reduced cognitive function [86]. This clinically perceptible difference in cognitive ability among reduced-function controls compared to high-function controls was associated with a two-fold increase in blood A$\beta$ levels [86]. A further five-fold increase in circulating A$\beta$ concentration was identified among individuals with a diagnosis of MCI/prodromal AD compared to reduced-function controls [86]. In profoundly demented AD patients, we identified a precipitous drop in circulating A$\beta$ levels compared to those individuals with MCI [86], and may indicate an accumulation of A$\beta$ in brain due to total or near-total collapse of the metabolic machinery necessary to rid the brain of the toxin.

Having found that there may be a relationship between circulating copper levels and the clearance of A$\beta$ from the brain, we turned our studies to an investigation of individuals annually evaluated clinically prior to demise [91]. We established cortical copper levels among these individuals and compared sub-populations grouped according to clinical performance and consensus clinical diagnosis of normal control ($N = 32$), MCI ($N = 7$), and AD ($N = 32$). The control population was further sub-grouped based on an individual's cognitive performance. We compared cortical copper levels and clinical performance for a group of 16 individuals considered high-function controls and a group of 16 individuals considered low-function controls. The interval between an individual's last clinical evaluation and demise was less than 12 months in all cases. Post-mortem measures of cortical copper levels were correlated to cognitive performance and diagnosis at an individual's most recent yearly clinical evaluation. We found that the mean copper levels in temporal cortex were significantly reduced in AD compared to cognitively normal controls, and levels in MCI were between

Table 7
Differences in mean copper levels (± SEM) in plasma (mg/l) and brain (mg/kg) produced by cholesterol diet (2%) and copper water (0.12 PPM) in rabbits

| Diet/water | N | Blood copper | Brain copper | Blood/brain ratio |
|---|---|---|---|---|
| Normal/distilled | 7 | 0.45 ± 0.03 | 2.27 ± 0.16 | 0.198 |
| Normal/copper | 5 | 0.50 ± 0.04 | 1.89 ± 0.07 | 0.264 |
| Cholesterol/distilled | 7 | 0.52 ± 0.05 | 2.41 ± 0.12 | 0.215 |
| Cholesterol/copper | 7 | 0.70 ± 0.06 | 2.22 ± 0.14 | 0.315 |

AD and controls [91]. Mean cortical copper levels were lower in low-function controls compared to high-function controls, but the difference did not achieve significance

Simultaneously we performed studies comparing copper levels Appendix between brain and blood among rabbits fed either a normal diet or a cholesterol diet and administered either unaltered distilled drinking water or distilled water supplemented with copper. We found that adding copper to the water of rabbits eating normal chow induced an increase in circulating copper levels and a reduction of copper levels in the brain (Table 7). A similar inverse relationship was identified in cholesterol-fed animals.

These results are clearly consistent with observations in humans where increasing levels of copper in the blood seemed to be associated with reduced levels in the brain. This supports the hypothesis suggesting that binding of copper by $A\beta$ is a scavenger mechanism by which the peptide aggregates with or flocculates copper, thus inactivating the metal ion and possibly $A\beta$ [59]. Increased production of $A\beta PP$ and $A\beta$ have been associated with reduce CNS copper levels, while increasing copper levels tend to reduce brain $A\beta$ levels [48]. In mice producing $A\beta$, there are reduced copper levels in brain and oral administration of copper returns brain copper levels to normal and reduces $A\beta$ production, and suggest that reduced levels of copper in AD may be overcome by supplementation of copper ion [40].

It is conceivable that initially clearance of $A\beta$ can keep up with production, while any level of accumulation occurring in the brain is below detection limits. Thereafter, with greater $A\beta$ exposure (aging) production may overwhelm clearance capacity and there is progressively more severe central accumulation. Recent studies in humans strongly indicate that a significant mechanism for accumulation of brain $A\beta$ may in fact be reduced clearance [4]. It is equally possible that prolonged sustained production of $A\beta$ directly produces metabolic "burn-out" at one or more points in the clearance pathway, or indirectly by inducing an alteration in the metabolic cascade or by deleteriously affecting an essential component required in the performance or control of $A\beta$ clearance.

Nevertheless, the continued ability of an individual to purge the brain of minimal $A\beta$ levels normally produced or pathologically elevated levels caused by a disease process (i.e., incipient AD) would reduce and/or delay the deleterious cognitive outcomes associated with accumulation and deposition of $A\beta$. The hypothesis that accumulation and subsequent disposition of $A\beta$ leading to progressively severe cognitive dysfunction is dependent on the capacity and continued ability to clear $A\beta$ from the brain, gains support from both human and animal model studies.

Based on the foregoing, it could easily be opined that if brain $A\beta$ clearance rate exceeds the rate of production, cognitive deterioration anticipated with central accumulation of the toxin may be delayed or even detained. The evidence clearly indicates that $A\beta$ overproduced in the brain can be transported to the blood and that reduced clearance may be pivotal in the onset and continued accumulation of $A\beta$ in the brain. It is equally clear that circulating copper may play a part in regulating $A\beta$ clearance from the brain.

These data indicate that in contrast to circulating blood levels, brain copper levels may gradually decrease with increasingly severe deterioration of cognitive performance, but bottoms out once an individual is clearly demented. It is of note that there is a copper binding site on $A\beta$ [9,105], and studies where $A\beta$ was co-injected (at physiologic concentrations found in plaques of AD) with iron, copper or zinc into rat cortex demonstrated that $A\beta$-Fe and $A\beta$-Zn were more toxic that $A\beta$ alone, while $A\beta$-copper was non-toxic [8]. Other studies support a relationship between increased copper and reduced $A\beta$ production in a transgenic mouse model of AD [5]. In the case of AD, it would seem possible that as brain copper levels wane, so does the capacity to inactivate $A\beta$s thus enhancing neurotoxicity. One might propose that increased central $A\beta$ accumulation predicates even more severe cognitive deterioration.

## EPILOGUE

The above does not constitute all that was uncovered while blazing the cholesterol trail. There are a few observations that will remain undisclosed until I am able to re-visit them. And to the others that have taken up the mantle or might, one must remember increased circulating cholesterol and water quality deleteriously affects most of the systemic organs also and no organ works in isolation.

## APPENDIX

### Sample preparation and Copper Determination

Triplicate wet samples of cortex (1 g) were excised from each temporal pole, freeze-dried and precisely weighed. To avoid any contamination, all glassware used for mineral analysis was first acid washed in 50% $HNO_3$ and thoroughly rinsed in distilled deionized water before use. Copper in serum was determined by atomic absorption spectrophotometric method as reported by Grosser [27]. A blank was run in parallel with each sample. The stock standard solutions were prepared daily from certified atomic standards (Fisher Scientific, NJ) and ranged from 0.0 $\mu$g/mL to 100 $\mu$g copper per 100 mL. Freeze-dried tissue was wet digested (1:10 W/V) overnight at 60°C in 5 ml nitric $HNO_3$ (AOAC, 1990 method 985-40; Official Methods of Analysis. $15^{th}$ ed. Association of Official Analytical Chemists. Arlington, VA.). Digested solution is measured directly or adjusted to suitable dilutions for copper analysis by atomic absorption spectrophotometer. Reagent blank using 70% $HNO_3$ was run in parallel to determine possible contamination. Copper is quantified by reference to standard curves made as indicated above using an AAnalyst 200 Pekin Elmer Atomic Absorption Spectrometer equipped with an air-acetylene flame. Values of copper are presented as $\mu$g/g wet tissue weight.

## ACKNOWLEDGMENTS

The studies were supported in part by the Arizona Biomedical Research Commission. Dr. Sparks receives consulting fees from Pfizer and Resverlogix Pharmaceuticals and owns stock in Sartoris, Inc.

## References

[1] J. Aarli, Role of cytokines in neurological disorders, *Curr Med Chem* **10** (2003), 1931–1937.
[2] N. Anitschkow, Uber die veranderungen der kaninchenaorta bie experimenteller cholesterinsteatose, *Beitr Pathol Anat* **56** (1913), 379–404.
[3] K.R. Bales, C. Fishman, C. DeLong, Y. Du, W. Jordan and S.M. Paul, Diet-induced hyperlipidemia accelerates amyloid deposition in the APPv717f transgenic mouse model of Alzheimer's disease, *Neurobiol Aging* **21** (2000), S139.
[4] R.J. Bateman, L.Y. Munsell, J.C. Morris, R. Swarm, K.E. Yarasheski and D.M. Holtzman, Human amyloid-beta synthesis and clearance rates as measured in cerebrospinal fluid in vivo, *Nat Med* **12** (2006), 856–861.
[5] T.A. Bayer, S. Schafer, A. Simons, A. Kemmling, T. Kamer, R. Tepest, A. Eckert, K. Schussel, O. Eikenberg, C. Sturchler-Pierrat, D. Abramowki, M. Staufenbiel and G. Multhaup, Dietary Cu stabilizes brain superoxide dismutase 1 activity and reduces amyloid AB production in APP23 transgenic mice, *Proc Natl Acad Sci USA* **100** (2003), 14187–14192.
[6] T. Beach, D.G. Walker, P. Potter, L. Sue and A. Fisher, Reduction of ceerbrospinal fluid amyloid beta after systemic administration of M1 muscarinic agonists, *Brain Res* **905** (2001), 220–223.
[7] C. Behl and B. Moosmann, Oxidative nerve cell death in Alzheimer's disease and stroke: antioxidants as neuroprotective compounds, *Biol Chem* **383** (2002), 521–536.
[8] G.M. Bishop and S.R. Robinson, The amyloid paradox: amyloid-beta-metal complexes can be neurotoxic and neuroprotective, *Brain Pathol* **14** (2004), 448–452.
[9] M. Chen, J. Durr and H.L. Fernandez, Possible role of calpain in normal processing of beta-amyloid precursor protein in human platelets, *Biochem Biophys Res Common* **273** (2000), 170–175.
[10] R. Deane, Z. Wu, A. Sagare, J. Davis, S. DuYan, K. Hamm, F. Xu, M. Parisi, B. LaRue, H.W. Hy, P. Spijkers, H. Guo, X. Song, P.J. Lenting, W.E. Van Nostrand and B.V. Zlokovic, LRP/amyloid beta-peptide interaction mediates differential brain efflux of Abeta isoforms, *Neuron* **43** (2004), 333–344.
[11] R. Deane, A. Sagare, M. Coma, M. Paris, R. Gelein, I. Singh and B. Zlokovic, A novel role for copper: Disruption of LRP-dependent brain A$\beta$ clearance, *Soc Neurosci* **37** (2007), 857.
[12] R.B. DeMattos, K.R. Bales, D.J. Cummins, S.M. Paul and D.M. Holtzman, Brain to plasma amyloi-beta efflux" A measure of brain amyloid burden in a mouse model of zlheimer's disease, *Science* **295** (2002), 2264–2267.
[13] R.B. DeMattos, K.R. Bales, M. Parsadanian, M.A. O'Dell, E.M. Foss, S.M. Paul and D.M. Holtzman, Plaque-associated disruption of CSF and plasma amyloid-beta (Abeta) equilibrium in a mouse model of Alzheimer's disease, *J Neurochem* **81** (2002), 229–236.
[14] D.W. Dickson, S.C. Lee, L.A. Mattiace, S.C. Yen and C. Brosnan, Microglia and cytokines in neurological disease, with special reference to AIDS and Alzheimer's disease, *Glia* **7** (1993), 75–83.
[15] J.M. Dopp, S.M. Breneman and J.A. Olschowka, Expression of ICAM-1, VCAM-1, L-selectin, and leukosialin in the mouse central nervous system during the induction and remission stages of experimental allergic encephalomyelitis, *J Neuroimmunol* **54** (1994), 129–144.
[16] R.A. Durham, C.A. Parker, M.R. Emmerling, C.L. Bisgaier and L.C. Walker, Effect of age and diet on the expression of

[16] beta-amyloid 1–40 and 1–42 in the brains of apolipoprotein-E-deficient mice, *Neurobiol Aging* **19** (1998), S281.
[17] S. Efthimiopoulos, D. Vassilacopoulou, J.A. Ripellino, N. Tezapsidis and N.K. Robakis, Cholinergic agonists stimulate secretion of soluble full-length amyloid precursor protein in neuroendocrine cells, *Proc Natl Acad Sci USA* **93** (1996), 8046–8050.
[18] B. Engelhardt, F.K. Conley and E.C. Butcher, Cell adhesion molecules on vessels during inflammation in the mouse central nervous system, *J Neuroimmunol* **51** (1994), 199–208.
[19] G. Engstrom, P. Lind, B. Hedblad, L. Stavenow, L. Janzon and F. Lindgarde, Effects of cholesterol and inflammation-sensitive plasma protein on incidence of myocardial infarction and stroke in men, *Circulation* **105** (2002), 2632–2637.
[20] C. Finch and D. Cohen, Aging, metabolism, and Alzheimer disease: review and hypothesis, *Exp Neurol* **143** (1997), 82–102.
[21] A. Fisher, R. Brandeis, I. Karton, Z. Pittel, D. Gurwitz, R. Haring, M. Sapir, A. Levy and E. Heldman, (Cis)-2-methyl-spiro(1,3-oxathiolane-5,3')quinuclidine, an M1 selective cholinergic agonist, attenuates cognitive dysfunctions in an animal model of Alzheimer's disease, *J Pharmacol Exptl Therap* **257** (1991), 392–403.
[22] C.E. Fishman, S.L. White, C.A. DeLong, D.J. Cummins, W.H. Jordan, K.R. Bales and S.M. Paul, High fat diet potentiates $\beta$-amyloid deposition in the APP V717F transgenic mouse model of Alzheimer's disease, *Soc Neurosci* **25** (1999), 1859.
[23] F. Fonnum, Radiochemical microassay for the determination of choline acetyltransferase and acetylcholinesterase activities, *Biochem J* **115** (1969), 465–472.
[24] M. Frampton, R. Harvey and V. Kirchner, Propentofylline for dementia, *Cochrane Database Syst Rev* **2** (2003), CD002853.
[25] J. Ghiso, m. Shayo, M. Calero, D. Ng, Y. Tomidokoro, S. Gandy, A. Rostagno and B. Franqione, Systemic catabolism of Alzheimer's Abeta40 and Abeta42, *J Biol Chem* **279** (2004), 45897–45908.
[26] M. Gotz, G. Kunig, P. Riederer and M. Youdim, Oxidative stress: free radical production in neural degeneration, *Pharmacol Ther* **63** (1994), 37–122.
[27] Z.A. Grosser, The determination of major and minor elements in serum by flame atomic absorption, *Atomic Spectroscopy* **17** (1996), 229–231.
[28] Z.-Z. Guan, W.-F. Uy and A. Nordberg, Dual effects of nicotine on oxidative stress & neuroprotection in PC12 cells, *Neurochem Int* **43** (2003), 243–249.
[29] U. Haas and D.L. Sparks, Cathepsin D: Activity and immunocytochemical localization in Alzheimer's disease and aging, *Mol Chem Neuropathol* **29** (1996), 1–14.
[30] E. Hellstrom-Lindahl, J. Court, J. Keverne, M. Svedberg, M. Lee, A. Marutle, A. Thomas, E. Perry, I. Bednar and A. Nordberg, Nicotine reduces A$\beta$ in the brain and cerebral vessels of APPsw mice, *Eur J Neurosci* **19** (2004), 2703–2710.
[31] E. Hirsch, T. Breidert, E. Rousselet, S. Hunot, A. Hartmann and P. Michel, The role of glial reaction and inflammation in Parkinson's disease, *Ann NY Acad Sci* **991** (2003), 214–228.
[32] P.R. Hof, C. Bouras, N. Mehal, D.L. Sparks, D.P. Perl and J.H. Morrison, Age-related distribution of neuropathologic changes in the cerebral cortex of Down Syndrome cases: quantitative regional analysis and comparison with Alzheimer's disease, *Arch Neurol* **52** (1995), 379–391.
[33] E. Hone, I.J. Martins, J. Fonte and R.N. Martins, Apolipoprotein E influences amyloid-beta clearance from the murine periphery, *J Alzheimers Dis* **5** (2003), 1–8.
[34] B.T. Hyman, K. Marzloff and P.V. Arriagada, The lack of accumulation of senile plaques or amyloid burden in Alzheimer's disease suggests a dynamic balance between amyloid deposition and resolution, *J Neuropath Exp Neurol* **52** (1993), 594–600.
[35] M. Iskra and W. Majewski, Copper and zinc concentration and the activities of ceruloplasmin and superoxide dismutatse in atherosclerosis obliterans, *Bil Trace Elem Res* **73** (2000), 55–65.
[36] R.N. Kalaria and S.N. Kroon, Expression of leukocyte antigen CD34 by brain capillaries in Alzheimer's disease and neurologically normal subjects, *Acta Neuropathol (Berl)* **82** (1992), 606–612.
[37] R.N. Kalaria and P. Hedera, Differential degeneration of the cerebral microvasculature in Alzheimer's disease, *Neuroreport* **6** (1995), 477–480.
[38] R.N. Kalaria, Cerebral vessels in ageing and Alzheimer's disease, *Pharmacol Ther* **72** (1996), 193–214.
[39] K.K. Kandimalla, G.L. Curran, S.S. Holasek, E.J. Gilles, T.M. Wengenack and J.F. Poduslo, Pharmacokinetic analysis of the blood-brain barrier transport of 1251-amyloid beta protein 40 in wild-type and Alzheimer's disease transgenic mice (APP, PS1) and its implications for amyloid plaque formation, *J Pharmacol Exp Ther* **313** (2005), 1370–1378.
[40] H. Kessler, F.G. Pajonk, T. supprian, P. Falkai, G. Multhaup and T.A. Bayer, Cerebrospinal fluid diagnositic markers correlate with lower plasma copper and ceruloplasmin in patients with AD, *J Neural Trans* **113** (2006), 581–585.
[41] T.N. Khavkin, Nikolai Nikolaevich Anitschkow. (In commemoration of the 90th anniversary of his birthday), *Beitr Pathol* **156** (1975), 301–312.
[42] S. Kim, Y. Kim, S. Jeong, C. Haass, Y. Kim and Y. Suh, Enhanced release of secreted form of Alzmeimer's amyloid precursor protein from PCI2 calls by nicotine, *Mol Pharmacol* **52** (1997), 430–436.
[43] D. Lahiri, D. Utsuki, D. Chen, M. Farlow, M. Shoaib, D. Ingram and N. Greig, Nicotine reduces the secretion of Alzheimer's beta-amyloid precursor protein containing beta-amyloid peptide in the rat without altering synaptic proteins, *Ann NY Acad Sci* **965** (2002), 365–372.
[44] L. Li, S. Zeigler, R.J. Lindsey and K. Fukuchi, Effects of an atherogenic diet on amyloidosis in transgenic mice overexpressing the C-terminal portion of $\beta$-amyloid precursor protein, *Soc Neurosci* **25** (1999), 1859.
[45] B. Liu and J. Hong, Role of microglia in inflammation-mediated neurodegenerative diseases: mechanisms and strategies for therapeutic intervention, *J Pharmacol Exp Ther* **304** (2003), 1–7.
[46] R.W. Mahley, Apolipoprotein E: Cholesterol transport protein with expanded role in cell biology, *Science* **240** (1988), 622–630.
[47] L. Martin, Neuronal cell death in nervous system development, disease, and injury, *Int J Mol Med* **7** (2001), 455–478.
[48] C.J. Maynard, A.I. Bush, C.L. Masters, R. Cappai and Q.X. Li, Metals and amyloid-beta in Alzheimer's disease, *Int J Exp Pathol* **86** (2005), 147–159.
[49] M.A. Mendez, M. Araya, M. Olivares, F. Pizarro and M. Gonzalez, Sex and ceruloplasmin modulate the response to copper exposure in healthy individuals, *Environ Health Perspect* **112** (2004), 1654–1657.

[50] M.C. Morris, D.A. Evans, C.C. Tangney, J.L. Bienias, J.A. Schneider, R.S. Wilson and P.A. Scherr, Dietary copper and high saturated and trans fat intakes associated with cognitive decline, *Arch Neurol* **63** (2006), 1085–1088.

[51] H. Nakanishi, Neuronal and microglial cathepsins in aging and age-related diseases, *Ageing Res Rev* **2** (2003), 367–381.

[52] M.B. Newman, G.W. Arendash, R.D. Shytle, P.C. Bickford, T. Tighe and P.R. Sanberg, Nicotine's oxidative and antioxidant properties in CNS, *Life Sciences* **71** (2002), 2807–2820.

[53] R.A. Nixon, A.M. Cataldo, P.A. Paskevich, D.A. Hamilton, T.R. Wheelock and L. Kanaley-Andrews, The lysosomal system in Neurons: Involvement at multiple stages of Alzheimer's disease pathogenesis, *Ann NY Acad Sci* **674** (1992), 65–88.

[54] A. Nordberg, E. Hellstrom-Lindahl, M. Lee, M. Johnson, M. Mousavi, R. Hall, E. Perry, I. Bednar and J. Court, Chronic nicotine treatment reduces beta-amyloidosis in the brain of a mouse model of Alzheimer's disease (APPsw), *J Neurochem* **81** (2002), 655–658.

[55] R. Ozcankaya and N. Delibas, Malondialdehyde, superoxide dismutase, melatonin, iron, copper and zinc blood concentrations in patients with Alzheimer's disease: Cross-sectional study, *Croatian Med J* **43** (2002), 28–32.

[56] G. Perry, M. Taddeo, A. Nunomura, X. Zhu, T. Zenteno-Savin, K. Drew, S. Shimohama, J. Avila, R. Castellani and M. Smith, Comparative biology and pathology of oxidative stress in Alzheimer and other neurodegenerative diseases: beyond damage and response, *Comp Biochem Physiol C Toxicol Pharmacol* **133** (2002), 507–513.

[57] M. Piorunska-Stolzman, M. Iskra and W. Majewski, The activity of cholesterol esterase and ceruloplasmin are inversely related in the serum of men with atherosclerosis obliterans, *Med Sci Monit* **7** (2001), 940–945.

[58] L.M. Refolo, M.A. Pappolla, B. Malester, J. LaFrancois, R. Bryant-Thomas Wang, K. Sambamurti and K. Duff, Hypercholesterolemia accelerates Alzheimer's amyloid pathology in a transgenic mouse model, *Neurobiol Dis* **7** (2000), 321–331.

[59] S.R. Robinson and G.M. Bishop, A$\beta$ as a bioflocculant: Implications for the amyloid hypothesis of Alzheimer's disease, *Neurobiol Aging* **23** (2002), 1051–1072.

[60] B.G. Schreurs, C.A. Smith-Bell, J. Lochhead and D.L. Sparks, Cholesterol modifies classical conditioning of the rabbit nictitating membrane response, *Behav Neurosci* **117** (2003), 1220–1232.

[61] S.A. Scott, S.A. Johnson, C. Zarow and L.S. Perlmutter, Inability to detect beta-amyloid protein precursor mRNA in Alzheimer plaque-associated microglia, *Exp Neurol* **121** (1993), 113–118.

[62] F.S. Shie, R.C. LeBoeuf, J.B. Leverenz and L.W. Jin, Effects of cholesterol feeding on histopathologic hallmarks of Alzheimer's disease in $\beta$-amyloid precursor protein (APP) transgenic mice, *Soc Neurosci* **25** (1999), 1859.

[63] J. Snaedal, J. Kristinsson, S. Gunnarsdottir, A. Olafsdottir, M. Baldvinsson and T. Johannesson, Copper, ceruloplasmin and superoxide dismutase in patients with Alzheimer's disease, *Dementia Geriatr Cog Dis* **9** (1998), 239–242.

[64] J. Snaedal, J. Kristinsson, S. Gunnarsdottir, R. Olafsdottir, M. Baldvinsson and T. Johannesson, Copper, ceruloplasmin and superoxide dismutase in patients with Alzheimer's disease. a case-control study, *Dement Geriatr Cogn Disord* **5** (1998), 239–242.

[65] D. Sparks, J. Hunsaker, 3rd, S. Scheff, R. Kryscio, J. Henson and W. Markesbery, Cortical senile plaques in coronary artery disease, aging and Alzheimer's disease, *Neurobiol Aging* **11** (1990), 601–607.

[66] D.L. Sparks, J.C. Hunsaker III and J.T. Slevin, Postmortem accumulation of 3-Methoxytyramine in the brain, *New Engl J Med* **311** (1984), 540.

[67] D.L. Sparks, J.T. Slevin and J.C. Hunsaker III, Putamenal 3-Methoxy-tyramine as a gauge of the postmortem interval, *J Forensic Sci* **31** (1986), 962–971.

[68] D.L. Sparks, P.R. Oeltgen, R.K. Kryscio and J.C. Hunsaker III, Comparison of chemical methods for determining postmortem interval, *J Forensic Sci* **34** (1989), 197–206.

[69] D.L. Sparks and J.C. Hunsaker, Down's Syndrome: Occurrence of ALZ-50 reactive neurons and the formation of senile plaques, *J Neurol Sci* **109** (1992), 77–82.

[70] D.L. Sparks, H. Liu, S.W. Scheff, C.M. Coyne and J.C. Hunsaker III, Temporal sequence of plaque formation in the cerebral cortex of non-demented individuals, *J Neuropath Exp Neurol* **52** (1993), 135–142.

[71] D.L. Sparks, S.W. Scheff, J.C. Hunsaker III, H. Liu, T. Landers and D.R. Gross, Induction of Alzheimer-like $\beta$-amyloid immunoreactivity in the brains of rabbits with dietary cholesterol, *Exp Neurol* **126** (1994), 88–94.

[72] D.L. Sparks, H. Liu, D.R. Gross and S.W. Scheff, Increased density of cortical Apolipoprotein E. Immunoreactive neurons in rabbit brain after dietary administration of cholesterol, *Neurosci Lett* **187** (1995), 142–144.

[73] D.L. Sparks, S.W. Scheff, H. Liu, T. Landers, C.M. Coyne and J.C. Hunsaker, Increased density of neurofibrillary tangles (NFT) in non-demented individuals with hypertension, *J Neurol Sci* **131** (1995), 162–169.

[74] D.L. Sparks, Intraneuronal $\beta$-amyloid immunoreactivity in the CNS, *Neurobiol Aging* **17** (1996), 291–299.

[75] D.L. Sparks, Dietary cholesterol induces Alzheimer-like - amyloid immunoreactivity in rabbit brain, *Nutr Metab Cardiovasc Dis* **7** (1997), 255–266.

[76] D.L. Sparks, Rabbit brain cholesterol content and free radical enzyme activity in brain and blood after dietary administration of cholesterol, *Nutr Metab Cardiovasc Dis* **7** (1997), 255–266.

[77] D.L. Sparks, Neuropathologic links between Alzheimer's disease and vascular disease, in: *Alzheimer's Disease and Related Disorders*, K. Iqbal, D.F. Swaab, B. Winblad and H.M. Wisniewski, eds, John Wiley & Sons Ltd, 1999, pp. 153–163.

[78] D.L. Sparks, Vascular related and mediated alterations in Alzheimer's disease, in: *Cerebral Cortex*, J. Morrison and A. Peters, eds, Raven Press, New York, NY, 1999, pp. 733–772.

[79] D.L. Sparks, D.J. Connor, D.R. Wasser, J.E. Lopez and M.N. Sabbagh, The Alzheimer's Disease Atorvastatin Treatment Trial: Scientific basis and position on the use of HMG-CoA reductase inhibitors (statins) that do or do not cross the blood-brain barrier, in: *Drug Discovery and Development for Alzheimer's Disease*, H. Fillit and A. O'Connell, eds, Springer Publishing Co., 2000, pp. 244–252.

[80] D.L. Sparks, Y.-M. Kou, A. Roher, T.A. Martin and R.J. Lukas, Alterations of Alzheimer's disease in the cholesterol-fed rabbit, including vascular inflammation. Preliminary observations, *Ann NY Acad Sci* **903** (2000), 335–344.

[81] D.L. Sparks, J. Lochhead, D. Horstman, T. Wagoner and T. Martin, Water quality has a pronounced effect on cholesterol-induced accumulation of Alzheimer amyloid $\beta$ (A$\beta$) in rabbit brain, *J Alzheimers Dis* **4** (2002), 523–529.

[82] D.L. Sparks, R. Martins and T. Martin, Cholesterol and cognition: Rationale for the AD Cholesterol-lowering Treatment Trial and sex-related differences in $\beta$-amyloid accumulation

in the brains of spontaneously hypercholesterolemic Watanabe rabbits, *Ann N Y Acad Sci* **977** (2002), 356–366.

[83] D.L. Sparks and B.G. Schreurs, Trace amounts of copper in water induce $\beta$-amyloid plaques and learning deficits in a rabbit model of alzheimer's disease, *Proc Natl Acad Sci USA* **100** (2003), 1065–1069.

[84] D.L. Sparks, Cholesterol, copper and accumulation of thioflavine S-reactive Alzheimer-like amyloid $\beta$ in rabbit brain, *J Mol Neurosci* **24** (2004), 97–104.

[85] D.L. Sparks, J. Lochhead, A. Fisher and T. Martin, Water quality & cholesterol-induced pathology: differential effects of the M-1 muscarinic receptor agonist AF267b on accumulation of Alzheimer-like Amyloid B in rabbit brain, in: *Recent Progress in Alzheimer's and Parkinson's Diseases*, I. Hanin, R. Cacabelos and A. Fisher, eds, Taylor & Francis, Boca Raton, FL, 2005, pp. 197–203.

[86] D.L. Sparks, S. Petanceska, M. Sabbagh, D. Connor, Soares H, C. Adler, J. Lopez, C. Ziolowski, J. Lochhead and P. Browne, Cholesterol, copper and A$\beta$ in controls, MCI, AD and the AD Cholesterol-Lowering Treatment trial (ADCLT), *Curr Alzheimer Res* **2** (2005), 527–539.

[87] D.L. Sparks, M.N. Sabbagh, D.J. Connor, J. Lopez, L.J. Launer, P. Browne, D. Wasser, S. Johnson-Traver, J. Lochhead and C. Ziolwolski, Atorvastatin for the treatment of mild-to-moderate Alzheimer's disease: preliminary results, *Arch Neurol* **62** (2005), 753–757.

[88] D.L. Sparks, D. Connor, M.N. Sabbagh, R.P. Petersen, J. Lopez and P. Browne, Circulating cholesterol levels, apolipoprotein E genotype and dementia severity influence the benefit of atorvastatin treatment in AD: Results of the Alzheimer's Disease Cholesterol-Lowering Treatment (AD-CLT) trial, *Acta Neuol Scand* **114** (2006), 3–7.

[89] D.L. Sparks, R. Friedland, S. Petanceska, B.G. Schreurs, J. Shi, G. Perry, M.A. Smith, A. Sharma, S. Derosa, C. Ziolkowski and G. Stankovic, Trace copper levels in drinking water, but not zinc or aluminum influence CNS Alzheimer's like pathology, *J Nutr Health Aging* **10** (2006), 247–254.

[90] D.L. Sparks, Cholesterol metabolism and brain amyloidosis: Evidence for a role of copper in the clearance of A$\beta$ through the liver, *Curr Alzheimer Res* **4** (2007), 165–169.

[91] D.L. Sparks, Copper and cognition in AD and PD, *INSERN* (2008), in press.

[92] A. Spector, Oxidative stress and disease, *J Ocul Pharmacol Ther* **16** (2000), 193–201.

[93] R. Squitti, D. Lupoi, P. Pasqualetti, G. Dal Forno, F. Vernieri, P. Chiovenda, L. Rossi, M. Cortesi, E. Cassetta and P.M. Rossini, Elevation of serum copper levels in Alzheimer's disease, *Neurology* **59** (2002), 1153–1161.

[94] R. Squitti, P. Pasqualetti, G. Dal Forno, S. Cesaretti, F. Pedance, A. Finazzi, A. and P.M. Rossini, Elevated serum copper levels discriminates Alzheimer's disease from vascular dementia, *Neurology* **60** (2003), 2013–2014.

[95] R. Squitti, P. Pasqualetti, G. Dal Forno, F. Moffa, E. Cassetta, D. Lupoi, F. Vernieri, L. Rossi, M. Baldassini and P.M. Rossini, Excess of serum copper not related to ceruloplasmin in Alzheimer disease, *Neurology* **64** (2005), 1040–1046.

[96] R. Squitti, G. Barbati, L. Rossi, M. Ventriglia, G. Dal Forno, S. Cesaretti, F. Moffa, I. Caridi, E. Cassetta, P. Pasqualetti, L. Calabrese, D. Lupoi and P.M. Rossini, Excess of nonceruloplasmin serum copper in AD correlates with MMSE, CSF B-amyloid, and h-tau, *AAN Enterprises, Inc.* (2006), 76–82.

[97] D. Strausak, J.F.B. Mercer, H.H. Dieter, W. Stremmel and G. Multhaup, Copper in disorders with neurological symptoms: Alzheimer's, Menkes, and Wilson diseases, *Brain Res Bull* **55** (2003), 175–185.

[98] W.J. Streit and D.L. Sparks, Appearance of microglia nodules coincides with presence of senile plaques in Down's syndrome, *J. Neuropath Exp Neurol* **52** (1993), 280.

[99] W.J. Streit and D.L. Sparks, Activation of microglia in the brains of humans with heart disease and hypercholesterolemic rabbits, *J Mol Med* **75** (1997), 130–138.

[100] H. Tago, H. Kimura and T. Maeda, Visualization of detailed acetylcholinesterase fiber and neuronal staining in rat brain by a sensitive histochemical procedure, *J Histochem Cytochem* **34** (1986), 1431–1438.

[101] C. Tamaki, S. Ohtsuki, T. Iwatsubo, T. Hashimoto, K. Yamada, C. Yabuki and T. Terasaki, Major involvement of low-density lipoprotein receptor-related protein 1 in the clearance of plasma free amyloid beta-peptide by the liver, *Pharm Res* **23** (2006), 1407–1416.

[102] T. Utsuki, M. Shoaib, H. Holloway, D. Ingram, W. Wallace, V. Haroutunian, K. Sambamurti, D. Lahiri and N. Greig, Nicotine lowers the secretion of the Alzheimer's amyloid beta-protein precursor that contains amyloid beta-peptide in rat, *J Alzheimers Dis* **4** (2002), 405–415.

[103] D.J. Waggoner, T.B. Bartnikas and J.D. Gitlin, The role of copper in neurodegenerative disease, *Neurobiol Dis* **6** (1999), 221–230.

[104] H. Wang, M. Yu, M. Ochani, C.A. Amella, M. Tanovic, S. Susarla, J.H. Li, H. Wang. H, Yang, L. Ulloa, Y. Al-Abed, C.J. Czura and K.J. Tracey, Nicotinic acetylcholine receptor alpha7 subunit is an essential regulator of inflammation, *Nature* **421** (2003), 384–388.

[105] A.R. White, R. Reyes, J.F. Mercer, J. Camakaris, H. Zheng, A.I. Bush, G. Multhaup, K. Beyreuther, C.L. Maters and R. Cappai, Copper levels are increased in the cerebral cortex and liver of APP and APLP2 knockout mice, *Brain Res* **842** (1999), 439–444.

[106] Q.-S. Xue, D.L. Sparks and W.J. Streit, Microglial activation in the hippocampus of hypercholesterolemic rabbits occurs independent of increased amyloid production, *J Neuroinflammation* **4** (2007), 20–25.

[107] C.V. Zerbinatti and B. G., LRP and Alzheimer's disease, *Rev Neurosci* **16** (2005), 123–135.

# APPENDIX: New Pilot Data

In our previous studies we administered 2% cholesterol and varying qualities of drinking water – ranging from local tap water to distilled water – and we were able to manipulate Alzheimer-like amyloid beta (Aβ) accumulation and memory deficits in less than 3 months but were unable to identify any changes in tau as the major component of the other characteristic lesion in AD β the neurofibrillary tangle (NFT). A recent publication indicated that administering lower levels of dietary cholesterol (1%) over longer periods of time (7 months) could produce NFT-like changes in the brain along with Aβ. deposition. We found that there was AD-like tau neuropathology produced by 5 months of 1% cholesterol diet in female rabbits administered distilled water, but not in animals administered normal chow and tap water to drink.

**Cholesterol influence on Tau**. Tau is the precursor for the other important neuropathologic feature of AD, the neurofibrillary tangle (NFT). Specifically phosphorylated tau is a main component of the NFT (1). Single strands of phosphorylated tau become paired into filaments called paired helical filaments (PHF), which then assemble into fibrillary complexes within a neuron eventually becoming a NFT. Antibodies have been developed to detect different stages of the NFT formation, including PHF-1 antibody identifying paired helical filaments and AT8 antibody highlighting phosphorylated tau (1). A recent study reported long duration low level cholesterol administration in female rabbits induced the presence of Aβ and NFT AD-like neuropathology. Seven months of 1% cholesterol diet caused increased PHF-1 immunoreactivity in hippocampus of female rabbits (2) as well plaque-like Aβ deposition in the hippocampus and cortical brain regions (3). The authors suggested that shorter periods of administration of higher levels of cholesterol were not long enough to produce both pathologic lesions of AD. These authors did not note the quality of water provided the animal in the manuscripts (2) (3). They eventually divulged that they administered the animals reverse osmosis (RO) purified water.

We have run a pilot study of 5 female rabbits - three fed 1% cholesterol and drinking distilled water for 5 months compared to two animals administered normal chow and tap water. In this very small study we focus on changes in tau by measuring total tau levels by ELISA and evaluating the brain for the presence of phosphorylated tau by performing AT8 immuno histochemistry. We found that circulating levels of tau were increased 2.4 fold in the animals on cholesterol compared to animals on normal chow. Similarly we found regional increases in the brain between the groups; there were 2.2, 1.8 and 1.4 fold increases in the frontal cortex, hippocampus and superior temporal cortex, respectively. We investigated the hippocampus for AT8 immunoreactivity; we found essentially no AT8 staining in the hippocampus of either of the rabbits fed normal chow and tap water and that 2 of the 3animals on 1% cho lesterol and distilled water showed AT8 immunoreactive neurons. It was of note that the animal with the highest levels of total tau in the hippocampus showed the greatest number of more intensely AT8 stained neurons and the animal with the lowest increase in total tau showed only the faintest of AT8 staining β with the animal having mid-level increase in total tau also having the mid-level of number and intensity of AT8 immunoreactive neurons in hippocampus. Furthermore the animal with the highest levels of tau in the hippocampus also showed AT8 immunoreative neurons in the superior temporal gyrus.

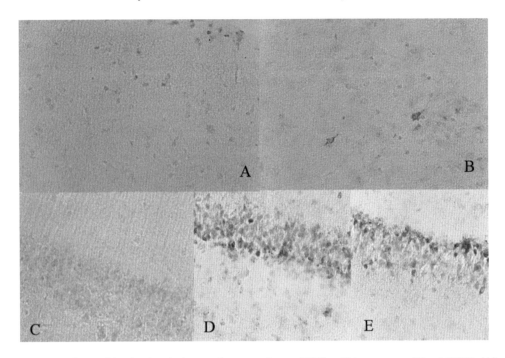

Figure 1. AT8 immunohistochemistry in the superior temporal gyrus (STG) and hippocampus of female NZW rabbits The STG in an animal fed normal chow and administered tap water to drink (A) compared to an animal fed 1% cholesterol diet drinking distilled water (B) for 5 months. The hippocampus in an animal fed normal chow and administered tap water (C) compared to the animals fed 1% cholesterol and drinking distilled water having a 2.5-fold increase in tau in the hippocampus (D) and a 4-fold increase in hippocampal tau levels (E). AT8 immunoreative neurons are seen in the cholesterol fed rabbit STG (B) and not in the animal fed normal chow. No darkly stained neurons are seen in the fascia dentate of the animal fed normal chow (C) and increasing number of stained neurons are seen in animals with increasing tau levels (D and E, respectively).

## References

[1] Mercken M., Vandermeeren M., Lubke U., et al., Monoclonal antibodies with selective specificity for Alzheimer Tau are directed against phosphatase-sensitive epitopes. *Acta Neuropathol* 1992; **84**:265-272.

[2] Ghribi O., Larsen B., Schrag M.,Herman M.M., High cholesterol content in neurons increases BACE, B-amyloid, and phosphory- lated tau levels in rabbit hippocampus. *Experimental Neurology* 2006; 200:460-467.

[3] Ghribi O., Golvovko M.Y., Larsen B., Schrag M.,Murphy E.J., Deposition of iron and B-amyloid plaques is associated with cortical cellular damage in rabbits fed with long-term cholesterol-enriched diets *Journal of Neurochemistry* 2006; 1-12.

# A Rabbit Model of Alzheimer's Disease: Valid at Neuropathological, Cognitive, and Therapeutic Levels

Diana S. Woodruff-Pak[a,*], Alexis Agelan[b], and Luis Del Valle[c]
[a]Department of Psychology, Temple University and Departments Neurology, & Radiology,
[b]University Laboratory Animal Resources, and
[c]Department of Neuroscience & Center for Neurovirology, Temple University School of Medicine

**Abstract.** Supplementing a rabbit's diet with 2% cholesterol alone or with a trace amount of copper created neuropathological changes that resembled those seen in Alzheimer's disease (AD). AD model rabbits were impaired in eyeblink classical conditioning; a form of learning severely impaired in AD. Our aim was to replicate AD rabbit model neuropathology, test eyeblink conditioning in this model, and determine if galantamine (Razadyne[TM]) would ameliorate impaired conditioning. In Experiment 1 rabbit chow with 2% cholesterol and drinking water with 0.12 mg/liter copper sulfate were administered for 10 weeks. Control rabbits received normal food and water. Rabbit brains were probed for neuropathology. AD model rabbits had significant neuronal loss in frontal cortex, hippocampus and cerebellum. Changes in neurons in the hippocampus were consistent with neurofibrillary degeneration and cytoplasmic immunoreactivity for β-amyloid and tau. In Experiment 2 AD model rabbits were injected daily with vehicle or 3.0 mg/kg galantamine and tested on 750 ms trace and delay eyeblink conditioning. Galantamine improved eyeblink conditioning significantly over vehicle. The AD rabbit model has validity from neuropathological to cognitive levels and offers a promising addition to the available animal models of AD. Galantamine ameliorated impaired eyeblink conditioning, extending the validity of the AD rabbit model to treatment modalities.

Keywords: Eyeblink classical conditioning, trace paradigm, delay paradigm, β-amyloid, tau, neuronal loss, hippocampus, cerebral cortex, cerebellum.

## INTRODUCTION

A limiting factor for rapid progress in research on Alzheimer's disease (AD) has been the small number of animal models of the disease. To bridge the wide gap between the molecular biology of AD and clinical therapeutics, it is essential to have valid non-human animal models to investigate disease mechanisms, test treatments, evaluate preventative strategies and cures. A promising and understudied animal model of AD is the cholesterol-fed rabbit. Cholesterol-fed rabbits develop a number of pathological indices of AD [38] that are accelerated when a trace amount of copper is added to the drinking water [36]. Deposition of tau as well as β-amyloid was seen in the brains of rabbits fed 1% cholesterol over seven months [12]. The major aim of this study was to replicate and examine further the utility of the AD model rabbit model and to determine whether galantamine (Razadyne[TM]), a drug approved by the FDA to treat mild to moderate AD, would ameliorate impaired learning in this animal model of AD.

*The Alzheimer's disease model rabbit*

The amino acid sequence of β-amyloid that forms amyloid plaques is identical in rabbits, humans, and some other mammals, but not rodents [19]. The cho-

---

[*]Corresponding author: Diana S. Woodruff-Pak, Ph.D., Department of Psychology, 1701 North 13th Street, Temple University, Philadelphia, PA 19122, USA. Tel.: 215 204 1258; Fax: 215 204 5539; Email: pak@temple.edu

lesterol-fed rabbit has been demonstrated to have validity on a number of levels. At a molecular level, the brain of this animal model has at least a dozen features similar to the pathology observed in the AD brain including elevated β-amyloid concentration, neuronal accumulation of β-amyloid immunoreactivity, extracellular Aβ plaques, elevated brain cholesterol, apolipoprotein E immunoreactivity, breaches in the blood brain barrier, microgliosis, and neuronal loss [35, 36, 37, 38]. Feeding rabbits over a long time period with cholesterol causes an increase in the cholesterol content in neurons, and this high cholesterol content in neurons is accompanied by an increase in BACE1 levels [12]. The accumulation of β-amyloid in the brain can be associated with BACE1, the enzyme that initially cleaves amyloid precursor protein (APP) to generate β-amyloid and cause the accumulation of $A\beta_{1-42}$ peptide. The accumulation of β-amyloid is associated with the phosphorylation of tau [e.g., 40]. Data demonstrating excessive cholesterol content in neurons following a seven-month diet adding 1% cholesterol to rabbit chow triggers a cascade of neuropathological changes. The increase in cholesterol content in neurons may underlie the increase in BACE1 and β-amyloid levels than in turn trigger the phosphorylation of tau [12].

Normal brain function involves the use of cholesterol and metals such as copper, but both may be involved in the etiology of AD. Recent findings from domains as divergent as epidemiology, cell biology, and genetics converge to indicate that cholesterol plays a central role in the biology of APP and its proteolytic product, β-amyloid [44]. Mounting evidence demonstrates roles for APP and β-amyloid in metal homeostasis [2]. The widespread replication of AD neuropathology in sites comparable to AD neuropathology coupled with cognitive impairment on a task similar to the impairment observed in AD makes this model promising for the evaluation of AD therapeutics as well as for the investigation of mechanisms of neuropathological development in AD.

*Transgenic mouse models of Alzheimer's disease*

Research on AD has made significant advances with the development of the promising models that have been created with transgenic mice. Mini-gene copies of human APP and Presenelin 1 and 2 (PS1 and PS2) genes implicated in familial AD have been used to create transgenic mouse models of AD [9, 16]. Mice doubly transgenic for APP + PS1 mutations show rapid progressive development of compact plaques beginning at 3 months of age and reaching a plateau at 12 months [15]. APP + PS1 mice are impaired in spatial learning tasks by the age of 15 months [1, 14]. This mouse model has proved useful in tests of immunization with synthetic $A\beta_{1-42}$ peptide [25]. However, this model has limitations. The APP + PS1 mouse model is a model for amyloidosis, and there is little neuronal loss or neurofibrillary tangle formation [26]. This mouse model does not have the full compliment of AD neuropathology. In most AD mouse models, other AD neuropathology such as neurofibrillary tangles are not produced. An exception is the triple-transgenic mouse model of AD that has both plaque and tangle neuropathology [29]. This model is proving invaluable in studies of AD mechanisms as well as therapeutics [2]. However, mouse models of AD have little or no neuronal loss, and the cholinergic system is not affected to the degree that it is compromised in AD. In addition to transgenic mouse models of AD, other experimental models are needed.

Animal models of AD must have cognitive as well as physiological and molecular validity. One of the many challenges that have made it so difficult to create valid animal models of AD is time and the length of human life expectancy. Age is the greatest risk factor for AD [4, 27]. Advanced age and long time periods extending far beyond the life expectancy of most mammals are associated with formation of AD neuropathology. Another major problem for validity in animal models is that AD is a cortical dementia affecting forms of cognition with few parallels in non-human mammals.

*Sensitivity of eyeblink classical conditioning in Alzheimer's disease*

Severe memory loss is the most prominent clinical symptom of AD, and this memory impairment has long been associated with impairment in acetylcholine neurotransmission. Disrupted cholinergic neurotransmission is not the single cause of memory loss in AD, but this deficit characteristic of AD clearly impairs memory. Disruption of the brain cholinergic system in AD links this dementing disease to the model system of eyeblink conditioning. Eyeblink conditioning impairment in AD may reflect medial-temporal lobe atrophy and associated central nervous system cholinergic dysfunction that occurs early in disease progression.

Demonstration that antagonists to acetylcholine impaired eyeblink conditioning in rabbits [24] and that this effect was mediated via the hippocampus [33] led to the hypothesis that patients with AD and a disrupted cholinergic system would be impaired in delay eyeblink classical conditioning. In the delay classical conditioning procedure, a neutral stimulus such as a tone or light is the conditioned stimulus (CS) that onsets and then is followed by a reflex-eliciting stimulus such as an air puff that causes an eyeblink called the unconditioned stimulus (US). What makes it a delay procedure is that there is a delay after CS onset and before US onset and that the CS and US overlap and coterminate. Disruption of the hippocampus during acquisition of delay conditioning prolongs acquisition, making the organism take hundreds more training trials to achieve learning criterion [33, 46, 49]. In the trace eyeblink classical conditioning procedure, the CS onsets and then turns off and a blank "trace" period ensues before the onset of the US. The CS and US never overlap. The hippocampus is essential for acquisition in the trace procedure if the trace interval exceeds a critical interval. For rabbits a trace greater than 300 ms makes the hippocampus essential [28].

Using the 400 ms delay eyeblink classical conditioning in 20 healthy older adults matched to 20 patients diagnosed with AD, we observed highly significant differences in eyeblink conditioning between the AD patients and age-matched, non-demented control subjects [47]. These results have been replicated in other laboratories [23, 32]. The 400 ms delay paradigm was better than the 750 ms trace paradigm at differentiating healthy older adults from adults diagnosed with AD [51]. An additional replication of the sensitivity of the delay eyeblink conditioning paradigm for AD indicated that 400 ms delay differentiated some cerebrovascular dementia patients from patients with probable AD [53]. Eyeblink conditioning in adults over the age of 35 with Down's syndrome and AD neuropathology was similar to conditioning in probable AD patients [54]. Conditioning in patients with Huntington's disease [52] and Parkinson's disease [6, 34] is relatively normal and clearly differentiated from eyeblink conditioning in AD.

Adding 2% cholesterol to the rabbit diet and adding a trace amount of copper to the drinking water for 8 to 10 weeks creates neuropathological changes in the brain of young adult male rabbits that resembles AD neuropathology in humans [35]. Rabbits treated in this manner are impaired in eyeblink classical conditioning [39]. The major aim of this study was to replicate this model and then to extend the model by determining whether it responded to AD therapeutics. A drug approved by the FDA to treat mild to moderate AD (galantamine [Razadyne™]) was tested to determine if this treatment would ameliorate impaired eyeblink classical conditioning in this rabbit model of AD.

## MATERIALS AND METHODS

### Subjects

Data are reported for a total of 28 of 30 male Specific Pathogen Free (SPF) New Zealand white rabbits that completed Experiment 1 or Experiment 2. Two AD model rabbits in the vehicle treatment group in Experiment 2 did not survive to complete behavioral testing. Rabbits were 4 months old at the beginning of the study and weighed a mean of 2.6 kg. All rabbits were purchased from Covance (Denver, PA). Rabbits were individually housed in stainless steel cages in temperature and humidity controlled rooms. The light/dark cycle was 12/12-h. The Institutional Animal Care and Use Committee (IACUC) at Temple University approved research procedures used in this study. This research was carried out in accordance with the Guide for the Care and Use of Laboratory Animals as adopted and promulgated by the National Institutes of Health. The minimum number of animals required to show an effect was used, and procedures were designed to minimize or completely eliminate pain and distress.

### Dietary treatment

For 10 weeks, 20 of the rabbits were fed 160 g/day of a commercially produced diet (Test Diet 7520) of 2% cholesterol added to Purina Mills High Fiber Diet and 0.12 mg/liter copper sulfate added to distilled drinking water. After 10 weeks on this diet, rabbits' mean weight was 3.1 kg. Rabbits were switched to the untreated Purina Mills High Fiber Diet and tap water for the duration of the experiment that involved 3 weeks of behavioral testing. Ten additional rabbits were fed 160 g/day of Purina Mills High Fiber Diet and tap water throughout the study.

*Experiment 1: Neuropathology*

This experiment was carried out to compare neuropathology in 10 AD model rabbits and 10 control rabbits maintained for the same period of time. For 10 weeks AD model rabbits were fed a diet of 2% cholesterol added to Purina Mills High Fiber Diet and 0.12 mg/liter copper sulfate added to distilled drinking water. During this period control rabbits were fed a diet of untreated Purina Mills High Fiber Diet and tap water. Rabbits were euthanized with an overdose of pentobarbital and brains were immediately removed and stored in 10% buffered formalin and embedded in paraffin.

Sections of 10 μm in thickness from representative areas of the brain, including the frontal and parietal cortex, the basal ganglia, the hippocampus, the cerebellum and the brainstem were cut in a microtome (Microm HM315) and placed on electromagnetically charged slides. Staining with Hematoxylin and Eosin was performed for routine histological evaluation. Immunohistochemistry was performed using the avidin-biotin-peroxidase system according to the manufacturer's instructions (Vector Laboratories). Our modified protocol includes deparaffination in xylene, re-hydration through alcohol up to water, non-enzymatic antigen retrieval with Citrate buffer pH 6.0 heated to 95°C for 30 minutes and endogenous peroxidase quenching with 6% $H_2O_2$ in methanol for 20 minutes. After rinsing with PBS, sections were blocked with normal horse serum for 1 hour at room temperature and primary antibodies were incubated over night in a humidified chamber. Primary antibodies utilized in this study included an mouse monoclonal anti-β-amyloid raised against aminoacids 672-714 of amyloid A4, which recognizes β-amyloid of human, mouse and rat origin (Clone B-4, 1:100 dilution, Santa Cruz Biotechnology), a goat polyclonal anti-tau (V-20, 1:250 dilution, Santa Cruz Biotechnology), and a mouse monoclonal anti-calbindin (Clone KR6, 1:100 dilution, NovoCastra Laboratories). After washing with PBS, biotinilated anti-mouse or anti-goat secondary antibodies were incubated for 1 hour at room temperature (1:250 dilution) and avidin biotin-peroxidase complexes were incubated also for 1 hour (ABC Kit, Vector Laboratories. Finally the peroxidase was developed with Diaminobencidine (DAB, Sigma Laboratories), and sections were counterstained with Hematoxylin, dehydrated with alcohol, cleared in xylene and mounted with Permount (Fischer Scientific).

*Experiment 2: Eyeblink classical conditioning and drug efficacy*

This experiment was carried out to explore further learning in AD model rabbits in the trace and delay eyeblink classical condition paradigms. Previous research with eyeblink conditioning in AD model rabbits tested a short-delay paradigm after a longer trace paradigm [35]. Our aim was to hold the interstimulus interval (ISI) in the delay and trace paradigms constant. We used a 750 ms ISI for both paradigms. The other aim was to test the effect of a cognition-enhancing drug, galantamine, on learning in animals with established AD neuropathology. After AD neuropathology was created in rabbits with 10 weeks of the cholesterol/copper regimen, three weeks of behavioral testing and treatment with galantamine ($n$ = 10 AD model rabbits) or vehicle ($n$ = 8 AD model rabbits) was carried out.

AD model rabbits were tested in 750 ms trace (250 ms 85 dB 1 KHz tone conditioned stimulus [CS], 500 ms trace, 100 ms corneal air puff unconditioned stimulus [US]) eyeblink classical conditioning for 10 days and 750 ms delay (850 ms 85 dB 1 KHz tone CS followed 750 ms after its onset with a 100 ms corneal air puff US) eyeblink classical conditioning for 5 days. Each daily session consisted of 90 paired trials of CS and US with a random inter-trial interval ranging between 20 and 30 ms with a mean of 25 ms. A single session lasted approximately 45 minutes.

Techniques were similar to published reports [e.g., 46]. The rabbits were adapted to Plexiglas restrainers and headpieces for one hour on each of the two consecutive days before behavioral testing began. On the second day of adaptation, a local ophthalmic anesthetic (proparacaine hydrochloride) was applied to the left eye so that a 6-0 nylon suture loop could be placed in the temporal margin of the nictitating membrane (NM).

The conditioning apparatus consisted of four separate sound-attenuating chambers, permitting four rabbits to be trained simultaneously. A speaker mounted to the wall of each chamber delivered a 1-kHz, 85-dB tone that was used as the CS. The headpiece, affixed behind the rabbit's ears and under its muzzle, held a plastic tube to deliver 3 psi corneal-directed air puff US and a minitorque potentiometer (San Diego Instruments, San Diego, CA) to measure the rabbit's NM/eyeblink response. Elastic eyelid retractors kept the rabbit's eye open. The potentiometer was secured to the NM via a lever and the nylon suture loop. Analog output from the potentiometer was digitized, stored and analyzed using an IBM PC-compatible system [3]. This system also controlled the timing and presentation of the stimuli.

Changes in the position of the NM detected by the potentiometer were processed and stored in 3-ms bins by the computer. The program recorded a response when the NM moved a minimum of 0.5 mm. A conditioned response (CR) was recorded in both the delay and trace procedures if the response occurred between 25 and 750 ms after the onset of the CS. An unconditioned response (UR) was recorded if the response took place more than 750 ms after the onset of the CS. In both paradigms, CR and UR amplitudes were taken as a measure of response magnitude. A trial was eliminated if NM activity crossed the response threshold within 100 ms prior to the onset of the CS. The criterion for learning was defined as 8 consecutive CRs in a block of 9 trials, with at least 40% of all trials within a training session as CRs.

*Drug administration*

AD model rabbits received 15 daily subcutaneous injections of either 3.0 mg/kg galantamine ($n = 10$) or vehicle ($n = 8$) 30 min before each eyeblink conditioning session. Previously, it had been demonstrated that a dose of 3.0 mg/kg galantamine was the optimal dose to ameliorate impaired eyeblink classical conditioning in older rabbits [55]. Galantamine was supplied by Janssen Pharmaceutica, N.V.

RESULTS

*Experiment 1: Neuropathology*

Histological examination of the brain of AD model rabbits revealed significant neuropathological changes, particularly in the cerebral cortex, hippocampus and cerebellum. Sections from the frontal and parietal cortex revealed a decrease in the number of neurons. The total number of cortical neurons in 25 medium magnification fields (400x) per rabbit was counted. In the control group an average of 1,350 neurons were accounted for, compared with 978 neurons in the cortex of AD model rabbits (Figure 1, Panels A, control and B, AD model). At higher magnification the nuclei of cortical neurons was condensed and their cytoplasm was slim and intensely eosinophillic (Figure 1, Panel D). Interestingly, immunohistochemical detection of β-amyloid and tau yielded negative results in these altered cortical neurons (Figure 1, Panels F and H, respectively).

Examination of the hippocampus revealed similar features, with well-populated Horns of Ammon in the control group (Figure 2, Panel A), and decreased number of neurons in thinner Horns of Ammon in AD model rabbits (Figure 2, Panel B). At higher magnification, hippocampal neurons also show condensed nuclei and scanted eosinophyllic cytoplasm with a filamentous pattern ((Panel D), similar to the neurofibrillary degeneration observed in AD in humans. Corroborating this notion, immunohistochemistry for β-amyloid and tau was positive in the cytoplasm of these altered neurons (Figure 2, Panels F and H, respectively).

Similar alterations were observed in the cerebellum. While the cerebellum of control animals was fully populated by round Purkinje cells with long dendritic processes (Figure 3, Panel A), AD model rabbits showed few Purkinje cells with, scant cytoplasm and a significant loss of processes (Panel B). Also noticeable in Figure 3 is the significantly thinner granule layer in the cerebellum of AD model rabbits. Immunohistochemical detection of Calbindin a neuronal marker specific for Purkinje cells shows at higher magnification the comparison between healthy neurons in the control group (Panel C) and the pathological appearance of the few remaining Purkinje cells in the AD model rabbits (Panel D). These abnormal Purkinje cells robustly expressed cytoplasmic β-amyloid (Panel F), but expression of tau was not present (Panel H).

*Experiment 2: Eyeblink classical conditioning*

*750 ms Trace Paradigm*
A 2 (Drug Treatment) by 10 (Training Sessions) repeated measures analysis of variance (ANOVA)

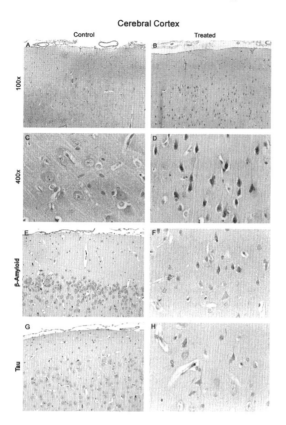

Fig. 1. A low magnification view shows a normally populated cerebral cortex in a control rabbit (Control; Panel A), compared with an Alzheimer's disease (AD) model rabbit (Treated; Panel B) in which fewer neurons are present. At higher magnification, neurons in the control rabbits show a normal pyramidal shape (Panel C), while groups of neurons in the AD model rabbits have condensed, intensely eosinophillic cytoplasm with a fibrillary appearance (Panel D). Immunohistochemistry for β-amyloid and tau were negative in both control (Panels E and G, respectively) and AD model rabbits (Panels F and H, respectively). Panels A to D stained with Hematoxylin & Eosin, A and B, original magnification x100. Panels C, D, F and H original magnification, x400, E and G, x200.

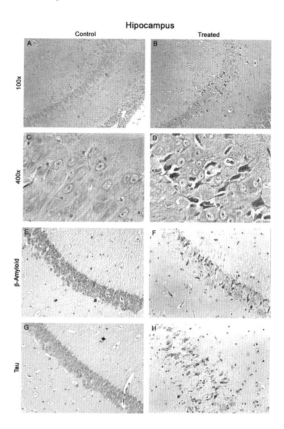

Fig. 2. A low magnification view of the hippocampus in control rabbits (Control; Panel A) shows a fully populated well-organized Horn of Ammon, whereas a disorganized Horn of Ammon with a lower number of neurons is apparent in AD model rabbits (Treated; Panel B). At higher magnification, normal neurons and nuclei are evident (Panel C), but neuronal bodies in the AD model rabbits have scant cytoplasm with an intensely eosinophylic fibrillary appearance, similar to neurofibrillary tangles of Alzheimer's disease (Panel D). These neurons in AD model rabbits were immunoreactive for both β-amyloid (Panel F), and tau (panel H). Hippocampal neurons in control rabbits were negative for β-amyloid (Panel E), and tau (panel G). Panels A to D, Hematoxylin & Eosin. Panels A, B, E and F original magnification x100. Panels C, D, F and H original magnification, x400, E and G, x200.

using the dependent measure of percentage of CRs in the trace conditioning paradigm indicated statistically significant main effects of Drug Treatment, $F(1, 16) = 5.30$, $p = 0.035$ and Training Sessions, $F(9, 144) = 17.40$, $p < 0.001$. There was also a significant Drug Treatment by Training Sessions interaction effect, $F(9, 144) = 3.86$, $p < 0.004$ (Figure 4). Post hoc analysis of the statistically significant interaction effect indicated that the differences between the two groups of AD model rabbits were significant in sessions 6 through 10. AD model rabbits treated with galantamine acquired CRs more rapidly and at a higher magnitude than did AD model rabbits treated with vehicle.

The pattern of results was similar in a repeated measures ANOVA for two other dependent measures, response latency (latency after CS onset to the first response of 0.5 mm or greater) and CR amplitude (peak amplitude of the conditioned response). For response latency, the effect of Drug Treatment was statistically significant, $F(1, 16) = 6.37$, $p = 0.023$, as was the effect of Training Sessions, $F(9, 144) = 13.54$, $p < 0.001$. The Drug Treatment by Training Sessions interaction effect was also significant, $F(9, 144) = 3.68$, $p < 0.001$. For CR amplitude, the effect of Training Sessions was significant,

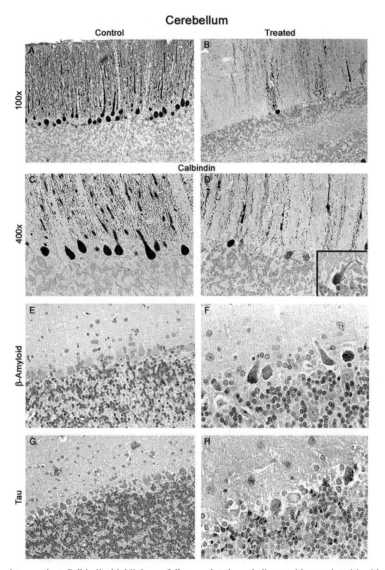

Fig. 3. Immunohistochemistry against Calbindin highlights a fully populated cerebellum, with round and healthy-looking Purkinje cells, which posses abundant cytoplasmic processes in a control rabbit (Control; Panel A), compared with smaller, less populated cerebella in an AD model rabbit (Treated; Panel B). Note the considerably smaller granule layer. At higher magnification the condensed fewer Purkinje cells and the loss of axons and dendrites is evident in the AD model rabbit (Panel D), especially in comparison to the control rabbit (Panel C). Note in Panel D the intensely eosinophillic, condensed and fibrillary cytoplasm (inset, Hematoxylin & Eosin). Purkinje cells show cytoplasmic expression of β-amyloid in AD model rabbits (Panel F), but not in control rabbits (Panel E). Immunoreactivity for tau was absent in control rabbits (Panel G) and was weak at best in AD model rabbits (Panel H). Panels A and B original magnification x100. Panels C to H original magnification x400.

$F(9, 144) = 7.72$, $p < 0.001$ as was the Drug Treatment by Training Sessions interaction, $F(9, 144) = 2.47$, $p = 0.012$. The main effect of Drug Treatment did not reach significance at the 0.05 level of confidence, $F(1, 16) = 3.06$, $p = 0.10$.

Trials to learning criterion were compared in a $t$-test. There was a statistically significant effect of Drug Treatment, $t(16) = 2.77$, $p = 0.014$. AD model rabbits treated with galantamine achieved learning criterion in significantly fewer trials that AD model rabbits treated with vehicle (Figure 5).

The unconditioned response (UR) is a measure of the reflexive motor response to the air puff unconditioned stimulus. A 2 (Drug Treatment) by 10 (Training Sessions) repeated measures ANOVA using the dependent measure of UR amplitude indicated that the reflexive motor response was not affected by the dietary treatment or drug. Neither the main effects

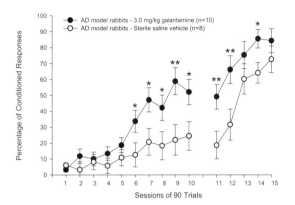

Fig. 4. Percentage of conditioned responses for 10 90-trial sessions in 750 ms trace eyeblink conditioning followed by 5 90-trial sessions in 750 ms delay eyeblink conditioning in 10 AD model rabbits treated with 3.0 mg/kg galantamine and 8 AD model rabbits treated with vehicle. Error bars are standard error of the mean. Asterisks indicate sessions in which there were statistically significant differences between AD model rabbits treated with galantamine and vehicle (* $p < 0.05$; ** $p < 0.01$).

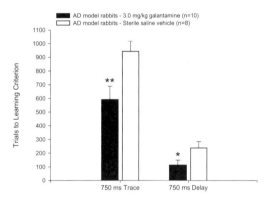

Fig. 5. Trials to a learning criterion of 8 conditioned responses (CRs) in 9 consecutive trials in a session of 40% or more CRs in AD model rabbits treated with 3.0 mg/kg galantamine and AD model rabbits treated with vehicle in 750 ms trace and 750 ms delay eyeblink classical conditioning. There were statistically difference between trials to learning criterion in both the delay and trace paradigms (* $p < 0.05$; ** $p < 0.01$)

nor the interaction effect were significantly different between the treatment groups over training sessions.

*750 ms Delay Paradigm*

A 2 (Drug Treatment) by 5 (Training Sessions) repeated measures ANOVA using the dependent measure of percentage of CRs demonstrated a statistically significant effect of Drug Treatment, (1, 16) = 6.75, $p = 0.019$, and a significant effect of Training Sessions, $F$ (4, 64) = 19.51, $p < 0.001$. Post hoc analysis indicated statistically significant differences between the groups in the first, second, and fourth session of delay conditioning (Sessions 11, 12, and 14 of trace and delay eyeblink conditioning). The interaction effect was not significant, $F$ (4, 64) = 1.31, $p = 0.28$ (Figure 4). The pattern of results was similar in a repeated measures ANOVA for response latency and CR amplitude. For response latency, the effect of Drug Treatment was statistically significant, (1, 16) = 5.20, $p = 0.037$, and the effect of Training Sessions was statistically significant, $F$ (4, 64) = 20.07, $p < 0.001$. The Drug Treatment by Training Sessions interaction was not significant, $F$ (4, 64) = 0.44, $p = 0.783$. For CR amplitude, there was a statistically significant effect of Drug Treatment, (1, 16) = 5.06, $p = 0.039$, and a significant effect of Training Sessions, $F$ (4, 64) = 8.64, $p < 0.001$. The Drug Treatment by Training Sessions interaction was not significant, $F$ (4, 64) = 1.56, $p = 0.20$. Trials to learning criterion were compared in a *t*-test. There was a statistically significant effect of Drug Treatment, $t$ (16) = 2.31, $p = 0.049$ (Figure 5).

A 2 (Drug Treatment) by 10 (Training Sessions) repeated measures ANOVA using the dependent measure of UR amplitude indicated that the reflexive motor response was not affected by the dietary treatment or drug. Neither the main effects nor the interaction effect were significantly different between the treatment groups over training sessions.

## DISCUSSION

A mammalian model of AD originated by Sparks [38] and repeatedly replicated [e.g., 12, 36, 39] was also replicated in this study by feeding young male rabbits for 10 weeks a commercially prepared rabbit chow diet to which 2% cholesterol was added and including a trace amount of copper sulfate in the distilled drinking water. Control rabbits were fed normal rabbit chow and water during this same time period. The brains of treated rabbits showed AD-like neuropathology in hippocampus, cerebral cortex, and cerebellum including apoptosis and positive staining for β-amyloid and tau. In the three weeks after the dietary regime, AD model rabbits were tested with 750 ms trace and then 750 ms delay eyeblink conditioning. Animals were injected 30 min before the daily conditioning session with either equal volumes of solution containing either sterile saline vehicle or 3.0 mg/kg galantamine. The administration of 3.0 mg/kg galantamine significantly ameliorated trace and delay eyeblink classical conditioning of AD model rabbits over conditioning in vehicle-treated AD model rabbits.

*AD neuropathology*

Neuropathological changes, similar to those observed in AD in humans, including neuronal loss and immunoreactivity for β-amyloid and tau were found in the cortex, hippocampus as well as the cerebellum of treated rabbits. Neuropathology was found in cerebral cortex, but not expression of β-amyloid and tau by immunohistochemistry.

We examined the cerebellum in AD model rabbits primarily because the cerebellum is the essential substrate for eyeblink classical conditioning. Although the cerebellum has traditionally been viewed as one of the last structures to be impacted by classical AD pathology in humans, evidence has mounted to indicate that the cerebellum is affected by AD. Antibodies to β-amyloid revealed the presence of diffuse plaques in the molecular layer of the AD and aged-Down's syndrome cerebellum [17, 20, 43]. Ultrastructural studies have revealed only rare amyloid fibrils in diffuse plaques, whereas the dystrophic neurites in these plaques contain membranous and vesicular dense bodies but no paired helical filaments [7, 57]. Purkinje cell loss and reactive astrocytosis of the cerebellum in familial AD was found to be more severe than in sporadic AD in a Japanese sample, but the β-amyloid deposition in the cerebellum in both familial and sporadic AD were similar [11]. In a British sample, there was lysosomal abnormality in Purkinje cells in AD. These differences may be associated with the absence of senile plaques and the presence of "diffuse" amyloid plaques in the cerebellum in AD [8]. Abnormalities in AD model rabbits in cerebellar Purkinje cells are consistent with findings in the cerebellum of humans with AD supporting the validity of the AD model rabbit.

*Eyeblink classical conditioning in AD model rabbits*

In the initial study of rabbits treated with the cholesterol/copper regimen and tested in eyeblink classical conditioning, the first 10 sessions were carried out with the 750 ms trace paradigm and the final 5 sessions used the 400 ms delay paradigm. AD model rabbits were impaired in 750 ms trace but not 400 ms delay eyeblink conditioning. The investigators concluded that the hippocampus-dependent trace paradigm was impaired, but the delay paradigm that does not require the hippocampus was not impaired. However, the two paradigms varied on the interval between the CS and the US, called the inter-stimulus interval (ISI) as well as on the contiguity of CS and US. Studies in rabbits [e.g., 30] and humans [e.g., 22] demonstrated the importance of the ISI in rate of conditioning. In rabbits, ISIs of 250-500 ms are optimal. In humans, this optimal range is wider – from 400-1800 ms. We controlled the ISI in the present study and compared a 750 ms trace to a 750 ms delay paradigm. When the ISIs were equal, AD model rabbits were impaired in delay as well as trace eyeblink classical conditioning.

Vehicle-treated rabbits were impaired in 750 delay eyeblink conditioning even though they had received 10 training sessions in 750 ms trace conditioning using the same 1 KHz tone CS and corneal air puff US. To compare further the impact of the cholesterol/copper regimen on the two eyeblink classical conditioning paradigms, delay should be tested first, followed by trace. Given the significant pathology we observed in the cerebellum as well as hippocampus, it is likely that delay eyeblink classical conditioning will be impaired when tested before trace eyeblink conditioning.

The model we first articulated for the cause of severe impairment of delay eyeblink conditioning in AD [48] was based on the scopolamine-injected rabbit [33]. Disruption of the septohippocampal cholinergic system perturbed the cerebellar circuitry in acquisition of CRs. The cholinergic system is impaired in AD model rabbits [36] and severely impaired in AD in humans [5]. Transgenic APP + PS1 mice are not impaired in delay or trace eyeblink classical conditioning [10]. We interpreted the results with eyeblink conditioning in APP + PS1 transgenic mice as consistent with the effects of amyloidogenic pathology on the cholinergic system in these mice at 12 months of age. The effects of the pathology are considerably lower than the cholinergic deficits reported in AD in humans [18, 45]. Learning was disrupted in PDAPP V717F mice in the trace eyeblink classical conditioning procedure [42]. These transgenic mice had reduced hippocampal and callosal brain volume [1] and some impairment in cholinergic neurotransmission [29].

*AD therapeutics*

Previous studies in retired breeder rabbits demonstrated the efficacy of 3.0 mg/kg galantamine in ameliorating impaired eyeblink classical conditioning in the delay [55, 56] and trace procedures [31, 41]. Even in young rabbits that were at ceiling performance on delay eyeblink conditioning, 3.0 mg/kg galantamine facilitated acquisition [56] and reversed antagonists to nicotinic acetylcholine receptors [50]. There is an association between upregulation of nicotinic acetylcholine receptors in galantamine-treated normal older rabbits and improved eyeblink classical conditioning [56]. This physiological change may have accounted for amelioration of impaired conditioning in galantamine-treated AD model rabbits. We did not expect that established AD neuropathology in rabbits treated for 10 weeks with the cholesterol/copper regimen would be changed by 15 daily injections of galantamine, and we observed no differences in AD model rabbit brains treated with galantamine or vehicle. Nevertheless, results in the present study demonstrate that the performance in delay and trace eyeblink conditioning of young male rabbits impaired with AD neuropathology can be ameliorated with 3.0 m/kg galantamine.

Acetylcholine neurotransmission plays a crucial role in learning and memory and has been the focus of pharmacological therapy for AD. The first four drugs approved by the Food and Drug Administration to treat memory impairment in mild to moderate AD were acetylcholinesterase inhibitors that prolonged acetylcholine action at the synapse. Of these compounds, galantamine is the only molecule approved in Europe with mechanisms of action that include both acetylcholinesterase inhibition and allosteric potentiating effects at the nicotinic acetylcholine receptor. Although these drugs approved to treat mild to moderate AD are all of demonstrated efficacy in treatment of the disease, none prevent the eventual decline of cognitive functions. Additional drugs with various therapeutic routes must be identified and tested for cognition enhancement in AD. These results with the AD model rabbit indicate that the model, with its demonstrated validity from molecular to cognitive levels, has utility as a preclinical test of drug efficacy in AD.

## ACKNOWLEDGEMENTS

We are grateful to Sue Seta for managing and testing the AD model rabbits and to Laura Scarlota and Mike Tobia for their assistance with animal training and data entry. D. Larry Sparks provided invaluable advice on the implementation of the AD rabbit model. This research was supported by grants from NIH (AG021925) and from Janssen Pharmaceutica, N.V. and Ortho-McNeil Neurologics. This project was also funded, in part, under a grant with the Pennsylvania Department of Health. The Department specifically disclaims responsibility for any analyses, interpretations or conclusions.

## References

[1] G.W. Arendash, D.L. King, M.N. Gordon, D. Morgan, J.M. Hatcher, C.E. Hope and D.M. Diamond, Progressive, age-related behavioral impairments in transgenic mice carrying both mutant amyloid precursor protein and presenilin-1 transgenes, *Brain Res.* **891** (2001) 42-53.

[2] A. Caccamo, S. Oddo, L.M. Billings, K.N. Green, H. Martinez-Coria, A. Fisher and F.M. LaFerla, M1 receptors play a central role in modulating AD-like pathology in transgenic mice, *Neuron* **9** (2004) 671-682.

[3] G. Chen and J.E. Steinmetz, A general-purpose computer system for behavioral conditioning and neural recording experiments, *Behav. Res. Meth. Instr. & Computers* **30** (1998) 384-391.

[4] C. Costantini, R.M.K. Kolasani and L. Puglielli, Ceramide and cholesterol: Possible connections between normal aging of the brain and Alzheimer's disease. Just hypotheses or molecular pathways to be identified?, *Alzheimer's & Dementia* **1** (2005) 43-50.

[5] J.T. Coyle, D.L. Price and M.R. DeLong, Alzheimer's disease: a disorder of cortical cholinergic innervation, *Science* **219** (1983) 1184-1190.

[6] I. Daum, M.M. Schugens, C. Breitenstein, H. Topka and S. Spieker, Classical eyeblink conditioning in Parkinson's disease, *Mov. Disord.* **11** (1996) 639-646.

[7] D.W. Dickson, A. Wertkin, L.A. Mattiace, E. Fier, Y. Kress, P. Davies and S-H. Yen, Ubiquitin immunoelectron microscopy of dystrophic neurites in cerebellar senile plaques of Alzheimer's disease, *Acta. Neuropathol.* **79** (1990) 486-493.

[8] J.H. Dowson, C.Q. Mountjoy, M.R. Cairns, H. Wilton-Cox and W. Bondareff, Lipopigment changes in Purkinje cells in Alzheimer's disease, *J. Alzheimers Dis.* **1** (1998) 71-79.

[9] K. Duff, C. Eckman, C. Zehr, X. Yu, C.M. Prada, J. Perez-Tur, M. Hutton, L. Buee, Y. Harigaya, D. Yager, D. Morgan, M.N. Gordon, L. Holcomb, L. Refolo, B. Zenk, J. Hardy and S. Younkin, Increased amyloid-beta 42(43) in brains of mice expressing mutant presenilin 1, *Nature* **383** (1996) 710-713.

[10] M. Ewers, D. Morgan, M. Gordon and D.S. Woodruff-Pak, Associative and motor learning in 12-month-old transgenic APP + PS1 mice, *Neurobiol. Aging.* (2006 in press).

[11] Y. Fukutani, N.J. Cairns, M.N. Rossor and P.L. Lantos, Cerebellar pathology in sporadic and familial Alzheimer's disease including APP 717 (Val→Ile) mutation

[12] O. Ghribi, B. Larsen, M. Schrag and M.M. Herman, High cholesterol content in neurons increases BACE, β-amyloid, phosphorylated tau levels in rabbit hippocampus, *Exp. Neurol.* **200** (2006) 460-467.

[13] F. Gonzalez-Lima, J.D. Berndt, J.E. Valla, D. Games and E.M. Reiman, Reduced corpus callosum, fornix and hippocampus in PDAPP transgenic mouse model of Alzheimer's disease, *Neuroreport* **12** (2001) 2375-2379.

[14] M.N. Gordon, D.L. King, D.M. Diamond, P.T. Jantzen, K.V. Boyett, C.E. Hope, J.M. Hatcher, G. DiCarlo, W.P. Gottschall, D. Morgan and G.W. Arendash, Correlation between cognitive deficits and Abeta deposits in transgenic APP+PS1 mice, *Neurobiol. Aging* **22** (2001) 377-385.

[15] L. Holcomb, M.N. Gordon, E. McGowan, X. Yu, S. Benkovic, P. Jantzen, K. Wright, I. Saad, R. Mueller, D. Morgan, S. Sanders, C. Zehr, K. O'Campo, J. Hardy, C.M. Prada, C. Eckman, S. Younkin, K. Hsiao and K. Duff, Accelerated Alzheimer-type phenotype in transgenic mice carrying both mutant amyloid precursor protein and presenilin 1 transgenes, *Nat. Med.* **4** (1998) 97-100.

[16] K. Hsiao, P. Chapman, S. Nilsen, C. Eckman, Y. Harigaya, S. Younkin, F. Yang and G. Cole, Correlative memory deficits, Abeta elevation, and amyloid plaques in transgenic mice, *Science* **274** (1996) 99-102.

[17] S.I. Ikeda, D. Allsop and G.G. Glenner, The morphology and distribution of plaque and related deposits in the brains of Alzheimer's disease and control cases: An immunohistochemical study using amyloid beta-protein antibody, *Lab. Invest.* **60** (1989) 113-122.

[18] S. Jaffar, S.E. Counts, S.Y. Ma, E. Dadko, M.N. Gordon, D. Morgan and E. J. Mufson, Neuropathology of mice carrying mutant APP(swe) and/or PS1(M146L) transgenes: alterations in the p75(NTR) cholinergic basal forebrain septohippocampal pathway, *Exp. Neurol.* **170** (2001) 227-243.

[19] E.M. Johnstone, M.O. Chaney, F.H. Norris, R. Pascual and S.P. Little, Conservation of the sequence of the Alzheimer's disease amyloid peptide in dog, polar bear and five other mammals by cross-species polymerase chain reaction analysis, *Brain. Res. Mol. Brain. Res.* **10** (1991) 299-305.

[20] Y-T. Li, D.S. Woodruff-Pak and J.Q. Trojanowski, Amyloid plaques in cerebellar cortex and the integrity of Purkinje cell dendrites, *Neurobiol. Aging* **15** (1994) 1-9.

[21] C.J. Maynard, A.I. Bush, C.L. Masters, R. Cappai and Q-X. Li, Metals and amyloid-β in Alzheimer's disease, *Int. J. Exp. Path.* **86** (2005) 147-159.

[22] W.R. McAllister, Eyeblink conditioning as a function of the CS-UCS interval, *J. Exp. Psychol.* **45** (1953) 417-423.

[23] A.B. Moore, M.W. Bondi, D.P. Salmon and C. Murphy, Eyeblink classical conditioning to auditory and olfactory stimuli: performance among older adults with and without the apolipoprotein E epsilon 4 allele, *Neuropsychol.* **19** (2005) 437-445.

[24] J.W. Moore, N.A. Goodell and P.R. Solomon, Central cholinergic blockade by scopolamine and habituation, classical conditioning, and latent inhibition of the rabbit's nictitating membrane response, *Physiol. Psychol.* **4** (1976) 395-399.

[25] D. Morgan and B.D. Gitter,. Evidence supporting a role for anti-Abeta antibodies in the treatment of Alzheimer's disease, *Neurobiol. Aging* **25** (2004) 605-608.

[26] D. Morgan and R.K. Keller, What evidence would prove the amyloid hypothesis? Toward rational drug treatments for Alzheimer's disease, *J. Alzheimer's Dis.* **4** (2002) 257-260.

[27] J.C. Morris, Is Alzheimers disease inevitable with age? Lessons from clinicopathologic studies of healthy aging and very mild Alzheimer's disease, *J. Clin. Invest.* **104** (1999) 1171-1173.

[28] J.R. Moyer, R.A. Deyo and J.F. Disterhoft, Hippocampectomy disrupts trace eye-blink conditioning in rabbits, *Behav. Neurosci.* **104** (1990) 243-252.

[29] S. Oddo, A. Caccamo, J.D. Shepherd, M.P. Murphy, T.E. Golde, R. Kayed, R. Metherate, M.P. Mattson, Y. Akbari and F.M. LaFerla, Triple-transgenic model of Alzheimer's disease with plaques and tangles: intracellular Abeta and synaptic dysfunction, *Neuron* **39** (2003) 409-421.

[30] N. Schneiderman, Interstimulus interval function of the nictitating membrane response of the rabbit under delay versus trace conditioning, *J. Comp. Physiol. Psychol.* **62** (1966) 397-402.

[31] B.B. Simon, B. Knuckley and D.A. Powell, Galantamine facilitates acquisition of a trace-conditioned eyeblink response in healthy, young rabbits, *Learn. Mem.* **11** (2004)116-122.

[32] P.R. Solomon, E. Levine, T. Bein and W.W. Pendlebury, Disruption of classical conditioning in patients with Alzheimer's disease, *Neurobiol. Aging* **12** (1991) 283-287.

[33] P.R. Solomon, S.D. Solomon, E. Vander Schaaf and H.E. Perry, Altered activity in the hippocampus is more detrimental to classical conditioning than removing the structure, *Science* **220** (1983) 329-331.

[34] M. Sommer, J. Grafman, K. Clark and M. Hallett, Learning in Parkinson's disease: eyeblink conditioning, declarative learning, and procedural learning, *J. Neurol. Neurosurg. Psychiatry* **67** (1999) 27-34.

[35] D.L. Sparks, Intraneuronal β-amyloid immunoreactivity in the CNS, *Neurobiol. Aging.* **17** (1996) 291-299.

[36] D.L. Sparks, Cholesterol, copper, and accumulation of thioflavine S-reactive Alzheimer's-like amyloid β in rabbit brain, *J. Mol. Neurosci.* **24** (2004) 97-104.

[37] D.L. Sparks, Y-M. Kou, A. Roher, T.A. Martin and R.J. Lukas, Alterations of Alzheimer's disease in the cholesterol-fed rabbit, including vascular inflammation. Preliminary observations, *Ann. NY Acad. Sci.* **903** (2000) 335-344.

[38] D.L. Sparks, S.W. Scheff, J.C. Hunsaker III, H. Liu, T. Landers and D.R. Gross, Induction of Alzheimer-like β-amyloid immunoreactivity in the brains of rabbits with dietary cholesterol. *Exp. Neurol.* **126** (1994) 88-94.

[39] D.L. Sparks and B. Schreurs, Trace amounts of copper in water induce beta-amyloid plaques and learning deficits in a rabbit model of Alzheimer's disease, *PNAS* **100** (2003) 11065-11069.

[40] A. Takashima, T. Honda, K. Yasutake, G. Michel, O. Murayama, M. Murayama, K. Ishiguro and H. Yamaguchi, Acitvation of tau protein kinase I/glycogen synthase kinase-3beta by amyloid beta peptide (25-35) enhances phosphorylation of tau in hippocampal neurons, *Neurosci. Res.* **31** (1998) 317-323.

[41] A.P. Weible, M.M. Oh, G. Lee and J.F. Disterhoft, Galantamine facilitates acquisition of hippocampus-dependent trace eyeblink conditioning in aged rabbits, *Learn. Mem.* **11** (2004) 108-115.

[42] C. Weiss, P.N. Venkatasubramanian, A.S. Aguado, J.M. Power, B.C. Tom, L. Li, K.S. Chen, J.F. Disterhoft and A.M. Wyrwicz, Impaired eyeblink conditioning and decreased hippocampal volume in PDAPP V717F mice, *Neurobiol. Dis.* **11** (2002) 425-433.

[43] H.M. Wisniewski, C. Bancher, M. Barcikowska, G.Y. Wen and J. Currie, Spectrum of morphological appearance of amyloid deposits in Alzheimer's disease, *Acta. Neuropath.* **78** (1989) 337-347.

[44] B. Wolozin, Cholesterol and the biology of Alzheimer's disease, *Neuron* **41** (2004) 7-10.

[45] T.P.Wong, T. Debeir, K. Duff and A.C. Cuello, Reorganization of cholinergic terminals in the cerebral cortex and hippocampus in transgenic mice carrying mutated presenilin-1 and amyloid precursor protein transgenes, *J. Neurosci.* **19** (1999) 2706-2716.

[46] D.S. Woodruff-Pak, Mecamylamine reversal by nicotine and by a partial alpha 7 nicotinic acetylcholine receptor agonist (GTS-21) in rabbits tested with delay eyeblink classical conditioning, *Behav. Brain Res.* **143** (2003) 159-167.

[47] D.S. Woodruff-Pak, R.G. Finkbiner and I. Katz, A model system demonstrating parallels in animal and human aging: Extension to Alzheimer's disease, in: *Novel approaches to the treatment of Alzheimer's disease*, E.M. Meyer, J.W. Simpkins and J. Yamamoto, eds., Plenum, New York, 1989, pp. 355-371.

[48] D.S. Woodruff-Pak, R. Finkbiner and D. Sasse, Eyeblink conditioning discriminates Alzheimer's patients from non-demented aged, *NeuroReport* **1** (1990) 45-48.

[49] D.S. Woodruff-Pak, Y-T. Li, A. Kazmi and W.R. Kem, Nicotinic cholinergic system involvement in eyeblink classical conditioning in rabbits, *Behav. Neurosci.* **108** (1994) 486-493.

[50] D.S. Woodruff-Pak, Y-T. Li and W.R. Kem, A nicotinic agonist (GTS-21), eyeblink conditioning, and nicotinic receptor binding in rabbit brain, *Brain Res.* **645** (1994) 309-317.

[51] D.S. Woodruff-Pak and M. Papka, Alzheimer's disease and eyeblink conditioning: 750 ms trace versus 400 ms delay paradigm, *Neurobiol. Aging* **17** (1996) 397-404.

[52] D.S. Woodruff-Pak and M. Papka, Huntington's disease and eyeblink classical conditioning: Normal learning but abnormal timing, *J. Int. Neuropsychol. Soc.* **2** (1996) 323-334.

[53] D.S. Woodruff-Pak, M. Papka, S. Romano and Y-T. Li, Eyeblink classical conditioning in Alzheimer's disease and cerebrovascular dementia, *Neurobiol. Aging* **17** (1996) 505-512.

[54] D.S. Woodruff-Pak, M. Papka and E.W. Simon, Eyeblink classical conditioning in Down's Syndrome, Fragile X Syndrome, and normal adults over and under age 35, *Neuropsychol.* **8** (1994) 14-24.

[55] D.S. Woodruff-Pak and I. Santos, Nicotinic modulation in an animal model of a form of associative learning impaired in Alzheimer's disease, *Behav. Brain. Res.* **113** (2000) 11-19.

[56] D.S. Woodruff-Pak, R.W. Vogel III and G.L.Wenk, Galantamine: Effect on nicotinic receptor binding, acetylcholinesterase inhibition, and learning, *PNAS* **98** (2001) 2089-2094.

[57] T. Yamazaki, H. Yamaguchi, Y. Nakazato, K. Ishiguro, T. Kawarabayashi and S. Hirai, Ultrastructural characterization of cerebellar diffuse plaques in Alzheimer's disease, *J. Neuropathol. Exp. Neurol.* **51** (1992) 281-286.

# Transgenic *Drosophila* Models of Alzheimer's Amyloid-β 42 Toxicity

Koichi Iijima[a,c,1] and Kanae Iijima-Ando[b,c,1]
[a]*Laboratory of Neurobiology and Genetics,*
[b]*Laboratory of Neurogenetics and Pathobiology,*
[c]*Department of Neuroscience, Farber Institute for Neurosciences, Department of Biochemistry and Molecular Biology, Thomas Jefferson University, Philadelphia, PA19107, USA*

**Abstract.** Alzheimer disease (AD) is the most common form of senile dementia, and a cure is desperately needed. The amyloid-β 42 (Aβ42) has been suggested to play a central role in the pathogenesis of AD. However, the mechanism by which Aβ42 causes AD remains unclear. To understand the pathogenesis and to develop therapeutic avenues, it is crucial to generate animal models of Aβ42 toxicity in genetically tractable organisms. *Drosophila* is a well-established model system for which abundant genetic tools are available, and recently emerges as a model for human neurodegenerative diseases. We and others have established transgenic flies that express human Aβ42 in the nervous system. These Aβ42 flies developed age-dependent short-term memory impairment and neurodegeneration accompanied by Aβ42 deposits. Here we will summarize key features of transgenic *Drosophila* models of Aβ42 toxicity and a number of insights into disease mechanisms as well as therapeutic implications gained from these models.

Keywords: Alzheimer disease, amyloid-β 42, transgenic animal models, *Drosophila*, learning and memory, neurodegeneration, protein misfolding, genetic screen

## INTRODUCTION

The establishment of animal models of human disease is crucial for understanding disease pathogenesis as well as for the discovery and evaluation of potential therapies. The fruit fly *Drosophila* has been used as a powerful genetic model to investigate many aspects of biology. Analysis of the *Drosophila* genome has revealed that approximately 70% of human disease-related genes have homologs in *Drosophila* [1, 2], suggesting that studies using *Drosophila* may provide insights into the pathomechanisms of human diseases.

Recently, transgenic *Drosophila* has emerged as a model to study human neurological and neurodegenerative disorders. Many late-onset neurodegenerative diseases including Alzheimer's disease (AD), tauopathies, Parkinson's disease, Huntington's disease and other polyglutamine diseases, amyotrophic lateral sclerosis (ALS), and prion disease are characterized by accumulation of misfolded proteins [3]. Molecular genetic studies revealed the genes associated with familial forms of such diseases, and a transgenic overexpression approach has been used to model those diseases caused by toxic gain-of-function mechanisms. Overexpression of some of those genes in *Drosophila* neurons has been shown to recapitulate key features of human neurodegenerative diseases [4-21].

With the short generation time and large number of progeny, one of the most important tools that *Drosophila* provides is the ability to carry out large-scale genetic screens for mutations that affect a given process. A number of forward genetic screens using chemical or transposon-based mutagenesis, in which genes are randomly disrupted, have been conducted to unravel the molecular mechanisms underlying various biological processes ranging from

---

[1]Corresponding authors: Koichi Iijima, 900 Walnut Street, JHN410, Philadelphia, PA19107, USA. E-mail: Koichi.Iijima@jefferson.edu, or Kanae Iijima-Ando, 900 Walnut Street, JHN411, Philadelphia, PA19107, USA. E-mail: Kanae.Iijima-Ando@jefferson.edu

development to behavior. These efforts have resulted in the availability of an amazing collection of loss of function mutations and transgenic lines that are available in the fly community. Moreover, the completion of a whole genome analysis in *Drosophila* [22] enables us to perform systematic RNAi-based knockouts of all fly genes [23, 24]. These tools allow us to efficiently test hypotheses derived from biochemical or histopathological analysis of human disease tissues, genetic association studies performed on particular disease phenotypes, or findings in *in vitro* model systems. In addition to hypothesis testing, the unbiased, genome-wide, forward genetic screens have great potential to identify novel genes and pathways that modify a given disease phenotype.

A number of excellent reviews regarding modeling of human neurodegenerative diseases in the fly have been published [25-31]. In this review, we will summarize key features of transgenic *Drosophila* models of AD and findings in pathogenic mechanisms and therapeutic implications gained from these models.

## ESTABLISHMENT OF TRANSGENIC *DROSOPHILA* AS A MODEL TO STUDY HUMAN AMYLOID-β 42 TOXICITY

*The role of the Aβ peptide in the pathogenesis of Alzheimer's disease*

AD is a fatal disorder and, in its later stages, global cognitive functions are disrupted and associated motor disabilities lead patients to become bedridden [32]. Short-term memory impairment is detected in the early stage of the disease along with other psychiatric problems such as sleep disorders and increased agitation, which distinguishes AD from other neurodegenerative conditions such as Parkinson's disease, tauopathies, or Huntington's disease [32, 33].

At the level of cellular pathology, extensive neuron loss and two characteristic hallmarks, senile plaques (SPs) and neurofibrillary tangles (NFTs), are observed in AD brains [34]. SPs are extracellularly deposited protein aggregates that are referred to as amyloid deposits. Biochemical studies have revealed that the major components of these SPs are 40 or 42 amino acid peptides called amyloid-β 40 or 42 (Aβ40 or Aβ42) [35, 36]. Although a low number of SPs can be detected in normal aged brains, this lesion is relatively specific to AD. In contrast, NFTs, which are intracellular protein inclusions composed of hyperphosphorylated microtubule-associated tau protein [37], are also observed in several other neurological diseases called tauopathies.

The majority of AD cases are sporadic with disease onset after 65 years of age. Less than 10% of all AD cases are inherited in an autosomal dominant manner [38]. Aβ peptides are physiological metabolites of the amyloid β-precursor protein (AβPP) and result from sequential cleavage by the β-secretase and γ-secretase complexes, whose catalytic subunits are Presenilin 1 (PS1) and Presenilin 2 (PS2) [39]. Molecular genetic studies of early-onset familial AD (EOFAD) patients have identified causative mutations in the *APP*, *PS1*, and *PS2* genes, and these mutations promote Aβ42 production, aggregation, and stability against clearance [40]. These results provide a strong causative link between Aβ42 and AD [41].

Tau mutations have not been associated with any known form of familial AD to date; however, tau haplotypes driving slightly higher tau expression increase AD risk [42, 43], suggesting that tau plays a role in the pathogenesis of AD as a modulator of disease progression. Studies in human AD cases following Aβ immunization have shown decreases in amyloid burden and in phosphorylated tau in neurites surrounding the amyloid plaques [44, 45]. In transgenic mice overproducing human Aβ and tau proteins, Aβ facilitates the abnormal phosphorylation of tau at AD-related sites and enhances the formation of NFTs [46-48]. Aβ immunization removes amyloid pathology as well as early-stage tau lesions [49] and ameliorated cognitive decline in transgenic mice that form plaques and tangles [50-52]. Knockdown of tau expression suppresses Aβ-induced neurotoxicity in cultured neurons [53, 54], and lowering or eliminating endogenous tau expression in transgenic mice suppresses Aβ-induced behavioral deficits [55]. These reports suggest that Aβ lies upstream of pathology and toxicity of tau in AD [56]. However, the mechanism by which the Aβ42 peptide initiates the complex pathogenesis of AD remains elusive.

*Drosophila models of Alzheimer's amyloid-β 42 toxicity*

*Drosophila* has a clear homolog of AβPP, AβPP-like protein (dAPPL) [57]. dAPPL shares about 30% overall sequence identity with human AβPP, however, the region in dAPPL that corresponds to the

Aβ peptide lacks significant homology with the human peptide [57]. Flies deficient for dAPPL exhibit behavioral abnormalities (phototaxis deficiency), which can be rescued by a human AβPP transgene, indicating a functional conservation between dAPPL and human AβPP [57]. dAPPL is involved in synaptic differentiation [58], synaptic development [59], and neurite arborization [60]. dAPPL overexpression causes axonal transport defects [61]. Similarly, overexpression of human AβPP induces axonal transport defects and increases cell death in the larval fly brain [62].

*Drosophila* has all the components of the γ-secretase complex [63, 64], but the Aβ peptide cannot be produced from human AβPP in the fly, because *Drosphila* has very low β-secretase activity [65, 66]. Recently, Carmine-Simmen et al. identified *Drosophila* β-secretase-like enzyme (dBACE), which has 25% identity to human BACE1 and 28% identity to human BACE2. dBACE does cleave human AβPP but not at β-site [14, 67]. Interestingly, dBACE overexpression cleaves dAPPL and produces a fragment containing the region in dAPPL that corresponds to the Aβ peptide, which aggregate into intracellular fibrils, amyloid deposits, and cause age-dependent behavioral deficits and neurodegeneration [67]. In the non-amyloidogenic pathway, AβPP is cleaved within the Aβ domain by α-secretase precluding deposition of intact Aβ peptide. Kuzbanian (Kuz) is the *Drosophila* ortholog of α-secretase ADAM10 [68, 69] and cleaves dAPPL [67]. These results suggest that not only the biological functions but also proteolytic processing of AβPP are conserved between *Drosophila* and humans.

To study human Aβ peptide-induced amyloid formation and neurodegeneration in *Drosophila*, transgenic flies have been generated by several approaches using the GAL4/UAS transgene expression system (Figure 1) [70]. To overproduce Aβ peptides in flies, Greeve et al. [14] generated triple transgenic flies expressing human AβPP, human β-secretase, and *Drosophila* presenilin (dPSn) or dPSn with point mutations that are found in the early-onset familial AD (EOFAD) patients, N141I, L235P, and E280A [71]. These flies produced modest levels of Aβ peptides, presumably including Aβ40 and Aβ42, and developed β-amyloid plaques and age-dependent neurodegeneration as well as semilethality. The neurodegeneration phenotype was enhanced in the flies expressing dPSn carrying EOFAD mutations, while it is suppressed by a genetic reduction in fly endogeneous dPSn, suggesting that neurodegeneration is dependent on γ-secretase activity.

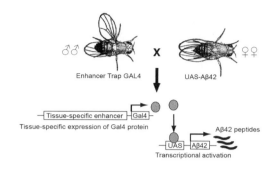

Fig. 1. Spatial targeting of transgene expression in *Drosophila*. GAL4/UAS system. *Driver lines* expressing the transcriptional activator GAL4 in a tissue-specific fashion are crossed to UAS-lines with genomic inserts of a target gene fused to five GAL4-binding sites arrayed in tandem (5 × UAS) (shown here as UAS-Aβ42). Adapted from Development, [70].

In contrast, we and others took more direct approach to overproduce human Aβ42 peptides in fly neurons. To express human Aβ40 or Aβ42 in the secretory pathway, Aβ peptide sequences were directly fused to a signal peptide at the N-terminus. This artificial construct produced intact Aβ40 or Aβ42 peptide in the fly brain [13, 15], which allowed us to evaluate the toxicity of Aβ40 and Aβ42 separately in the brain for the first time in an animal model. Both the Aβ40 and Aβ42 peptides accumulated during aging in the fly brain but only Aβ42 formed amyloid deposits [13]. This result is consistent with previous observations of amyloid deposits in AD or Down syndrome patient brains, which have shown that Aβ42 first accumulates in amyloid plaques in the brain parenchyma [72].

Using the Pavlovian olfactory classical conditioning assay [73, 74], we detected an age-dependent short-term memory defect in Aβ42 flies. High levels of Aβ40 expression also caused short-term memory defects, which suggest that an excess of Aβ40 was also toxic to synaptic plasticity. Sensory motor activity was not significantly affected in Aβ flies, suggesting that the observed defects were attributable to short-term memory defects.

At later stage, Aβ42, but not Aβ40, caused locomotor defects, premature death and an age-dependent neurodegeneration in the brains. An electron microscopic analysis of Aβ42 fly brains revealed that most of degenerating neurons showed necrotic cell death. Neither amyloid fibril, nor NFTs were detected in this model [13]. Using the same model, Finelli et al. showed that Aβ42 peptides expressed in fly eyes developed amyloid deposits and caused retinal degeneration [15].

Using a similar approach, Crowther et al. also established the transgenic flies expressing the wild-type and EOFAD-related Arctic mutant (E22G) Aβ42 peptides in *Drosophila* neural tissue, and showed that neuronal dysfunction and degeneration induced by the Arctic Aβ42 were more severe than that induced by wild-type Aβ42 [16].

*Comparison of the Aβ42 flies to mouse models of Alzheimer's disease*

There are many animal models of AD including nematodes, fly, mouse, rat, rabbit, canine, and non-human primates [75]. Interestingly, each model seems to recapitulate somewhat different aspects of AD. Most transgenic mice that overproduce Aβ successfully recapitulate several pathological lesions including extracellular amyloid deposits, behavioral deficits and memory defects, but they do not exhibit global neuronal loss [76, 77]. In contrast, progressive neurodegeneration was induced, and extensive intracellular accumulation of Aβ42 within neurons was observed in the transgenic Aβ42 flies [13, 15, 16, 78]. Human AD brains show significant accumulation of intraneuronal Aβ42 [79, 80], and, a few mouse models of AD have shown intraneuronal accumulation of Aβ42 that was associated with memory defects [48], significant neuron loss [81, 82], and axonopathy [83]. These observations suggest that the Aβ42 fly model recapitulate neuronal dysfunction and degeneration caused by intraneuronal Aβ42.

Aβ42 accumulation has been detected in intracellular compartments of neurons, including the endoplasmic reticulum (ER) [84], Golgi apparatus [84], endosomal-lysosomal system [85], multivesicular bodies [80], autophagolysosomes [86], mitochondria [87], and nucleus [88] in AD patients and/or in the brains of transgenic AD model mice. In addition, traumatic brain injury, which is a risk factor of AD, induces intraaxonal accumulation of Aβ in the brains of humans and animal models [89, 90]. Under normal conditions, the endosomal system is a major cellular compartment for Aβ generation [91]. However, several lines of evidence suggest that Aβ can be generated intracellularly from the early secretory pathway [92]. For example, Aβ42, but not Aβ40, can be produced in the ER [93-96]. In addition, a series of recent reports showed that defects in axonal transport induced intraaxonal generation of Aβ peptide [97, 98] and facilitated amyloid deposit formation in mouse models of AD [97]. Moreover, Aβ secreted into the extracellular space can be taken up by neurons and internalized into intracellular pools [99]. Therefore, under abnormal conditions, Aβ42 may be generated, retained, or recycled back to several intracellular locations, thus facilitating pathological intraneuronal accumulation, which may result in Aβ42-mediated disruption of various cellular processes [100].

Another unique feature of our fly model is that the toxicity of Aβ40 and Aβ42 can be dissected *in vivo*. Since most of the mouse models of AD overexpress AβPP, which produces a series of Aβ species including Aβ40 and Aβ42 after proteolysis, it is difficult to study the toxicity of either species individually. In contrast, we found that while both Aβ40 and Aβ42 could cause memory defects, only Aβ42 caused neurodegeneration with amyloid deposits. Recently established transgenic mice separately expressing Aβ40 or Aβ42 have shown that extracellular amyloid deposit formation is induced by Aβ42, but not by Aβ40. However, no neuronal loss has been observed in Aβ42 mice [101]. The Aβ42 fly model thus provides a unique tool for studying the mechanism underlying intraneuronal Aβ42 accumulation and neurodegeneration.

## IMPLICATIONS FOR ANTI-Aβ THERAPIES FROM FLY MODELS

An increase in Aβ42 levels above a given threshold in the brain is generally regarded to be the primary event in AD pathogenesis, and approaches to develop disease-modifying therapies have focused on lowering Aβ42 levels. In contrast to EOFAD, in which Aβ42 production and/or aggregation is enhanced as a result of genetic factors, the mechanisms by which Aβ42 reaches pathological levels in the brains of late-onset AD (LOAD) patients are not well understood. The steady state level of Aβ42 reflects the balance between production and clearance of Aβ42, and an imbalance of these activities could be sufficient to raise Aβ42 levels. Thus, reducing Aβ42 levels can be achieved either by attenuating production, or by facilitating degradation and/or clearance of Aβ42 from the brain [102, 103].

## Targeting β- and γ-secretases

In order to reduce the Aβ42 level in the brain, the prevention of amyloidogenic processing of AβPP by application of inhibitors of β-secretase (BACE) and γ-secretase is a viable option [104]. Both β- and γ-secretase inhibitor treatments ameliorate Aβ-induced toxicity, including semi-lethality and premature death, in triple transgenic flies expressing human AβPP, human β-secretase, and fly presenilin (dPsn) [14]. Recently, Rajendran et al. developed a β-secretase inhibitor peptide conjugated to a sterol moiety as a membrane anchor, which markedly increased the potency of the inhibitor [105]. Feeding sterol-linked β-secretase inhibitor to the triple transgenic flies increases the survival rates, indicating a reduction in toxicity in vivo.

Drosophila has been used as a functional in vivo system to search for genetic and pharmacological modifiers of γ-secretase activity. Drosophila has all the components of the γ-secretase complex, which is responsible for intramembranous cleavage of several transmembrane proteins, including AβPP, DCC, ErbB4, N-Cadherin, E-Cadherin, Notch, and Delta [63, 64, 106]. Psn is the enzymatic component of the γ-secretase complex, and EOFAD mutations in Psn affect the function of the γ-secretase complex in Drosophila [71]. Ubiquilin 1 (UBQLN1) is a Psn interactor that promotes the accumulation of PS1 and regulates its endoproteolysis [107]. The Drosophila ortholog of human UBQLN1, dUbqln, modifies the eye phenotype induced by overexpression of dPsn, indicating that the proteins have a functional interaction in vivo [108, 109]. Using a transgenic reporter of γ-secretase-mediated AβPP processing [110], Gross et al. showed that dUbqln and the Drosophila homolog of X11 [111], an AβPP interacting protein that modulates Aβ production [112], affects AβPP processing [113].

In Drosophila, Notch signaling is required during development and functions in many cell fate specification events. The modifier screens of Psn-dependent Notch-related phenotypes identified many genes, and some of these modifiers genetically interact with AβPP [114, 115]. Inhibitors of γ-secretase induce developmental defects in Drosophila that are remarkably similar to those caused by genetic reduction of Notch, indicating that the three-dimensional structure of the drug-binding site(s) in Drosophila γ-secretase is remarkably conserved vis-à-vis the same site(s) in the mammalian enzyme [116, 117]. In summary, these reports suggest that fly models may be a sensitive and functional in vivo system for the validation of newly developed and the prescreening of drug candidates for β- and γ-secretases.

## Enhancing Aβ-degrading enzymes

Several peptidases have been identified as candidate Aβ-degrading enzymes, including neprilysin (NEP) [118], insulin-degrading enzyme (IDE) [119], endothelin-converting enzyme 1 (ECE1) [120], cathepsin D [121], and serine protease-α2-macroglobulin [122]. Interestingly, the expression of NEP is reduced in both LOAD patient brains and normally aging brains [123, 124], suggesting that a reduction in Aβ-degrading activity may contribute to the onset and/or progression of LOAD.

Among the Aβ42 degrading enzymes, NEP has been identified as one of the major rate-limiting Aβ-degrading enzymes in the brain [118, 125]. A deficiency in NEP accelerates formation of extracellular amyloid deposits [126], amyloid angiopathy [126], synaptic dysfunctions [127], and memory defects [127] caused by human Aβ in transgenic mice. Transgenic [128, 129], viral [130, 131], or ex vivo [132] delivery of NEP to brains of AβPP transgenic mice reduces extracellular amyloid deposits, synaptic dysfunction, and premature death. Thus, activation of NEP in the brain could be a potential disease modifying therapy for AD by reducing extracellular Aβ.

In addition to extracellular Aβ, intraneuronal Aβ42 may contribute to AD pathogenesis. However, the protective effects of neuronal NEP expression on intraneuronal Aβ42 accumulation and neurodegeneration were not clear. To investigate the effects of NEP on intraneuronal Aβ42 accumulation and neuron loss induced by Aβ42 in vivo, we established transgenic flies expressing human NEP and examined the effects of NEP on Aβ42-induced toxicity in the Aβ42 fly model [133].

Expression of NEP significantly reduced the level of Aβ42 in both the detergent soluble and insoluble fractions of Aβ42 brains. Additionally, neuronal expression of NEP prevented formation of intraneuronal Thioflavin S-positive Aβ42 deposits in the cell body. Furthermore, NEP dramatically suppressed neuron loss induced by Aβ42. These protective effects were not observed in transgenic flies expressing an inactive mutant form of human NEP (NEP-mut) with the amino acid substitution E585V in the zinc-binding motif of NEP [134, 135], suggesting that the observed effects are attributable to the enzymatic activity of NEP [133]. These results are consistent with a recent study using AβPP transgenic mice [136].

Interestingly, neuronal expression of NEP significantly shortened the lifespan of flies, in part through reduced cAMP-responsive element binding protein (CREB) activity. NEP expression also caused axon degeneration in the aged fly brain. These results suggest that the transgenic fly model may also be used to detect potential side effects of a given manipulation [133].

*Targeting pyroglutamate-modification of Aβ*

N-terminally truncated and, in particular, pyroglutamate (pE)-modified Aβ42 peptides have been suggested to be important in the initiation of pathological cascades because of their abundance, resistance to proteolysis, rapid aggregation, and neurotoxicity. The pE-modification of Aβ42 is catalyzed by glutaminyl cyclase. A recent report showed that transgenic *Drosophila* expressing pE-modified Aβ42 (Aβ3(pE)-42), which were fed an inhibitor of glutaminyl cyclase, exhibited a reduced accumulation of Aβ3(pE)-42 [137].

*Targeting Aβ aggregation*

Aβ is a hydrophobic and self-aggregation prone peptide [138]. Aggregation of the Aβ42 peptide in the brain parenchyma is a hallmark of AD pathology, and the prevention of Aβ aggregation has been proposed as a therapeutic intervention in AD. The azo-dye Congo Red binds to β-amyloid and inhibits fibrillization of Aβ42 *in vitro*. Feeding Congo Red to flies resulted in a significant increase in longevity, reduction in plaque formation, and delay in neurodegeneration [16]. In a recent study, ligands were designed to specifically target aggregation of Aβ by binding and stabilizing the α-helical conformation of amino acids 13–26 in Aβ. These inhibitors reduced fibril formation of Aβ40 *in vitro*, and ameliorated Aβ42 toxicity to cells in culture and to hippocampal slice preparations. Additionally, feeding these inhibitors to flies expressing Aβ42 in the central nervous system prolonged lifespan, increased locomotor activity, and reduced neurodegeneration [139]. Also, expression of a small engineered binding protein (Z(Abeta3)) that binds with nanomolar affinity to Aβ, facilitated degradation of Aβ42 and abolished their neurotoxic effects in transgenic *Drosophila* expressing human Aβ42 [140].

In addition to mature amyloid fibrils, Aβ can form a variety of misfolded structures *in vitro*, including multiple monomer conformers, different types of prefibrillar assemblies including small oligomers [141], higher molecular weight complexes known as Aβ-derived diffusible ligands (ADDLs) [142], oligomers composed of 15–20 monomers [143], dodecameric oligomers (Aβ*56) [144], and protofibrils [145]. Some of these intermediate Aβ species can also be found in the cerebrospinal fluid and brains of AD patients [146, 147]. Since aggregation of Aβ42 in the brain parenchyma is a hallmark of AD pathology [148], the neurotoxicity of Aβ42 was initially correlated with the tendency of Aβ42 to form insoluble mature amyloid fibrils [149, 150]. However, recent experimental data indicates that the soluble prefibrillar assemblies cause more severe synaptic dysfunctions, cognitive defects, and neurodegeneration than do the mature fibrils [151].

A study using the *Drosophila* model in combination with computational predictions of Aβ42 aggregation has further demonstrated that there is a strong positive correlation between the magnitude of neurotoxicity, as manifested in locomotor deficits and reduced lifespan, and the propensity of Aβ42 to aggregate into protofibrils [152]. These results indicate that soluble prefibrillar assemblies, but not mature amyloid fibrils, of Aβ42 are the primary neurotoxic species, which is consistent with the observation that the level of soluble Aβ is better correlated with the severity of cognitive impairment than the density of insoluble amyloid deposits [153, 154].

*The complexity of Aβ42 aggregation and toxicity*

In addition to the "quantitative" changes in toxicity, the structural diversity of Aβ42 aggregates may also "qualitatively" influence the pathogenicity of Aβ42, as has been well established in the pathogenesis of prion disease mediated by various PrP species [155]. Deshpande et al. reported that both Aβ oligomers and ADDLs induced rapid and massive neuronal death, with ADDLs exhibiting their effects with a slightly slower time course. In contrast, Aβ fibrils induced progressive dystrophy and modest

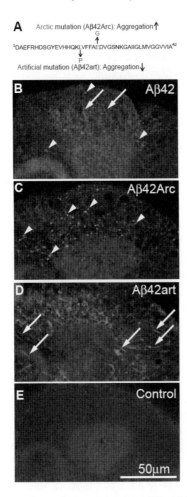

Fig. 2. Distribution and aggregation of Aβ42, Aβ42Arc, and Aβ42art peptides in fly brains. (A) Sequences of Aβ42, Aβ42Arc, and Aβ42art. (B-E) Thioflavin S staining of brains of 25 days-after-eclosion flies. Arrowheads and arrows indicate Thioflavin S-positive deposits and neurites, respectively. No signal was detected in the control (E). Adopted from [78]

cell death [156]. Recently, Chiang et al. reported that Aβ42 oligomers and larger aggregates have different effects on synaptic transmission and LTD at the neuromuscular junction of *Drosophila* larvae [157].

We investigated the correlation between levels of aggregation of Aβ42 and memory defects and neurodegeneration through genetic manipulation of Aβ42 aggregation in the *Drosophila* model. We have established transgenic fly lines carrying human Aβ42 mutants with differing tendencies to aggregate [78]. Human Aβ42 with the Arctic mutation (E22G, Aβ42Arc), which causes EOFAD [158], is more aggregation-prone than wild-type Aβ42 [159-162]. In contrast, an artificial mutation L17P (Aβ42art) suppresses amyloid fibril formation [150, 163, 164] (Figure 2). Aβ42Arc accumulated in the insoluble fraction more readily than Aβ42. In contrast, Aβ42art accumulated primarily in the soluble fraction and was greatly reduced in the insoluble fraction [78].

The severity of locomotor dysfunction and premature death positively correlated with the aggregtion tendencies of Aβ peptides. Surprisingly, however, Aβ42art caused earlier onset of memory defects than Aβ42. More remarkably, each Aβ induced qualitatively different pathologies. Aβ42Arc caused greater neuron loss than did Aβ42, while Aβ42art flies showed the strongest neurite degeneration. This pattern of degeneration coincides with the distribution of Thioflavin S-stained Aβ aggregates: Aβ42Arc formed large deposits in the cell body, Aβ42art accumulated preferentially in the neurites, while Aβ42 accumulated in both location (Figure 2) [78]. These results indicate that changes in the intrinsic aggregation tendencies of Aβ42 can modify the pathogenicity of Aβ42 *in vivo* and result in distinct pathological phenotypes in the fly brain.

Interestingly, although the accumulation patterns of Aβ42Arc (more aggregation-prone than Aβ42) and Aβ42art (less aggregation-prone than Aβ42) in the fly brain are dissimilar, the intracellular distribution of Aβ42 overlaps with the profiles of both Aβ42Arc and Aβ42art (Figure 2). This observation leads us to hypothesize that Aβ42Arc and Aβ42art may mimic particular conformations of Aβ42. Thus, a fraction of Aβ42 may induce neuron loss, while a different fraction of Aβ42 may induce neurite degeneration depending on the particular misfolded structure. As a result, the pathology observed in the Aβ42 fly brain may be the sum of qualitatively different pathologies induced by different misfolded structures of Aβ42 (Figure 3). These results also suggest that the partial prevention of Aβ42 amyloidgenesis by aggregation inhibitors may result in qualitative shifts in the pathogenic effects of Aβ42 [165]. Furthermore, the tendency of Aβ42, which is normally an unfolded polypeptide consisting primarily of a random-coil structure in the native, soluble state [166, 167], to aggregate may be affected by a combination of genetic [168], environmental [169], and aging factors [170], and the resultant Aβ42 conformers or species may contribute to the heterogeneous pathogenesis of AD [171].

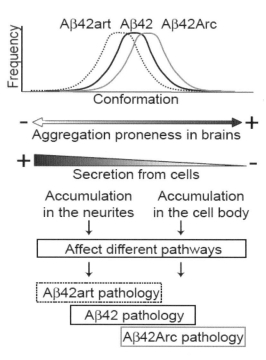

Fig 3. The hypothetical model that explains pathologies induced by Aβ42 in the fly brains. The aggregation propensity of Aβ42 in brains may be influenced by host factors, and a mixture of Aβ42 with different misfolding structures may be formed. The frequency of appearance of each Aβ42 conformation would distribute as shown in the top graph (Aβ42, depicted as black). Each misfolding structure has different secretion efficiency and cellular distribution pattern and/or may affect different molecular pathways. Thus, the sum of qualitatively different pathologies induced by different misfolding structures of Aβ42 would be observed as brain pathology in Aβ42 flies. An increase (Aβ42Arc, depicted as gray) or a decrease (Aβ42art, dotted line) of the intrinsic aggregation proneness of Aβ42 by mutations shifts the distribution of frequency of misfolding structures of Aβ, resulting in qualitative as well as quantitative shifts of the brain pathology

## INSIGHTS INTO PATHOMECHANISMS OF Aβ42-INDUCED TOXICITY FROM FLY AD MODELS

The power of genetics in *Drosophila* provides *in vivo* experimental platforms to examine whether gain or loss of function of a given gene can modify disease phenotypes in a time efficient and cost effective manner [27]. Since *Drosophila* generally does not possess as many redundant gene families as mammals, studying the consequence of a single gene disruption is easier in the fly. This enables us to systematically test hypotheses derived from biochemical or histopathological analysis of human disease tissues, genetic association studies performed on particular disease phenotypes, or findings in *in vitro* model systems. In addition, one of the most powerful advantages of using a *Drosophila* model is the ability to conduct large scale, genome-wide forward genetic screens to identify novel genes and pathways that modify given disease phenotypes. The findings from this unbiased approach have great potential to open novel research areas. As summarized in this section, several groups have demonstrated that this system can lead to powerful insights into the pathophysiological mechanisms underlying Aβ42 toxicity.

### Oxidative stress

Oxidative stress occurs early in the progression of Alzheimer disease [172]. Rival et al. conducted a genome wide gene-expression analysis and complementary genetic screen in Aβ42 flies. Microarray analysis identified changes in the expression of oxidative stress-related genes induced by Aβ42 expression, and a subsequent genetic screen confirmed the importance of oxidative stress. The iron-binding protein ferritin and the $H_2O_2$ scavenger catalase are the most potent suppressors of wild-type and Arctic (E22G) Aβ42 toxicity. Likewise, treatment with the iron-binding compound clioquinol increased the lifespan of flies expressing Arctic Aβ42 [173].

### Abnormality in mitochondrial distribution

Mitochondrial dysfunctions are implicated in AD brains [174-177], and recent studies have shown that mitochondria are reduced in neuronal processes in AD neurons [86, 174-181]. We recently reported that Aβ42 induces mitochondrial mislocalization, which contributes to Aβ42-induced neuronal dysfunction in a transgenic *Drosophila* model. In the Aβ42 fly brain, mitochondria were reduced in axons and dendrites, and accumulated in the somata without severe mitochondrial damage or neurodegeneration (Figure 4A) [182]. In contrast, organization of microtubule or global axonal transport was not significantly altered at this stage [182].

Mitochondria are linked to motors by the mitochondrial membrane GTPase Miro, which is linked to kinesin by milton to allow transport in axons and dendrites [183]. Null mutations in *milton* and *Miro* have been reported to disrupt axonal and dendritic transport of mitochondria in neurons [184, 185]. Using RNAi-mediated knockdown of *milton* and a heterozygous *Miro* mutation, Aβ42-induced locomotor deficits was enhanced by genetic reductions of mitochondrial transport (Figure 4B). Importantly, perturbations in mitochondrial transport in neurons induced late-onset behavioral deficits. These results indicate that mitochondrial mislocalization contributes to Aβ42-induced behavioral deficits [182].

Using a different transgenic *Drosophila* model [16], Zhao et al. [186] have performed extensive time course analysis of motor neurons expressing Aβ42, and found depletion of presynaptic mitochondria, slowdown of bi-directional transports of axonal mitochondria, decreased synaptic vesicles, increased large vacuoles, and elevated synaptic fatigue. The depletion of presynaptic mitochondria was the earliest detected phenotype and was not caused by the change in axonal transport of mitochondria. Moreover, axonal mitochondria exhibited a dramatic reduction in number but a significant

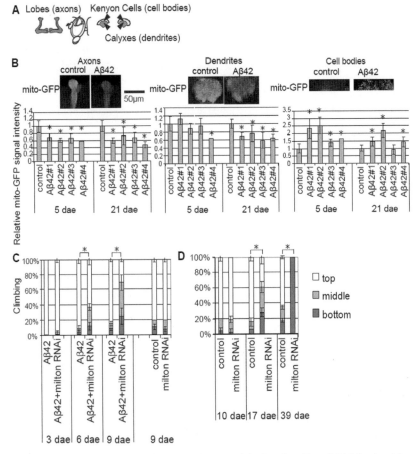

Fig. 4. Mitochondrial mislocalization underlies Aβ42-induced neuronal dysfunction. (A and B) Mitochondria are reduced in the axons and dendrites in Aβ42 fly brain. (A) A schematic view of the mushroom body. (B) Signal intensities of mito-GFP in axon bundle tips, dendrites, and cell bodies of the mushroom body structure in control and Aβ42 flies at 5 days after eclosion (dae) and 21 dae. Ratios relative to control are shown (mean ± SD, n=6-10; *, p<0.05, Student's t-test). Representative images at 21 dae are shown at the top. (C) Aβ42-induced locomotor deficits are enhanced by genetic reductions of mitochondrial transport. Enhancement of Aβ42-induced locomotor defects by neuronal knockdown of milton was detected by climbing assay [13]. In this assay, flies were placed in an empty plastic vial and tapped to the bottom. The number of flies at the top, middle, or bottom of the vial was scored after 10 seconds. The average percentage of flies at the top (white), middle (light gray), or bottom (dark gray) of assay vials is shown (mean±SD, n = 5). Asterisks indicate the significant difference in the percentage of the flies stayed at the bottom (p<0.05, Student's t-test). (D) Disruption of mitochondrial transport causes age-dependent behavioral deficits. Age-dependent locomotor defects in the flies with a neuronal knockdown of milton using UAS-milton RNAi transgenic flies. Asterisks indicate the significant difference in the percentage of the flies stayed at the bottom (mean ± SD, n=5, p<0.05, Student's t-test). Adopted from [182].

increase in size in aged Aβ42-expressing flies. From these results, they propose that Aβ42 induces a global depletion of mitochondria in the axons due to an impairment of mitochondrial fission.

*Autophagy*

The macroautophagy (autophagy) pathway is thought to be involved in a variety of neurodegenerative diseases, including AD [86]. Recent study in a *Drosophila* model showed that Aβ42 caused an extensive accumulation of autophagic vesicles that become increasingly dysfunctional with age [187]. Aβ42 impairs the degradative function, as well as the structural integrity, of post-lysosomal autophagic vesicles, which triggers a neurodegenerative cascade that can be enhanced by autophagy activation or partially rescued by autophagy inhibition. Compromise and leakage from post-lysosomal vesicles result in cytosolic acidification, additional damage to membranes and organelles, and erosive destruction of cytoplasm leading to eventual neuron death. They propose that neuronal autophagy initially appears to play a pro-survival role that changes in an age-dependent way to a pro-death role in the context of Aβ42 expression [187].

*Copper homeostasis and FK506 binding proteins*

Altered copper homeostasis has clearly been shown to have a role in Alzheimer's disease [188]. FK506 binding proteins (FKBPs) are prolyl-isomerases (PPIases), which have diverse cellular functions. Among FKBPs, FKBP52 is abundantly expressed in neurons and plays a role in the regulation of intracellular copper metabolism by interacting with a copper transporter Atox1 [189]. Sanokawa-Akakura et al. showed that the Aβ42 pathology was enhanced by feeding copper and by *Atox1* mutation. FKBP52 overexpression suppressed Aβ42-induced neurodegeneration and increased lifespan in Aβ42 flies, whereas loss of function of FKBP52 exacerbated these Aβ42 phenotypes. These studies suggest a novel role for FKBP52 in modulating toxicity of Aβ42 peptides through copper homeostasis [190].

*Retromer sorting pathway*

Deficiencies in the retromer sorting pathway have been linked to late-onset Alzheimer's disease [91]. Muhammad et al. tested the effects of retromer deficiency on memory formation, synaptic functions and generation of Aβ peptides using mouse and *Drosophila* models. They found that retromer-deficient mice develop hippocampal-dependent memory and synaptic dysfunction, which was associated with elevations in mouse endogenous Aβ peptide. Using transgenic flies expressing human AβPP and BACE, they found that retromer-deficiency enhanced neuronal loss and formation of Aβ aggregates in the flies expressing human AβPP and human BACE [191]. These results suggest that retromer deficiency observed in late-onset AD may contribute to the disease pathogenesis.

*PI3 kinase signaling*

Multiple intracellular signals are altered in AD brain tissues, including the PI3K/Akt pathway. Chiang et al. recently demonstrated that Aβ42 expression induced PI3K hyperactivity, and genetic silencing or pharmacological inhibition of PI3K functions suppressed Aβ42-induced memory loss, as well as Aβ42 accumulation, in transgenic *Drosophila*. Their results suggest that Aβ42 stimulates PI3K, which in turn causes memory loss in association with an increase in accumulation of Aβ42 aggregates [192].

*cAMP/Protein kinase A signaling*

Impaired regulation of the cAMP/PKA pathway has been reported in the brains of AD patients. We recently demonstrated that Aβ42-induced behavioral defects were modulated by cAMP levels and PKA activity [182]. Levels of putative PKA substrate phosphoproteins were reduced in the Abeta42 fly brains. Perturbations in mitochondrial transport in neurons disrupted PKA signaling. Since Aβ42 induces mitochondrial mislocalization, these results suggest that a disruption of PKA signaling is one of the mechanisms whereby mitochondrial mislocalization contributes to Aβ42-induced neuronal dysfunction [182].

*An unbiased genetic screen in Aβ42 flies*

Retinal toxicity has been widely used in genetic screens in *Drosophila* models to identify modifiers of neurodegenerative disease. Expression of Aβ42 in the fly eye results in a visible reduction in eye size and a roughened eye surface due to the death of intrinsic photoreceptors and supporting cells [193]. Using eye degeneration as the read-out phenotype, Cao et al. have tested the effects of 1,963 EP transposon insertions in the fly genome. The EP transposon has a GAL4 activated promoter and, depending upon the site and orientation of the insertion, the transposon will upregulate or downregulate gene activity [194]. Using this approach, 23 modifiers were identified, including genes involved in the secretory pathway (e.g., the human orthologs of carboxypeptidase D and the AP3 subunit δ-1), cholesterol homeostasis (AMP kinase γ-subunit), copper transport (ATP7), and ubiquitin/proteolysis (ubiquitin-conjugating enzyme E2Q and NEP 2), which have been implicated in AD pathogenesis. This screen also identified genes related to the regulation of chromatin structure as a potential pathway mediating Aβ42 toxicity. Loss-of-function mutations of several components of the Sin3A complex, which regulates transcription and chromatin remodeling through interactions with other transcription factors, including SAP130, and the histone deacetylases HDAC1 and 4, all enhanced the rough eye phenotype in Aβ42 flies [193]. Another screen for loss-of-function mutations that enhance or suppress the eye degeneration phenotype induced by Aβ42 identified that the Toll/NFκB signaling pathway as mediating human Aβ42 toxicity, suggesting the involvement of innate immunity/inflammatory pathway components in Aβ42 toxicity [195].

## CONCLUSION

*Drosophila* have proven to be excellent models for human neurodegenerative diseases as a powerful tool for both hypothesis-testing approaches and non-biased genome-wide screens [25-31]. The fly models of neurodegenerative disease can be used to systematically evaluate the effects of known genetic and environmental risk factors, and candidates of susceptibility genes identified by genetic association studies with humans or that revealed by functional genomics with mouse models. Also, the genes discovered in the forward genetic approach in fly models may help reveal disease susceptibility genes in humans. Furthermore, because of the lack of a stringent blood–brain barrier in the fly, which allows compounds to easily gain access to the nervous system, fly models are also excellent *in vivo* model for the testing and screening of therapeutic compounds [196]. In summary, *Drosophila* is a powerful genetic model system for the study of complex human diseases such as AD. Along with the genetic tools that are currently available, as well as those constantly being developed, we believe that the *Drosophila* system will make a significant a contribution to AD research, as it has in many other biological contexts.

## ACKNOWLEDGEMENT

Our works have been supported by start-up funds from the Farber Institute for Neurosciences, a pilot research grant from the Thomas Jefferson University, and grants from Gilbert Foundation/ the American Federation for Aging Research, the Alzheimer's Association (NIRG-08-91985) and the National Institute of Health (R01AG032279-A1).

## References

[1] Reiter LT, Potocki L, Chien S, Gribskov M, Bier E (2001) A systematic analysis of human disease-associated gene sequences in Drosophila melanogaster. *Genome Res* **11**, 1114-1125.

[2] Fortini ME, Skupski MP, Boguski MS, Hariharan IK (2000) A survey of human disease gene counterparts in the Drosophila genome. *J Cell Biol* **150**, F23-30.

[3] Soto C (2003) Unfolding the role of protein misfolding in neurodegenerative diseases. *Nat Rev Neurosci* **4**, 49-60.

[4] Raeber AJ, Muramoto T, Kornberg TB, Prusiner SB (1995) Expression and targeting of Syrian hamster prion protein induced by heat shock in transgenic Drosophila melanogaster. *Mech Dev* **51**, 317-327.

[5] Deleault NR, Dolph PJ, Feany MB, Cook ME, Nishina K, Harris DA, Supattapone S (2003) Post-transcriptional suppression of pathogenic prion protein expression in Drosophila neurons. *J Neurochem* **85**, 1614-1623.

[6] Warrick JM, Paulson HL, Gray-Board GL, Bui QT, Fischbeck KH, Pittman RN, Bonini NM (1998) Expanded polyglutamine protein forms nuclear inclusions and causes neural degeneration in Drosophila. *Cell* **93**, 939-949.

[7] Jackson GR, Salecker I, Dong X, Yao X, Arnheim N, Faber PW, MacDonald ME, Zipursky SL (1998) Polyglutamine-expanded human huntingtin transgenes induce degeneration of Drosophila photoreceptor neurons. *Neuron* **21**, 633-642.

[8] Feany MB, Bender WW (2000) A Drosophila model of Parkinson's disease. *Nature* **404**, 394-398.

[9] Wittmann CW, Wszolek MF, Shulman JM, Salvaterra PM, Lewis J, Hutton M, Feany MB (2001) Tauopathy in Drosophila: neurodegeneration without neurofibrillary tangles. *Science* **293**, 711-714.

[10] Jackson GR, Wiedau-Pazos M, Sang TK, Wagle N, Brown CA, Massachi S, Geschwind DH (2002) Human wild-type tau interacts with wingless pathway components and produces neurofibrillary pathology in Drosophila. *Neuron* **34**, 509-519.

[11] Ratnaparkhi A, Lawless GM, Schweizer FE, Golshani P, Jackson GR (2008) A Drosophila model of ALS: human ALS-associated mutation in VAP33A suggests a dominant negative mechanism. *PLoS ONE* **3**, e2334.

[12] Watson MR, Lagow RD, Xu K, Zhang B, Bonini NM (2008) A Drosophila model for amyotrophic lateral sclerosis reveals motor neuron damage by human SOD1. *J Biol Chem*.

[13] Iijima K, Liu HP, Chiang AS, Hearn SA, Konsolaki M, Zhong Y (2004) Dissecting the pathological effects of human Abeta40 and Abeta42 in Drosophila: a potential model for Alzheimer's disease. *Proc Natl Acad Sci U S A* **101**, 6623-6628.

[14] Greeve I, Kretzschmar D, Tschape JA, Beyn A, Brellinger C, Schweizer M, Nitsch RM, Reifegerste R (2004) Age-dependent neurodegeneration and Alzheimer-amyloid plaque formation in transgenic Drosophila. *J Neurosci* **24**, 3899-3906.

[15] Finelli A, Kelkar A, Song HJ, Yang H, Konsolaki M (2004) A model for studying Alzheimer's Abeta42-induced toxicity in Drosophila melanogaster. *Mol Cell Neurosci* **26**, 365-375.

[16] Crowther DC, Kinghorn KJ, Miranda E, Page R, Curry JA, Duthie FA, Gubb DC, Lomas DA (2005) Intraneuronal Abeta, non-amyloid aggregates and neurodegeneration in a Drosophila model of Alzheimer's disease. *Neuroscience* **132**, 123-135.

[17] Li Y, Ray P, Rao EJ, Shi C, Guo W, Chen X, Woodruff EA, 3rd, Fushimi K, Wu JY (2010) A Drosophila model for TDP-43 proteinopathy. *Proc Natl Acad Sci U S A* **107**, 3169-3174.

[18] Feiguin F, Godena VK, Romano G, D'Ambrogio A, Klima R, Baralle FE (2009) Depletion of TDP-43 affects Drosophila motoneurons terminal synapsis and locomotive behavior. *FEBS Lett* **583**, 1586-1592.

[19] Lu Y, Ferris J, Gao FB (2009) Frontotemporal dementia and amyotrophic lateral sclerosis-associated disease protein TDP-43 promotes dendritic branching. *Mol Brain* **2**, 30.

[20] Hanson KA, Kim SH, Wassarman DA, Tibbetts RS (2010) Ubiquilin modifies toxicity of the 43 kilodalton TAR-DNA binding protein (TDP-43) in a Drosophila model of amyotrophic lateral sclerosis (ALS). *J Biol Chem*.

[21] Tsuda H, Han SM, Yang Y, Tong C, Lin YQ, Mohan K, Haueter C, Zoghbi A, Harati Y, Kwan J, Miller MA, Bellen HJ (2008) The amyotrophic lateral sclerosis 8 protein VAPB is cleaved, secreted, and acts as a ligand for Eph receptors. *Cell* **133**, 963-977.

[22] Adams MD, Celniker SE, Holt RA, Evans CA, Gocayne JD, Amanatides PG, Scherer SE, Li PW, Hoskins RA, Galle RF, George RA, Lewis SE, Richards S, Ashburner M, Henderson SN, Sutton GG, Wortman JR, Yandell MD, Zhang Q, Chen LX, Brandon RC, Rogers YH, Blazej RG, Champe M, Pfeiffer BD, Wan KH, Doyle C, Baxter EG, Helt G, Nelson CR, Gabor GL, Abril JF, Agbayani A, An HJ, Andrews-Pfannkoch C, Baldwin D, Ballew RM, Basu A, Baxendale J, Bayraktaroglu L, Beasley EM, Beeson KY, Benos PV, Berman BP, Bhandari D, Bolshakov S, Borkova D, Botchan MR, Bouck J, Brokstein P, Brottier P, Burtis KC, Busam DA, Butler H, Cadieu E, Center A, Chandra I, Cherry JM, Cawley S, Dahlke C, Davenport LB, Davies P, de Pablos B, Delcher A, Deng Z, Mays AD, Dew I, Dietz SM, Dodson K, Doup LE, Downes M, Dugan-Rocha S, Dunkov BC, Dunn P, Durbin KJ, Evangelista CC, Ferraz C, Ferriera S, Fleischmann W, Fosler C, Gabrielian AE, Garg NS, Gelbart WM, Glasser K, Glodek A, Gong F, Gorrell JH, Gu Z, Guan P, Harris M, Harris NL, Harvey D, Heiman TJ, Hernandez JR, Houck J, Hostin D, Houston KA, Howland TJ, Wei MH, Ibegwam C, Jalali M, Kalush F, Karpen GH, Ke Z, Kennison JA, Ketchum KA, Kimmel BE, Kodira CD, Kraft C, Kravitz S, Kulp D, Lai Z, Lasko P, Lei Y, Levitsky AA, Li J, Li Z, Liang Y, Lin X, Liu X, Mattei B, McIntosh TC, McLeod MP, McPherson D, Merkulov G, Milshina NV, Mobarry C, Morris J, Moshrefi A, Mount SM, Moy M, Murphy B, Murphy L, Muzny DM, Nelson DL, Nelson DR, Nelson KA, Nixon K, Nusskern DR, Pacleb JM, Palazzolo M, Pittman GS, Pan S, Pollard J, Puri V, Reese MG, Reinert K, Remington K, Saunders RD, Scheeler F, Shen H, Shue BC, Siden-Kiamos I, Simpson M, Skupski MP, Smith T, Spier E, Spradling AC, Stapleton M, Strong R, Sun E, Svirskas R, Tector C, Turner R, Venter E, Wang AH, Wang X, Wang ZY, Wassarman DA, Weinstock GM, Weissenbach J, Williams SM, Woodage T, Worley KC, Wu D, Yang S, Yao QA, Ye J, Yeh RF, Zaveri JS, Zhan M, Zhang G, Zhao Q, Zheng L, Zheng XH, Zhong FN, Zhong W, Zhou X, Zhu S, Zhu X, Smith HO, Gibbs RA, Myers EW, Rubin GM, Venter JC (2000) The genome sequence of Drosophila melanogaster. *Science* **287**, 2185-2195.

[23] Ueda R (2001) Rnai: a new technology in the postgenomic sequencing era. *J Neurogenet* **15**, 193-204.

[24] Dietzl G, Chen D, Schnorrer F, Su KC, Barinova Y, Fellner M, Gasser B, Kinsey K, Oppel S, Scheiblauer S, Couto A, Marra V, Keleman K, Dickson BJ (2007) A genome-wide transgenic RNAi library for conditional gene inactivation in Drosophila. *Nature* **448**, 151-156.

[25] Bilen J, Bonini NM (2005) Drosophila as a model for human neurodegenerative disease. *Annu Rev Genet* **39**, 153-171.

[26] Sang TK, Jackson GR (2005) Drosophila models of neurodegenerative disease. *NeuroRx* **2**, 438-446.

[27] Shulman JM, Shulman LM, Weiner WJ, Feany MB (2003) From fruit fly to bedside: translating lessons from Drosophila models of neurodegenerative disease. *Curr Opin Neurol* **16**, 443-449.

[28] Marsh JL, Thompson LM (2004) Can flies help humans treat neurodegenerative diseases? *Bioessays* **26**, 485-496.

[29] Fortini ME, Bonini NM (2000) Modeling human neurodegenerative diseases in Drosophila: on a wing and a prayer. *Trends Genet* **16**, 161-167.

[30] Lu B, Vogel H (2009) Drosophila models of neurodegenerative diseases. *Annu Rev Pathol* **4**, 315-342.

[31] Khurana V (2008) Modeling Tauopathy in the fruit fly Drosophila melanogaster. *J Alzheimers Dis* **15**, 541-553.

[32] Cummings JL (2003) *The neuropsychiatry of Alzheimer's disease and other dementias*, Martin Dunitz, London.

[33] Selkoe DJ (2002) Alzheimer's disease is a synaptic failure. *Science* **298**, 789-791.

[34] Selkoe DJ (2001) Alzheimer's disease: genes, proteins, and therapy. *Physiol Rev* **81**, 741-766.

[35] Glenner GG, Wong CW (1984) Alzheimer's disease and Down's syndrome: sharing of a unique cerebrovascular amyloid fibril protein. *Biochem Biophys Res Commun* **122**, 1131-1135.

[36] Masters CL, Simms G, Weinman NA, Multhaup G, McDonald BL, Beyreuther K (1985) Amyloid plaque core protein in Alzheimer disease and Down syndrome. *Proc Natl Acad Sci U S A* **82**, 4245-4249.

[37] Lee VM, Balin BJ, Otvos L, Jr., Trojanowski JQ (1991) A68: a major subunit of paired helical filaments and derivatized forms of normal Tau. *Science* **251**, 675-678.

[38] Bertram L, Tanzi RE (2005) The genetic epidemiology of neurodegenerative disease. *J Clin Invest* **115**, 1449-1457.

[39] Gandy S (2005) The role of cerebral amyloid beta accumulation in common forms of Alzheimer disease. *J Clin Invest* **115**, 1121-1129.

[40] Tanzi RE, Bertram L (2005) Twenty years of the Alzheimer's disease amyloid hypothesis: a genetic perspective. *Cell* **120**, 545-555.

[41] Hardy J, Selkoe DJ (2002) The amyloid hypothesis of Alzheimer's disease: progress and problems on the road to therapeutics. *Science* **297**, 353-356.

[42] Myers AJ, Kaleem M, Marlowe L, Pittman AM, Lees AJ, Fung HC, Duckworth J, Leung D, Gibson A, Morris CM, de Silva R, Hardy J (2005) The H1c haplotype at the MAPT locus is associated with Alzheimer's disease. *Hum Mol Genet* **14**, 2399-2404.

[43] Kauwe JS, Cruchaga C, Mayo K, Fenoglio C, Bertelsen S, Nowotny P, Galimberti D, Scarpini E, Morris JC, Fagan AM, Holtzman DM, Goate AM (2008) Variation in MAPT is associated with cerebrospinal fluid tau levels in the presence of amyloid-beta deposition. *Proc Natl Acad Sci U S A* **105**, 8050-8054.

[44] Nicoll JA, Wilkinson D, Holmes C, Steart P, Markham H, Weller RO (2003) Neuropathology of human Alzheimer disease after immunization with amyloid-beta peptide: a case report. *Nat Med* **9**, 448-452.

[45] Ferrer I, Boada Rovira M, Sanchez Guerra ML, Rey MJ, Costa-Jussa F (2004) Neuropathology and pathogenesis of encephalitis following amyloid-beta immunization in Alzheimer's disease. *Brain Pathol* **14**, 11-20.

[46] Lewis J, Dickson DW, Lin WL, Chisholm L, Corral A, Jones G, Yen SH, Sahara N, Skipper L, Yager D, Eckman C, Hardy J, Hutton M, McGowan E (2001) Enhanced neurofibrillary degeneration in transgenic mice expressing mutant tau and APP. *Science* **293**, 1487-1491.

[47] Gotz J, Chen F, van Dorpe J, Nitsch RM (2001) Formation of neurofibrillary tangles in P301l tau transgenic mice induced by Abeta 42 fibrils. *Science* **293**, 1491-1495.

[48] Oddo S, Caccamo A, Shepherd JD, Murphy MP, Golde TE, Kayed R, Metherate R, Mattson MP, Akbari Y, LaFerla FM (2003) Triple-transgenic model of Alzheimer's disease with plaques and tangles: intracellular Abeta and synaptic dysfunction. *Neuron* **39**, 409-421.

[49] Oddo S, Billings L, Kesslak JP, Cribbs DH, LaFerla FM (2004) Abeta immunotherapy leads to clearance of early, but not late, hyperphosphorylated tau aggregates via the proteasome. *Neuron* **43**, 321-332.

[50] Oddo S, Caccamo A, Smith IF, Green KN, LaFerla FM (2006) A dynamic relationship between intracellular and extracellular pools of Abeta. *Am J Pathol* **168**, 184-194.

[51] Oddo S, Caccamo A, Tran L, Lambert MP, Glabe CG, Klein WL, LaFerla FM (2006) Temporal Profile of Amyloid-beta (Abeta) Oligomerization in an in Vivo Model of Alzheimer Disease: A LINK BETWEEN Abeta AND TAU PATHOLOGY. *J. Biol. Chem.* **281**, 1599-1604.

[52] Oddo S, Vasilevko V, Caccamo A, Kitazawa M, Cribbs DH, LaFerla FM (2006) Reduction of soluble Abeta and tau, but not soluble Abeta alone, ameliorates cognitive decline in transgenic mice with plaques and tangles. *J Biol Chem* **281**, 39413-39423.

[53] Rapoport M, Dawson HN, Binder LI, Vitek MP, Ferreira A (2002) Tau is essential to beta -amyloid-induced neurotoxicity. *Proc Natl Acad Sci U S A* **99**, 6364-6369.

[54] King ME, Kan HM, Baas PW, Erisir A, Glabe CG, Bloom GS (2006) Tau-dependent microtubule disassembly initiated by prefibrillar beta-amyloid. *J Cell Biol* **175**, 541-546.

[55] Roberson ED, Scearce-Levie K, Palop JJ, Yan F, Cheng IH, Wu T, Gerstein H, Yu GQ, Mucke L (2007) Reducing endogenous tau ameliorates amyloid beta-induced deficits in an Alzheimer's disease mouse model. *Science* **316**, 750-754.

[56] Ballatore C, Lee VM, Trojanowski JQ (2007) Tau-mediated neurodegeneration in Alzheimer's disease and related disorders. *Nat Rev Neurosci* **8**, 663-672.

[57] Luo L, Tully T, White K (1992) Human amyloid precursor protein ameliorates behavioral deficit of flies deleted for Appl gene. *Neuron* **9**, 595-605.

[58] Torroja L, Packard M, Gorczyca M, White K, Budnik V (1999) The Drosophila beta-amyloid precursor protein homolog promotes synapse differentiation at the neuromuscular junction. *J Neurosci* **19**, 7793-7803.

[59] Ashley J, Packard M, Ataman B, Budnik V (2005) Fasciclin II signals new synapse formation through amyloid precursor protein and the scaffolding protein dX11/Mint. *J Neurosci* **25**, 5943-5955.

[60] Leyssen M, Ayaz D, Hebert SS, Reeve S, De Strooper B, Hassan BA (2005) Amyloid precursor protein promotes post-developmental neurite arborization in the Drosophila brain. *EMBO J* **24**, 2944-2955.

[61] Torroja L, Chu H, Kotovsky I, White K (1999) Neuronal overexpression of APPL, the Drosophila homologue of the amyloid precursor protein (APP), disrupts axonal transport. *Curr Biol* **9**, 489-492.

[62] Gunawardena S, Goldstein LS (2001) Disruption of axonal transport and neuronal viability by amyloid precursor protein mutations in Drosophila. *Neuron* **32**, 389-401.

[63] Takasugi N, Tomita T, Hayashi I, Tsuruoka M, Niimura M, Takahashi Y, Thinakaran G, Iwatsubo T (2003) The role of presenilin cofactors in the gamma-secretase complex. *Nature* **422**, 438-441.

[64] Periz G, Fortini ME (2004) Functional reconstitution of gamma-secretase through coordinated expression of pre-

[65] Fossgreen A, Bruckner B, Czech C, Masters CL, Beyreuther K, Paro R (1998) Transgenic Drosophila expressing human amyloid precursor protein show gamma-secretase activity and a blistered-wing phenotype. *Proc Natl Acad Sci U S A* **95**, 13703-13708.

[66] Yagi Y, Tomita S, Nakamura M, Suzuki T (2000) Overexpression of human amyloid precursor protein in Drosophila. *Mol Cell Biol Res Commun* **4**, 43-49.

[67] Carmine-Simmen K, Proctor T, Tschape J, Poeck B, Triphan T, Strauss R, Kretzschmar D (2009) Neurotoxic effects induced by the Drosophila amyloid-beta peptide suggest a conserved toxic function. *Neurobiol Dis* **33**, 274-281.

[68] Allinson TM, Parkin ET, Turner AJ, Hooper NM (2003) ADAMs family members as amyloid precursor protein alpha-secretases. *J Neurosci Res* **74**, 342-352.

[69] Rooke J, Pan D, Xu T, Rubin GM (1996) KUZ, a conserved metalloprotease-disintegrin protein with two roles in Drosophila neurogenesis. *Science* **273**, 1227-1231.

[70] Brand AH, Perrimon N (1993) Targeted gene expression as a means of altering cell fates and generating dominant phenotypes. *Development* **118**, 401-415.

[71] Ye Y, Fortini ME (1999) Apoptotic activities of wild-type and Alzheimer's disease-related mutant presenilins in Drosophila melanogaster. *J Cell Biol* **146**, 1351-1364.

[72] Iwatsubo T, Odaka A, Suzuki N, Mizusawa H, Nukina N, Ihara Y (1994) Visualization of A beta 42(43) and A beta 40 in senile plaques with end-specific A beta monoclonals: evidence that an initially deposited species is A beta 42(43). *Neuron* **13**, 45-53.

[73] Quinn WG, Harris WA, Benzer S (1974) Conditioned behavior in Drosophila melanogaster. *Proc Natl Acad Sci U S A* **71**, 708-712.

[74] Tully T, Quinn WG (1985) Classical conditioning and retention in normal and mutant Drosophila melanogaster. *J Comp Physiol [A]* **157**, 263-277.

[75] Woodruff-Pak DS (2008) Animal models of Alzheimer's disease: therapeutic implications. *J Alzheimers Dis* **15**, 507-521.

[76] Gotz J, Ittner LM (2008) Animal models of Alzheimer's disease and frontotemporal dementia. *Nat Rev Neurosci* **9**, 532-544.

[77] McGowan E, Eriksen J, Hutton M (2006) A decade of modeling Alzheimer's disease in transgenic mice. *Trends Genet*.

[78] Iijima K, Chiang HC, Hearn SA, Hakker I, Gatt A, Shenton C, Granger L, Leung A, Iijima-Ando K, Zhong Y (2008) Abeta42 mutants with different aggregation profiles induce distinct pathologies in Drosophila. *PLoS ONE* **3**, e1703.

[79] Gouras GK, Tsai J, Naslund J, Vincent B, Edgar M, Checler F, Greenfield JP, Haroutunian V, Buxbaum JD, Xu H, Greengard P, Relkin NR (2000) Intraneuronal Abeta42 accumulation in human brain. *Am J Pathol* **156**, 15-20.

[80] Takahashi RH, Milner TA, Li F, Nam EE, Edgar MA, Yamaguchi H, Beal MF, Xu H, Greengard P, Gouras GK (2002) Intraneuronal Alzheimer abeta42 accumulates in multivesicular bodies and is associated with synaptic pathology. *Am J Pathol* **161**, 1869-1879.

[81] Casas C, Sergeant N, Itier JM, Blanchard V, Wirths O, van der Kolk N, Vingtdeux V, van de Steeg E, Ret G, Canton T, Drobecq H, Clark A, Bonici B, Delacourte A, Benavides J, Schmitz C, Tremp G, Bayer TA, Benoit P, Pradier L (2004) Massive CA1/2 neuronal loss with intraneuronal and N-terminal truncated Abeta42 accumulation in a novel Alzheimer transgenic model. *Am J Pathol* **165**, 1289-1300.

[82] Oakley H, Cole SL, Logan S, Maus E, Shao P, Craft J, Guillozet-Bongaarts A, Ohno M, Disterhoft J, Van Eldik L, Berry R, Vassar R (2006) Intraneuronal beta-amyloid aggregates, neurodegeneration, and neuron loss in transgenic mice with five familial Alzheimer's disease mutations: potential factors in amyloid plaque formation. *J Neurosci* **26**, 10129-10140.

[83] Wirths O, Weis J, Szczygielski J, Multhaup G, Bayer TA (2006) Axonopathy in an APP/PS1 transgenic mouse model of Alzheimer's disease. *Acta Neuropathol (Berl)* **111**, 312-319.

[84] Rodrigo J, Fernandez-Vizarra P, Castro-Blanco S, Bentura ML, Nieto M, Gomez-Isla T, Martinez-Murillo R, MartInez A, Serrano J, Fernandez AP (2004) Nitric oxide in the cerebral cortex of amyloid-precursor protein (SW) Tg2576 transgenic mice. *Neuroscience* **128**, 73-89.

[85] Cataldo AM, Petanceska S, Terio NB, Peterhoff CM, Durham R, Mercken M, Mehta PD, Buxbaum J, Haroutunian V, Nixon RA (2004) Abeta localization in abnormal endosomes: association with earliest Abeta elevations in AD and Down syndrome. *Neurobiol Aging* **25**, 1263-1272.

[86] Nixon RA, Wegiel J, Kumar A, Yu WH, Peterhoff C, Cataldo A, Cuervo AM (2005) Extensive involvement of autophagy in Alzheimer disease: an immuno-electron microscopy study. *J Neuropathol Exp Neurol* **64**, 113-122.

[87] Lustbader JW, Cirilli M, Lin C, Xu HW, Takuma K, Wang N, Caspersen C, Chen X, Pollak S, Chaney M, Trinchese F, Liu S, Gunn-Moore F, Lue LF, Walker DG, Kuppusamy P, Zewier ZL, Arancio O, Stern D, Yan SS, Wu H (2004) ABAD directly links Abeta to mitochondrial toxicity in Alzheimer's disease. *Science* **304**, 448-452.

[88] Ohyagi Y, Asahara H, Chui DH, Tsuruta Y, Sakae N, Miyoshi K, Yamada T, Kikuchi H, Taniwaki T, Murai H, Ikezoe K, Furuya H, Kawarabayashi T, Shoji M, Checler F, Iwaki T, Makifuchi T, Takeda K, Kira J, Tabira T (2005) Intracellular Abeta42 activates p53 promoter: a pathway to neurodegeneration in Alzheimer's disease. *Faseb J* **19**, 255-257.

[89] Chen XH, Siman R, Iwata A, Meaney DF, Trojanowski JQ, Smith DH (2004) Long-term accumulation of amyloid-beta, beta-secretase, presenilin-1, and caspase-3 in damaged axons following brain trauma. *Am J Pathol* **165**, 357-371.

[90] Uryu K, Chen XH, Martinez D, Browne KD, Johnson VE, Graham DI, Lee VM, Trojanowski JQ, Smith DH (2007) Multiple proteins implicated in neurodegenerative diseases accumulate in axons after brain trauma in humans. *Exp Neurol* **208**, 185-192.

[91] Small SA, Gandy S (2006) Sorting through the cell biology of Alzheimer's disease: intracellular pathways to pathogenesis. *Neuron* **52**, 15-31.

[92] Busciglio J, Gabuzda DH, Matsudaira P, Yankner BA (1993) Generation of beta-amyloid in the secretory

[93] pathway in neuronal and nonneuronal cells. *Proc Natl Acad Sci U S A* **90**, 2092-2096.
[93] Cook DG, Forman MS, Sung JC, Leight S, Kolson DL, Iwatsubo T, Lee VM, Doms RW (1997) Alzheimer's A beta(1-42) is generated in the endoplasmic reticulum/intermediate compartment of NT2N cells. *Nat Med* **3**, 1021-1023.
[94] Lee SJ, Liyanage U, Bickel PE, Xia W, Lansbury PT, Jr., Kosik KS (1998) A detergent-insoluble membrane compartment contains A beta in vivo. *Nat Med* **4**, 730-734.
[95] Skovronsky DM, Doms RW, Lee VM (1998) Detection of a novel intraneuronal pool of insoluble amyloid beta protein that accumulates with time in culture. *J Cell Biol* **141**, 1031-1039.
[96] Wild-Bode C, Yamazaki T, Capell A, Leimer U, Steiner H, Ihara Y, Haass C (1997) Intracellular generation and accumulation of amyloid beta-peptide terminating at amino acid 42. *J Biol Chem* **272**, 16085-16088.
[97] Stokin GB, Lillo C, Falzone TL, Brusch RG, Rockenstein E, Mount SL, Raman R, Davies P, Masliah E, Williams DS, Goldstein LS (2005) Axonopathy and transport deficits early in the pathogenesis of Alzheimer's disease. *Science* **307**, 1282-1288.
[98] Araki Y, Kawano T, Taru H, Saito Y, Wada S, Miyamoto K, Kobayashi H, Ishikawa HO, Ohsugi Y, Yamamoto T, Matsuno K, Kinjo M, Suzuki T (2007) The novel cargo Alcadein induces vesicle association of kinesin-1 motor components and activates axonal transport. *Embo J* **26**, 1475-1486.
[99] Clifford PM, Zarrabi S, Siu G, Kinsler KJ, Kosciuk MC, Venkataraman V, D'Andrea MR, Dinsmore S, Nagele RG (2007) Abeta peptides can enter the brain through a defective blood-brain barrier and bind selectively to neurons. *Brain Res* **1142**, 223-236.
[100] Cruz JC, Kim D, Moy LY, Dobbin MM, Sun X, Bronson RT, Tsai LH (2006) p25/cyclin-dependent kinase 5 induces production and intraneuronal accumulation of amyloid beta in vivo. *J Neurosci* **26**, 10536-10541.
[101] McGowan E, Pickford F, Kim J, Onstead L, Eriksen J, Yu C, Skipper L, Murphy MP, Beard J, Das P, Jansen K, Delucia M, Lin WL, Dolios G, Wang R, Eckman CB, Dickson DW, Hutton M, Hardy J, Golde T (2005) Abeta42 is essential for parenchymal and vascular amyloid deposition in mice. *Neuron* **47**, 191-199.
[102] Iwata N, Higuchi M, Saido TC (2005) Metabolism of amyloid-beta peptide and Alzheimer's disease. *Pharmacol Ther* **108**, 129-148.
[103] Tanzi RE, Moir RD, Wagner SL (2004) Clearance of Alzheimer's Abeta peptide: the many roads to perdition. *Neuron* **43**, 605-608.
[104] Citron M (2004) Strategies for disease modification in Alzheimer's disease. *Nat Rev Neurosci* **5**, 677-685.
[105] Rajendran L, Schneider A, Schlechtingen G, Weidlich S, Ries J, Braxmeier T, Schwille P, Schulz JB, Schroeder C, Simons M, Jennings G, Knolker HJ, Simons K (2008) Efficient inhibition of the Alzheimer's disease beta-secretase by membrane targeting. *Science* **320**, 520-523.
[106] Brunkan AL, Goate AM (2005) Presenilin function and gamma-secretase activity. *J Neurochem* **93**, 769-792.
[107] Mah AL, Perry G, Smith MA, Monteiro MJ (2000) Identification of ubiquilin, a novel presenilin interactor that increases presenilin protein accumulation. *J Cell Biol* **151**, 847-862.
[108] Li A, Xie Z, Dong Y, McKay KM, McKee ML, Tanzi RE (2007) Isolation and characterization of the Drosophila ubiquilin ortholog dUbqln: in vivo interaction with early-onset Alzheimer disease genes. *Hum Mol Genet* **16**, 2626-2639.
[109] Ganguly A, Feldman RM, Guo M (2008) ubiquilin antagonizes presenilin and promotes neurodegeneration in Drosophila. *Hum Mol Genet* **17**, 293-302.
[110] Guo M, Hong EJ, Fernandes J, Zipursky SL, Hay BA (2003) A reporter for amyloid precursor protein gamma-secretase activity in Drosophila. *Hum Mol Genet* **12**, 2669-2678.
[111] Hase M, Yagi Y, Taru H, Tomita S, Sumioka A, Hori K, Miyamoto K, Sasamura T, Nakamura M, Matsuno K, Suzuki T (2002) Expression and characterization of the Drosophila X11-like/Mint protein during neural development. *J Neurochem* **81**, 1223-1232.
[112] Sano Y, Syuzo-Takabatake A, Nakaya T, Saito Y, Tomita S, Itohara S, Suzuki T (2006) Enhanced amyloidogenic metabolism of the amyloid beta-protein precursor in the X11L-deficient mouse brain. *J Biol Chem* **281**, 37853-37860.
[113] Gross GG, Feldman RM, Ganguly A, Wang J, Yu H, Guo M (2008) Role of X11 and ubiquilin as in vivo regulators of the amyloid precursor protein in Drosophila. *PLoS One* **3**, e2495.
[114] Mahoney MB, Parks AL, Ruddy DA, Tiong SY, Esengil H, Phan AC, Philandrinos P, Winter CG, Chatterjee R, Huppert K, Fisher WW, L'Archeveque L, Mapa FA, Woo W, Ellis MC, Curtis D (2006) Presenilin-based genetic screens in Drosophila melanogaster identify novel notch pathway modifiers. *Genetics* **172**, 2309-2324.
[115] van de Hoef DL, Hughes J, Livne-Bar I, Garza D, Konsolaki M, Boulianne GL (2009) Identifying genes that interact with Drosophila presenilin and amyloid precursor protein. *Genesis* **47**, 246-260.
[116] Micchelli CA, Esler WP, Kimberly WT, Jack C, Berezovska O, Kornilova A, Hyman BT, Perrimon N, Wolfe MS (2003) Gamma-secretase/presenilin inhibitors for Alzheimer's disease phenocopy Notch mutations in Drosophila. *FASEB J* **17**, 79-81.
[117] Groth C, Alvord WG, Quinones OA, Fortini ME (2010) Pharmacological analysis of Drosophila melanogaster gamma-secretase with respect to differential proteolysis of Notch and APP. *Mol Pharmacol* **77**, 567-574.
[118] Iwata N, Tsubuki S, Takaki Y, Watanabe K, Sekiguchi M, Hosoki E, Kawashima-Morishima M, Lee HJ, Hama E, Sekine-Aizawa Y, Saido TC (2000) Identification of the major Abeta1-42-degrading catabolic pathway in brain parenchyma: suppression leads to biochemical and pathological deposition. *Nat Med* **6**, 143-150.
[119] Qiu WQ, Walsh DM, Ye Z, Vekrellis K, Zhang J, Podlisny MB, Rosner MR, Safavi A, Hersh LB, Selkoe DJ (1998) Insulin-degrading enzyme regulates extracellular levels of amyloid beta-protein by degradation. *J Biol Chem* **273**, 32730-32738.
[120] Eckman EA, Reed DK, Eckman CB (2001) Degradation of the Alzheimer's amyloid beta peptide by endothelin-converting enzyme. *J Biol Chem* **276**, 24540-24548.
[121] Hamazaki H (1996) Cathepsin D is involved in the clearance of Alzheimer's beta-amyloid protein. *FEBS Lett* **396**, 139-142.
[122] Qiu WQ, Borth W, Ye Z, Haass C, Teplow DB, Selkoe DJ (1996) Degradation of amyloid beta-protein by a serine protease-alpha2-macroglobulin complex. *J Biol Chem* **271**, 8443-8451.

[123] Yasojima K, Akiyama H, McGeer EG, McGeer PL (2001) Reduced neprilysin in high plaque areas of Alzheimer brain: a possible relationship to deficient degradation of beta-amyloid peptide. *Neurosci Lett* **297**, 97-100.

[124] Yasojima K, McGeer EG, McGeer PL (2001) Relationship between beta amyloid peptide generating molecules and neprilysin in Alzheimer disease and normal brain. *Brain Res* **919**, 115-121.

[125] Iwata N, Tsubuki S, Takaki Y, Shirotani K, Lu B, Gerard NP, Gerard C, Hama E, Lee HJ, Saido TC (2001) Metabolic regulation of brain Abeta by neprilysin. *Science* **292**, 1550-1552.

[126] Farris W, Schutz SG, Cirrito JR, Shankar GM, Sun X, George A, Leissring MA, Walsh DM, Qiu WQ, Holtzman DM, Selkoe DJ (2007) Loss of neprilysin function promotes amyloid plaque formation and causes cerebral amyloid angiopathy. *Am J Pathol* **171**, 241-251.

[127] Huang SM, Mouri A, Kokubo H, Nakajima R, Suemoto T, Higuchi M, Staufenbiel M, Noda Y, Yamaguchi H, Nabeshima T, Saido TC, Iwata N (2006) Neprilysin-sensitive synapse-associated amyloid-beta peptide oligomers impair neuronal plasticity and cognitive function. *J Biol Chem* **281**, 17941-17951.

[128] Leissring MA, Farris W, Chang AY, Walsh DM, Wu X, Sun X, Frosch MP, Selkoe DJ (2003) Enhanced proteolysis of beta-amyloid in APP transgenic mice prevents plaque formation, secondary pathology, and premature death. *Neuron* **40**, 1087-1093.

[129] Poirier R, Wolfer DP, Welzl H, Tracy J, Galsworthy MJ, Nitsch RM, Mohajeri MH (2006) Neuronal neprilysin overexpression is associated with attenuation of Abeta-related spatial memory deficit. *Neurobiol Dis* **24**, 475-483.

[130] Marr RA, Rockenstein E, Mukherjee A, Kindy MS, Hersh LB, Gage FH, Verma IM, Masliah E (2003) Neprilysin gene transfer reduces human amyloid pathology in transgenic mice. *J Neurosci* **23**, 1992-1996.

[131] Iwata N, Mizukami H, Shirotani K, Takaki Y, Muramatsu S, Lu B, Gerard NP, Gerard C, Ozawa K, Saido TC (2004) Presynaptic localization of neprilysin contributes to efficient clearance of amyloid-beta peptide in mouse brain. *J Neurosci* **24**, 991-998.

[132] Hemming ML, Patterson M, Reske-Nielsen C, Lin L, Isacson O, Selkoe DJ (2007) Reducing amyloid plaque burden via ex vivo gene delivery of an Abeta-degrading protease: a novel therapeutic approach to Alzheimer disease. *PLoS Med* **4**, e262.

[133] Iijima-Ando K, Hearn SA, Granger L, Shenton C, Gatt A, Chiang HC, Hakker I, Zhong Y, Iijima K (2008) Overexpression of neprilysin reduces alzheimer amyloid-beta42 (Abeta42)-induced neuron loss and intraneuronal Abeta42 deposits but causes a reduction in cAMP-responsive element-binding protein-mediated transcription, age-dependent axon pathology, and premature death in Drosophila. *J Biol Chem* **283**, 19066-19076.

[134] Shirotani K, Tsubuki S, Iwata N, Takaki Y, Harigaya W, Maruyama K, Kiryu-Seo S, Kiyama H, Iwata H, Tomita T, Iwatsubo T, Saido TC (2001) Neprilysin degrades both amyloid beta peptides 1-40 and 1-42 most rapidly and efficiently among thiorphan- and phosphoramidon-sensitive endopeptidases. *J Biol Chem* **276**, 21895-21901.

[135] Hama E, Shirotani K, Masumoto H, Sekine-Aizawa Y, Aizawa H, Saido TC (2001) Clearance of extracellular and cell-associated amyloid beta peptide through viral expression of neprilysin in primary neurons. *J Biochem* **130**, 721-726.

[136] Spencer B, Marr RA, Rockenstein E, Crews L, Adame A, Potkar R, Patrick C, Gage FH, Verma IM, Masliah E (2008) Long-term neprilysin gene transfer is associated with reduced levels of intracellular Abeta and behavioral improvement in APP transgenic mice. *BMC Neurosci* **9**, 109.

[137] Schilling S, Zeitschel U, Hoffmann T, Heiser U, Francke M, Kehlen A, Holzer M, Hutter-Paier B, Prokesch M, Windisch M, Jagla W, Schlenzig D, Lindner C, Rudolph T, Reuter G, Cynis H, Montag D, Demuth HU, Rossner S (2008) Glutaminyl cyclase inhibition attenuates pyroglutamate Abeta and Alzheimer's disease-like pathology. *Nat Med* **14**, 1106-1111.

[138] Lansbury PT, Lashuel HA (2006) A century-old debate on protein aggregation and neurodegeneration enters the clinic. *Nature* **443**, 774-779.

[139] Nerelius C, Sandegren A, Sargsyan H, Raunak R, Leijonmarck H, Chatterjee U, Fisahn A, Imarisio S, Lomas DA, Crowther DC, Stromberg R, Johansson J (2009) Alpha-helix targeting reduces amyloid-beta peptide toxicity. *Proc Natl Acad Sci U S A* **106**, 9191-9196.

[140] Luheshi LM, Hoyer W, de Barros TP, van Dijk Hard I, Brorsson AC, Macao B, Persson C, Crowther DC, Lomas DA, Stahl S, Dobson CM, Hard T (2010) Sequestration of the Abeta peptide prevents toxicity and promotes degradation in vivo. *PLoS Biol* **8**, e1000334.

[141] Walsh DM, Klyubin I, Fadeeva JV, Rowan MJ, Selkoe DJ (2002) Amyloid-beta oligomers: their production, toxicity and therapeutic inhibition. *Biochem Soc Trans* **30**, 552-557.

[142] Lambert MP, Barlow AK, Chromy BA, Edwards C, Freed R, Liosatos M, Morgan TE, Rozovsky I, Trommer B, Viola KL, Wals P, Zhang C, Finch CE, Krafft GA, Klein WL (1998) Diffusible, nonfibrillar ligands derived from Abeta1-42 are potent central nervous system neurotoxins. *Proc Natl Acad Sci U S A* **95**, 6448-6453.

[143] Kayed R, Head E, Thompson JL, McIntire TM, Milton SC, Cotman CW, Glabe CG (2003) Common Structure of Soluble Amyloid Oligomers Implies Common Mechanism of Pathogenesis. *Science* **300**, 486-489.

[144] Lesne S, Koh MT, Kotilinek L, Kayed R, Glabe CG, Yang A, Gallagher M, Ashe KH (2006) A specific amyloid-beta protein assembly in the brain impairs memory. *Nature* **440**, 352-357.

[145] Harper JD, Wong SS, Lieber CM, Lansbury PT (1997) Observation of metastable Abeta amyloid protofibrils by atomic force microscopy. *Chem Biol* **4**, 119-125.

[146] Kuo YM, Emmerling MR, Vigo-Pelfrey C, Kasunic TC, Kirkpatrick JB, Murdoch GH, Ball MJ, Roher AE (1996) Water-soluble Abeta (N-40, N-42) oligomers in normal and Alzheimer disease brains. *J Biol Chem* **271**, 4077-4081.

[147] Georganopoulou DG, Chang L, Nam JM, Thaxton CS, Mufson EJ, Klein WL, Mirkin CA (2005) Nanoparticle-based detection in cerebral spinal fluid of a soluble pathogenic biomarker for Alzheimer's disease. *Proc Natl Acad Sci U S A* **102**, 2273-2276.

[148] Thal DR, Capetillo-Zarate E, Del Tredici K, Braak H (2006) The development of amyloid beta protein depos-

its in the aged brain. *Sci Aging Knowledge Environ* **2006**, re1.

[149] Yankner BA, Dawes LR, Fisher S, Villa-Komaroff L, Oster-Granite ML, Neve RL (1989) Neurotoxicity of a fragment of the amyloid precursor associated with Alzheimer's disease. *Science* **245**, 417-420.

[150] Murakami K, Irie K, Morimoto A, Ohigashi H, Shindo M, Nagao M, Shimizu T, Shirasawa T (2003) Neurotoxicity and physicochemical properties of Abeta mutant peptides from cerebral amyloid angiopathy: implication for the pathogenesis of cerebral amyloid angiopathy and Alzheimer's disease. *J Biol Chem* **278**, 46179-46187.

[151] Haass C, Selkoe DJ (2007) Soluble protein oligomers in neurodegeneration: lessons from the Alzheimer's amyloid beta-peptide. *Nat Rev Mol Cell Biol* **8**, 101-112.

[152] Luheshi LM, Tartaglia GG, Brorsson AC, Pawar AP, Watson IE, Chiti F, Vendruscolo M, Lomas DA, Dobson CM, Crowther DC (2007) Systematic in vivo analysis of the intrinsic determinants of amyloid Beta pathogenicity. *PLoS Biol* **5**, e290.

[153] Lue LF, Kuo YM, Roher AE, Brachova L, Shen Y, Sue L, Beach T, Kurth JH, Rydel RE, Rogers J (1999) Soluble amyloid beta peptide concentration as a predictor of synaptic change in Alzheimer's disease. *Am J Pathol* **155**, 853-862.

[154] Naslund J, Haroutunian V, Mohs R, Davis KL, Davies P, Greengard P, Buxbaum JD (2000) Correlation between elevated levels of amyloid beta-peptide in the brain and cognitive decline. *Jama* **283**, 1571-1577.

[155] Prusiner SB (2004) *Prion biology and diseases*, Cold Spring Harbor Laboratory Press, Cold Spring Harbor, N.Y.

[156] Deshpande A, Mina E, Glabe C, Busciglio J (2006) Different conformations of amyloid beta induce neurotoxicity by distinct mechanisms in human cortical neurons. *J Neurosci* **26**, 6011-6018.

[157] Chiang HC, Iijima K, Hakker I, Zhong Y (2009) Distinctive roles of different beta-amyloid 42 aggregates in modulation of synaptic functions. *FASEB J* **23**, 1969-1977.

[158] Nilsberth C, Westlind-Danielsson A, Eckman CB, Condron MM, Axelman K, Forsell C, Stenh C, Luthman J, Teplow DB, Younkin SG, Naslund J, Lannfelt L (2001) The 'Arctic' APP mutation (E693G) causes Alzheimer's disease by enhanced Abeta protofibril formation. *Nat Neurosci* **4**, 887-893.

[159] Johansson AS, Berglind-Dehlin F, Karlsson G, Edwards K, Gellerfors P, Lannfelt L (2006) Physiochemical characterization of the Alzheimer's disease-related peptides A beta 1-42Arctic and A beta 1-42wt. *Febs J* **273**, 2618-2630.

[160] Whalen BM, Selkoe DJ, Hartley DM (2005) Small nonfibrillar assemblies of amyloid beta-protein bearing the Arctic mutation induce rapid neuritic degeneration. *Neurobiol Dis* **20**, 254-266.

[161] Cheng IH, Palop JJ, Esposito LA, Bien-Ly N, Yan F, Mucke L (2004) Aggressive amyloidosis in mice expressing human amyloid peptides with the Arctic mutation. *Nat Med* **10**, 1190-1192.

[162] Lord A, Kalimo H, Eckman C, Zhang XQ, Lannfelt L, Nilsson LN (2006) The Arctic Alzheimer mutation facilitates early intraneuronal Abeta aggregation and senile plaque formation in transgenic mice. *Neurobiol Aging* **27**, 67-77.

[163] Morimoto A, Irie K, Murakami K, Masuda Y, Ohigashi H, Nagao M, Fukuda H, Shimizu T, Shirasawa T (2004) Analysis of the secondary structure of beta-amyloid (Abeta42) fibrils by systematic proline replacement. *J Biol Chem* **279**, 52781-52788.

[164] Fay DS, Fluet A, Johnson CJ, Link CD (1998) In vivo aggregation of beta-amyloid peptide variants. *J Neurochem* **71**, 1616-1625.

[165] Iijima K, Iijima-Ando K (2008) Drosophila models of Alzheimer's amyloidosis: the challenge of dissecting the complex mechanisms of toxicity of amyloid-beta 42. *J Alzheimers Dis* **15**, 523-540.

[166] Kelly JW (2005) Attacking amyloid. *N Engl J Med* **352**, 722-723.

[167] Rochet JC, Lansbury PT, Jr. (2000) Amyloid fibrillogenesis: themes and variations. *Curr Opin Struct Biol* **10**, 60-68.

[168] DeMattos RB, Cirrito JR, Parsadanian M, May PC, O'Dell MA, Taylor JW, Harmony JA, Aronow BJ, Bales KR, Paul SM, Holtzman DM (2004) ApoE and clusterin cooperatively suppress Abeta levels and deposition: evidence that ApoE regulates extracellular Abeta metabolism in vivo. *Neuron* **41**, 193-202.

[169] Cherny RA, Atwood CS, Xilinas ME, Gray DN, Jones WD, McLean CA, Barnham KJ, Volitakis I, Fraser FW, Kim Y, Huang X, Goldstein LE, Moir RD, Lim JT, Beyreuther K, Zheng H, Tanzi RE, Masters CL, Bush AI (2001) Treatment with a copper-zinc chelator markedly and rapidly inhibits beta-amyloid accumulation in Alzheimer's disease transgenic mice. *Neuron* **30**, 665-676.

[170] Cohen E, Bieschke J, Perciavalle RM, Kelly JW, Dillin A (2006) Opposing Activities Protect Against Age-Onset Proteotoxicity. *Science* **313**, 1604-1610.

[171] Cummings JL (2000) Cognitive and behavioral heterogeneity in Alzheimer's disease: seeking the neurobiological basis. *Neurobiol Aging* **21**, 845-861.

[172] Moreira PI, Santos MS, Oliveira CR, Shenk JC, Nunomura A, Smith MA, Zhu X, Perry G (2008) Alzheimer disease and the role of free radicals in the pathogenesis of the disease. *CNS Neurol Disord Drug Targets* **7**, 3-10.

[173] Rival T, Page RM, Chandraratna DS, Sendall TJ, Ryder E, Liu B, Lewis H, Rosahl T, Hider R, Camargo LM, Shearman MS, Crowther DC, Lomas DA (2009) Fenton chemistry and oxidative stress mediate the toxicity of the beta-amyloid peptide in a Drosophila model of Alzheimer's disease. *Eur J Neurosci* **29**, 1335-1347.

[174] Mattson MP, Gleichmann M, Cheng A (2008) Mitochondria in neuroplasticity and neurological disorders. *Neuron* **60**, 748-766.

[175] Lin MT, Beal MF (2006) Mitochondrial dysfunction and oxidative stress in neurodegenerative diseases. *Nature* **443**, 787-795.

[176] Reddy PH (2009) Amyloid beta, mitochondrial structural and functional dynamics in Alzheimer's disease. *Exp Neurol*.

[177] Wang X, Su B, Zheng L, Perry G, Smith MA, Zhu X (2009) The role of abnormal mitochondrial dynamics in the pathogenesis of Alzheimer's disease. *J Neurochem* **109 Suppl 1**, 153-159.

[178] Wang J, Xiong S, Xie C, Markesbery WR, Lovell MA (2005) Increased oxidative damage in nuclear and mitochondrial DNA in Alzheimer's disease. *J Neurochem* **93**, 953-962.

[179] Hirai K, Aliev G, Nunomura A, Fujioka H, Russell RL, Atwood CS, Johnson AB, Kress Y, Vinters HV, Tabaton

M, Shimohama S, Cash AD, Siedlak SL, Harris PL, Jones PK, Petersen RB, Perry G, Smith MA (2001) Mitochondrial abnormalities in Alzheimer's disease. *J Neurosci* **21**, 3017-3023.
[180] Wang X, Su B, Lee HG, Li X, Perry G, Smith MA, Zhu X (2009) Impaired balance of mitochondrial fission and fusion in Alzheimer's disease. *J Neurosci* **29**, 9090-9103.
[181] Knott AB, Perkins G, Schwarzenbacher R, Bossy-Wetzel E (2008) Mitochondrial fragmentation in neurodegeneration. *Nat Rev Neurosci* **9**, 505-518.
[182] Iijima-Ando K, Hearn SA, Shenton C, Gatt A, Zhao L, Iijima K (2009) Mitochondrial Mislocalization Underlies Aβ42-Induced Neuronal Dysfunction in a Drosophila Model of Alzheimer's Disease. *PLoS ONE* **4**, e8310.
[183] Glater EE, Megeath LJ, Stowers RS, Schwarz TL (2006) Axonal transport of mitochondria requires milton to recruit kinesin heavy chain and is light chain independent. *J Cell Biol* **173**, 545-557.
[184] Stowers RS, Megeath LJ, Gorska-Andrzejak J, Meinertzhagen IA, Schwarz TL (2002) Axonal transport of mitochondria to synapses depends on milton, a novel Drosophila protein. *Neuron* **36**, 1063-1077.
[185] Guo X, Macleod GT, Wellington A, Hu F, Panchumarthi S, Schoenfield M, Marin L, Charlton MP, Atwood HL, Zinsmaier KE (2005) The GTPase dMiro is required for axonal transport of mitochondria to Drosophila synapses. *Neuron* **47**, 379-393.
[186] Zhao XL, Wang WA, Tan JX, Huang JK, Zhang X, Zhang BZ, Wang YH, YangCheng HY, Zhu HL, Sun XJ, Huang FD (2010) Expression of beta-amyloid Induced age-dependent presynaptic and axonal changes in Drosophila. *J Neurosci* **30**, 1512-1522.
[187] Ling D, Song HJ, Garza D, Neufeld TP, Salvaterra PM (2009) Abeta42-induced neurodegeneration via an age-dependent autophagic-lysosomal injury in Drosophila. *PLoS ONE* **4**, e4201.
[188] Barnham KJ, Bush AI (2008) Metals in Alzheimer's and Parkinson's diseases. *Curr Opin Chem Biol* **12**, 222-228.
[189] Sanokawa-Akakura R, Dai H, Akakura S, Weinstein D, Fajardo JE, Lang SE, Wadsworth S, Siekierka J, Birge RB (2004) A novel role for the immunophilin FKBP52 in copper transport. *J Biol Chem* **279**, 27845-27848.
[190] Sanokawa-Akakura R, Cao W, Allan K, Patel K, Ganesh A, Heiman G, Burke R, Kemp FW, Bogden JD, Camakaris J, Birge RB, Konsolaki M (2010) Control of Alzheimer's amyloid beta toxicity by the high molecular weight immunophilin FKBP52 and copper homeostasis in Drosophila. *PLoS One* **5**, e8626.
[191] Muhammad A, Flores I, Zhang H, Yu R, Staniszewski A, Planel E, Herman M, Ho L, Kreber R, Honig LS, Ganetzky B, Duff K, Arancio O, Small SA (2008) Retromer deficiency observed in Alzheimer's disease causes hippocampal dysfunction, neurodegeneration, and Abeta accumulation. *Proc Natl Acad Sci U S A* **105**, 7327-7332.
[192] Chiang HC, Wang L, Xie Z, Yau A, Zhong Y (2010) PI3 kinase signaling is involved in A{beta}-induced memory loss in Drosophila. *Proc Natl Acad Sci U S A*.
[193] Cao W, Song HJ, Gangi T, Kelkar A, Antani I, Garza D, Konsolaki M (2008) Identification of novel genes that modify phenotypes induced by Alzheimer's beta amyloid overexpression in Drosophila. *Genetics*.
[194] Rorth P, Szabo K, Bailey A, Laverty T, Rehm J, Rubin GM, Weigmann K, Milan M, Benes V, Ansorge W, Cohen SM (1998) Systematic gain-of-function genetics in Drosophila. *Development* **125**, 1049-1057.
[195] Tan L, Schedl P, Song HJ, Garza D, Konsolaki M (2008) The Toll-->NFkappaB signaling pathway mediates the neuropathological effects of the human Alzheimer's Abeta42 polypeptide in Drosophila. *PLoS One* **3**, e3966.
[196] Marsh JL, Thompson LM (2006) Drosophila in the Study of Neurodegenerative Disease. *Neuron* **52**, 169-178.

# Section 2
# Using Animal Models to Understand Mechanisms of Disease

# Estrogen, Progesterone and Hippocampal Plasticity in Rodent Models

Michael R. Foy[a,*], Michel Baudry[b], Roberta Diaz Brinton[b,c] and Richard F. Thompson[b]
[a] Department of Psychology, Loyola Marymount University, Los Angeles, CA, USA
[b] Neuroscience Program, University of Southern California, Los Angeles, CA, USA
[c] Department of Pharmacy, University of Southern California, Los Angeles, CA, USA

**Abstract.** Accumulating evidence indicates that ovarian hormones regulate a wide variety of non-reproductive functions in the central nervous system by interacting with several molecular and cellular processes. A growing animal literature using both adult and aged rodent models indicates that 17β-estradiol (E2), the most potent of the biologically relevant estrogens, and progesterone (P4), a major naturally occurring progestogen, facilitate some forms of learning and memory, in particular those that involve hippocampal-dependent tasks. A recently developed triple-transgenic mouse (3xTg-AD) has been widely used as an animal model of Alzheimer's disease (AD), as this mouse exhibits an age-related and progressive neuropathological phenotype that includes both plaque and tangle pathology mainly restricted to hippocampus, amygdala and cerebral cortex. In this report, we examine recent studies that compare the effects of ovarian hormones on synaptic transmission and synaptic plasticity in adult and aged rodents. A better understanding of the non-reproductive functions of ovarian hormones has far-reaching implications for hormone therapy to maintain health and function within the nervous system throughout aging.

Keywords: 3xTg AD, aging, estrogen, progesterone, hippocampus, LTP, mouse, synaptic plasticity

## INTRODUCTION

Within the past decade, there has been increasing interest in the effects of ovarian hormones on neural function. This effort is driven, in part, by the results of clinical studies suggesting that estrogen therapy administered after menopause may delay or prevent the onset of Alzheimer's disease (AD) in older women. Other, still controversial, research indicates that ovarian hormones may enhance or diminish certain forms of memory in postmenopausal women. Much of the most current research related to ovarian hormones and brain function is focused in two primary directions. First, clinical studies have examined the potential effect estrogen might offer in protecting against cognitive decline during normal aging and against AD (neuroprotection). Second, laboratory studies have examined the mechanisms by which estrogen and progesterone can modify the structure of nerve cells and alter the way neurons communicate with other cells in the brain (neuroplasticity). Rodent models have been used extensively to study a variety of behavioral, cognitive and anatomical changes linked to important features of AD found in humans. In this chapter, we examine recent evidence from experimental research using rodent models on the effects of estrogen and progesterone on neuroplasticity and neuroprotection in both adult and aged rats, as well as information regarding the recent triple transgenic mouse model of AD.

AD is an age-related, irreversible and neurodegenerative disorder that causes a progressive deterioration of cognitive function, including a profound loss of memory [56]. Neuropathologically, AD is characterized by the accumulation of amyloid-β (Aβ) deposits in brain, neurofibrillary tangles, which consist of hyperphosphorylated tau aggregates [103], and progressive neuronal loss in cortical structures. Aβ has been found to be the primary component of amyloid plaques and is generated from the amyloid-β protein precursor (APP) by sequential proteolytic

---

[*] Corresponding Author.

cleavage at the β and γ sites on the peptide [56]. Neurofibrillary tangles are composed of hyperphosphorylated tau, a microtubule-associated protein. While tau protein promotes the assembly of microtubules, its hyperphosphorylation interferes with its normal biological functions by reducing its ability to bind to and stabilize microtubules [123]. The occurrence of both amyloid plaques and neurofibrillary tangles are necessary to confirm a diagnosis of AD. Epidemiological studies indicate that women develop a higher risk of AD even after adjusting for age [25,63], suggesting a genetic or hormonal cause. The depletion of the sex steroid hormones estrogen and progesterone at menopause appears to be a significant risk factor for the development of AD in women [69,95,96,119,142]. Prospective and case-control studies have demonstrated that hormone therapy (HT) can reduce the risk of AD in women [12,69,96]. However, the relationship between the therapeutic benefits of estrogen- and progesterone-based HT, and both normal cognitive decline and development of AD have been the recent topics of heated debate and controversy (see [31]). Furthermore, clinical findings from the massive Women's Health Initiative Memory Study (WHIMS) demonstrating a higher incidence of dementia in subjects receiving estrogen-based HT [100,106] was quite unexpected, given the early background of estrogen and AD. The WHIMS findings raised many important points, perhaps the most important being the need to better understand the role of estrogen and progesterone in AD pathogenesis, and to optimize HT.

*Estrogen and cognition*

In humans, the issue of hormone therapy has yielded much debate with many clinical studies supporting a neuroprotective role of estrogen, along with those including the use of progesterone to counteract estrogen-induced proliferation of the endometrium [65,104]. The depletion of estrogen that occurs after menopause increases the susceptibility of women to AD [59,96], whereas estrogen replacement in postmenopausal women improves verbal memory [2,101]. Healthy postmenopausal women with estrogen replacement scored significantly higher on tests of immediate and delayed paragraph recall compared with healthy postmenopausal women not taking estrogen replacement [68]. Other evaluations of estrogen-replacement therapy in AD patients indicated that estrogen does not alleviate cognitive impairment associated with the disease [86], but it does seem to have a beneficial effect as a preventive treatment [119], which is most apparent in younger, postmenopausal women [60]. The WHIMS study [106] was designed to address this seemingly contradictory literature. A possible confound in many of the earlier estrogen studies of cognition was the use of combinations of estrogens and progestins. The WHIMS reported a non-significant increase in the number of women with mild cognitive impairment who were using estrogen (conjugated equine estrogens), but a highly significant increase in the number of women with mild cognitive impairment who were using combined estrogens and progestins.

In animal models, ovarian hormones have been shown to influence memory via actions on neurons, particularly in hippocampus [83,87,118,131]. A neuroprotective role of estrogen has been established in both *in vivo* and *in vitro* animal models of neurodegeneration [43,44,113,115]. Additionally, E2 has been found to facilitate some forms of learning and memory function in rodents, particularly for hippocampal-dependent tasks [8,9]. Post-training E2 injection facilitated retention in the Morris water maze [111], and a cholinergic agonist enhanced this effect [94]. In other studies, the effects of E2 and raloxifene, a selective estrogen-receptor modulator, have been evaluated on the acquisition of a delayed matching to position in a T-maze task and on hippocampal acetylcholine release in ovariectomized rats. E2, but not raloxifene, enhanced the T-maze task performance, and E2 and a high dose of raloxifene increased potassium-stimulated acetylcholine release in hippocampus [53]. By contrast, some studies found little or no effect of the estrous cycle, and thereby of endogenous E2, on tasks involving spatial memory [12,49,130,138].

A recent review of sex differences in rodent models of learning and memory function suggests that male rats have advantages in some forms of memory, but this finding was found not to be as strong in mouse models of memory [67]. Most research suggests that E2, and perhaps P4, influence learning and memory, but do so in a task-dependent manner [34]. In a series of rat studies, P4 was found to impair cognitive function [17,18]. Gonadally-intact aged female rats performed much more poorly on a reference memory task (radial arm maze) than gonadally-intact young adult female rats. Interestingly, if the aged animals were ovariectomized, they performed as well as the young animals. It appears that in the gonadally-intact aged animals, E2 levels are virtually the same as in young animals, but P4 levels are much higher in the aged animals. Indeed, if aged ovariectomized animals were chronically implanted with P4 pellets, they performed the radial arm maze task as poorly as gonadally-intact aged female rats. In this cognitive task, P4 supplementation appears to reverse the somewhat cognitive enhancing effects of ovariectomy. Despite a vast literature on ovarian hormones and cognition, the effects of ovarian hormones on learning and memory functions still remain unclear. What emerges from many of these reports is that E2 and P4 enhance learning and memory in some instances, but impair, or have no effect on learning in others.

*Estrogen and hippocampus*

For more than 30 years, electrophysiological investigations have found estrogen to promote changes in synaptic plasticity within the nervous system. In a pioneering study, decreased hippocampal seizure thresholds were found in animals primed with estrogen and also during proestrus, the time of the estrous cycle when estrogen levels are at their highest levels [120]. In humans, changes in electrical activity of nervous system tissue correlate with hormonal factors that appear to play a role in catamenial epilepsy, a form of epilepsy in which the likelihood of seizures varies during the menstrual cycle. Many women with catamenial epilepsy experience a sharp increase in seizure frequency immediately before menstruation, when estrogen concentrations relative to those of progesterone are also at their highest levels [3]. Changes in hippocampal responsiveness correlate with estrogen activity, as induction of long-term potentiation (LTP), an induced form of synaptic plasticity (see below), is maximal in female rats during the afternoon of proestrus, when endogenous estrogen concentrations are highest [129]. Furthermore, induction of hippocampal LTP is facilitated in ovariectomized rats treated with E2 compared to untreated ovariectomized rats [30].

The development of *in vitro* models to study the mechanisms of neuronal plasticity has provided researchers better tools to investigate how estrogen regulates synaptic excitability in the nervous system, and, in particular, in hippocampus. It should be stressed, however, that the binding of $^3$H-E2 in hippocampus does not approach that seen in hypothalamus and related diencephalic structures [81,82]. Nonetheless, studies by Teyler and colleagues using *in vitro* hippocampal slice preparations have shown that gonadal steroids dramatically affect neuronal excitability in specific pathways of the rodent hippocampus [121,126]. In the initial series of experiments, extracellular monosynaptic population field responses recorded from area CA1 of hippocampal slices from male and female rats were monitored before and after the addition of E2 (100 pM) to the slice incubation medium (artificial cerebrospinal fluid; aCSF). In male rats, E2 produced a rapid (< 10 min) enhancement of population field responses evoked by stimulation of the afferents to CA1 pyramidal cells (Figure 1). This was the first published report demonstrating that picomolar concentrations of the gonadal steroid E2 directly enhanced glutamatergic synaptic transmission in hippocampus [121].

Although the mechanism of action of gonadal steroids in hippocampus is not entirely understood, it is likely to be receptor-mediated. There was no facilitation of field responses when the inactive estrogen, 17α-estradiol, was added to hippocampal slice medium [42], and the further addition of E2 no longer resulted in an increased response, as observed in the presence of E2 alone [42,132,133]. Similar results were found when the estrogen receptor antagonist tamoxifen was applied to hippocampal slices before E2 addition [41]. The ability of 17α-estradiol and tamoxifen to block E2 effects on hippocampal excitability provides strong evidence that the rapid physiological modulation of gonadal hormones is most likely due to the activation of a plasma membrane receptor.

*Synaptic plasticity and long-term potentiation*

Long-term potentiation of synaptic transmission in hippocampus and neocortex is considered to be a cellular model of memory trace formation in the brain, at least for certain forms of memory [9,19,74]. Although there is a large body of work regarding the molecular and synaptic mechanisms underlying LTP [10,52], there is a relative paucity of studies demonstrating the critical role of LTP in behavioral learning and memory [105]. Nonetheless, whether LTP is or not the substrate of the synaptic modifications which occur during learning in forebrain structures of vertebrates, studies of its mechanisms have re-

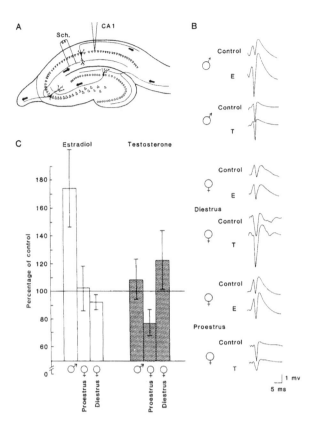

Fig. 1. (A) Diagram of transverse hippocampal slice, (B) representative field potentials from slice preparations in the various experimental conditions (artificial cerebrospinal fluid = control, 17β-estradiol = E, testosterone = T), and (C) bar graphs summarizing major experimental outcomes in hippocampal slices from male rats, and female rats in proestrus and diestrus. Reprinted with permission from [121]. Copyright 2010 American Association for the Advancement of Science.

vealed the existence of a number of processes that undoubtedly play critical roles in memory formation [13]. In area CA1 of hippocampus, the most widely studied form of LTP requires NMDA receptor activation for its induction, and an increase in the number of α-amino-3-hydroxy-5-methyl-4-isoxazoleproprianate (AMPA) receptors for its expression and maintenance. In addition, Teyler and associates have demonstrated a second form of tetanus-induced LTP in CA1 that is independent of NMDA receptors, and involves voltage-dependent calcium channels [54].

*Estrogen and non-genomic mechanism of action*

The genomic mechanism of action of estrogen has been the traditional framework for interpreting the effect of estrogen on cell function, but an increasing number of reports document the effects of acute applications of estrogenic steroids that are too rapid (occurring ≤ 10 min) to be accounted for exclusively by a genomic pathway. In particular, the existence of rapid estrogenic steroid-induced changes in neuronal excitability suggests other, non-genomic mechanisms involving direct interactions with sites on the plasma membrane that alter or regulate a variety of ion channels and neurotransmitter transporters [97,135].

*In vitro* intracellular recordings of CA1 neurons from adult ovariectomized female rats have shown that E2 increases synaptic excitability in part by enhancing the magnitude of AMPA receptor-mediated responses [133]. The rapid onset of the increased excitability, and its blockade by 6-cyano-7-nitroquinaxaline (CNQX, an AMPA receptor antagonist) but not by D-2-amino-5-phosphonovalerate (D-APV a competitive NMDA receptor antagonist), supported a postsynaptic membrane site of action resulting in enhanced non-NMDA glutamate receptors. Later studies using whole cell recordings found that acute E2 application potentiated kainate-induced currents in a subpopulation of CA1 cells [55], although a direct interaction between E2 and the receptor channel was not excluded [134].

*Estrogen and NMDA receptor regulation*

A large body of evidence demonstrates that E2-mediated regulation of synapse formation is dependent on NMDA receptor activation. Morphological studies during the course of neuronal development conducted in cultured neurons prepared from embryonic day 18 rat fetuses have shown that estrogenic steroids exert a growth-promoting, neurotrophic effect on hippocampal and cortical neurons via a mechanism that requires NMDA receptor activation [23,24]. *In vivo* studies using adult ovariectomized female rats have also revealed a proliferation of dendritic spines in hippocampal CA1 pyramidal cells after E2 treatment that could be prevented by blockade of NMDA receptors, but not by AMPA or muscarinic receptor antagonists [136]. Other reports using adult ovariectomized female rats provided evidence that chronic E2 treatment increased the number of NMDA receptor binding sites and NMDA receptor-mediated responses [50,137]. These studies indicate that estrogen and NMDA receptors are heavily involved in synapse formation.

The possibility of direct regulation of NMDA receptor-mediated synaptic transmission by E2 may not have been detected previously (e.g., [133]) because tests of this hypothesis had not been conducted under optimal conditions. Because of the voltage-dependent blockade of the NMDA receptor channel by $Mg^{2+}$ and the slow kinetics of the channel opening relative to that of the AMPA receptor, there is only a minor NMDA receptor-mediated component of the excitatory postsynaptic potential (EPSP) evoked by low-frequency stimulation of glutamatergic afferents. This NMDA receptor component can be enhanced with low $Mg^{2+}$ concentrations or high-frequency stimulation patterns used to induce the depolarization accompanying the summation of overlapping EPSPs [140]. In experiments using low $Mg^{2+}$ concentrations and in the presence of the AMPA receptor antagonist 6,7-dinitroquinoxal-ine-2,3-dione (DNQX), an acute application of E2 in adult male rat hippocampal slices resulted in a rapid increase in the amplitude of NMDA receptor-mediated EPSPs evoked by stimulation of the Schaffer collaterals [43]. The effect of E2 on pharmacologically isolated NMDA receptor-mediated synaptic responses was such that concentrations of E2 greater than 10 nM induced seizure activity in hippocampal neurons, and lower concentrations (1 nM) markedly increased the amplitude of NMDA receptor-mediated EPSPs.

*Estrogen and hippocampal LTP in male rats*

To investigate the effect of estrogen on synaptic plasticity associated with learning and memory function, E2 was applied to hippocampal slices from adult male rats before the slices were exposed to high-frequency stimulation designed to induce LTP. When LTP was assessed after high-frequency stimulation, fEPSP values were increased significantly for E2-treated slices compared to control-aCSF slices (Figure 2). fEPSP mean increases in slope was 192% (experimental) vs. 154% (control). Thus, hippocampal slices from adult male rats treated with E2 exhibited a pronounced, persisting and significant increase in LTP as measured by both population fEPSP slope and fEPSP amplitude recordings [43,47].

To further evaluate the effects of E2 on the magnitude of hippocampal LTP, the intensity of afferent stimulation to Schaffer collaterals in slices perfused with E2 was decreased in order to produce baseline values to pre-E2 levels immediately before the delivery of the high-frequency stimulation train used to elicit LTP [14]. Under these conditions, E2 still produced an increase in the amplitude of LTP from adult male rat hippocampal slices compared to that obtained in control (aCSF) slices. These findings indicate that E2-induced enhancement of hippocampal LTP is not due to simply a change in basal EPSP level, but is more likely due to biochemical activation of an intracellular cascade, presumably mediated by activation of a src tyrosine pathway that enhances NMDA receptor function. Two recent studies indicate that E2 activates a cellular cascade that consists of m-calpain and LIM kinase resulting in cofilin phosphorylation, increased actin polymerization and increased number of synaptic AMPA receptors [71,141].

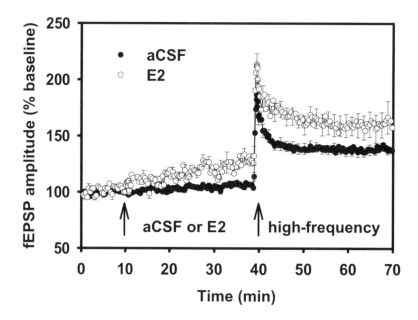

Fig. 2. Field EPSP (f-EPSP) recordings in area CA1 from male hippocampal slices perfused with either artificial cerebrospinal fluid (aCSF) or 17β-estradiol (E2) before and after high-frequency stimulation used to induce LTP. Reprinted with permission from [47]. Copyright 2010 American Psychological Association.

*Estrogen and hippocampal LTP in female rats*

In another series of animal studies, estrous cycle changes in rats were correlated with changes in synaptic plasticity. Hippocampal slices from cycling female rats in diestrus (low estrogen concentration) and proestrus (high estrogen concentration) were perfused with aCSF, and LTP was elicited by high-frequency stimulation. The difference in LTP magnitude between these groups following high-frequency stimulation was dramatic: slices from rats in proestrus exhibited LTP representing about a 50% increase over baseline, whereas slices from rats in diestrus had LTP values representing about a 25% increase over baseline [15] (Figure 3). These findings support the original interpretation of Teyler et al (1980) who identified changes in baseline synaptic transmission that were correlated with the phase of the estrus cycle in female rats at the time of hippocampal slice preparation.

Since the electrophysiological study above has shown that female rats in proestrus exhibited an increased magnitude of hippocampal LTP as compared to female in diestrus, results of a current study examining the effect of E2 on hippocampal LTP during the two critical time periods in the rat estrous cycle, proestrus and diestrus, are reported here. Estrous cycles of adult (3-5 mo) Sprague-Dawley rats were monitored for 10 days prior to any physiological experiments, and hippocampal slices were prepared from rats that were either in proestrus or diestrus. Recording and stimulating electrodes were positioned in the dendrites of area CA1 and Schaffer collaterals, respectively. Baseline stimulation (0.05 Hz, 100 μsec) was adjusted to elicit 50% of the maximum fEPSP amplitude. After 10 min of stable baseline stimulation, aCSF (control group) or E2 at a concentration of 100 pM (experimental group) was perfused in the slices for 30 min, and LTP was induced by a brief period of high-frequency stimulation (5 trains of 20 pulses at 100 Hz). Subsequent synaptic responses were monitored for 30 min post-LTP induction. The magnitude of LTP induced in area CA1 was larger in slices from proestrus rats, compared to slices from diestrus rats, as previously reported [15]. However, addition of E2 increased LTP in slices from diestrus rats, while it decreased LTP in slices from proestrus rats (Figure 4) [40]. These observations suggest that E2 alters hippocampal LTP in female rats, depending on the state of their estrous cycle (i.e., on the levels of circulating E2). In cycling female rats, when endogenous circu-

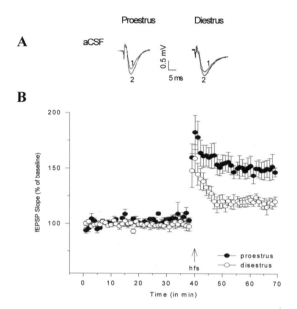

Fig. 3. (A) Representative waveforms from female rats in proestrus and diestrus for pre-high-frequency stimulation (1) and post-high frequency stimulation (2) periods, and (B) experimental outcomes of fEPSP activity before and after high-frequency stimulation (hfs) used to induce LTP. All slices were perfused with aCSF. Reprinted by permission from [15]. Copyright 2010 National Academy of Sciences, USA.

lating levels of E2 are at their highest levels (i.e., proestrus), LTP magnitude is increased, and exogenously applied E2 during proestrus decreases LTP magnitude, possibly through the activation of inhibitory or ceiling effect. When endogenous circulating levels of E2 are at their lowest levels (i.e., diestrus), the situation is completely reversed from that observed in the proestrus state. Here, LTP magnitude is decreased, and exogenously applied E2 increases LTP magnitude under this condition.

These results suggest that cyclic changes in estrogen levels occurring during the estrous cycle in female rats are associated with changes in the magnitude of LTP recorded from hippocampal CA1 cells. They also corroborate work mentioned earlier indicating facilitation of LTP induction by estrogen in ovariectomized female rats [30], increased LTP in the afternoon of proestrus of female rats [129], and support the results of a study showing improved memory performance with high estrogen levels in female rats [76].

### Estrogen, synaptic plasticity and aging in rats

It has been reported that during aging, when memory function declines, the processes of synaptic plasticity in hippocampus are altered. Specifically, LTP is impaired and the opposite process of long-term depression (LTD) is enhanced [5-7,38,39,51,72,73,91,92]. We recently replicated this effect of aging on LTD and discovered a profound action of estrogen on this process in aged male rats [47,127]. LTD was induced in CA1 region of hippocampal slices using standard conditions (stimulation of Schaffer collaterals at 1 Hz for 15 min) in adult (3-5 mo) and aged (18-24 mo) Sprague-Dawley rats. In agreement with earlier studies, the standard protocol for inducing LTD resulted in little or no LTD in slices from adult animals, but marked LTD in slices from aged animals (Figure 5A) [38,39]. Infusion of E2 in slices caused a slight increase in synaptic transmission (baseline), as in previous studies. It had little effect on LTD in slices from adult animals, but markedly attenuated LTD in slices from aged animals (Figure 5B). Thus, the prevention by E2 of age-related LTD enhancement may account, in part, for the protective effects of estrogen on memory functions in aged organisms reported in some studies (see below).

### Estrogen and two forms of LTP

As we noted earlier, Teyler and associates discovered a form of LTP in CA1 pyramidal neurons that is independent of NMDA receptors and involves voltage-dependent calcium channels [54]. This form of LTP is optimally induced by very high frequency

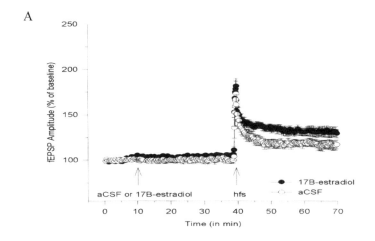

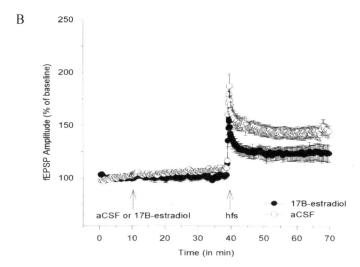

Fig. 4. (A) Changes in LTP in field CA1 of hippocampal slices from female rats in diestrus. 17β-estradiol (filled circles) enhanced LTP relative to control aCSF (open circles). (B) Changes in LTP in field CA1 of hippocampal slices from female rats in proestrus. 17β-estradiol impaired LTP relative to control. Reprinted by permission from [40]. Copyright 2010 Cambridge University Press.

stimulation of Schaffer collaterals (e.g., 200 Hz for 1 sec) and is blocked by nifedipine (L-type calcium channel blocker), but not by the NMDA receptor antagonist D-APV. In contrast, the NMDA receptor-dependent form of LTP in CA1 is best induced by lower frequencies of stimulation (e.g., 25 Hz) and is blocked by D-APV but not by nifedipine. The standard stimulation paradigm used for LTP induction (i.e., 100 Hz for 1 sec) therefore induces both forms of LTP [28,85].

We evaluated the effects of acute E2 application in hippocampal slices from male rats on both forms of LTP in CA1. Using 25 Hz tetanus of Schaffer collaterals, E2 and nifedipine were infused and extracellular field EPSPs recorded. E2 caused the expected increase in synaptic transmission and pronounced enhancement of LTP whereas nifedipine had no effect on either process, implying that under this condition, E2 was acting only on NMDA receptor-dependent LTP [143].

We then used a 100 Hz tetanus of Schaffer collaterals (in CA1 region of hippocampal slices from adult male rats), and both extracellular field EPSPs and intracellular EPSPs were recorded from pyramidal neurons and E2 and nifedipine were infused. E2 alone caused the expected increase in synaptic trans-

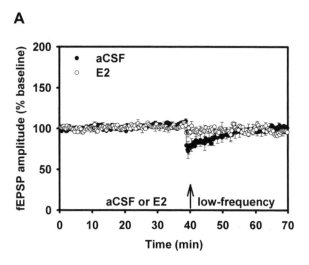

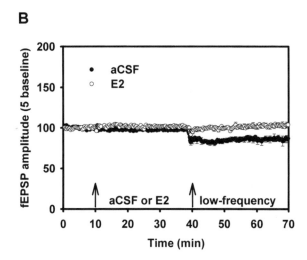

Fig. 5. Long-term depression (LTD) in (A) aCSF vs. 17β-estradiol (E2) in hippocampal slices recorded from young adult (3-5 mo) male Sprague-Dawley rats. (B) LTD recorded from aged (18-24 mo) male Sprague-Dawley rats. Low-frequency stimulation (used to induce LTD) delivered to aged rat slices perfused with 17β-estradiol failed to induce robust LTD. Reprinted with permission from [47]. Copyright 2010 American Psychological Association.

mission and pronounced enhancement of LTP, but both effects of E2 were reduced in magnitude by nifedipine. Therefore, under this condition, it would seem that E2 is acting by modulating both L-type voltage gated calcium channels and NMDA receptors. Intracellularly recorded EPSPs in response to paired subthreshold stimuli with a short interstimulus interval (50 ms) in the presence of E2 indicated an increase in EPSP amplitude to both stimuli without changes in paired-pulse ratio, strongly supporting a postsynaptic origin of E2 effects [1].

The possibility that E2 may modulate calcium influx through L-type calcium channels is consistent with the effects of aging on synaptic transmission and plasticity in hippocampus. Thus, it has been reported that aging is associated with enhanced activity of voltage-gated calcium channels in hippocampal CA1 neurons [26], and that blocking calcium influx through L-type calcium channels inhibits LTD induction and enhances LTP in aged animals in the CA1 region of hippocampal slices [92]. Blocking L-type calcium channels in hippocampus has been reported to enhance memory in several paradigms and particularly to enhance learning and memory processes in aged animals [33,98,99].

*Effects of estrogen in the 3xTg-AD mouse model of Alzheimer's disease*

A triple-transgenic mouse model of AD (called 3xTg-AD) has been developed in LaFerla's laboratory at the University of California, Irvine [93]. Mutations in three genes linked to AD and frontotemporal dementia were utilized. Human $APP_{SWE}$ and $tau_{P301L}$ transgenes were co-microinjected into single-cell embryos harvested from homozygous mutant $PS1_{M146V}$ knock-in mice. There are several advantages in using this model. First, the reported tight APP and tau linkage paired with the 'knock-in' PS1 approach yielded homozygous mice that breed readily, thus facilitating rapid, straightforward and cost-effective generation of a study colony. Second, and more importantly, the 3xTg-AD mouse exhibits age-related neuropathological phenotypes that includes both Aβ accumulation and hyperphosphorylated tau that exhibit a regional pattern similar to AD. Specifically, intracellular Aβ accumulates first in cortical regions (around 4 mo) and later in hippocampus, while tau hyperphosphorylation develops after Aβ accumulation (between 12-15 mo), beginning in limbic structures and progressing to cortical regions [16,93]. An *in vitro* study examining synaptic dysfunction in 3xTg-AD mice at 6 months found lowered levels of hippocampal basal synaptic transmission and reduced levels of LTP compared to non-transgenic mice [93]. In this report, it was suggested that the synaptic dysfunction found in transgenic mice might represent an early change preceding the accumulation of the hallmark pathological lesions that accompany AD. Another study using 3xTg-AD mice found that in ovariectomized transgenic mice, P4 blocked the beneficial effect on Aβ peptide accumulation provided by E2 treatment alone, but did not affect spontaneous alternation behavior in a Y-maze, a rodent model of working memory [27]. Significantly higher amounts of Aβ peptide was also found in female 3xTg-AD mice compared to male 3xTg-AD mice [62]. Research utilizing the 3xTg-AD mouse model has indicated that this mouse also shows hypothalamic-pituitary-adrenal (HPA) axis hyperactivity in both an age- and sex-dependent fashion [122]. 3xTg-AD mice exhibit HPA hyperactivity in response to stress, which is more pronounced in 9 month-old female 3xTg-AD mice compared to age-matched non-transgenic female mice and 3xTg-AD male mice [29].

In an *in vitro* electrophysiological study, synaptic plasticity was examined comparing the effects of E2 in hippocampal slices prepared from both gonadally-intact and gonadectomized 6 month-old male 3xTg-AD and wild-type (wt: 129/C57BL/6 F1 hybrid) mice [45]. E2 induced an increase in LTP in both 3xTg-AD groups (intact and gonadectomized) compared to their respective vehicle controls [45]. However, in the wild-type groups, E2 produced a smaller enhancement of LTP. These findings suggest a differential effect of E2 on hippocampal synaptic plasticity in 3xTg-AD and wild-type mice.

In a related behavioral study, the impact of allopregnanolone, a metabolite of P4 that promotes proliferation of neural progenitor cells derived from rat hippocampus and cerebral cortex [128], was evaluated on the hippocampal-dependent trace eyeblink conditioning task in 3xTg-AD and wild-type 3 month old mice [110]. The 3xTg-AD mice showed significantly lower levels of neurogenesis in the dentate gyrus than wild-type mice before training. In delay eyeblink conditioning, the conditioned stimulus (CS; e.g., auditory tone) onset precedes the unconditioned stimulus (US; e.g., an air puff to the cornea of the eye) onset, and the two CS and US stimuli overlap and coterminate with one another. In trace eyeblink conditioning, the CS precedes the US, and there is a short stimulus-free period (trace interval) between the CS offset and the US onset. We previously reported that both delay and trace eyeblink conditioning require the cerebellum, while the trace procedure also requires involvement of the hippocampus [114,117]. The 3x-Tg-AD mice were significantly impaired in learning the trace procedure compared to the wild-type mice. Mice were injected with 10 mg/kg allopregnanolone (s.c.) and 100 mg/kg BrdU to label dividing cells. Seven days later, mice were trained in the trace eyeblink-conditioning paradigm (250 ms tone followed by 100 ms, 60 Hz shock, 30 trials, 2 sessions/day for 5 days, trace interval of 250 ms). Following the learning trials, mice were returned to their home cages for another seven days, and subsequently tested for memory of the learned association. Allopregnanolone treatments markedly and significantly enhanced neurogenesis in the 3xTg-AD mice but not in the wild-type mice. Allopregnanolone treatment also enhanced the rate of learning in 3xTg-AD mice, increased the magnitude of the learning performance to the level of the wild-type mice and reversed their memory deficit. Importantly, in the 3xTg-AD mice there was a high and significant correlation (r = 0.80) between number of conditioned responses and number of new surviving neurons [110]. These results suggest that allopregnanolone is a potent cognitive enhancer in a mouse model of AD, and that allopregnanolone could be a potential therapeutic to prevent or delay cognitive deficits associated with AD.

In summary, the neuropathology of AD exhibited in the 3xTg-AD mouse (development of Aβ and tau pathology) provides an exciting model with which to study the molecular, cellular and behavioral interactions between the many processes underlying learning and memory function, and their modifications during the post-menopausal period and aging-related disorders.

*Estrogen and cellular neuroprotection*

In a well-established model of estrogen-induced neuroprotection, primary cultures of hippocampal neurons were exposed to the excitatory amino acid, glutamic acid, and neuronal injury was assessed by measuring lactate dehydrogenase (LDH) release in the culture medium. A 5 min treatment with 100 μM glutamate caused significant cell death compared to control conditions; neuronal cell death was significantly decreased following pre-treatment with E2 [89]. Maximal neuroprotection of approximately 18% was provided by a concentration of 10 ng/ml E2 [89].

To investigate the mechanism underlying E2-mediated neuroprotection, and in particular, the contributions of changes in intracellular calcium concentration, calcium concentration was determined by imaging techniques and microfluorescence in cultured hippocampal neurons. Surprisingly, in neurons pretreated with E2, changes in intracellular calcium concentration elicited by glutamate application were increased by about 70% [88]. E2 itself induced a rapid increase in intracellular calcium concentration within minutes of exposure, an effect that was blocked by an L-type calcium channel antagonist suggesting that E2 directly or indirectly regulates some properties of these voltage-gated calcium channels [139]. E2-induced calcium influx was required for E2-mediated activation of a biochemical signaling pathway involving Src, ERK, CREB, and Bcl-2; a schematic diagram of the molecular signaling cascade leading to E2-induced neuroprotection was described in [139]. These results demonstrate that at the single-cell level, an E2 membrane-associated receptor mediates rapid E2 effects in cultured neurons. Estrogen-induced neuroprotection against excitotoxic glutamate also requires the mitogen-activated protein kinase (MAPK) cascade in primary cortical neuron cultures [90,109]. A similar neuroprotective effect of E2 against NMDA-mediated neurotoxicity was reported in cultured hippocampal slices, and this effect also involved the activation of the MAP kinase pathway [15].

Collectively, these reports indicate that although significant progress has been made regarding the identification of the cellular mechanisms involved in E2-mediated neuroprotection, there is a need to further elucidate a number of unresolved issues regarding the complex and mostly indirect ways in which estrogen interacts with numerous cellular signaling pathways regulating cell survival and death.

*Molecular mechanisms of estrogen effects in brain*

Recent results from several laboratories have provided a general framework to understand the mechanisms underlying the multiple effects of E2 on synaptic structure and function (Figure 6) [75]. Physiological concentrations of E2 (10 pM-1 nM) interact with Erα and Erβ receptors to produce both direct and indirect genomic effects. The direct genomic effects are due to interactions between E2 and traditional cytoplasmic receptors followed by tran location of the hormone-receptor complex to the nucleus and the regulation of transcription of specific genes, through interactions with ERE regulatory elements present in these genes. In neurons, these include anti-apoptotic genes of the bcl-2 family, which are probably responsible for the neuroprotective effects of E2 observed in a number of models of neuronal death. In astrocytes, they include the GFAP (down regulation) and laminin (up regulation) genes, which might be involved in regulating sprouting responses following lesions, as well as in normal astrocyte activation in brains from aged animals [70]. The indirect genomic effects of E2 might be linked to the stimulation of the phosphoinositol-3 (PI3) kinase/Akt system [32,108] and/or of a G protein, and/or of Src tyrosine kinase and ERK/MAP kinase pathways [112].

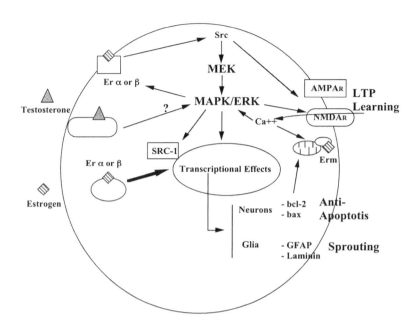

Fig. 6. Schematic representation of the general hypotheses linking estrogen/testosterone with the MAPK/ERK pathway. Reprinted with permission from [40]. Copyright 2010 Cambridge University Press.

The MAP kinase pathway occupies a central place in the regulation of synaptic plasticity [80,116]. Pharmacological manipulations directed at blocking this pathway have consistently produced impairments in synaptic plasticity and learning and memory, and this pathway is activated with LTP-inducing tetanus or in different learning paradigms [11,20,22,102]. We have shown that endogenous estrogen levels in cycling female rats produce a tonic phosphorylation/activation of extracellular signal-regulated kinase 2 (ERK2)/MAP kinase [15]. In addition, we have shown that this activation of the MAP kinase pathway is also linked to the regulation of glutamate ionotropic receptors and might be involved in the "cognitive enhancing" effects of E2. Indeed, the acute estrogen-mediated enhancement of LTP is mediated by activation of a src tyrosine kinase pathway [14]. Thus, acute application of the src inhibitor PP2 in the perfusing medium of hippocampal slices from adult male rats abolished E2-mediated enhancement of both synaptic transmission and LTP, but had no effect on LTP itself. Similarly, this pathway might also be involved in the neuroprotective effects of E2, as MAP kinase inhibitors have consistently been shown to block the neuroprotective effects of E2 in a variety of models of neurodegeneration. Moreover, growth factors and other factors providing neuroprotection, such as PDGF and BDNF, also use the MAP kinase pathway for their neuroprotective effects.

Interestingly, it appears that Erα stimulation is critically involved in the neuroprotective effect of estrogen, as Erα knock-out mice are not protected by 17β-estradiol against ischemia-induced neuronal damage [35]. Furthermore, recent results obtained from the same knock-out mice suggest the possible existence of novel E2 receptors responsible for the activation of the ERK/MAP kinase pathway [112]. These results indicate that several steps described in Figure 6 remain to be elucidated.

*P4 regulation of memory and neuronal excitability*

Much of the literature cited above suggests that E2 exerts a variety of effects on neural structures and function, particularly within hippocampus [83,136,137]. Electrophysiological studies have shown that E2 enhances hippocampal CA1 synaptic transmission and plasticity by increasing NMDA and AMPA receptor activity, resulting in neuronal excitation [43,121,132,133]. While these studies have focused exclusively on the effects of E2 on brain structure and function, more recent studies have investigated the effects of P4 and its neuroactive metabolites, allopregnanolone (APα) and pregnanolone, on cognitive function and neural excitability.

P4 and APα are known to regulate cognitive function, particularly those related to mood and/or associated with changes in the menstrual cycle (e.g., postpartum depression, major depression, epilepsy) [4]. The GABAergic system also participates in major depression (for review, see [21]). The GABA$_A$ receptor mediates the majority of rapid (1-100 ms) synaptic inhibition in the mammalian brain, and APα and pregnanolone exert both anxiolytic and anesthetic effects by enhancing GABA-stimulated chloride conductance. This enhanced conductance serves to hyperpolarize postsynaptic membranes and results in neuronal inhibition [77,78]. More recent evidence suggests that specific neurosteroids 'fine-tune' neuronal inhibition via the GABAergic system [84]. Another recent study has shown the ability of P4 to influence cognition and memory of biologically salient stimuli within healthy young women [125]. Here, a single dose of P4 was orally administered to women who were then asked to memorize and recognize faces while undergoing functional magnetic resonance imaging. The results revealed that P4 decreases recognition accuracy without affecting reaction times. P4 also decreases amygdala and fusiform gyrus activity elicited by faces during memory encoding, supporting the conclusion that P4 alters memory function by influencing amygdala activity [125].

In animals, P4 and its metabolites severely impair learning and memory performance immediately following administration in the Morris water maze test [66,107]. Although the mechanism underlying this impairment is unknown, a recent study demonstrated that pretreatment of rats with APα induces a partial tolerance against the acute effects of APα in the Morris water maze test [124]. These authors suggest that prolonged exposure to APα in women (e.g., pregnancy, postmenopausal hormone replacement therapy, menstrual cycle) may alter cognitive behavior such as learning and memory processing, possibly through a GABAergic-dependent mechanism.

*P4 and synaptic plasticity*

Few studies have examined the acute effects of P4 on synaptic transmission and plasticity, with the results of these early studies being mostly contradictory. P4 (10 μM) reportedly has no effect on LTP in CA1 slices from 4-week-old rats, but no non-drug control was used [64]. In another study, P4 (8-10 μM in CA1 slices) significantly enhanced synaptic transmission, as seen by increased field potential and population spike amplitude; however, following seizure-induced tetanus, P4 decreased field potential, population spike responses, and duration of afterdischarges [36]. In whole cell patch clamp of pyramidal neurons from slices of prelimbic cortex, P4 (100 μM) had no effect on the frequency of excitatory postsynaptic currents (EPSCs), but inhibited dopamine-induced increases in EPSCs [37]. P4 dose-response functions were not obtained in any of these studies.

To address some of the concerns raised above, we evaluated the effects of several concentrations of P4 on basal synaptic transmission and two forms of synaptic plasticity in rat hippocampal slices: LTP and LTD. P4 (1 μM) acts relatively quickly (~30 min) to cause a decrease in CA1 fEPSPs [46]. From the dose-response experiments reported in this study, the highest concentration of P4 tested (1 μM) resulted in a decreased level of basal synaptic transmission that persisted for at least 45 min following initial P4 application. It was also found that not only can P4 cause a decrease in hippocampal synaptic transmission, but also markedly decrease LTP in CA1 neurons from adult, ovariectomized rats. Furthermore, we found no evidence that P4 causes any change in synaptic activity following low-frequency stimulation used to induce LTD. In contrast to P4's effect on LTP, the same concentrations of P4 did not affect LTD, suggesting that high-frequency stimulation could significantly increase inhibitory action of P4 on excitatory synaptic transmission in hippocampus [46] (Figure 7). In a most recent study, we examined the effects of E2 and P4 treatment on hippocampal synaptic plasticity in ovariectomized rats. Young adult rats (3-5 mo) were ovariectomized, and after one week of recovery, were implanted with either E2, P4, or E2 + P4 hormone pellets. At the end of 60 days of hormonal treatment, hippocampal slices were prepared from the experimental animals, and using extracellular recording techniques from CA1 hippocampal neurons, tested for baseline synaptic transmission, LTP and LTD. Our results indicate that aCSF-treated slices from ovariectomized animals implanted with P4 exhibited the lowest amount of LTP (~110% of baseline), while aCSF-treated slices prepared from E2 + P4 animals exhibited the highest amount of LTP (~140% of baseline),

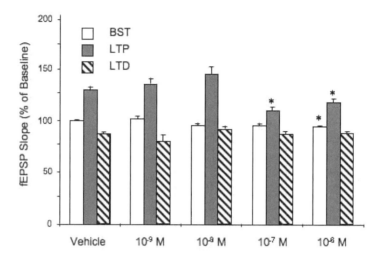

Fig. 7. Bar graph representing fEPSP slope values (vehicle and P4 at different concentrations) during baseline synaptic transmission (BST), LTP and LTD periods. At the highest concentration tested, P4 attenuated BST and LTP, but had no effect on LTD. Reprinted with permission from [46]. Copyright 2010 Cold Spring Harbor Laboratory Press.

similar to that observed in gonadally intact animals [48]. Finally, intracellular studies suggest that the effects of P4 on hippocampal synaptic transmission may be mediated, at least in part, by activation of $GABA_A$ receptors, as isolated currents were completely blocked by the $GABA_A$-receptor antagonist picrotoxin (100 μM) [46]. This is consistent with another study linking $GABA_A$ receptors with P4 action [61]. In sum, we find that P4 at concentrations greater than 0.1 μM significantly decreases LTP, while having no effect on LTD, most likely due to the young age of the subjects. These findings suggest that P4 action is dependent on high-frequency stimulation and a compensatory recruitment of $GABA_A$ receptors, either by direct action of P4, or its metabolites. Furthermore, *in vivo* treatment with E2 + P4 in ovariectomized rats restores baseline synaptic transmission and LTP levels to those found in intact female rats.

*Summary of estrogen's and progesterone's effects on synaptic plasticity*

The studies reviewed above establish several fundamental characteristics of the effects of E2 and P4 on synaptic transmission in mammalian central nervous system (CNS). E2 acts rapidly via presumed membrane mechanisms to enhance both NMDA and AMPA receptor/channel responses elicited by glutamate released from excitatory presynaptic terminals. P4 acts rapidly as well, presumably by enhancing $GABA_A$ activity.

E2 rapidly and markedly enhances hippocampal LTP in CA1 neurons of adult, male rats, whereas P4 decreases hippocampal LTP. E2-mediated LTP enhancement is due to an increase in NMDA receptor and AMPA receptor functions. Both possibilities are consistent with intracellular data, which also suggest the activation of $GABA_A$ receptor activity by P4 to decrease LTP. Changes in E2 levels in cycling female rats have also been correlated with changes in synaptic plasticity, as measured by changes in LTP magnitude. Furthermore, E2 has also been found to enhance LTP from male hippocampal slices prepared from 3xTg-AD mice. These findings suggest a mechanism by which naturally fluctuating endogenous hormone levels can impact a cellular model associated with important aspects of learning and/or memory in mammalian CNS.

To the extent that LTP is a mechanism involved in processes of coding and storage of information, i.e., in memory formation, E2 appears to enhance these processes while P4 appears to diminish them. Indeed, E2-mediated LTP enhancement suggests a possible mechanism by which E2 can exert its facilitatory effects on memory processes in humans. Clinical evidence indicates that estrogenic steroids can enhance cognitive functions in humans, particularly in postmenopausal women [57,58,69]; however, some prospective observational studies have yet to find a protective effect of E2 on either cognition, or the incidence of dementia [8,79]. Understanding the mechanisms underlying the changes in some of the effects of E2 and P4 associated with aging will represent a significant advance in understanding the mechanisms involved in age-related decline in cognitive function.

## ACKNOWLEDGMENTS

This work was supported by the National Institute of Health Grants NIA P01 AG014751 (Caleb E. Finch), Project 3 (RFT); P01 AG026572 (RDB), Project 2 (RFT); and R01 AG023742 (Diana Woodruff-Pak & RFT).

## References

[1] G. Akopian, M.R. Foy and R.F. Thompson, 17B-estradiol enhancement of LTP involves activation of voltage-dependent calcium channels, *Society for Neuroscience Abstracts* **255.2** (2003).

[2] S. Asthana, L.D. Baker, S. Craft, F.Z. Stanczyk, R.C. Veith, M.A. Raskind and S.R. Plymate, High-dose estradiol improves cognition for women with AD: results of a randomized study, *Neurology* **57** (2001), 605-612.

[3] T. Backstrom, Epileptic seizures in women related to plasma estrogen and progesterone during the menstrual cycle, *Acta Neurol Scand* **54** (1976), 321-347.

[4] T. Backstrom, L. Andreen, V. Birzniece, I. Bjorn, I.M. Johansson, M. Nordenstam-Haghjo, S. Nyberg, I. Sundstrom-Poromaa, G. Wahlstrom, M. Wang and D. Zhu, The role of hormones and hormonal treatments in premenstrual syndrome, *CNS Drugs* **17** (2003), 325-342.

[5] C.A. Barnes, Memory deficits associated with senescence: a neurophysiological and behavioral study in the rat, *J Comp Physiol Psychol* **93** (1979), 74-104.

[6] C.A. Barnes, G. Rao, T.C. Foster and B.L. McNaughton, Region-specific age effects on AMPA sensitivity: electrophysiological evidence for loss of synaptic contacts in hippocampal field CA1, *Hippocampus* **2** (1992), 457-468.

[7] C.A. Barnes, Normal aging: regionally specific changes in hippocampal synaptic transmission, *Trends Neurosci* **17** (1994), 13-18.

[8] E. Barrett-Connor and D. Kritz-Silverstein, Estrogen replacement therapy and cognitive function in older women, *Jama* **269** (1993), 2637-2641.

[9] M. Baudry, J.L. Davis and R.F. Thompson, *Advances in Synaptic Plasticity*, The MIT Press, Cambridge, MA, 2000.

[10] M.F. Bear and R.C. Malenka, Synaptic plasticity: LTP and LTD, *Curr Opin Neurobiol* **4** (1994), 389-399.

[11] D.E. Berman, S. Hazvi, K. Rosenblum, R. Seger and Y. Dudai, Specific and differential activation of mitogen-activated protein kinase cascades by unfamiliar taste in the insular cortex of the behaving rat, *J Neurosci* **18** (1998), 10037-10044.

[12] B. Berry, R. McMahan and M. Gallagher, Spatial learning and memory at defined points of the estrous cycle: effects on performance of a hippocampal-dependent task, *Behav Neurosci* **111** (1997), 267-274.

[13] G. Bi and M. Poo, Synaptic modification by correlated activity: Hebb's postulate revisited, *Annu Rev Neurosci* **24** (2001), 139-166.

[14] R. Bi, G. Broutman, M.R. Foy, R.F. Thompson and M. Baudry, The tyrosine kinase and mitogen-activated protein kinase pathways mediate multiple effects of estrogen in hippocampus, *Proc Natl Acad Sci U S A* **97** (2000), 3602-3607.

[15] R. Bi, M.R. Foy, R.M. Vouimba, R.F. Thompson and M. Baudry, Cyclic changes in estradiol regulate synaptic plasticity through the MAP kinase pathway, *Proc Natl Acad Sci U S A* **98** (2001), 13391-13395.

[16] L.M. Billings, S. Oddo, K.N. Green, J.L. McGaugh and F.M. LaFerla, Intraneuronal Abeta causes the onset of early Alzheimer's disease-related cognitive deficits in transgenic mice, *Neuron* **45** (2005), 675-688.

[17] H.A. Bimonte-Nelson, R.S. Singleton, C.L. Hunter, K.L. Price, A.B. Moore and A.C. Granholm, Ovarian hormones and cognition in the aged female rat: I. Long-term, but not short-term, ovariectomy enhances spatial performance, *Behav Neurosci* **117** (2003), 1395-1406.

[18] H.A. Bimonte-Nelson, R.S. Singleton, B.J. Williams and A.C. Granholm, Ovarian hormones and cognition in the aged female rat: II. progesterone supplementation reverses the cognitive enhancing effects of ovariectomy, *Behav Neurosci* **118** (2004), 707-714.

[19] T.V. Bliss and G.L. Collingridge, A synaptic model of memory: long-term potentiation in the hippocampus, *Nature* **361** (1993), 31-39.

[20] S. Blum, A.N. Moore, F. Adams and P.K. Dash, A mitogen-activated protein kinase cascade in the CA1/CA2 subfield of the dorsal hippocampus is essential for long-term spatial memory, *J Neurosci* **19** (1999), 3535-3544.

[21] P. Brambilla, J. Perez, F. Barale, G. Schettini and J.C. Soares, GABAergic dysfunction in mood disorders, *Mol Psychiatry* **8** (2003), 721-737, 715.

[22] R. Brambilla, N. Gnesutta, L. Minichiello, G. White, A.J. Roylance, C.E. Herron, M. Ramsey, D.P. Wolfer, V. Cestari, C. Rossi-Arnaud, S.G. Grant, P.F. Chapman, H.P. Lipp, E. Sturani and R. Klein, A role for the Ras signalling pathway in synaptic transmission and long-term memory, *Nature* **390** (1997), 281-286.

[23] R.D. Brinton, P. Proffitt, J. Tran and R. Luu, Equilin, a principal component of the estrogen replacement therapy premarin, increases the growth of cortical neurons

[24] via an NMDA receptor-dependent mechanism, *Exp Neurol* **147** (1997), 211-220.
[24] R.D. Brinton, J. Tran, P. Proffitt and M. Montoya, 17 beta-Estradiol enhances the outgrowth and survival of neocortical neurons in culture, *Neurochem Res* **22** (1997), 1339-1351.
[25] R. Brookmeyer, S. Gray and C. Kawas, Projections of Alzheimer's disease in the United States and the public health impact of delaying disease onset, *Am J Public Health* **88** (1998), 1337-1342.
[26] L.W. Campbell, S.Y. Hao, O. Thibault, E.M. Blalock and P.W. Landfield, Aging changes in voltage-gated calcium currents in hippocampal CA1 neurons, *J Neurosci* **16** (1996), 6286-6295.
[27] J.C. Carroll, E.R. Rosario, L. Chang, F.Z. Stanczyk, S. Oddo, F.M. LaFerla and C.J. Pike, Progesterone and estrogen regulate Alzheimer-like neuropathology in female 3xTg-AD mice, *J Neurosci* **27** (2007), 13357-13365.
[28] I. Cavus and T. Teyler, Two forms of long-term potentiation in area CA1 activate different signal transduction cascades, *J Neurophysiol* **76** (1996), 3038-3047.
[29] L.K. Clinton, L.M. Billings, K.N. Green, A. Caccamo, J. Ngo, S. Oddo, J.L. McGaugh and F.M. LaFerla, Age-dependent sexual dimorphism in cognition and stress response in the 3xTg-AD mice, *Neurobiol Dis* **28** (2007), 76-82.
[30] D.A. Cordoba Montoya and H.F. Carrer, Estrogen facilitates induction of long term potentiation in the hippocampus of awake rats, *Brain Res* **778** (1997), 430-438.
[31] J. Couzin, Estrogen research. The great estrogen conundrum, *Science* **302** (2003), 1136-1138.
[32] S.R. Datta, A. Brunet and M.E. Greenberg, Cellular survival: a play in three Akts, *Genes Dev* **13** (1999), 2905-2927.
[33] J.F. Disterhoft, W.W. Wu and M. Ohno, Biophysical alterations of hippocampal pyramidal neurons in learning, ageing and Alzheimer's disease, *Ageing Res Rev* **3** (2004), 383-406.
[34] G.P. Dohanich, Gonadal steroids, learning and memory, in: *Hormones, brain and behavior*, R.T. Rubin, eds., Academic Press, San Diego, 2002, pp. 265-327.
[35] D.B. Dubal, H. Zhu, J. Yu, S.W. Rau, P.J. Shughrue, I. Merchenthaler, M.S. Kindy and P.M. Wise, Estrogen receptor alpha, not beta, is a critical link in estradiol-mediated protection against brain injury, *Proc Natl Acad Sci U S A* **98** (2001), 1952-1957.
[36] H.E. Edwards, T. Epps, P.L. Carlen and J.M. N, Progestin receptors mediate progesterone suppression of epileptiform activity in tetanized hippocampal slices in vitro, *Neuroscience* **101** (2000), 895-906.
[37] X.Q. Feng, Y. Dong, Y.M. Fu, Y.H. Zhu, J.L. Sun, Z. Wang, F.Y. Sun and P. Zheng, Progesterone inhibition of dopamine-induced increase in frequency of spontaneous excitatory postsynaptic currents in rat prelimbic cortical neurons, *Neuropharmacology* **46** (2004), 211-222.
[38] T.C. Foster and C.M. Norris, Age-associated changes in Ca(2+)-dependent processes: relation to hippocampal synaptic plasticity, *Hippocampus* **7** (1997), 602-612.
[39] T.C. Foster, Involvement of hippocampal synaptic plasticity in age-related memory decline, *Brain Res Brain Res Rev* **30** (1999), 236-249.
[40] M. Foy, M. Baudry and R. Thompson, Estrogen and hippocampal synaptic plasticity, *Neuron Glia Biol* **1** (2004), 327-338.
[41] M.R. Foy, Neuromodulation: Effects of estradiol and THC on brain excitability, *Doctoral Dissertation, Kent State University, Kent, OH* (1983).
[42] M.R. Foy and T.J. Teyler, 17-alpha-Estradiol and 17-beta-estradiol in hippocampus, *Brain Res Bull* **10** (1983), 735-739.
[43] M.R. Foy, J. Xu, X. Xie, R.D. Brinton, R.F. Thompson and T.W. Berger, 17beta-estradiol enhances NMDA receptor-mediated EPSPs and long-term potentiation, *J Neurophysiol* **81** (1999), 925-929.
[44] M.R. Foy, 17beta-estradiol: effect on CA1 hippocampal synaptic plasticity, *Neurobiol Learn Mem* **76** (2001), 239-252.
[45] M.R. Foy, G. Akopian and R.F. Thompson, 17B-estradiol enhances LTP in hippocampal slices from 3xTg-AD mice, *Society for Neuroscience Abstracts* **536.15** (2006).
[46] M.R. Foy, G. Akopian and R.F. Thompson, Progesterone regulation of synaptic transmission and plasticity in rodent hippocampus, *Learn Mem* **15** (2008), 820-822.
[47] M.R. Foy, M. Baudry, J.G. Foy and R.F. Thompson, 17beta-estradiol modifies stress-induced and age-related changes in hippocampal synaptic plasticity, *Behav Neurosci* **122** (2008), 301-309.
[48] M.R. Foy, G.K. Akopian, Y.K. Kim, S. Lin, T. Morgan and R.F. Thompson, Effects of estrogen and progesterone treatment on hippocampal synaptic plasticity in ovariectomized rats, *Society for Neuroscience Abstracts* **499.1** (2009).
[49] L.A. Galea, M. Kavaliers, K.P. Ossenkopp and E. Hampson, Gonadal hormone levels and spatial learning performance in the Morris water maze in male and female meadow voles, Microtus pennsylvanicus, *Horm Behav* **29** (1995), 106-125.
[50] A.H. Gazzaley, N.G. Weiland, B.S. McEwen and J.H. Morrison, Differential regulation of NMDAR1 mRNA and protein by estradiol in the rat hippocampus, *J Neurosci* **16** (1996), 6830-6838.
[51] Y. Geinisman, L. Detoledo-Morrell, F. Morrell and R.E. Heller, Hippocampal markers of age-related memory dysfunction: behavioral, electrophysiological and morphological perspectives, *Prog Neurobiol* **45** (1995), 223-252.
[52] Y. Geinisman, Structural synaptic modifications associated with hippocampal LTP and behavioral learning, *Cereb Cortex* **10** (2000), 952-962.
[53] R.B. Gibbs, D. Wu, L.B. Hersh and D.W. Pfaff, Effects of estrogen replacement on the relative levels of choline acetyltransferase, trkA, and nerve growth factor messenger RNAs in the basal forebrain and hippocampal formation of adult rats, *Exp Neurol* **129** (1994), 70-80.
[54] L.M. Grover and T.J. Teyler, Two components of long-term potentiation induced by different patterns of afferent activation, *Nature* **347** (1990), 477-479.
[55] Q. Gu and R.L. Moss, 17 beta-Estradiol potentiates kainate-induced currents via activation of the cAMP cascade, *J Neurosci* **16** (1996), 3620-3629.
[56] J. Hardy and D.J. Selkoe, The amyloid hypothesis of Alzheimer's disease: progress and problems on the road to therapeutics, *Science* **297** (2002), 353-356.

[57]  V.W. Henderson, Estrogen replacement therapy for the prevention and treatment of Alzheimer's disease, *CNS Drugs* **8** (1997), 343-351.

[58]  V.W. Henderson, *Hormone Therapy and the Brain: A Clinical Perspective on the Role of Estrogen*, Parthenon Publishing, New York, 2000.

[59]  V.W. Henderson, A. Paganini-Hill, B.L. Miller, R.J. Elble, P.F. Reyes, D. Shoupe, C.A. McCleary, R.A. Klein, A.M. Hake and M.R. Farlow, Estrogen for Alzheimer's disease in women: randomized, double-blind, placebo-controlled trial, *Neurology* **54** (2000), 295-301.

[60]  V.W. Henderson, K.S. Benke, R.C. Green, L.A. Cupples and L.A. Farrer, Postmenopausal hormone therapy and Alzheimer's disease risk: interaction with age, *J Neurol Neurosurg Psychiatry* **76** (2005), 103-105.

[61]  M.B. Herd, D. Belelli and J.J. Lambert, Neurosteroid modulation of synaptic and extrasynaptic GABA(A) receptors, *Pharmacol Ther* **116** (2007), 20-34.

[62]  C. Hirata-Fukae, H.F. Li, H.S. Hoe, A.J. Gray, S.S. Minami, K. Hamada, T. Niikura, F. Hua, H. Tsukagoshi-Nagai, Y. Horikoshi-Sakuraba, M. Mughal, G.W. Rebeck, F.M. Laferla, M.P. Mattson, N. Iwata, T.C. Saido, W.L. Klein, K.E. Duff, P.S. Aisen and Y. Matsuoka, Females exhibit more extensive amyloid, but not tau, pathology in an Alzheimer transgenic model, *Brain Res* (2008),

[63]  L.X. Hy and D.M. Keller, Prevalence of AD among whites: a summary by levels of severity, *Neurology* **55** (2000), 198-204.

[64]  K. Ito, K.L. Skinkle and T.P. Hicks, Age-dependent, steroid-specific effects of oestrogen on long-term potentiation in rat hippocampal slices, *J Physiol* **515 ( Pt 1)** (1999), 209-220.

[65]  K. Ito, H. Utsunomiya, N. Yaegashi and H. Sasano, Biological roles of estrogen and progesterone in human endometrial carcinoma--new developments in potential endocrine therapy for endometrial cancer, *Endocr J* **54** (2007), 667-679.

[66]  I.M. Johansson, V. Birzniece, C. Lindblad, T. Olsson and T. Backstrom, Allopregnanolone inhibits learning in the Morris water maze, *Brain Res* **934** (2002), 125-131.

[67]  Z. Jonasson, Meta-analysis of sex differences in rodent models of learning and memory: a review of behavioral and biological data, *Neurosci Biobehav Rev* **28** (2005), 811-825.

[68]  D.L. Kampen and B.B. Sherwin, Estrogen use and verbal memory in healthy postmenopausal women, *Obstet Gynecol* **83** (1994), 979-983.

[69]  C. Kawas, S. Resnick, A. Morrison, R. Brookmeyer, M. Corrada, A. Zonderman, C. Bacal, D.D. Lingle and E. Metter, A prospective study of estrogen replacement therapy and the risk of developing Alzheimer's disease: the Baltimore Longitudinal Study of Aging, *Neurology* **48** (1997), 1517-1521.

[70]  S.G. Kohama, J.R. Goss, C.E. Finch and T.H. McNeill, Increases of glial fibrillary acidic protein in the aging female mouse brain, *Neurobiol Aging* **16** (1995), 59-67.

[71]  E.A. Kramar, L.Y. Chen, N.J. Brandon, C.S. Rex, F. Liu, C.M. Gall and G. Lynch, Cytoskeletal changes underlie estrogen's acute effects on synaptic transmission and plasticity, *J Neurosci* **29** (2009), 12982-12993.

[72]  P.W. Landfield and G. Lynch, Impaired monosynaptic potentiation in in vitro hippocampal slices from aged, memory-deficient rats, *J Gerontol* **32** (1977), 523-533.

[73]  P.W. Landfield, T.A. Pitler and M.D. Applegate, The effects of high Mg2+-to-Ca2+ ratios on frequency potentiation in hippocampal slices of young and aged rats, *J Neurophysiol* **56** (1986), 797-811.

[74]  P.W. Landfield and S.A. Deadwyler, *Long-Term Potentiation: From Biophysics to Behavior*, Alan R. Liss, Inc., New York, 1988.

[75]  S.J. Lee and B.S. McEwen, Neurotrophic and neuroprotective actions of estrogens and their therapeutic implications, *Annu Rev Pharmacol Toxicol* **41** (2001), 569-591.

[76]  B. Leuner, S. Mendolia-Loffredo and T.J. Shors, High levels of estrogen enhance associative memory formation in ovariectomized females, *Psychoneuroendocrinology* **29** (2004), 883-890.

[77]  M.D. Majewska, N.L. Harrison, R.D. Schwartz, J.L. Barker and S.M. Paul, Steroid hormone metabolites are barbiturate-like modulators of the GABA receptor, *Science* **232** (1986), 1004-1007.

[78]  M.D. Majewska, Neurosteroids: endogenous bimodal modulators of the GABAA receptor. Mechanism of action and physiological significance, *Prog Neurobiol* **38** (1992), 379-395.

[79]  K. Matthews, J. Cauley, K. Yaffe and J.M. Zmuda, Estrogen replacement therapy and cognitive decline in older community women, *J Am Geriatr Soc* **47** (1999), 518-523.

[80]  C. Mazzucchelli and R. Brambilla, Ras-related and MAPK signalling in neuronal plasticity and memory formation, *Cell Mol Life Sci* **57** (2000), 604-611.

[81]  B. McEwen, J. Gerlach and D. Micco, Putative glucocorticoid receptors in hippocampus and other regions of the rat brain, in: *The Hippocampus, Vol 2: Neurophysiology and behavior*, R.I.K. Pribram, eds., Plenum, 1975, pp.

[82]  B. McEwen and S.E. Alves, Estrogen actions in the central nervous system, *Endocr Rev* **20** (1999), 279-307.

[83]  B. McEwen, K. Akama, S. Alves, W.G. Brake, K. Bulloch, S. Lee, C. Li, G. Yuen and T.A. Milner, Tracking the estrogen receptor in neurons: implications for estrogen-induced synapse formation, *Proc Natl Acad Sci U S A* **98** (2001), 7093-7100.

[84]  E.A. Mitchell, M.B. Herd, B.G. Gunn, J.J. Lambert and D. Belelli, Neurosteroid modulation of GABAA receptors: molecular determinants and significance in health and disease, *Neurochem Int* **52** (2008), 588-595.

[85]  S.L. Morgan, C.M. Coussens and T.J. Teyler, Depotentiation of vdccLTP requires NMDAR activation, *Neurobiol Learn Mem* **76** (2001), 229-238.

[86]  R.A. Mulnard, C.W. Cotman, C. Kawas, C.H. van Dyck, M. Sano, R. Doody, E. Koss, E. Pfeiffer, S. Jin, A. Gamst, M. Grundman, R. Thomas and L.J. Thal, Estrogen replacement therapy for treatment of mild to moderate Alzheimer disease: a randomized controlled trial. Alzheimer's Disease Cooperative Study, *Jama* **283** (2000), 1007-1015.

[87]  D.D. Murphy, N.B. Cole, V. Greenberger and M. Segal, Estradiol increases dendritic spine density by reducing GABA neurotransmission in hippocampal neurons, *J Neurosci* **18** (1998), 2550-2559.

[88] J. Nilsen and R.D. Brinton, Impact of progestins on estradiol potentiation of the glutamate calcium response, *Neuroreport* **13** (2002), 825-830.

[89] J. Nilsen and R.D. Brinton, Impact of progestins on estrogen-induced neuroprotection: synergy by progesterone and 19-norprogesterone and antagonism by medroxyprogesterone acetate, *Endocrinology* **143** (2002), 205-212.

[90] J. Nilsen and R.D. Brinton, Divergent impact of progesterone and medroxyprogesterone acetate (Provera) on nuclear mitogen-activated protein kinase signaling, *Proc Natl Acad Sci U S A* **100** (2003), 10506-10511.

[91] C.M. Norris, D.L. Korol and T.C. Foster, Increased susceptibility to induction of long-term depression and long-term potentiation reversal during aging, *J Neurosci* **16** (1996), 5382-5392.

[92] C.M. Norris, S. Halpain and T.C. Foster, Reversal of age-related alterations in synaptic plasticity by blockade of L-type Ca2+ channels, *J Neurosci* **18** (1998), 3171-3179.

[93] S. Oddo, A. Caccamo, J.D. Shepherd, M.P. Murphy, T.E. Golde, R. Kayed, R. Metherate, M.P. Mattson, Y. Akbari and F.M. LaFerla, Triple-transgenic model of Alzheimer's disease with plaques and tangles: intracellular Abeta and synaptic dysfunction, *Neuron* **39** (2003), 409-421.

[94] M.G. Packard and L.A. Teather, Posttraining estradiol injections enhance memory in ovariectomized rats: cholinergic blockade and synergism, *Neurobiol Learn Mem* **68** (1997), 172-188.

[95] A. Paganini-Hill and V.W. Henderson, Estrogen deficiency and risk of Alzheimer's disease in women, *Am J Epidemiol* **140** (1994), 256-261.

[96] A. Paganini-Hill and V.W. Henderson, Estrogen replacement therapy and risk of Alzheimer disease, *Arch Intern Med* **156** (1996), 2213-2217.

[97] D.W. Pfaff and B.S. McEwen, Actions of estrogens and progestins on nerve cells, *Science* **219** (1983), 808-814.

[98] J.M. Power, W.W. Wu, E. Sametsky, M.M. Oh and J.F. Disterhoft, Age-related enhancement of the slow outward calcium-activated potassium current in hippocampal CA1 pyramidal neurons in vitro, *J Neurosci* **22** (2002), 7234-7243.

[99] J. Quevedo, M. Vianna, D. Daroit, A.G. Born, C.R. Kuyven, R. Roesler and J.A. Quillfeldt, L-type voltage-dependent calcium channel blocker nifedipine enhances memory retention when infused into the hippocampus, *Neurobiol Learn Mem* **69** (1998), 320-325.

[100] S.R. Rapp, M.A. Espeland, S.A. Shumaker, V.W. Henderson, R.L. Brunner, J.E. Manson, M.L. Gass, M.L. Stefanick, D.S. Lane, J. Hays, K.C. Johnson, L.H. Coker, M. Dailey and D. Bowen, Effect of estrogen plus progestin on global cognitive function in postmenopausal women: the Women's Health Initiative Memory Study: a randomized controlled trial, *Jama* **289** (2003), 2663-2672.

[101] S.M. Resnick and P.M. Maki, Effects of hormone replacement therapy on cognitive and brain aging, *Ann N Y Acad Sci* **949** (2001), 203-214.

[102] J.C. Selcher, C.M. Atkins, J.M. Trzaskos, R. Paylor and J.D. Sweatt, A necessity for MAP kinase activation in mammalian spatial learning, *Learn Mem* **6** (1999), 478-490.

[103] D.J. Selkoe, Alzheimer's disease: genes, proteins, and therapy, *Physiol Rev* **81** (2001), 741-766.

[104] N. Shabani, I. Mylonas, U. Jeschke, A. Thaqi, C. Kuhn, T. Puchner and K. Friese, Expression of estrogen receptors alpha and beta, and progesterone receptors A and B in human mucinous carcinoma of the endometrium, *Anticancer Res* **27** (2007), 2027-2033.

[105] T.J. Shors and L.D. Matzel, Long-term potentiation: what's learning got to do with it?, *Behav Brain Sci* **20** (1997), 597-614; discussion 614-555.

[106] S.A. Shumaker, C. Legault, S.R. Rapp, L. Thal, R.B. Wallace, J.K. Ockene, S.L. Hendrix, B.N. Jones, 3rd, A.R. Assaf, R.D. Jackson, J.M. Kotchen, S. Wassertheil-Smoller and J. Wactawski-Wende, Estrogen plus progestin and the incidence of dementia and mild cognitive impairment in postmenopausal women: the Women's Health Initiative Memory Study: a randomized controlled trial, *Jama* **289** (2003), 2651-2662.

[107] J.M. Silvers, S. Tokunaga, R.B. Berry, A.M. White and D.B. Matthews, Impairments in spatial learning and memory: ethanol, allopregnanolone, and the hippocampus, *Brain Res Brain Res Rev* **43** (2003), 275-284.

[108] T. Simoncini, A. Hafezi-Moghadam, D.P. Brazil, K. Ley, W.W. Chin and J.K. Liao, Interaction of oestrogen receptor with the regulatory subunit of phosphatidylinositol-3-OH kinase, *Nature* **407** (2000), 538-541.

[109] C.A. Singer, X.A. Figueroa-Masot, R.H. Batchelor and D.M. Dorsa, The mitogen-activated protein kinase pathway mediates estrogen neuroprotection after glutamate toxicity in primary cortical neurons, *J Neurosci* **19** (1999), 2455-2463.

[110] C. Singh, J.M. Wang, B. Guerrero, R.D. Brinton and R.F. Thompson, Allopregnanolone reverses the learning and memory deficits in the triple transgenic Alzheimer disease mouse model, *Society for Neuroscience Abstracts* **49.10** (2007).

[111] M. Singh, E.M. Meyer, W.J. Millard and J.W. Simpkins, Ovarian steroid deprivation results in a reversible learning impairment and compromised cholinergic function in female Sprague-Dawley rats, *Brain Res* **644** (1994), 305-312.

[112] M. Singh, G. Setalo, Jr., X. Guan, D.E. Frail and C.D. Toran-Allerand, Estrogen-induced activation of the mitogen-activated protein kinase cascade in the cerebral cortex of estrogen receptor-alpha knock-out mice, *J Neurosci* **20** (2000), 1694-1700.

[113] M. Singh, N. Sumien, C. Kyser and J.W. Simpkins, Estrogens and progesterone as neuroprotectants: what animal models teach us, *Front Biosci* **13** (2008), 1083-1089.

[114] L.R. Squire, C.E. Stark and R.E. Clark, The medial temporal lobe, *Annu Rev Neurosci* **27** (2004), 279-306.

[115] S. Suzuki, C.M. Brown and P.M. Wise, Mechanisms of neuroprotection by estrogen, *Endocrine* **29** (2006), 209-215.

[116] J.D. Sweatt, The neuronal MAP kinase cascade: a biochemical signal integration system subserving synaptic plasticity and memory, *J Neurochem* **76** (2001), 1-10.

[117] K. Takehara, S. Kawahara and Y. Kirino, Time-dependent reorganization of the brain components underlying memory retention in trace eyeblink conditioning, *J Neurosci* **23** (2003), 9897-9905.

[118] P. Tanapat, N.B. Hastings and E. Gould, Ovarian steroids influence cell proliferation in the dentate gyrus of

[119] M.X. Tang, D. Jacobs, Y. Stern, K. Marder, P. Schofield, B. Gurland, H. Andrews and R. Mayeux, Effect of oestrogen during menopause on risk and age at onset of Alzheimer's disease, *Lancet* **348** (1996), 429-432.

[120] E. Terasawa and P.S. Timiras, Electrical activity during the estrous cycle of the rat: cyclic changes in limbic structures, *Endocrinology* **83** (1968), 207-216.

[121] T.J. Teyler, R.M. Vardaris, D. Lewis and A.B. Rawitch, Gonadal steroids: effects on excitability of hippocampal pyramidal cells, *Science* **209** (1980), 1017-1018.

[122] C. Touma, O. Ambree, N. Gortz, K. Keyvani, L. Lewejohann, R. Palme, W. Paulus, K. Schwarze-Eicker and N. Sachser, Age- and sex-dependent development of adrenocortical hyperactivity in a transgenic mouse model of Alzheimer's disease, *Neurobiol Aging* **25** (2004), 893-904.

[123] J.Q. Trojanowski and V.M. Lee, Phosphorylation of paired helical filament tau in Alzheimer's disease neurofibrillary lesions: focusing on phosphatases, *Faseb J* **9** (1995), 1570-1576.

[124] S. Turkmen, M. Lofgren, V. Birzniece, T. Backstrom and I.M. Johansson, Tolerance development to Morris water maze test impairments induced by acute allopregnanolone, *Neuroscience* **139** (2006), 651-659.

[125] G. van Wingen, F. van Broekhoven, R.J. Verkes, K.M. Petersson, T. Backstrom, J. Buitelaar and G. Fernandez, How progesterone impairs memory for biologically salient stimuli in healthy young women, *J Neurosci* **27** (2007), 11416-11423.

[126] R.M. Vardaris and T.J. Teyler, Sex differences in the response of hippocampal CA1 pyramids to gonadal steroids: Effects of testosterone and estradiol on the in vitro slice preparation, *Society for Neuroscience Abstracts* **A153.9** (1980),

[127] R.M. Vouimba, M.R. Foy, J.G. Foy and R.F. Thompson, 17beta-estradiol suppresses expression of long-term depression in aged rats, *Brain Res Bull* **53** (2000), 783-787.

[128] J.M. Wang, P.B. Johnston, B.G. Ball and R.D. Brinton, The neurosteroid allopregnanolone promotes proliferation of rodent and human neural progenitor cells and regulates cell-cycle gene and protein expression, *J Neurosci* **25** (2005), 4706-4718.

[129] S.G. Warren, A.G. Humphreys, J.M. Juraska and W.T. Greenough, LTP varies across the estrous cycle: enhanced synaptic plasticity in proestrus rats, *Brain Res* **703** (1995), 26-30.

[130] S.G. Warren and J.M. Juraska, Spatial and nonspatial learning across the rat estrous cycle, *Behav Neurosci* **111** (1997), 259-266.

[131] P.M. Wise, D.B. Dubal, M.E. Wilson, S.W. Rau, M. Bottner and K.L. Rosewell, Estradiol is a protective factor in the adult and aging brain: understanding of mechanisms derived from in vivo and in vitro studies, *Brain Res Brain Res Rev* **37** (2001), 313-319.

[132] M. Wong and R.L. Moss, Electrophysiological evidence for a rapid membrane action of the gonadal steroid, 17 beta-estradiol, on CA1 pyramidal neurons of the rat hippocampus, *Brain Res* **543** (1991), 148-152.

[133] M. Wong and R.L. Moss, Long-term and short-term electrophysiological effects of estrogen on the synaptic properties of hippocampal CA1 neurons, *J Neurosci* **12** (1992), 3217-3225.

[134] M. Wong and R.L. Moss, Patch-clamp analysis of direct steroidal modulation of glutamate receptor-channels, *J Neuroendocrinol* **6** (1994), 347-355.

[135] M. Wong, T.L. Thompson and R.L. Moss, Nongenomic actions of estrogen in the brain: physiological significance and cellular mechanisms, *Crit Rev Neurobiol* **10** (1996), 189-203.

[136] C.S. Woolley and B.S. McEwen, Estradiol regulates hippocampal dendritic spine density via an N-methyl-D-aspartate receptor-dependent mechanism, *J Neurosci* **14** (1994), 7680-7687.

[137] C.S. Woolley, N.G. Weiland, B.S. McEwen and P.A. Schwartzkroin, Estradiol increases the sensitivity of hippocampal CA1 pyramidal cells to NMDA receptor-mediated synaptic input: correlation with dendritic spine density, *J Neurosci* **17** (1997), 1848-1859.

[138] C.S. Woolley, Estrogen-mediated structural and functional synaptic plasticity in the female rat hippocampus, *Horm Behav* **34** (1998), 140-148.

[139] T.W. Wu, J.M. Wang, S. Chen and R.D. Brinton, 17Beta-estradiol induced Ca2+ influx via L-type calcium channels activates the Src/ERK/cyclic-AMP response element binding protein signal pathway and BCL-2 expression in rat hippocampal neurons: a potential initiation mechanism for estrogen-induced neuroprotection, *Neuroscience* **135** (2005), 59-72.

[140] X. Xie, T.W. Berger and G. Barrionuevo, Isolated NMDA receptor-mediated synaptic responses express both LTP and LTD, *J Neurophysiol* **67** (1992), 1009-1013.

[141] S. Zadran, Q. Qin, X. Bi, H. Zadran, Y. Kim, M.R. Foy, R. Thompson and M. Baudry, 17-Beta-estradiol increases neuronal excitability through MAP kinase-induced calpain activation, *Proc Natl Acad Sci U S A* **106** (2009), 21936-21941.

[142] R.F. Zec and M.A. Trivedi, The effects of estrogen replacement therapy on neuropsychological functioning in postmenopausal women with and without dementia: a critical and theoretical review, *Neuropsychol Rev* **12** (2002), 65-109.

[143] Y. Zeng, M.R. Foy, T.J. Teyler and R.F. Thompson, 17B-estradiol enhancement of NMDA LTP in hippocampus, *Society for Neuroscience Abstracts* **972.2** (2004).

# Apparent Behavioral Benefits of Tau Overexpression in P301L Tau Transgenic Mice

Dave Morgan[a,*], Sanjay Munireddy[a], Jennifer Alamed[a], Jason DeLeon[a], David M. Diamond[b,e], Paula Bickford[c,e], Michael Hutton[d], Jada Lewis[d], Eileen McGowan[d] and Marcia N. Gordon[a]

[a] Alzheimer Research Laboratory, Department of Pharmacology, University of South Florida, Tampa, FL, USA
[b] Departments of Psychology, Molecular Pharmacology and Center for Preclinical and Clinical Research on PTSD, University of South Florida, Tampa, FL, USA
[c] Center of Excellence for Aging and Brain Repair and Department of Neurosurgery, University of South Florida, Tampa, FL, USA
[d] Department of Neuroscience, Mayo Clinic College of Medicine, Jacksonville, FL, USA
[e] James A. Haley Veterans Administration Medical Center, Tampa, FL, USA

**Abstract**. Transgenic mice expressing human tau containing the P301L tau mutation (JNPL3; tau mice) develop motor neuron loss, paralysis and death between 7 and 12 months. Surprisingly, at 5 and 7 months of age, tau transgenic mice were superior to other genotypes in the rotarod task, and had near perfect scores on the balance beam and coat hanger tests. One tau transgenic mouse was performing at a superior level in the rotarod one day prior to developing paralysis. Cognitive function was also normal in the tau mice evaluated in the radial arm water maze and the Y-maze tasks. We also crossed the tau transgenic mice with Tg2576 amyloid-$\beta$ protein precursor (A$\beta$PP) transgenic mice. Although A$\beta$PP mice were deficient in the radial arm maze task, A$\beta$PP + tau mice were not impaired, implying a benefit of the tau transgene. Some mice were homozygous for the retinal degeneration mutation (rd/rd) and excluded from the genotype analysis. Only the water maze task discriminated the rd/rd mice from nontransgenic mice. In conclusion, it seems that the modest tau overexpression or the presence of mutant tau in the JNPL3 tau mice may provide some benefit with respect to motor and cognitive performance before the onset of paralysis.

Keywords: Amyloid, rotarod, tau, transgenic mice, water maze, Y-maze

## INTRODUCTION

The histopathology of Alzheimer's disease reveals two prominent features, amyloid plaques derived from the amyloid-$\beta$ (A$\beta$) peptide and neurofibrillary tangles derived from hyperphosphorylated tau. Mutations in either the amyloid-$\beta$ protein precursor (A$\beta$PP) or in tau lead to neurodegeneration and dementia syndromes (with different clinical presentations [12,15]). The tau transgenic line JNPL3 expresses 4 repeat human tau (without N terminal inserts) harboring the FTDP-17 mutation P301L under control of the prion promoter [17]. These mice have little phenotype at birth, but develop a progressive motor abnormality associated with loss of lower motor neurons. These motor abnormalities begin as early as 6.5 months, and have a prevalence of 90% by 10 months, although the phenotype is delayed on certain strain backgrounds. When bred with the Tg2576 A$\beta$PP transgenic mouse, there is an increase in the formation of filamentous tau inclusions in forebrain regions [18].

---

*Corresponding author: Dave Morgan, PhD, Alzheimer Research Laboratory, MDC Box 9, University of South Florida, Tampa, FL 33612-4799, USA. Tel.: +1 813 974 3949; Fax: +1 813 974 2565; E-mail: dmorgan@hsc.usf.edu.

Given that the greatest degeneration in the JNPL3 tau transgenic line is in the spinal cord, we wished to investigate when the earliest motor deficits might be detected. As a first approach, we examined mice at 5 months of age, an age prior to the onset of overt movement abnormalities, using several tasks effective at detecting motor deficits in mice. We tested the same cohort a second time at 7 months, closer to the age of motor disturbance onset and finished with cognitive function testing at 8 months of age. We further compared the tau mice with $A\beta PP$ +tau mice derived from a cross between Tg2576 and JNPL3 lines to determine if any early motor deficits might be accelerated by the presence of increased $A\beta$, as was the case for abnormally phosphorylated tau isoforms in prior work [11,18].

## METHODS

### Mouse breeding and procedures

Mice were obtained from a cross between Tg2576 mice and JNPL3 mice in the University of South Florida Alzheimer Research Laboratory colony. This cross results in four genotypes, tau only, $A\beta PP$ only, $A\beta PP$ +tau and nontransgenic. We started the study with 10 mice in each group except the $A\beta PP$ mice which had a sample size of 12. All mice were born from the same litters and had a mixed genetic background (contributed by the founder mice) of 54% C57BL/6; 22% DBA/2; 18% SW and 6% SJL. During the study, 1 nontransgenic, 4 $A\beta PP$, 4 tau and 1 $A\beta PP$+tau mouse died resulting in final sample sizes of 9 nontrangenic, 8 $A\beta PP$, 6 tau and 9 $A\beta PP$+tau mice at the end of the study. In addition to these mice indicated above, there were 10 mice homozygous for the retinal degeneration (rd) mutation. 4 were nontransgenic, 2 were $A\beta PP$ and 4 were tau genotype. This mutation was contributed by the SW and SJL backgrounds of the parental lines. Although they were tested at the same time as the sighted mice, their data were analyzed separately. All mice were genotyped by PCR for $A\beta PP$, tau and *rd* both at weaning and at necropsy. The *rd* mutation genotyping used the primer pairs suggested on the JAX Laboratories web site.

Mice were tested at three ages for behavioral effects of genotype. The first test occurred at 5 months and consisted of the motor test battery of coat hanger test, balance beam and rotarod (see below). A second motor function test was performed at 7 months of age. Finally, cognitive function testing followed the motor function tests at 8 months of age. Brain tissue was then collected for histopathologic analysis of tau tangles (at this age, $A\beta PP$ mice do not yet have histologically detectable amyloid deposits).

### Behavioral testing

#### Coat hanger task

Strength and coordination were tested in the coat hanger test [16]. In this test, the mouse was suspended by its forepaws from the center of the horizontal portion of a wire coat hanger. The goal was for the mouse to raise its hind legs to the bar, move to the end of the bar and right itself onto the diagonal portion of the coat hanger. Mice are administered 3 trials of up to 60 seconds duration. Each trial was separated by at least 10 minutes. The task was scored by average time on the hanger without dropping (total hang time; presented and Table 1), and an agility score from 1–5 based upon how successful the mice were in reaching the goal (no differences between groups; data not presented).

#### Balance beam task

As a test of vestibular and general motor function, each animal was carefully placed at the center of a suspended beam and released. The balance time was averaged from 3 successive trials, separated by at least 10 minutes. Maximum trial length was 60 seconds. Because the intention of the balance beam and coat hanger tests was to detect deficits in motor function, the conditions chosen were designed to detect impairments, with most nontransgenic mice performing near maximal performance.

#### Rotating rod

Mice were placed onto the round portion of a motorized circular rod (Ugo Basile Rota-rod model 7750) which was slowly accelerated starting at 2.5 RPM and accelerated to 34 RPM over 5 minutes with an increase in speed every 30 seconds. Mice were required to walk at the speed of rod rotation to keep from falling. The time until falling was recorded for each mouse. Mice were given three trials each day, separated by at least 10 minutes, for 5 consecutive days. The maximum possible time on the rod was 300 seconds. The average time for each day was recorded. Only very rarely do mice reach the maximum time possible for this task.

Table 1
Motor Performance Testing

| Motor test | Nontransgenic | A$\beta$PP | Tau | A$\beta$PP + tau | rd/rd |
|---|---|---|---|---|---|
| 5 month coat hanger | 54 ± 4 | 46 ± 4 | 52 ± 4 | 49 ± 4 | 47 ± 3 |
| 5 month balance beam | 47 ± 6 | 44 ± 6 | 50 ± 5 | 36 ± 7 | 40 ± 7 |
| 7 month coat hanger | 47 ± 3 | 44 ± 4 | 45 ± 5 | 39 ± 6 | 45 ± 2 |
| 7 month balance beam | 38 ± 6 | 44 ± 6 | 40 ± 7 | 18 ± 4* | 53 ± 3 |

*Significantly lower than A$\beta$PP, tau and nontransgenic mice, $P < 0.01, 0.02$ and $0.05$, respectively.

*Y-maze*

Each animal was placed in a walled Y-maze for a single 5 minute trial. The sequence of arm entries and total number of arm choices were recorded. Spontaneous alternation (entering all three arms sequentially without repetition) was expressed as a percentage, as calculated according to the method of [3]. If an animal made the following sequence of arm selections (1,2,3,2,1,3,1,2), the total alternation opportunities (triads) was 6 (total entries minus 2) and the percentage alternation would be scored as 67% (4 out of 6).

*Radial arm water maze*

The radial arm water maze has the spatial complexity and ease of performance measurement comparable to the dry radial arm maze combined with the rapid learning normally observed in the Morris water maze without requiring footshock or food deprivation as motivating factors [7]. The radial arm water maze for these studies contained 6 swim paths (arms) radiating from an open central area, with a submerged escape platform located at the end of one of the arms (the goal arm; see [20]). On each trial, the mouse was initially placed in the center of a randomly selected start arm and allowed to swim in the maze for up to 60 seconds to find the escape platform. The platform was located in the same arm on each trial. On day one, mice were given 15 trials alternating between a visible platform (above the water) and a hidden platform (below the water). The next day they were given 15 additional trials with all the trials using a hidden platform. Entry into an incorrect arm (all four limbs within the arm) was scored as an error. If a mouse failed to make an arm entry within 20 seconds, this also was scored as an error. The errors for blocks of 3 consecutive trials were averaged for data analysis.

Mice were organized into cohorts of 4 mice and administered trials in a "spaced" (as opposed to "massed") manner designed to minimize fatigue. All mice in a cohort received trial 1, then all mice received trial 2, etc. Each cohort was given 3 trials (1 block) sequentially then returned to their home cages while a second cohort was tested. The first cohort was tested then for the second block, alternating with a second cohort until 15 trials (5 blocks) were completed for day 1. The start arm was varied for each trial so that mice rely upon spatial cues to solve the task instead of learning motor rules (i.e., second arm on the right). The goal arm for each successive mouse was different to avoid use of odor cues to locate the goal arm. Group averages of less than 1 error indicate learning of platform location. A detailed description of the general procedure with a few modifications has been published, complete with goal arm assignments and scoring sheets [1].

*Histopathology*

Mice were euthanatized shortly after completion of radial arm water maze testing by overdose with pentobarbital (200 mg/kg) followed by transcardial perfusion with saline followed by perfusion with neutral buffered paraformaldehyde. Brains were postfixed for 24 hours and cryopreserved in 10%, 20% and then 30% sucrose solutions. Frozen sections at 25 $\mu$m were collected in the horizontal plane through the brain. Every 12th section was stained for abnormally phosphorylated tau using antibody R145d directed against tau phospho-epitope S422 (Biosource International). We chose this antibody originally because Gotz and colleagues [11] found it was the phospho-form of tau that was most sensitive to stimulation by the presence of A$\beta$. Earlier work performed in our laboratory with a younger cohort of mice from this breeding had found a correlation between the number of phospho-tau cells per slide and performance on spatial navigation and visible platform tasks [4]. We used this same antibody again because we were attempting to confirm this observation. We counted all positively stained profiles in the hippocampus and cortex of these mice (this analysis was performed by the same individual as in [4]). Because the A$\beta$PP mice were collected at an age before deposits appeared, we did not stain for amyloid load in these mice.

## Statistics

Genotype effects were analyzed by ANOVA followed by means testing with Fisher's LSD analysis, using the statistical analysis program Statview (SAS, Chicago IL). For the rotarod and radial arm water maze tasks, days or blocks were included in a 2 way ANOVA as a repeated measure. $P < 0.05$ was selected as the criterion for statistical significance.

## RESULTS

### Motor function testing

Mice were given a motor function battery at two ages, in anticipation of detecting early signs of lower motor neuron dysfunction in the tau transgenic mice. The first age was 5 months, an age well in advance of the onset of hindlimb paralysis in prior work. A second age of 7 months was chosen as the earliest time when mice might exhibit paralysis. Mice were subjected to a short motor test battery consisting of a coat hanger test on one day, a balance beam task on a second day, and 5 consecutive days of rotarod testing.

All mice did well on the coat hanger test at 5 months. The mean hang times ranged from 46 to 54 (out of 60) seconds, and no differences were detected between the four genotypes (Table 1). A similar outcome was obtained for the balance beam task in 5 month old mice (Table 1). The rotarod testing revealed that all groups of mice improved performance over days, and mouse genotype had an effect on the time spent on the rod (Fig. 1; repeated measures 2 way ANOVA; significant effect of genotype, $P < 0.05$, and a significant effect of days, $P < 0.001$). Somewhat unexpectedly, the best performing mice were those expressing the tau transgene. When averaged across all 5 days, the tau mice performed significantly better than the nontransgenic mice ($P < 0.05$) and the A$\beta$PP mice ($P < 0.01$). The A$\beta$PP, nontransgenic and A$\beta$PP + tau groups did not differ from each other.

At 7 months of age, the coat hanger test again indicated comparable performance among the 4 genotypes (Table 1). In the balance beam test, the A$\beta$PP+tau mice were inferior to the other three groups (Table 1). At 7 months, rotarod testing again indicated a significant effect of days of training ($P < 0.001$) and of genotype ($P < 0.005$; Fig. 1). Analysis of the average of all 5 days indicated that the tau mice performed better than either the nontransgenic mice ($P < 0.05$) or

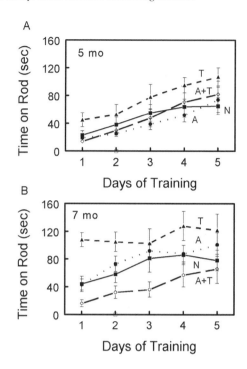

Fig. 1. Enhanced rotarod performance in P301L tau transgenic mice. Four genotypes of mice, tau (T; solid triangles $n = 9$), A$\beta$PP+tau (A + T; open diamonds; $n = 9$), nontransgenic mice (sold squares; solid line; $n = 9$) and A$\beta$PP mice (A; solid circles; $n = 11$) were tested in the rotarod for 3 trials each day for 5 consecutive days. Panel A is testing performed when mice were 5 months of age and panel B represents testing of the same mice at 7 months of age. Tau mice differed from nontransgenic mice at both ages (see text for detailed statistical results).

the A$\beta$PP +tau mice ($P < 0.001$). In addition, the A$\beta$PP + tau mice performed significantly worse on the rotarod than the A$\beta$PP only mice ($P < 0.05$), but not nontransgenic mice (Fig. 1).

Given the fact that transgenic tau expression in this mouse line ultimately caused paralysis, it seemed surprising that these mice exhibited superior performance on the rotarod task. We did note that one mouse became paralyzed during the testing procedure which led us to examine the performance of individual mice over days. In Fig. 2, individual mouse values are plotted in comparison to the mean and SEM for nontransgenic animals (solid line). Two mice with typical individual performances for tau transgenic mice are mouse 4 (solid circles) and mouse 34 (open circles). Their times were above the nontransgenic mean value on the first day, and progressively improved over the 5 days of the trial. Conversely, mouse #41 (open diamonds) started the first 3 days of rotarod testing considerably above the average for nontransgenic mice. Remarkably, on the fourth day the performance of mouse 41 had deteri-

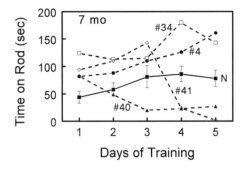

Fig. 2. Patterns of individual mouse performances in the rotarod testing. The mean ± standard error for nontransgenic (N) mice is represented by the solid line. Individual values for tau mice numbers 4 (solid circles), 34 (open squares), 41 (open diamonds) and 40 (solid triangles) are shown. Mouse # 41 developed hindlimb paralysis during the testing procedure.

orated and by the end of the session it was noticed that hind limb paralysis had set in. This mouse was unable to perform the rotarod task on day 5 and was euthanatized shortly thereafter. Another tau animal, mouse 40 (solid triangles), showed a progressive deterioration of performance over 5 days (an unusual response), but did not develop hindlimb paralysis before tissue was collected after the cognitive function tests. One potential explanation for such rapid changes in behavioral performance might be explained by increased neuronal activity, which might both improve motor performance, but also increase vulnerability to excitotoxic forms of damage.

*Cognitive function testing*

Mice were evaluated in the Y maze and in a 2 day version of the radial arm water maze shortly after completing the motor function testing at 7–8 months of age. No differences were detected in Y-maze alternation performance among the 4 genotype groups (Fig. 3A). All groups had mean alternation percentages around 60%, typical for control mice in our prior studies [5,10,13,14, 20,30]. However, the mice overexpressing the A$\beta$PP transgene exhibited an increased number of arm entries (Fig. 3B), an observation consistent with our prior work (cited above).

The 2 day radial arm water maze was used to assess spatial navigation learning and memory. In this task, nontransgenic mice showed continuous improvement in the task, achieving the less than 1 error criterion for learning by block 7 (Fig. 4). A$\beta$PP mice were deficient in learning the platform location in this task. Repeated measures ANOVA indicated a significant effect of blocks ($P < 0.001$) indicating that overall learning oc-

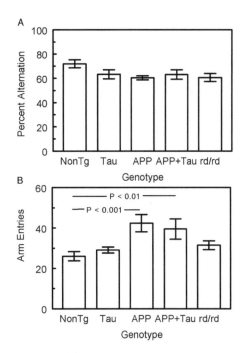

Fig. 3. Increased activity in A$\beta$PP transgenic mice. Mice were tested in the Y maze for 5 minutes and number of arm entries and % alternation recorded. As seen in a number of prior studies, mice with the A$\beta$PP transgene made a larger number of arm entries compared to nontransgenic animals. All mice exhibited normal alternation performance.

curred during the 2 days and a significant effect of genotype ($P < 0.005$). Means analysis at individual data points indicated that A$\beta$PP mice performed significantly worse that all three other groups during blocks 6, 8 and 10 ($P < 0.01$ or greater for all comparisons). The important observation is that the A$\beta$PP + tau mice performed significantly better than the A$\beta$PP only transgenic mice, suggesting that the two transgenes not only failed to synergize, but the tau transgene rescued the effect of the A$\beta$PP transgene on memory performance.

We also counted the number of neurons in tau and A$\beta$PP+tau transgenic mice that were labeled with antisera against phospho-ser 422 of human tau. Not all mice carrying a tau transgene exhibited staining for this abnormally phosphorylated form of tau in forebrain, and the numbers of such neurons in cortex or hippocampus were small in number, limiting the analysis to manual counts. We found a trend towards higher numbers of tau positive neurons per section in hippocampus of A$\beta$PP +tau mice (50 ± 27) versus tau only mice (3.1 ± 1.6) but this effect was not significant ($P < 0.09$). Still this trend is consistent with earlier work combining these lines where the strongest effects were observed in older mice than those examined here [18]. In

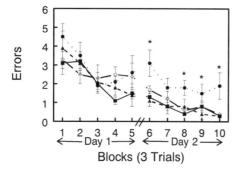

Fig. 4. Radial arm water maze performance deficits in AβPP transgenic mice are reversed by the presence of the tau transgene. Mice were run in the radial arm water maze to measure spatial navigation learning and performance. AβPP mice (solid circles) performed significantly worse than nontransgenic (solid squares), tau transgenic (solid triangles) or AβPP+tau (open diamonds) mice in several blocks on day 2 of training (see text for detailed statistical analyses).

cortex, a similar trend was present (8.9 ± 4.9 versus 3.8 ± 1.7; AβPP+tau versus tau). None of the AβPP only or nontransgenic mice had cells positive for this form of phospho-tau. We then attempted to correlate the number of phospho-tau positive neurons with cognitive performance on the radial arm water maze task. We correlated cognition and pathology in tau mice and AβPP+tau mice separately and also correlated behavior and pathology with the tau plus AβPP+tau mice combined. We compared the phospho-tau neuron numbers with several indices of learning task performance. None of these showed a correlation above 0.15, indicating these phenomena were unlinked.

*Effects of homozygosity for the retinal degeneration mutation*

Mice on this mixed genetic background, integrating the P301L tau transgenic mouse background and the Tg2576 AβPP transgenic mouse background, resulted in an unexpectedly large number of mice that were homozygous for the *rd*, retinal degeneration mutation. Although initially included during the behavioral testing of the mice, these blind mice were excluded from the data analyses presented above. We noticed, however, that these mice were largely indistinguishable from sighted mice on the balance beam task, the coat hanger task (Table 1), and the Y maze task (Fig. 3). Even on the rotarod, they performed as well as nontransgenic mice (Fig. 5A). The only task capable of distinguishing these blind mice from sighted mice was the radial arm water maze task. In Fig. 5B, it is clear that these mice do not improve over blocks on this task, yet the num-

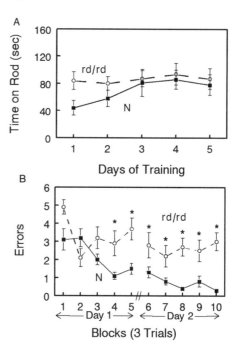

Fig. 5. Performance of blind (rd/rd) mice on rotarod and radial arm water maze tasks. Mice homozygous for the retinal degeneration mutation rd/rd performed comparably to the nontransgenic mice on the rotarod task (panel A; 7 month measurement). In the behavioral tests performed here, the only deficiency detected in the rd/rd mice was in the radial arm water maze (panel B) where the mice failed to improve significantly over the 2 days of testing.

ber of errors overall is not substantially different than sighted mice that are unaware of the platform location (3–4 errors is typically under these circumstances).

## DISCUSSION

The key observation in this study is that P301L tau transgenic animals, destined to develop hind limb motor paralysis secondary to straight or wavy tau filament formation [19] and death of lower motor neurons [17], do not exhibit antecedent deficits in motor behavior. Neither at 5 nor 7 months of age do these mice exhibit impairments on several tasks often used to detect motor abnormalities in transgenic mice. Remarkably, the performance of these mice on the rotarod task is actually superior to that of the other genotypes. However, the benefits of the tau transgene did not extend to the AβPP+tau transgenic mice, as their performance was not significantly different from the nontransgenic mice.

These results were certainly unexpected, given the anticipation that tau transgenic mice of the JNPL3 line invariably develop hindlimb paralysis. This is associated with oligodendrocytic apoptosis and axonal degeneration in the spinal cord, with 50% loss of lower motor neurons in mice exhibiting motor abnormalities [17,19, 31]. Thus, we designed these studies to detect early *deficits* in motor performance. It is uncertain if the enhancements observed with the rotarod task would generalize to other motor abilities using tasks sensitive to improvements in performance. A recent study indicated that JNPL3 tau mice surviving to 12 months and exhibiting motor abnormalities are impaired in the rotarod task [22], but earlier testing was not reported. In the THY-tau22 mouse model, Schindowski and colleagues [28] report that the transgenic mice stayed on the rotarod 30% longer on average than nontransgenic mice, although the effect did not reach statistical significance. Certainly, the balance beam and coat hanger task, often used successfully to detect motor deficits, failed to indicate any impairments in the tau animals. However, the task conditions were designed to detect deficits and most mice perform near the ceiling. Thus, they are relatively insensitive to improvements in motor performance. The rotarod is probably the only task in this series that has room for mice to improve, and, thus, is uniquely sensitive to detect enhancements in motor capability.

By inference, the cognitive function of the A$\beta$PP mice was improved by the presence of the tau transgene, as the A$\beta$PP+tau mice performed better than A$\beta$PP only mice in the last blocks of the radial arm water maze task. There was no difference in performance between tau mice and nontransgenic littermates. It should be noted that the apparent learning deficit in the A$\beta$PP + tau mice is not an artifact of failing to make arm choices, as mice failing to enter arms would receive a score of 3 (see Methods). Thus, on balance, the tau transgene effect on cognition tended towards being beneficial, rather than detrimental. These data are very similar to those obtained earlier with mice derived from the same breeding pairs as the present study. Arendash et al. [4] found impaired radial arm water maze function in A$\beta$PP mice, but no effect of the tau transgene compared to nontransgenic littermates. Unfortunately, this cohort lacked a sufficient number of tau+A$\beta$PP mice for a separate analysis, so they were grouped with the tau only animals for data analysis. In spite of the absence of a main effect of tau genotype on spatial navigation behavior, Arendash and colleagues [4] reported three significant correlations between a behavioral measure and the numbers of phospho-tau positive cells in one of the 5 brain regions evaluated (data collected by our research group). Our laboratory had concerns regarding the reliability of these correlations due to the extensive number of behavioral tests used (32) and the separate correlation with each of 5 brain regions (150 possible correlations). It seemed likely to us that some correlations would be positive by chance alone. Our study, using a larger dataset, restricted number of behavioral test values, and focused analysis on cortical and hippocampal cell numbers, failed to confirm these correlations. Although minor differences in the behavioral procedure and analysis of a different cohort of mice (albeit litters from the same transgenic parents) might explain the differences between these studies, we suggest that the initial observation resulted by chance due to the large number of correlations performed without error correction procedures.

At face value, these data suggest that modest overexpression of tau, or the presence of mutant tau, can be beneficial. However, it is important to recognize that in the absence of data on levels of tau expression within individual populations of neurons in JNPL3 mice, especially lower motor neurons, it is not possible to determine what levels of expression correspond to this apparent benefit. In a different P301L tau transgenic line, Boekhoorn et al. [6] reported that the tau mice at young ages (3 months) had improved object recognition performance and object recognition memory compared to nontransgenic mice. In the young THY-tau22 model, there is also evidence for increased neurogenesis [29]. Intriguing support for this argument also derives from work examining a different human tau transgenic line. These mice express a human PAC, termed 8c mice [8], but develop very little tau pathology. It is only when bred onto the tau knockout background that these mice develop neurofibrillary tangles and neuron loss [2], implying the endogenous mouse tau was somehow protective against the neuropathology. However, the opposite conclusion was reached in a well conducted study breeding A$\beta$PP mice onto a tau null background [25]. The tau null background had no impact on amyloid deposition in these mice, but rescued the memory deficits exhibited by these mice when bred on a tau sufficient background.

Certainly extensive overexpression of tau can have deleterious consequences. SantaCruz and coworkers [27]. demonstrated that a mouse expressing human P301L tau under control of the tetracycline response element (rTg4510) developed early and severe neuropathology, neuron loss and memory disruption. The memory loss was reversed when the transgene was suppressed with doxycycline, even though the neurofibrillary tangles continued to develop. Intriguingly, the doxycycline suppression reduced the expression of the tau transgene from 15 fold overexpression to 2.5 fold overexpression relative to endogenous mouse tau. In contrast, the average overxpression of the tau transgene in the JNPL3 is estimated as 2 fold endogenous murine expression. However, tau expression levels within spinal cord neurons that degenerate selectively are not known. Several of the tau mouse lines also exhibit memory deficits, but in general these do not impact all cognitive domains, nor are the effects as profound as those seen in A$\beta$PP transgenic mice unless severe neuron loss is present [21,24,26,28].

One unexpected outcome from the breeding between the Tg2576 A$\beta$PP mice and the JNPL3 tau mice was a large number of mice homozygous for the retinal degeneration mutation. This was most likely introduced through the SJL mice on the Tg2576 line and the SW background of the JNPL3 mice. Perhaps most remarkable is that the behavior of these mice was indiscernible from the behavior of mice with intact vision on most tasks, indicating that mice compensate for visual losses quite effectively. In fact, in prior work these mice even "passed" the visual cliff task intended to detect visual impairments in a neurological test battery [9]. Only the spatial navigation task revealed the considerable visual deficit. Importantly, the mice typically made 3–4 errors, a number consistent with mice skilled in the task, but without knowledge of the location of the platform. These mice appear to use an efficient search strategy, establishing a mental map of the maze and not repeating arms. It is simply that they cannot use the extramaze cues to orient their starting position within the maze, and thus are unable to determine the platform location when started in different arms each trial.

These rd/rd mice perform similarly to several inbred mouse lines previously demonstrated incapable of learning the Morris water maze, a spatial navigation task relying upon the same visuospatial skills used in the radial arm water maze [23]. The poor learning strains included FVB, C3H, SJL, A, BUB/BnJ and Balb/c. Of these lines, FVB, C3H, SJL, and BUB/BnJ are known to be homozygous for the rd mutation (see www.jax.org/jaxmice). In studies by Owen and colleagues of F1 hybrids [23], all hybrids performed well, even when one parent was from a line carrying the rd mutation, indicating heterozygosity at this allele lacks a major behavioral phenotype. Importantly, Owen et al. [23] predicted there was a sensory deficit in these poorly performing mouse lines, as these mice also did poorly on the visible platform test. This highlights the use of the visible platform as a task to detect mice lacking the sensory apparatus necessary to learn spatial navigation tasks.

In conclusion, these behavioral results fail to confirm the initial hypothesis that JNPL3 mice would have antecedent motor impairments at early ages as phosphotau species accumulate in their lower motor neurons. Surprisingly, we find the opposite effect, of enhanced running time on the rotarod task in tau transgenic mice. Coupled with other observations reported recently, these data suggest that modest levels of tau overexpression may provide some benefits to neurons, that are only overcome when levels of specific toxic tau species reach critical concentrations, and cause neurodegeneration. The basis for this benefit is unclear, however, the well established role of tau in stabilizing microtubules may provide a potential explanation in that ultra-stable microtubules may enable improved or more robust axonal function. This effect would likely be prominent in motor neurons due to their long axons which may in turn explain the impact on motor function in the rotarod test.

## ACKNOWLEDGMENTS

We thank Karen Hsiao Ashe for providing us with the Tg2576 A$\beta$PP mouse line. These results were supported by the National Institutes of Health (AG15490, AG18478, AG25509, and AG04418 to MG and DM). Dr. Morgan has served as a consultant for and/or received lecture fees from the following companies: AstraZeneca, Baxter, Bristol Myers Squibb, Forest, Merck, NeurImmune, Pfizer, Wyeth.

N.B. Disclosure statements were not received from Drs. Munireddy, Alamed, DeLeon, Hutton and Gordon.

## References

[1] J. Alamed, D.M. Wilcock, D.M. Diamond, M.N. Gordon and D. Morgan, Two-day radial-arm water maze learning and memory task; robust resolution of amyloid-related memory deficits in transgenic mice, *Nat Protoc* **1** (2006), 1671–1679.
[2] C. Andorfer, C.M. Acker, Y. Kress, P.R. Hof, K. Duff and P. Davies, Cell-cycle reentry and cell death in transgenic mice expressing nonmutant human tau isoforms, *J Neurosci* **25** (2005), 5446–5454.
[3] H. Anisman, Time-dependent variations in aversively motivated behaviors: nonassociative effects of cholinergic and catecholaminergic activity, *Psychol Rev* **82** (1975), 359–385.
[4] G.W. Arendash, J. Lewis, R.E. Leighty, E. McGowan, J.R. Cracchiolo, M. Hutton and M.F. Garcia, Multi-metric behavioral comparison of APPsw and P301L models for Alzheimer's disease: linkage of poorer cognitive performance to tau pathology in forebrain, *Brain Res* **1012** (2004), 29–41.

[5] L. Austin, G.W. Arendash, M.N. Gordon, D.M. Diamond, G. DiCarlo, C. Dickey, K. Ugen and D. Morgan, Short-term beta-amyloid vaccinations do not improve cognitive performance in cognitively impaired APP + PS1 mice, *Behav Neurosci* **117** (2003), 478–484.

[6] K. Boekhoorn, D. Terwel, B. Biemans, P. Borghgraef, O. Wiegert, G.J. Ramakers, K. de Vos, H. Krugers, T. Tomiyama, H. Mori, M. Joels, F. van Leuven and P.J. Lucassen, Improved long-term potentiation and memory in young tau-P301L transgenic mice before onset of hyperphosphorylation and tauopathy, *J Neurosci* **26** (2006), 3514–3523.

[7] D.M. Diamond, C.R. Park, K.L. Heman and G.M. Rose, Exposing rats to a predator impairs spatial working memory in the radial arm water maze, *Hippocampus* **9** (1999), 542–551.

[8] K. Duff, H. Knight, L.M. Refolo, S. Sanders, X. Yu, M. Picciano, B. Malester, M. Hutton, J. Adamson, M. Goedert, K. Burki and P. Davies, Characterization of pathology in transgenic mice over-expressing human genomic and cDNA tau transgenes, *Neurobiol Dis* **7** (2000), 87–98.

[9] M.F. Garcia, M.N. Gordon, M. Hutton, J. Lewis, E. McGowan, C.A. Dickey, D. Morgan and G.W. Arendash, The retinal degeneration (rd) gene seriously impairs spatial cognitive performance in normal and Alzheimer's transgenic mice, *NeuroReport* **15** (2004), 73–77.

[10] M.N. Gordon, D.L. King, D.M. Diamond, P.T. Jantzen, K.L. Boyett, C.E. Hope, J.M. Hatcher, G. DiCarlo, P. Gottschal, D. Morgan and G.W. Arendash, Correlation between cognitive deficits and Aá deposits in transgenic APP+PS1 mice, *Neurobiol Aging* **22** (2001), 377–385.

[11] J. Gotz, F. Chen, J. Van Dorpe and R.M. Nitsch, Formation of neurofibrillary tangles in P301l tau transgenic mice induced by Abeta 42 fibrils, *Science* **293** (2001), 1491–1495.

[12] J. Hardy and D.J. Selkoe, The amyloid hypothesis of Alzheimer's disease: progress and problems on the road to therapeutics, *Science* **297** (2002), 353–356.

[13] L.A. Holcomb, M.N. Gordon, E. McGowan, X. Yu, S. Benkovic, P. Jantzen, K. Wright, I. Saad, R. Mueller, D. Morgan, S. Sanders, C. Zehr, K. O'Campo, J. Hardy, C.M. Prada, C. Eckman, S. Younkin, K. Hsiao and K. Duff, Accelerated Alzheimer-type phenotype in transgenic mice carrying both mutant amyloid precursor protein and presenilin 1 transgenes, *Nat Med* **4** (1998), 97–100.

[14] L.A. Holcomb, M.N. Gordon, P. Jantzen, K. Hsiao, K. Duff and D. Morgan, Behavioral changes in transgenic mice expressing both amyloid precursor protein and presenilin-1 mutations: Lack of association with amyloid deposits, *Behav Gen* **29** (1999), 177–185.

[15] M. Hutton, Missense and splice site mutations in tau associated with FTDP-17: multiple pathogenic mechanisms, *Neurology* **56** (2001), S21–S25.

[16] R. Lalonde and S. Thifault, Absence of an association between motor coordination and spatial orientation in lurcher mutant mice, *Behav Genet* **24** (1994), 497–501.

[17] J. Lewis, E. McGowan, J. Rockwood, H. Melrose, P. Nacharaju, M. Van Slegtenhorst, K. Gwinn-Hardy, M.M. Paul, M. Baker, X. Yu, K. Duff, J. Hardy, A. Corral, W.L. Lin, S.H. Yen, D.W. Dickson, P. Davies and M. Hutton, Neurofibrillary tangles, amyotrophy and progressive motor disturbance in mice expressing mutant (P301L) tau protein, *Nat Genet* **25** (2000), 402–405.

[18] J. Lewis, D.W. Dickson, W.L. Lin, L. Chisholm, A. Corral, G. Jones, S.H. Yen, N. Sahara, L. Skipper, D. Yager, C. Eckman, J. Hardy, M. Hutton and E. McGowan, Enhanced neurofibrillary degeneration in transgenic mice expressing mutant tau and APP, *Science* **293** (2001), 1487–1491.

[19] W.L. Lin, J. Lewis, S.H. Yen, M. Hutton and D.W. Dickson, Ultrastructural neuronal pathology in transgenic mice expressing mutant (P301L) human tau, *J Neurocytol* **32** (2003), 1091–1105.

[20] D. Morgan, D.M. Diamond, P.E. Gottschall, K.E. Ugen, C. Dickey, J. Hardy, K. Duff, P. Jantzen, G. DiCarlo, D. Wilcock, K. Connor, J. Hatcher, C. Hope, M. Gordon and G.W. Arendash, A beta peptide vaccination prevents memory loss in an animal model of Alzheimer's disease, *Nature* **408** (2000), 982–985.

[21] T. Murakami, E. Paitel, T. Kawarabayashi, M. Ikeda, M.A. Chishti, C. Janus, E. Matsubara, A. Sasaki, T. Kawarai, A.L. Phinney, Y. Harigaya, P. Horne, N. Egashira, K. Mishima, A. Hanna, J. Yang, K. Iwasaki, M. Takahashi, M. Fujiwara, K. Ishiguro, C. Bergeron, G.A. Carlson, K. Abe, D. Westaway, P. St George-Hyslop and M. Shoji, Cortical neuronal and glial pathology in TgTauP301L transgenic mice: neuronal degeneration, memory disturbance, and phenotypic variation, *Am J Pathol* **169** (2006), 1365–1375.

[22] W. Noble, E. Planel, C. Zehr, V. Olm, J. Meyerson, F. Suleman, K. Gaynor, L. Wang, J. LaFrancois, B. Feinstein, M. Burns, P. Krishnamurthy, Y. Wen, R. Bhat, J. Lewis, D. Dickson and K. Duff, Inhibition of glycogen synthase kinase-3 by lithium correlates with reduced tauopathy and degeneration in vivo, *Proc Natl Acad Sci USA* **102** (2005), 6990–6995.

[23] E.H. Owen, S.F. Logue, D.L. Rasmussen and J.M. Wehner, Assessment of learning by the morris water task and fear conditioning in inbred mouse strains and F1 hybrids: Implications of genetic background for single gene mutations and quantitative trait loci analyses, *Neurosci* **80** (1997), 1087–1099.

[24] L. Pennanen, D.P. Wolfer, R.M. Nitsch and J. Gotz, Impaired spatial reference memory and increased exploratory behavior in P301L tau transgenic mice, *Genes Brain Behav* **5** (2006), 369–379.

[25] E.D. Roberson, K. Scearce-Levie, J.J. Palop, F. Yan, I.H. Cheng, T. Wu, H. Gerstein, G.Q. Yu and L. Mucke, Reducing endogenous tau ameliorates amyloid beta-induced deficits in an Alzheimer's disease mouse model, *Science* **316** (2007), 750–754.

[26] H. Rosenmann, N. Grigoriadis, H. Eldar-Levy, A. Avital, L. Rozenstein, O. Touloumi, L. Behar, T. Ben-Hur, Y. Avraham, E. Berry, M. Segal, I. Ginzburg and O. Abramsky, A novel transgenic mouse expressing double mutant tau driven by its natural promoter exhibits tauopathy characteristics, *Exp Neurol* **212** (2008), 71–84.

[27] K. Santacruz, J. Lewis, T. Spires, J. Paulson, L. Kotilinek, M. Ingelsson, A. Guimaraes, M. DeTure, M. Ramsden, E. McGowan, C. Forster, M. Yue, J. Orne, C. Janus, A. Mariash, M. Kuskowski, B. Hyman, M. Hutton and K.H. Ashe, Tau suppression in a neurodegenerative mouse model improves memory function, *Science* **309** (2005), 476–481.

[28] K. Schindowski, A. Bretteville, K. Leroy, S. Begard, J.P. Brion, M. Hamdane and L. Buee, Alzheimer's disease-like tau neuropathology leads to memory deficits and loss of functional synapses in a novel mutated tau transgenic mouse without any motor deficits, *Am J Pathol* **169** (2006), 599–616.

[29] K. Schindowski, K. Belarbi, A. Bretteville, K. Ando and L. Buee, Neurogenesis and cell cycle-reactivated neuronal death during pathogenic tau aggregation, *Genes Brain Behav* **7**(Suppl 1) (2008), 92–100.

[30] D.M. Wilcock, A. Rojiani, A. Rosenthal, S. Subbarao, M.J. Freeman, M.N. Gordon and D. Morgan, Passive immunother-

apy against Abeta in aged APP-transgenic mice reverses cognitive deficits and depletes parenchymal amyloid deposits in spite of increased vascular amyloid and microhemorrhage, *J Neuroinflammation* **1** (2004), 24.

[31] C. Zehr, J. Lewis, E. McGowan, J. Crook, W.L. Lin, K. Godwin, J. Knight, D.W. Dickson and M. Hutton, Apoptosis in oligodendrocytes is associated with axonal degeneration in P301L tau mice, *Neurobiol Dis* **15** (2004), 553–562.

# Activation of Cell Cycle Proteins in Transgenic Mice in Response to Neuronal Loss But Not Aβ and Tau Pathology

Joao P. Lopes[a], Mathew Blurton-Jones[b], Tritia R. Yamasaki[b], Paula Agostinho[a] and Frank M. LaFerla[b]
[a]Center for Neuroscience and Cell Biology, Faculty of Medicine, Biochemistry Institute, University of Coimbra, 3004 Coimbra, Portugal
[b]Department of Neurobiology and Behavior, 1109 Gillespie Neuroscience Facility, University of California, Irvine, Irvine, CA 92697-4545, USA

**Abstract**. Cell cycle proteins are elevated in the brain of patients and in transgenic models of Alzheimer's disease (AD), suggesting that aberrant cell cycle re-entry plays a key role in this disorder. However, the precise relationship between cell cycle reactivation and the AD hallmarks, amyloid-beta (Aβ) plaques and tau-laden neurofibrillary tangles, remains unclear. We sought to determine whether cell cycle reactivation initiates in direct response to Aβ and tau accumulation or whether it occurs as a downstream consequence of neuronal death pathways. Therefore, we used a triple transgenic mouse model of AD (3xTg-AD) that develops plaques and tangles, but does not exhibit extensive neuronal loss, whereas to model hippocampal neuronal death a tetracycline-regulatable transgenic model of neuronal ablation (CaM/Tet-DT$_A$ mice) was used. Cell-cycle proteins activation was determined in these two models of neurodegeneration, using biochemical and histological approaches. Our findings indicate that Cdk4, PCNA and phospho-Rb are significantly elevated in CaM/Tet-DT$_A$ mice following neuronal death. In contrast, no significant activation of cell-cycle proteins occurs in 3xTg-AD mice versus non-transgenic controls. Taken together, our data indicate that neuronal cell cycle reactivation is not a prominent feature induced by Aβ or tau pathology, but rather appears to be triggered by acute neuronal loss.

Keywords: Cell cycle, Alzheimer's disease, transgenic mice, tetracycline-inducible, amyloid-beta, PCNA, Cdk4, phospho-Rb, phospho-histone H3, apoptosis

## INTRODUCTION

Alzheimer's disease (AD) is an incapacitating and eventually fatal disorder, characterized by a progressive impairment of memory and cognition. Pathologically, AD is characterized by the accumulation of amyloid plaques and tau-laden neurofibrillary tangles (NFTs), as well as significant neuronal death, especially within the cortex and hippocampus [1, 2]. The amyloid-beta peptide (Aβ), the main component of amyloid plaques, is considered a key molecule in the pathogenesis of AD, triggering the onset of the disease and leading to tau pathology, synapse loss and neuronal death [3].

Studies suggest that during the progression of AD, a subset of neurons shows evidence of cell cycle re-activation [4-11]. In mature neurons, the cell cycle is normally arrested at the G0 phase [12]. Hence, neurons must be subjected to a potent stimulus to trigger the re-expression of cell cycle proteins and progression into the G1 phase [6]. Activation of complex-forming cyclin-dependent kinases (Cdks) leads to the phosphorylation of the retinoblastoma protein (Rb), releasing it from the E2F-1/Rb transcription-repressor complex, and allowing progression at the G1/S checkpoint [13]. Hypoxia, ischemia, oxidative stress, DNA damage or Aβ peptides are amongst a variety of stimuli that can cause cell cycle re-entry in neurons [7, 14-18]. Notably, it has been shown that, although cell cycle activation may contribute to the formation of NFTs [19], re-expression of cell cycle proteins in AD can also occur prior to tau hyper-

---

Correspondence address: Joao Pedro Lopes, Center for Neuroscience and Cell Biology, Faculty of Medicine, Biochemistry Institute, University of Coimbra, 3004 Coimbra, Portugal. Tel.: 351 239 820190; Fax: 351 239 822776; E-mail: jpplopes@gmail.com

phosphorylation [6, 11]. Furthermore, in one AD mouse model (R1.40-YAC), upregulation of the cell cycle markers preceded the appearance of amyloid plaques by months. Although this YAC model develops Aβ pathology and exhibits some tau phosphorylation, as with most AD models, there is no formation of mature NFTs [11, 20]. However, neuronal cyclin activation and cell cycle re-entry also occurs simply as a component of neuronal death [21].

To determine whether cell cycle re-entry occurs in direct response to Aβ and tau accumulation or instead, if reactivation occurs as a downstream consequence of neuronal death pathways, we examined two different transgenic mouse models. The triple transgenic mouse model of Alzheimer's disease (3xTg-AD) develops cognitive impairment, Aβ plaques, and NFTs in an age-dependent, hierarchical fashion, but like many other transgenic AD models, 3xTg-AD mice show little evidence of neuronal death [22, 23]. Consequently, we developed a novel transgenic model that allows the regulatable induction of diphtheria toxin (CaM/Tet-DT$_A$ mice), leading to robust hippocampal neuronal cell death [24]. Hence, these mice provide us with a method to determine whether cell cycle activation occurs as a downstream component of cell death pathways. We find no evidence of abortive cell cycle reactivation in 3xTg-AD mice, either prior to, or in response to Aβ or NFT pathology. In contrast, Cdk4, PCNA, and the retinoblastoma protein are significantly activated with the induction of neuronal death in CaM/Tet-DT$_A$ mice. The levels of tau and Aβ were also found altered in this mouse model. Our data indicates that neuronal cell cycle reactivation does not occur as a direct cause or consequence of the A and tau pathology, but is instead triggered as a component of neuronal death.

## MATERIAL AND METHODS

### Antibodies

Antibodies used include monoclonal anti-PCNA (PC10, 1:1000), polyclonal anti-Cdk4 (C-22, 1:1000), and anti-GAPDH (FL335, 1:10000) (Santa Cruz Biotechnology, CA, USA). Antibodies against the phosphorylated form of Rb (Ser807/811, 1:500) and Histone H3 (RR002, 1:15000) were from Cell Signalling (Denvers, MA, USA) and Upstate (Lake Placid, NY, USA), respectively. Anti- GFAP (1:500) and anti-NeuN (1:500) antibodies were from Chemicon-Milipore (Temecula, CA, USA). Antibodies against human PHF-tau (AT8, 1:250) and β-amyloid (6E10, 1:500) were from Pierce Endogen (Rockford, IL, USA) and Covance (Emeryville, CA, USA), respectively. Alexa-fluor conjugated secondary antibodies were from Invitrogen (Carlsbad, CA, USA).

### Mice

Generation and characterization of 3xTg-AD and CaM/Tet-DT$_A$ mice have been previously described [23, 24]. For this study, we used 1-, 6-, 12-, 18-, 24-month-old 3xTg-AD and age-matched nontransgenic (NonTg) mice. 4-month old CaM/Tet-DT$_A$ mice were subjected to neuronal lesions by removing doxycycline from their diet for 30 days. We previously established that 30 days of doxycycline withdrawal leads to marked neuronal death within CA1 of the hippocampus, and a more restricted loss of neurons within the hippocampal granule cell layer and the neocortex [24].

### Western Blot

Brains were homogenized in tissue protein extraction reagent (T-PER) (Pierce, Rockford, IL) in the presence of protease inhibitors (Roche Applied Science, Indianapolis, IN) and phosphatase inhibitors (5 mM sodium fluoride and 50 μM sodium orthovanadate) and centrifuged at 100,000 x $g$ for 1 h at 4°C. Supernatants were collected as the detergent-soluble fraction. Samples were normalized to protein concentration and denatured for 10 minutes at 95°C in sample buffer and NuPAGE LDS sample buffer (Invitrogen, Carlsbad, CA) and then resolved by SDS/PAGE under reducing conditions and transferred to a nitrocellulose membrane (Pierce, Rockford, IL). The membrane was incubated in a 5% solution of bovine serum albumin (BSA) for 1 h at room temperature and then overnight at 4°C with primary antibody. Reprobing with anti-GAPDH antibody was used as a loading control. Blots were washed in Tween 20-TBS (T-TBS) (0.02% Tween 20/100 mM Tris, pH 7.5/150 nM NaCl) for 20 min and incubated with HRP-conjugated secondary antibody (1hr RT). Blots were washed in T-TBS for 20

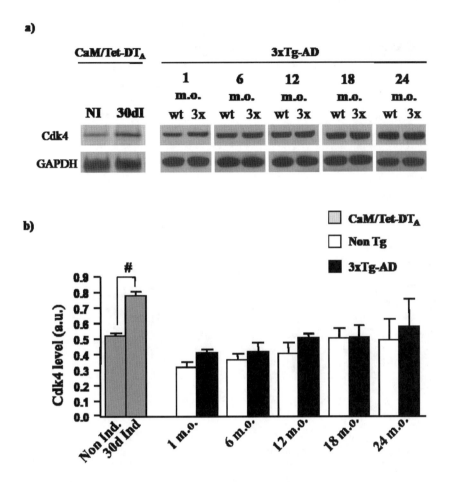

Fig. 1. Cdk4 is elevated in response to neuronal death but not Aβ and tau pathology. a) Lysates from whole brains of CaM/Tet-DT$_A$, 3xTg-AD and NonTg mice were analyzed by western Blot for Cdk4 and GAPDH as a loading control. b) Quantitative analysis of Cdk4 expression relative to GAPDH levels reveals a significant elevation in CaM/Tet-DT$_A$ mice subjected to neuronal ablation versus non-Induced controls. In contrast, 3xTg-AD mice exhibit no differences in Cdk4 expression versus age-matched NonTg controls. The data in each bar of the graph represents means means ±SEM of 4-6 experiments (1 animal per experiment) and are expressed as arbitrary units (a.u.). # $p<0.05$ compared with non-induced CaM/Tet-DT$_A$.

min, incubated for 5 min with SuperSignal (Pierce), washed, and exposed to film. Films (Pierce, Rockford, IL) were digitally scanned and immunoreactive bands quantified with Quantity One software (Hercules, CA, USA).

*Immunofluorescence*

Mice were killed by $CO_2$ asphyxiation, and brains were fixed for 48 h in 4% paraformaldehyde in TBS. Free-floating sections (50 μm thick) were sectioned on a vibratome (Pelco, Redding, CA), and stored in 0.02% sodium azide in phosphate-buffered saline (PBS) (Sigma, St. Louis, MO). Fluorescent immunohistochemistry followed standard protocols as previously described [25]. Briefly, sections were permeabilized with 0.2% Triton X-100/PBS for 15 min at RT, and blocked with 5% goat serum before incubation with primary antibody overnight at 4°C. Afterwards, sections were incubated with appropriate Alexa Fluor secondary antibody for 1 h at room temperature. Subsequently, sections were washed in PBS, mounted and cover slipped with Vectashield (Vector Laboratories). Representative images obtained randomly from the neocortical area of brain sections were visualized with a BioRad 2100 confocal microscope using lambda-strobing mode to prevent non-specific cross-excitation or cross-detection.

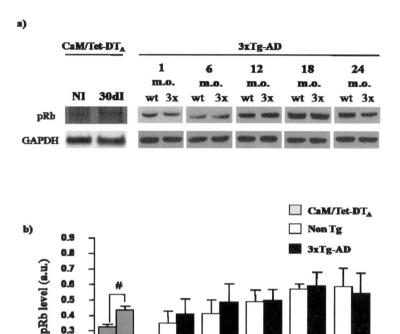

Fig. 2. Retinoblastoma protein is phosphorylated following neuronal ablation but shows no alteration in response to Aβ plaque and neurofibrillary tangle pathology. a) Lysates from whole brains of CaM/Tet-DT$_A$, 3xTg-AD and NonTg mice were resolved by SDS-PAGE and analyzed by western Blot with an antibody for pRb. GAPDH was used as a positive control. b) Quantitative analysis revealed a significant increase in Rb phosphorylation in lesion-induced CaM/Tet-DT$_A$ mice, whereas no changes in pRb were detected between 3xTg-AD and age-matched non-transgenic controls. The data in each bar of the graph represents means ±SEM of 4-6 experiments (1 animal per experiment) and are expressed as arbitrary units (a.u.). # p<0.05 compared with non-induced CaM/Tet-DT$_A$.

## Statistical analysis

Results are expressed as means ± SEM. Statistical analysis was performed with Graphpad Prism software. Significance was determined using an analysis of variance (ANOVA), followed by Bonferroni's posthoc tests for comparisons between 3xTg-AD and non-transgenic mice, across 1, 6, 12, 18, and 24-month ages. For comparisons between induced and non-induced CaM/Tet-DT$_A$ mice, a two-tailed Students' t-test was used.

## RESULTS

AD brains exhibit increased levels of Cdk4, a kinase that mediates one of the initial steps of cell cycle reactivation: Rb phosphorylation [5, 26, 27]. Consequently, we examined Cdk4 expression in CaM/Tet-DT$_A$, 3xTg-AD and NonTg mice by western blot. Neuronal ablation was induced in CaM/Tet-DT$_A$ mice by doxycycline withdrawal for 30 days and led to a highly significant 50% increase in Cdk4 levels (Fig. 1a). In contrast, Cdk4 expression was unchanged between 3xTg-AD and NonTg mice at any of the ages examined (1-, 6-, 12-, 18-, and 24-months). Immunohistochemical analysis of brain sections revealed perinuclear localization of Cdk4 in all groups, but no obvious changes in the subcellular distribution of this kinase (data not shown).

Since cell cycle activation proceeds via phosphorylation of the retinoblastoma protein by Cdk4 [28, 29], we determined the levels of phospho-Rb (pRb) in our two mouse models. Both western blot and immunofluorescence analysis revealed a signifi-

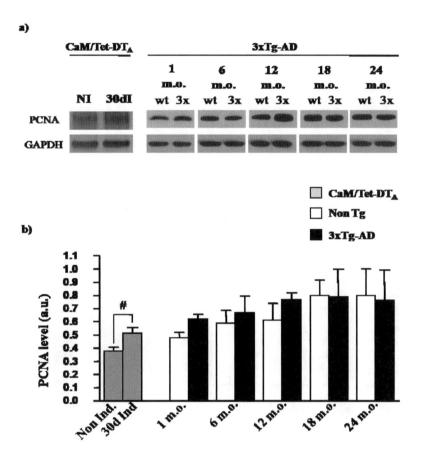

Fig. 3. PCNA expression is also elevated in CaM/Tet-DT$_A$ mice following neuronal ablation. a) Lysates from whole brains of CaM/Tet-DT$_A$, 3xTg-AD and NonTg mice were resolved by SDS-PAGE and analyzed by western Blot with an antibody for PCNA, revealing a significant increase in PCNA in response to neuronal ablation. b) Quantitative analysis of PCNA levels relative to GAPDH (loading control) reveals a significant elevation in CaM/Tet-DT$_A$ mice subjected to neuronal ablation versus non-Induced controls. In contrast, 3xTg-AD mice exhibit no differences in PCNA expression when compared with age-matched NonTg controls. The data in each bar of the graph represents means ±SEM of 4-6 experiments (1 animal per experiment) and are expressed as arbitrary units (a.u.). # $p<0.05$ compared with non-induced CaM/Tet-DT$_A$.

cant increase of phosphorylated Rb in response to neuronal ablation in CaM/Tet-DT$_A$ mice versus non-induced controls (Figs. 2 and 5a C, D -arrows), whereas no differences were detected between 3xTg-AD and NonTg mice at any time point (Figs. 2 and 5a A, B). Rb phosphorylation causes the disruption of the E2F1-Rb complex, with consequent progression past the G1/S checkpoint; hence, we evaluated the levels of proliferating cell nuclear antigen (PCNA), a component of the DNA replication machinery [13, 18]. As with the other cell cycle markers, we again found that PCNA levels did not significantly change in response to Aβ or tau pathology, at any age (Figs. 3 and 5a E, F). However, neuronal ablation in CaM/Tet-DT$_A$ animals was associated with a considerable increase in PCNA levels by both western blot and immunohistochemistry (Fig.3 and 5a G, H). Although being mainly a neuron-related phenomenon (Fig. 5b B, C), we also observed the presence of astrocytes labelled positive for these markers, indicating the existence of some gliosis (Fig. 5b A). The significant changes in the levels of proteins associated with early stages of cell cycle, occurred following the induction of neuronal death in CaM/Tet-DT$_A$ animals, suggested that these insulted neurons are able to advance past the G1/S

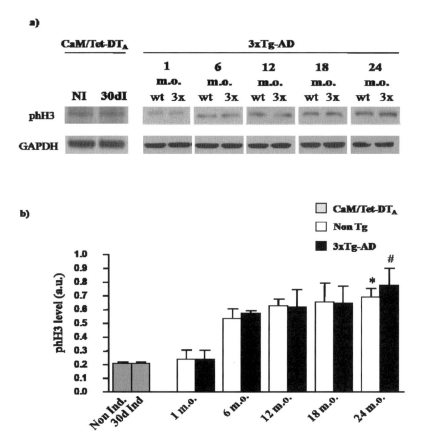

Fig. 4. phH3 levels are unchanged in transgenic models of neuronal ablation and AD. a) b) Western blot analysis of CaM/Tet-DT$_A$ mice revealed no changes in phosphorylated histone H3, suggesting that ablated neurons likely die prior to passing the G2/M checkpoint. Likewise, phH3 is unaltered in 3xTg-AD versus age-matched Non-transgenic mice. However, phH3 does exhibit an age-dependant increase in both transgenic and NonTg animals, thus some activation of the cell cycle appears to occur with brain aging, although whether this is indicative of simply increased gliosis remains unclear. The data in each bar of the graph represents means ±SEM of 4-6 experiments (1 animal per experiment) and are expressed as arbitrary units (a.u.). * $p<0.05$, # $p<0.05$, versus 1 month-old 3xTg-AD.

checkpoint, reaching the S or even the G2 phase. Consequently, we also assessed the levels of phosphorylated histone H3 (phH3) to determine whether re-cycling neurons are able to pass the G2/M cell cycle checkpoint. Interestingly, analysis of CaM/Tet-DT$_A$ mice revealed no changes in phosphorylated histone H3, indicating that ablated neurons die prior to passing the G2/M checkpoint (Fig. 4). Similarly, no changes in the phosphorylation of histone H3 were observed between NonTg and 3xTg-AD age-matched animals (Fig. 4). However, and in agreement with what can be seen in the other markers, we detected a trend towards an age-dependant increase in phH3 in both 3xTg-AD and Non-Tg mice, probably resulting from increased gliosis (Fig. 4).

Since neurons suffering excitotoxic damage can re-enter the cell cycle and later express AD-associated proteins [6], we determined the levels of Aβ and hyperphosphorylated tau in CaM/Tet-DT$_A$ animals. A significant augment of hyperphosphorylated tau was observed after the induction of neuronal death (Fig. 6a). Increased levels of Aβ were also found in 30-day Induced mice when compared with the Non-induced littermates (Fig. 6b).

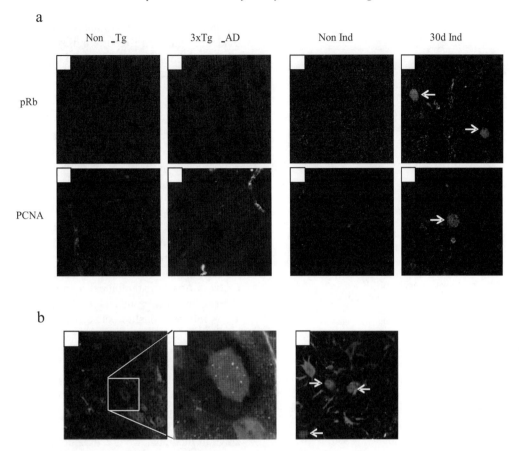

Fig. 5. Cell cycle activation occurs in a neuronal ablation transgenic model but not in 3xTg-AD a) Immunofluorescent analysis of the neocortex area in non-transgenic, 3xTg-AD, non-induced and 30d-induced CaM/Tet-DT$_A$ mice. pRb revealed a nuclear distribution of pRb (D) and PCNA (H) in 30d-induced animals, whereas no immunoreactivity was detected in non-induced controls (C, G), 18 m.o. non-transgenic (A, E) or 3xTg-AD animals (B, F). b) Colocalization of NeuN with pRb labeling confirmed the existence of cell cycle-active neurons (A, B). Positive immunoreactivity for GFAP was also found in some of the nuclei stained for PCNA, indicating the presence of gliosis (C). Images collected randomly by confocal microscopy using 400x magnification (Fig. 5a, A-G; Fig.5b, A,C) and 630x magnification (Fig. 5b, B).

## DISCUSSION

Aberrant cell cycle activation has been observed in several different neurodegenerative disease models and is closely associated with stimuli and pathways that culminate in neuronal death [10, 16, 18, 27, 30]. In the present work, we aimed to determine the relationship between anomalous cell cycle reentry, Aß and tau pathology, and neuronal death. Cell cycle activation was examined in the brains of two different transgenic models: 3xTg-AD mice and CaM/Tet-DT$_A$ mice, and we specifically focused our attention on several key proteins closely correlated with two important phase transitions: the G1/S and G2/M checkpoints [16, 31-33]. Interestingly, we detected no evidences of cell cycle activation in 3xTg-AD transgenic mice versus Non-Tg controls, suggesting that Aß and tau pathology are not directly causing cell cycle dysregulation. In contrast, we detected highly significant activation of cell cycle proteins in an inducible transgenic model of neuronal ablation, the CaM/Tet-DT$_A$ mice [24]. Taken together, our data suggest that AD-related cell cycle dysregulation occurs, not as a cause of, but rather as an event in apoptotic neuronal death.

Phosphorylation of the retinoblastoma protein (Rb) is considered to be an initial step in the triggering of the cell cycle at the G1 phase and several Cdks are implicated in this event, including Cdk2, Cdk4 and Cdk6 [16, 34, 35]. Increased expression of cyclin D1, the Cdk4 activator for the late G1 phase, has also been shown to occur in re-cycling cells [18]. In human AD brains, however, elevation of cyclin D1 is minimal in comparison to the robust elevation of Cdk4 observed, suggesting that phosphorylation of Rb in AD likely occurs primarily via ele-

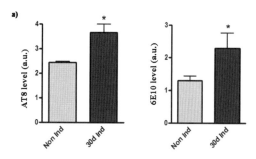

Fig. 6. Hyperphosphorylated tau and Aβ levels are increased in a transgenic model of neuronal ablation. a) b) Lysates from whole brains of CaM/Tet-DT$_A$ mice were resolved by SDS-PAGE and analyzed by western Blot with antibodies for hyperphosphorylated tau (AT8) and Aβ (6E10). Quantitative analysis revealed a significant increase in the levels of tau in CaM/Tet-DT$_A$ mice subjected to neuronal ablation versus non-Induced controls, whereas changes in Aβ were of lower extent. GAPDH was used as a loading control. The data in each bar of the graph represents means ± SEM of 4 experiments (1 animal per experiment) and are expressed as arbitrary units (a.u.). * p<0.05 compared with non-induced CaM/Tet-DT$_A$.

vated Cdk4 expression, not via increased cyclin D1 [5, 26]. Hence, we specifically examined Cdk4 expression in our transgenic models, whereas prior studies have not examined this kinase in AD transgenics [11, 30]. Our data revealed two different profiles for Cdk4 levels in the models used. Whereas 30-day induced CaM/Tet-DT$_A$ mice had a significant increase as compared with non-induced animals, no disparity was evident between 3xTg-AD and Non-Tg mice. Because the phosphorylation state of Rb can be altered via Cdks other than Cdk4 [35], the absence of changes in Cdk4 levels must also be combined with an analysis of pRb to more fully determine whether the neuronal cell cycle remains inactive. Indeed, analysis of pRb in 3xTg-AD versus NonTg mice supports just such a conclusion, since no changes were observed in pRb at any of the time points studied. Furthermore, the absence of alterations in pRb is corroborated by the observation that PCNA, a polymerase-associated protein synthesized in early G1 and S phases, also remained unaltered in 3xTg-AD mice.

The absence of cell cycle reactivation even in older 3xTg-AD animals was somewhat unexpected, since studies in different neurodegeneration-related animal models, like hypoxia/ischemia, amyotrophic lateral sclerosis or Parkinson's disease have documented ectopic cell cycle reactivation and an increase in the levels of PCNA [16, 18, 27]. However, although cell cycle re-entry has also been described in some transgenic AD models, there is conflicting evidence. There are reports indicating neuronal cell cycle activation in APP23 mice [11] whereas others have found cell cycle activation only within glia in this model [30]. Interestingly, APP23 mice are one of the few transgenic AD lines reported to exhibit significant neuronal cell death[36]. Thus, it remains unclear whether the conflicting reports of cell cycle activation in APP23 mice may relate to the degree to which neuronal death occurs between different colonies of these mice. For example, it is well established that some mouse background strains exhibit neuronal excitotoxic sensitivity [37], whereas others do not. Hence, slight variation in background strain or mixed-strain colonies could influence the degree to which neuronal death occurs in a given mouse model of AD. Therefore, separate examination of each set of key pathological features of AD, by using two transgenic models, allowed us to determine a possible relationship between cell cycle activation and AD pathology.

Despite the several studies in different models [5-10, 17, 19, 21], determination of whether cell cycle activation in AD occurs as a cause versus a consequence of pathology or neuronal death remains unclear. Our data showed two different behaviours in the models used: 3xTg-AD mice display a heavy amyloid load and strong tau pathology, but no apoptosis or cell cycle reactivation, whereas CaM/Tet-DT$_A$ animals have increased levels of Aβ and hyperphosphorylated tau, following the induction of apoptotic cell death and cell cycle re-entry. These results not only suggest that cell cycle reactivation in neurons can occur as a part of the apoptotic pathway, but also support recent observations that cell cycle re-entry may lead to AD-like changes [38].

Taken together our data suggest that cell cycle reactivation in Alzheimer's disease is not a direct consequence of Aβ deposition or tau pathology, but rather results from the downstream activation of a neuronal death pathway.

## ACKNOWLEDGEMENTS

This work was supported by FCT fellowship SFRH/BD/16381/2004 (JPL), NIH/NIA grant R01AG027544 (FML), and a California Institute for Regenerative Medicine Postdoctoral Scholar Award (MBJ).

# References

[1] Golde TE (2003) Alzheimer disease therapy: can the amyloid cascade be halted? *J Clin Invest* 111, 11-18.

[2] Selkoe DJ (2001) Alzheimer's disease: genes, proteins, and therapy. *Physiol Rev* 81, 741-766.

[3] Hardy J, Selkoe DJ (2002) The amyloid hypothesis of Alzheimer's disease: progress and problems on the road to therapeutics. *Science* 297, 353-356.

[4] Arendt T (2000) Alzheimer's disease as a loss of differentiation control in a subset of neurons that retain immature features in the adult brain. *Neurobiol Aging* 21, 783-796.

[5] Busser J, Geldmacher DS, Herrup K (1998) Ectopic cell cycle proteins predict the sites of neuronal cell death in Alzheimer's disease brain. *J Neurosci* 18, 2801-2807.

[6] Hernandez-Ortega K, Ferrera P, Arias C (2007) Sequential expression of cell-cycle regulators and Alzheimer's disease-related proteins in entorhinal cortex after hippocampal excitotoxic damage. *J Neurosci Res* 85, 1744-1751.

[7] Malik B, Currais A, Soriano S (2008) Cell cycle-driven neuronal apoptosis specifically linked to amyloid peptide Abeta1-42 exposure is not exacerbated in a mouse model of presenilin-1 familial Alzheimer's disease. *J Neurochem* 106, 912-916.

[8] Nagy Z, Esiri MM, Cato AM, Smith AD (1997) Cell cycle markers in the hippocampus in Alzheimer's disease. *Acta Neuropathol* 94, 6-15.

[9] Smith MZ, Nagy Z, Esiri MM (1999) Cell cycle-related protein expression in vascular dementia and Alzheimer's disease. *Neurosci Lett* 271, 45-48.

[10] Yang Y, Mufson EJ, Herrup K (2003) Neuronal cell death is preceded by cell cycle events at all stages of Alzheimer's disease. *J Neurosci* 23, 2557-2563.

[11] Yang Y, Varvel NH, Lamb BT, Herrup K (2006) Ectopic cell cycle events link human Alzheimer's disease and amyloid precursor protein transgenic mouse models. *J Neurosci* 26, 775-784.

[12] Nagy Z (2000) Cell cycle regulatory failure in neurones: causes and consequences. *Neurobiol Aging* 21, 761-769.

[13] Park DS, Morris EJ, Bremner R, Keramaris E, Padmanabhan J, Rosenbaum M, Shelanski ML, Geller HM, Greene LA (2000) Involvement of retinoblastoma family members and E2F/DP complexes in the death of neurons evoked by DNA damage. *J Neurosci* 20, 3104-3114.

[14] Katchanov J, Harms C, Gertz K, Hauck L, Waeber C, Hirt L, Priller J, von Harsdorf R, Bruck W, Hortnagl H, Dirnagl U, Bhide PG, Endres M (2001) Mild cerebral ischemia induces loss of cyclin-dependent kinase inhibitors and activation of cell cycle machinery before delayed neuronal cell death. *J Neurosci* 21, 5045-5053.

[15] Kruman, II, Wersto RP, Cardozo-Pelaez F, Smilenov L, Chan SL, Chrest FJ, Emokpae R, Jr., Gorospe M, Mattson MP (2004) Cell cycle activation linked to neuronal cell death initiated by DNA damage. *Neuron* 41, 549-561.

[16] Kuan CY, Schloemer AJ, Lu A, Burns KA, Weng WL, Williams MT, Strauss KI, Vorhees CV, Flavell RA, Davis RJ, Sharp FR, Rakic P (2004) Hypoxia-ischemia induces DNA synthesis without cell proliferation in dying neurons in adult rodent brain. *J Neurosci* 24, 10763-10772.

[17] Majd S, Zarifkar A, Rastegar K, Takhshid MA (2008) Different fibrillar Abeta 1-42 concentrations induce adult hippocampal neurons to reenter various phases of the cell cycle. *Brain Res* 1218, 224-229.

[18] Nguyen MD, Boudreau M, Kriz J, Couillard-Despres S, Kaplan DR, Julien JP (2003) Cell cycle regulators in the neuronal death pathway of amyotrophic lateral sclerosis caused by mutant superoxide dismutase 1. *J Neurosci* 23, 2131-2140.

[19] Illenberger S, Zheng-Fischhofer Q, Preuss U, Stamer K, Baumann K, Trinczek B, Biernat J, Godemann R, Mandelkow EM, Mandelkow E (1998) The endogenous and cell cycle-dependent phosphorylation of tau protein in living cells: implications for Alzheimer's disease. *Mol Biol Cell* 9, 1495-1512.

[20] Kulnane LS, Lamb BT (2001) Neuropathological characterization of mutant amyloid precursor protein yeast artificial chromosome transgenic mice. *Neurobiol Dis* 8, 982-992.

[21] Padmanabhan J, Park DS, Greene LA, Shelanski ML (1999) Role of cell cycle regulatory proteins in cerebellar granule neuron apoptosis. *J Neurosci* 19, 8747-8756.

[22] Billings LM, Oddo S, Green KN, McGaugh JL, LaFerla FM (2005) Intraneuronal Abeta causes the onset of early Alzheimer's disease-related cognitive deficits in transgenic mice. *Neuron* 45, 675-688.

[23] Oddo S, Caccamo A, Shepherd JD, Murphy MP, Golde TE, Kayed R, Metherate R, Mattson MP, Akbari Y, LaFerla FM (2003) Triple-transgenic model of Alzheimer's disease with plaques and tangles: intracellular Abeta and synaptic dysfunction. *Neuron* 39, 409-421.

[24] Yamasaki TR, Blurton-Jones M, Morrissette DA, Kitazawa M, Oddo S, LaFerla FM (2007) Neural stem cells improve memory in an inducible mouse model of neuronal loss. *J Neurosci* 27, 11925-11933.

[25] Blurton-Jones M, Tuszynski MH (2006) Estradiol-induced modulation of estrogen receptor-beta and GABA within the adult neocortex: a potential transsynaptic mechanism for estrogen modulation of BDNF. *J Comp Neurol* 499, 603-612.

[26] Biswas SC, Shi Y, Vonsattel JP, Leung CL, Troy CM, Greene LA (2007) Bim is elevated in Alzheimer's disease neurons and is required for beta-amyloid-induced neuronal apoptosis. *J Neurosci* 27, 893-900.

[27] Hoglinger GU, Breunig JJ, Depboylu C, Rouaux C, Michel PP, Alvarez-Fischer D, Boutillier AL, Degregori J, Oertel WH, Rakic P, Hirsch EC, Hunot S (2007) The pRb/E2F cell-cycle pathway mediates cell death in Parkinson's disease. *Proc Natl Acad Sci U S A* 104, 3585-3590.

[28] Dyson N (1998) The regulation of E2F by pRB-family proteins. *Genes Dev* 12, 2245-2262.

[29] Rashidian J, Iyirhiaro G, Aleyasin H, Rios M, Vincent I, Callaghan S, Bland RJ, Slack RS, During MJ, Park DS (2005) Multiple cyclin-dependent kinases signals are critical mediators of ischemia/hypoxic neuronal death in vitro and in vivo. *Proc Natl Acad Sci U S A* 102, 14080-14085.

[30] Gartner U, Bruckner MK, Krug S, Schmetsdorf S, Staufenbiel M, Arendt T (2003) Amyloid deposition in APP23 mice is associated with the expression of cyclins in astrocytes but not in neurons. *Acta Neuropathol* 106, 535-544.

[31] Liu DX, Greene LA (2001) Neuronal apoptosis at the G1/S cell cycle checkpoint. *Cell Tissue Res* 305, 217-228.

[32] Ogawa O, Zhu X, Lee HG, Raina A, Obrenovich ME, Bowser R, Ghanbari HA, Castellani RJ, Perry G, Smith MA (2003) Ectopic localization of phosphorylated histone H3 in Alzheimer's disease: a mitotic catastrophe? *Acta Neuropathol* 105, 524-528.

[33] Wen Y, Yang S, Liu R, Simpkins JW (2005) Cell-cycle regulators are involved in transient cerebral ischemia induced neuronal apoptosis in female rats. *FEBS Lett* 579, 4591-4599.

[34] Copani A, Condorelli F, Caruso A, Vancheri C, Sala A, Giuffrida Stella AM, Canonico PL, Nicoletti F, Sortino MA (1999) Mitotic signaling by beta-amyloid causes neuronal death. *Faseb J* 13, 2225-2234.

[35] Nguyen MD, Mushynski WE, Julien JP (2002) Cycling at the interface between neurodevelopment and neurodegeneration. *Cell Death Differ* 9, 1294-1306.

[36] Bondolfi L, Calhoun M, Ermini F, Kuhn HG, Wiederhold KH, Walker L, Staufenbiel M, Jucker M (2002) Amyloid-associated neuron loss and gliogenesis in the neocortex of amyloid precursor protein transgenic mice. *J Neurosci* 22, 515-522.

[37] McLin JP, Steward O (2006) Comparison of seizure phenotype and neurodegeneration induced by systemic kainic acid in inbred, outbred, and hybrid mouse strains. *Eur J Neurosci* 24, 2191-2202.

[38] McShea A, Lee HG, Petersen RB, Casadesus G, Vincent I, Linford NJ, Funk JO, Shapiro RA, Smith MA (2007) Neuronal cell cycle re-entry mediates Alzheimer disease-type changes. *Biochim Biophys Acta* 1772, 467-472.

# Electron Microscopic 3D Reconstruction Analysis of Amyloid Deposits in 3xTg-AD Mice and Aged Canines

Paworn Nuntagij[a,b], Naiphinich Kotchabhakdi[b] and Reidun Torp[a]
[a]*Centre for Molecular Biology and Neuroscience and Department of Anatomy, Institute of Basic Medical Sciences, University of Oslo, Norway*
[b]*Neuro-Behavioural Biology Centre, Institute of Molecular Biosciences, Mahidol University, Thailand*

**Abstract.** Little is known about how amyloid-β (Aβ) is deposited in relation to the complex ultrastructure of the brain. Here we combined serial section immunoelectron microscopy with 3D reconstruction to elucidate the spatial relationship between Aβ deposits and ultrastructurally identified cellular compartments. The analysis was performed in a transgenic mouse model with mutant presenilin-1, and mutant amyloid-β protein precursor (AβPP) and tau transgenes (3xTg-AD mice) and in aged dogs that develop Aβ plaques spontaneously. Reconstructions based on serial ultrathin sections of hippocampus (mice) or neocortex (dogs) that had been immunolabeled with Aβ (Aβ1−42) antibodies showed that the organization of extracellular Aβ deposits is more complex than anticipated from light microscopic analyses. In both species, deposits were tightly associated with plasma membranes of pyramidal cell bodies and major dendrites. The deposits typically consisted of thin sheets as well as slender tendrils that climbed along the large caliber dendritic stems of pyramidal neurons. No preferential association was observed between Aβ deposits and thin dendritic branches or spines, nor was there any evidence of preferential accumulation of Aβ around synaptic contacts or glial processes. Our data suggest that plaque formation is a precisely orchestrated process that involves specialized domains of dendrosomatic plasma membranes.

**Keywords.** Aged dogs, amyloid-β, electron microscopy, 3D images, immunocytochemistry, 3xTg-AD mice

## INTRODUCTION

Recent years have seen major advances in our understanding of the molecular mechanisms that underlie the generation of amyloid-β (Aβ) [1]. It is now well established that amyloidogenic fibrils are formed by abnormal cleavage of amyloid-β protein precursor (AβPP) and that the protofibrils are secreted by neurons, possibly at synaptic sites [2–5]. Much less is known about the subsequent steps in the pathophysiological process. Notably, we still lack an understanding of the factors that dictate fibrillogenesis, i.e., the process by which protofibrils combine to form deposits that can be recognized at the light or electron microscopic level [6,7]. Similarly, knowledge is scarce as to how Aβ deposits interact with specific neuronal or glial compartments in vivo.

It is important to fill this void as it was recently reported that the formation of Aβ deposits is a rapid event that precedes other types of pathology typically associated with Alzheimer's disease [8]. An essential task is to resolve how Aβ deposits develop at the ultrastructural level. Most previous studies of plaques have relied on light microscopic analyses, which fail to provide insight in how deposits relate to specific cellular compartments in the complex neuropil of the brain. By exploiting the possibilities offered by serial section immunoelectron microscopy and novel procedures for 3D reconstruction analysis, the present study aimed to establish the precise spatial relationship between extracellular Aβ deposits and the cellular compartments with which Aβ might interact. Our analysis was not centered on large, compact plaques but rather focused on small deposits, whose structure is likely to reflect the initial stage of fibrillogenesis. Our data show that the deposits contain intricately interwoven and delicate threads whose dimensions are beyond the resolution power of the light microscope. These deposits exhibit a striking preference for discrete domains of dendrosomatic plasma membranes. Our findings indicate that fibrillogenesis must be understood on the basis of specific molecular properties of these membrane domains.

## MATERIALS AND METHODS

### Tissue preparation

The transgenic model used here was derived by injecting two independent transgenes containing the human $APP_{Swe}$ and human $tau^{P301L}$ sequences into single-cell embryos from homozygous mutant $PS1^{M146V}$ knock-in mice [9]. Six animals aged 28–30 months were transcardially perfused with 2% dextran in phosphate buffer for 15 s followed by 4% formaldehyde (freshly depolymerized from paraformaldehyde) and 0.1% glutaraldehyde in 0.1M phosphate buffer, pH 7.4. The whole brains were cut at 500 $\mu$m on a Vibratome and stored in perfusion solution to which 0.2% sodium azide had been added as preservative. The whole hippocampi were dissected and embedded. Dog brain tissues were obtained from selections of naturally aging dogs with a behavioral profile indicating mild mental deficit [10,11]. Dogs were perfused transcardially with 4% formaldehyde and tissue specimens from dorsolateral prefrontal cortex were dissected out. Both mouse and dog tissues were embedded in Lowicryl LM20 as described [12,13]. There are two main steps in this procedure, cryoprotection and cryosubstitution. Cryoprotection was done by immersing the tissues into phosphate buffered glucose at increasing concentrations (10, 20, and 30%) prior to plunging the tissue specimens into liquid propane at −190°C in a liquid nitrogen-cooled cryofixation unit KF80 (Reichert, Vienna, Austria). Cryosubstitution was undertaken in 0.5% uranyl acetate in anhydrous methanol at −90°C for 24 h in a cryosubstitution unit AFS (Reichert). The temperature was stepwise increased to −45°C and Lowicryl HM20 was gradually substituted for methanol. Polymerization was performed under UV light for 48 h at −45°C.

### Serial section electron microscopy

After polymerization, sections through the whole tissue specimens were obtained and investigated in the light microscope (using cresyl violet staining). Areas of interest were identified and the blocks trimmed down with a razor blade assisted by a light microscope mounted on the ultramicrotome (Leica EM UC6; Wetzlar, Germany). After trimming, the block face had a rectangular shape of less than 0.5 mm in width and covered the area of interest identified in the light microscope. Serial sections (<70 nm) were cut with a diamond knife (Ultra 45°, Diatome, Biel, Switzerland) and placed on formvar-coated single-hole grids. Uninterrupted series of 80–160 consecutive tissue sections were divided between 10–15 grids with 8–12 sections on each grid.

### Immunocytochemistry

A$\beta$ deposits appeared as dark bundles of thread-like fibrils in the extracellular space of brain parenchyma. They can be easily seen if they are cut in the longitudinal plane. However, many A$\beta$ deposits could not be identified in standard uranyl/lead stained sections because they were too small or cut in an oblique or transverse plane. To ensure identification, each individual serial section was subjected to postembedding immunogold labeling with antibodies to A$\beta$. Polyclonal A$\beta$1−42 antibody [10,11,14] and monoclonal anti-A$\beta_{1-17}$ (6E10; Chemicon) were used in this study. Even though 6E10 recognizes most forms of A$\beta$, we mostly used polyclonal A$\beta_{1-42}$ because of its high sensitivity and selectivity for the fibrillary form of A$\beta$ aggregations. Also, the 3xTg-AD mice overexpress A$\beta$PP, which is one of the epitopes recognized by 6E10 [15]. The astroglial processes were labeled with polyclonal antibody against glutamine synthetase (GS; Sigma).

We followed the protocol for postembedding immunocytochemistry described by Matsubara et al. [12]. The whole series of sections were blocked by 50 mM glycine in 5 mM Tris buffer with 0.3% NaCl and 0.1% Triton X-100 (TBST) followed by 2% human serum albumin (HSA) in TBST. The primary antibodies in 2% HSA TBST were applied to the sections for 2 h (anti-A$\beta_{1-42}$ 1:300, 6E10 1:200, anti-GS 1:300). The sections were then rinsed twice by

TBST before being incubated with goat anti-rabbit or anti-mouse Fab fragments coupled to 10 nm gold particles (BBI) in 2% HSA TBST for 1 h. Double labeling with the same types of antibodies was achieved by incubating tissue section in formaldehyde vapor at 80°C for 1 h before applying the next round of immunoincubation. For enhancing the contrast, uranyl acetate (Fluorochem) in double distilled water and lead citrate were used. Sections were left to dry overnight before being investigated under electron microscope.

*3D reconstruction*

Analysis and 3D reconstruction were performed by the freeware Reconstruct [16] specifically developed for serial section electron microscopy [17]. Three areas of interest were chosen for 3D reconstruction: areas containing pyramidal cell bodies, areas containing primary dendrites in stratum radiatum, and neuropil regions in stratum lacunosum moleculare. Electron micrographs were obtained digitally from a transmission microscope (Tecnai 10 Philips) with a camera (SIS Mega View III) in an 8-bit TIF format at a magnification of 16,500x for 3D reconstruction and higher magnifications for corresponding illustrations. We chose these magnifications because gold particles are not visible under lower magnification. The process of reconstruction was initiated from the middle section in the series and was then extended in either direction. Fusion of images was done by automated software AnalySIS from 9–25 pictures. When the area of interest exceeded that size, fusion was done manually on Adobe Photoshop for Windows. Image aligning and 3D reconstruction were done on the software Reconstruct. For determining the surface of structures of interest, the membrane boundaries of the cellular structures were outlined and the immunoreactive Aβ deposits were defined as either 1) the outlines of highly contrasted structure of or 2) gold particles marked and shown as spheres in 3D images.

The structure of extracellular Aβ deposits and their spatial relationship to primary dendrites were analyzed in reconstructions of sections where the apical dendrites had been transversely or longitudinally cut. The spatial relationship between deposits and dendritic spines was demonstrated in 3D reconstructions from stratum lacunosum moleculare. The same protocols were used to analyze Aβ deposits and associated cell compartments in the naturally aged dog.

## RESULTS

We chose to study 28- to 30-month-old mice as these mice had a relatively high density of plaques, yet showed little evidence of tissue damage. However, dystrophic neurites containing multi-lamellar vesicles typical of autophagosomes [18,19] were associated with some of the Aβ deposits (Figure 1a,b). In single labeled sections, both antibodies (anti-Aβ$_{42}$ and 6E10) were found to produce distinct clusters of labeling in the extracellular space (Figure 1). In addition, the latter antibody (6E10) gave rise to dispersed labeling in the cytosol, reflecting the broader substrate specificity of this antibody (see Materials and Methods).

In double-labeled preparations, extracellular clusters of gold particles produced by the anti-Aβ$_{42}$ antibody invariably coincided with clusters of gold particles signaling binding of the anti-Aβ 6E10 antibody (Figure 1h). At high magnification, the gold particle clusters were found to overlie fibrillar structures with ultrastructural characteristics of Aβ deposits (e.g., Figure 1f,g). However, many of the fibrillar aggregates were so small as to go unnoticed in the absence of immunogold labeling (Figure 1a,d). Immunogold labeling of each individual section thus proved indispensable for an efficient detection and 3D reconstruction of the Aβ deposits.

The current electron microscopic investigation revealed that Aβ deposits are much more complex than would be surmised from standard light microscopic analyses. Rather than being compact and rather well demarcated structures, the deposits took the shape of branched and intertwined threads (Figure 2). Typically, the branches could be traced back to an area with particularly high densities of gold particles (Figure 2b) – as if each deposit had a "center of gravity".

A major aim of the present study was to resolve whether the Aβ deposits are randomly distributed in the extracellular space or whether they show a preferential association with particular membrane domains. From the 3D reconstructions of Aβ deposits in the CA1 stratum radiatum, it was clear that many of the thread-like extensions were in close proximity to stem dendrites and large caliber dendritic branches of pyramidal cells (Figure 3). These den-

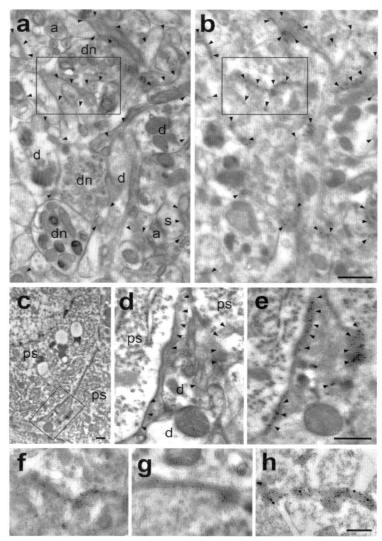

Fig. 1. Comparison of unstained sections (a, c, d) with immunolabeled sections (b, e-h). (a, b) The Aβ positive deposits identified by gold particles (arrowheads) are not readily identified in the unstained section (a; frames indicate corresponding areas in a and b shown at high magnification in f). Consecutive sections from stratum radiatum. (c-e) Perisomatic Aβ positive deposits (immunogold labeled in e, arrowheads) appear as electron dense zones in (c, d). From stratum pyramidale. Framed area in (c) is enlarged in (d). (f-h) Comparison of immunolabeling patterns with Aβ$_{42}$ and 6E10 antibodies. The sections in (f) and (g) were single labeled for Aβ and 6E10, respectively, whereas the section in (h) was double labeled (Aβ: small particles; 6E10: large particles). Note precise colocalization of the two particle sizes. From stratum radiatum. Abbreviations: dn, dystrophic neurites; d, dendrites; a, axons; ps, pyramidal cell somata; Scale bars:,0.5 μm (a-e); 0.25 μm (f-h).

and bore normal-looking spines with normal spacing (Figure 3b). No association was observed between Aβ deposits and finer dendritic branches in stratum radiatum (Figure 5c), nor was such an association found with the rather slender dendritic profiles that predominate in stratum lacunosum moleculare (not shown).

For a detailed analysis of the spatial relation between Aβ deposits and large caliber dendrites, we reconstructed deposits from series of more than 60 consecutive sections oriented roughly parallel to the long axes of the dendrites. The reconstructions showed that the threadlike extensions of the Aβ deposits were wrapped around the dendrites (Figure 3), often for a distance of more than 5 μm. For part of this distance (60–80%), the deposits were found to be in direct contact with the dendritic plasma membrane (Figure 3a,c).

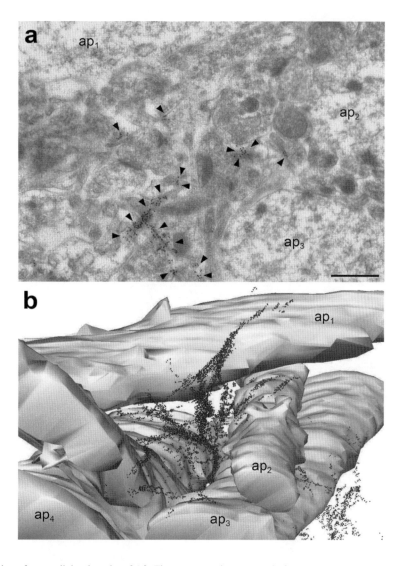

Fig. 2. 3D reconstruction of extracellular deposits of Aβ. The reconstruction was made from •100 consecutive sections and covered 400 $\mu m^2$ in each section. Amyloid deposits (arrowheads in *a*); dots representing gold particles in (*b*) form continuous meshworks composed of bundles of fibrils varying in size and direction. Some of the deposits are associated with dendrites (*ap1-ap5*; *ap4* and *ap5* not indicated in *a*). Scale bar, 0.5 μm. From stratum radiatum.

Pyramidal cell somata show a pattern of Aβ deposition similar to that of large caliber dendrites (Figure 4). In some cases Aβ deposits could be traced from neuronal somata onto the apical and basal stem dendrites (Figure 4b,c). This finding raised the issue whether there are any unique features that distinguish the dendrosomatic plasma membrane domain that is associated with Aβ deposits. The 3D reconstructions did not reveal any specific spatial relation between peridendritic Aβ deposits and spines (Figure 5d-g) or between peridendritic Aβ deposits and intracellular organelles such as mitochondria (Figure 5a,b). Moreover, there was no obvious association between Aβ deposits and astrocytic processes (Figure 6).

For comparison, we reconstructed Aβ plaques in the aged dog, using the same procedure as above. The 3D structure was strikingly similar to that described for deposits in mice. Thus the deposits took the shape of thin thread-like structures that twined around major dendrites (Figure 7). Compared with the situation in the 3xTg-AD model, the deposits in canine cortex were even more intimately attached to the dendritic plasma membranes, often forming

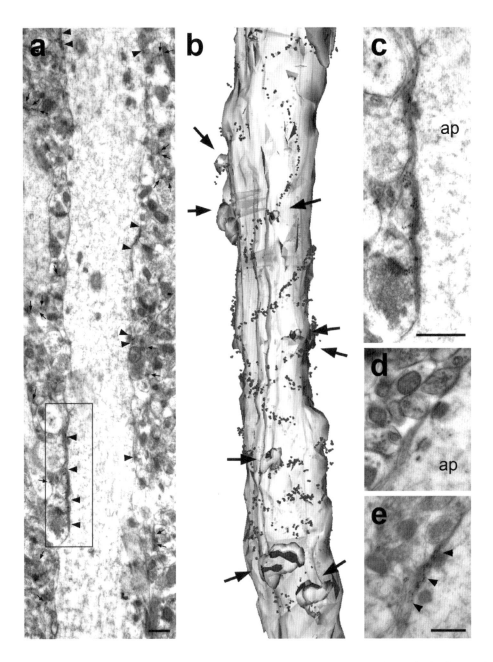

Fig. 3. Aβ deposits apposed to spiny dendrite in stratum radiatum. A series of electron micrographs of a longitudinally cut apical dendrite was chosen for 3D reconstruction. (*a*) Aβ deposits reside mainly in the extracellular space (arrows). Framed area in (*a* containing peridendritic deposit; arrowheads) is shown at higher magnification in (*c*). (*b*) Reconstructed dendrite (same as in *a*) shows several Aβ immunopositive threads (dots) entwining the dendrites. Spines (arrows) on the apical dendrites do not show any preferential association with Aβ deposits. (*c*) Immunolabeled, peridendritic amyloid deposit shown at low magnification in (*a*). (*d, e*) Aβ deposits are directly apposed to the dendritic plasma membranes with no space in between (non-immunolabeled section in d is adjacent to immunolabeled section in *e*). ap, Apical dendrite. Scale bars, 0.5 μm (*a*), 0.25 μm (*d, e*).

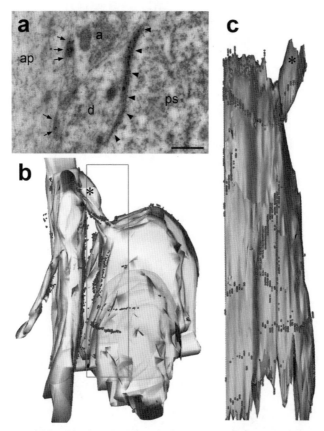

Fig. 4. Aβ deposits apposed to pyramidal cell body and primary dendrite. (a) Immunolabeled deposit (arrowheads, perisomatic; arrows, peridendritic) in direct contact with plasma membrane of pyramidal cell body (ps) and dendrite (ap). (b) 3D reconstruction of perisomatic deposit. Note that the deposit extends onto one of the primary basal dendrites (asterisk). A dendrite from a different cell passes close to the cell body. This dendrite is also associated with Aβ deposits. Part of cell body in frame is shown at a different angle in (c). Scale bar, 0.5 μm.

rather broad contact zones (Figure 7a,b). Those few deposits that were clearly separated from the neuronal surface in an individual section were found to be continuous with membrane-associated deposits when traced through several consecutive sections. Aβ deposits on the primary dendrites could be traced back to the cell somata (Figure 7c-e) where they formed thread-like structures similar to those found on the dendrites.

## DISCUSSION

This study represents the first detailed 3D reconstruction of Aβ deposits, based on Aβ$_{42}$ immunogold labeling of large series of consecutive ultrathin sections (each series comprising up to 160 sections, corresponding to >10 μm). In the CA1 of the 3xTg-AD mouse model, the deposits are tightly associated with the plasma membranes of pyramidal cell bodies and dendrites. The deposits show a clear preference for large caliber dendrites but do not seem to bear any spatial association to spines or to structures inside the cells (like mitochondria). A closely similar 3D structure of Aβ deposits was observed in aged dogs, which develop Aβ plaques "naturally" as part of the ageing process. Our finding that the two types of experimental animal produced the same 3D structure of Aβ deposits indicates that the structure observed is not dictated by the specific genetic manipulation in the 3xTg-AD mice but rather reflects the generic pattern of Aβ deposits in brain. Unfortunately, the fine structure of Aβ deposits rapidly breaks down postmortem so correlative studies in humans are not within the realm of present day technology.

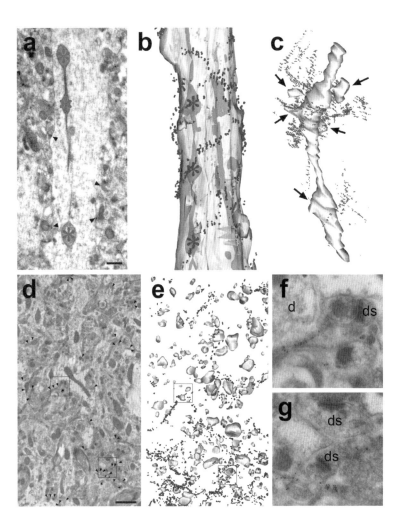

Fig. 5. Aβ deposits show no preferential association to mitochondria (a, b), thin dendrites (c), spines (d, e), or spine synapses (f, g). (a, b) Elongated mitochondria in apical dendrite (asterisks in a) are recon- structed in (b) along with peri- and extracellular Aβ deposits (dots). (c) Thin dendritic branch with spines (arrows) and reconstructed extracellular amyloid deposit. (d) Multiple Aβ deposits (arrowheads) in stratum lacunosum moleculare. (e) Reconstruction of deposits and dendritic spines in (d). Framed areas in (d, e) are shown at larger magnification in (f, g). d, dendrite; ds, dendritic spines. Scale bar, 0.5 μm.

Only a single study has been published so far on reconstructions of Aβ deposits at the EM level [20]. This study, conducted in aged rhesus monkeys, was not based on immunogold analysis, which in our hands was indispensable for a reliable detection of the fine caliber deposits in the extracellular space. Also, the latter study focused on late stage deposits whose structure is likely to be less informative in regard to the mechanisms that are at play when the deposits are initially formed. Here we chose to analyze plaques whose spatial relation to tissue elements is not confounded by the gross morphological changes that are typically seen in the vicinity of late stage plaques. In their EM study, Fiala et al. [20] described rather compact plaques and focused particularly on the relationship of these plaques to dystrophic neurites. No attempt was made to reconstruct the plaques along with dendrites or cell somata.

Numerous analyses have been made of plaques at the light microscopic level [21,22]. Typically such analyses leave the impression that plaques are roundish and rather compact structures randomly distributed through the neuropil. It is probable that this picture emerges simply because the fine threadlike deposits escape detection in the light microscope (their dimension often being less than 50 nm in width) so

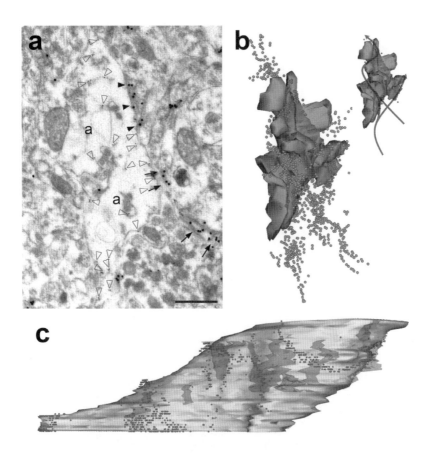

Fig. 6. Aβ deposits touch but do not extend along astrocyte processes. (*a*) Section double labeled for glutamine synthetase (small particles; open arrowheads) and Aβ (large particles; closed arrowheads). Glutamine synthetase is a marker of astrocytic processes (*a*). (*b*) Reconstruction of deposits shown in (*a*). Some of the gold particles signaling Aβ are found in close proximity to astrocytic processes (dark particles) while most particles (light particles) are more distant from the astrocyte plasma membrane. Arrows indicate general direction of the threadlike Aβ deposits. (*c*) Part of apical dendrite with areas of astrocyte apposition indicated. The peridendritic Aβ deposits (identified by gold particles) cross areas of astrocyte apposition (shaded areas) but show no preferential association with these. Scale bar, 0.5 μm.

that the accumulation of Aβ near the deposit's "center of gravity" comes to dominate the light microscopic appearance of the plaques.

The reported distribution of Aβ plaques at the light microscopic level is consistent with our finding of a close association of nascent plaques with dendrites. In the neocortex, plaques first form in layers IV and V (which contain large pyramidal cell dendrites) and develop much later in layer I (where thin, apical dendritic branches predominate) [9]. Similarly, in the hippocampus, light microscopic analyses indicate that Aβ is first laid down close to the pyramidal cell layer of CA1 and CA3 [9].

What does the present study tell us about the mechanisms underlying the formation of Aβ deposits? There is now strong evidence that soluble forms of Aβ are produced and then released from neurons prior to their being deposited in the form of plaques in the extracellular space [5,23–25]. Notably, previous light microscopic analyses indicate that neuronally derived AβPP is the source for Aβ deposits [26,27] and that such deposits are unlikely to form secondary to abnormal leakage from microvessels [28,29]. In fact, recent tracer studies suggest that perivascular deposits of Aβ (observed also in the present material) may reflect accumulation of brain derived Aβ along perivascular drainage routes [30,31]. The particular 3D appearance presently observed is likely to reflect the pattern of Aβ fibrillization. The most parsimonious explanation is that fibrillization is initiated at the surface of large caliber dendrites and that the process then spreads along the dendrites and also into the extracellular space off the dendritic plasma membrane. The dogs, even more so

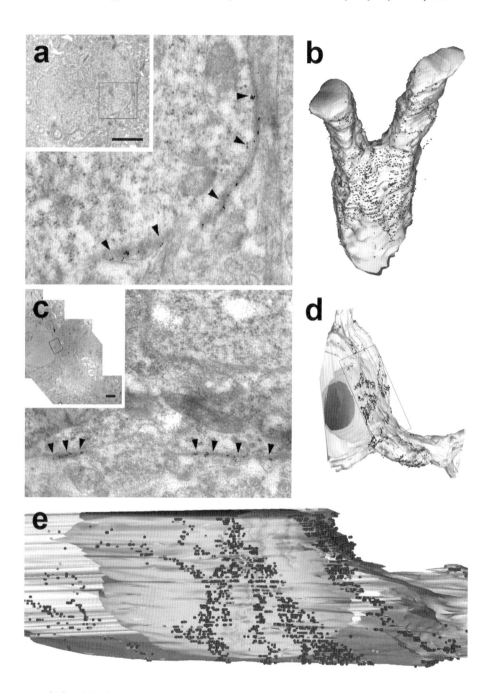

Fig. 7. Ultrastructure of Aβ positive deposits in aged dog (neocortex). Immunolabeling protocol was identical to that used in *Figures 1-6*. As in mouse, Aβ deposits were localized mostly on membranes of large dendrites and cell somata (arrowheads in *a, c*). Most deposits were tightly apposed to plasma membranes and did not fill the extracellular space. (*b, d*) 3D reconstructions of peridendritic deposits in (*a*) and perisomatic deposits in (*c*) revealed that Aβ deposits formed both sheath-like and thread-like structures although intermediate forms were also observed. Inset in (*d*) shows low magnification image of one of the sections used for reconstruction. Framed area in (*d*) is enlarged in (*e*) and shown at a different angle. Scale bar, 0.5 μm.

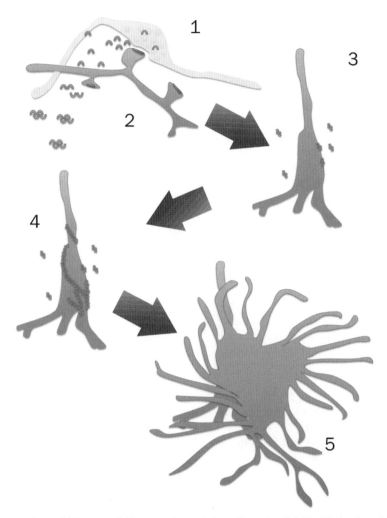

Fig. 8. Summary diagram showing multiple steps of Aβ aggregation and plaque formation. Soluble Aβ$_{42}$ is released from neurons, presumably at the synaptic terminals (*1*). Soluble Aβ oligomers form in the extracellular space and exert their effects on synapses, receptors, and membrane proteins (*2*). Initial deposition of Aβ oligomers occurs on the lipid membrane of the dendrosomatic compartment (*3*). Fibrillation is promoted by the presence of certain membrane components in the specialized domains (*4*). More Aβ aggregates and plaques are formed from an existing deposition on the plasma membrane compartment (*5*).

than the transgenic mice, showed an intimate association of plaque material with dendritic plasma membranes. In both species, when in one individual section extracellular deposits were found to be detached from the dendritic plasma membrane compartment they could be traced back to this compartment when followed through adjacent sections.

The molecular mechanisms favoring fibrillization at the dendritic surface remain to be identified. One possibility is that the lipid composition of the membrane is a critical factor and that fibrillization occurs predominantly at membrane domains with a specific lipid signature [3,32,33]. Indirect support for such a mechanism was provided in a recent study by Oksman and colleagues [34] demonstrating that the Aβ load can be modified by feeding animals with an excess of docosahexenoic acid (DHA). It should also be recalled that cholesterol decreasing drug statins are known to affect plaque formation in experimental and clinical studies [35,36]. Our finding that deposits appear to develop in contiguity with the lipid bilayer of the plasma membrane may provide a possible mechanistic basis for the latter observation. Unfortunately, antibodies that work at the EM level and that can distinguish lipid rafts or other specializations in the lipid bilayer are not available [37,38].

Of direct relevance to the present findings are previous *in vitro* data indicating that GM1 ganglioside-bound Aβ precipitates the formation of Aβ aggregates [39,40]. As gangliosides are plasma membrane molecules, these *in vitro* data would predict the intimate spatial relation presently found between plasma membranes and nascent plaques.

Although we consider preferential fibrillization to be the most likely explanation of the observed association between Aβ and dendrosomatic membranes, we cannot exclude the possibility that the complex pattern of peridendritic Aβ deposits reflects the distribution of vesicular release sites in the dendritic compartment [41]. It is now well established that signal molecules (such as glutamate and endocannabinoids) are released not only from the synaptic terminals but also from dendrites [42–44]. If the dendritic secretory pathway is used for protofibril secretion, this could provide an alternative explanation for the 3D pattern presently observed.

Based on the detailed 3D structure of Aβ deposits revealed in the present study, inferences can be made with respect to the mechanisms that promote the laying down of Aβ deposits in the intact brain (Figure 8). This important step in the pathogenesis of Alzheimer's disease has received little attention thus far as efforts have been focused on the molecular mechanisms underlying cleavage of AβPP to Aβ. A recent study shows that plaque formation is a rapid event that precedes other types of pathology including microglial recruitment and appearance of abnormal neurites [8]. This underlines the need for a better understanding of the mechanisms that govern the formation of plaques. The hitherto unrecognized complexity of the Aβ deposits indicates that plaque formation should be seen as a process that is dictated by molecular interactions with specific plasma membrane domains rather than as a simple polymerization process in the extracellular space.

## ACKNOWLEDGMENTS

This chapter is based on a paper previously published in the *Journal of Alzheimer's Disease* [45]. Paworn Nuntagij was supported through a grant from the Medical Scholars Program, Mahidol University. Thanks to Drs S. Oddo, University of Texas Health Science Center, TX and F. LaFerla, University of Irvine, CA for providing us with the 3xTg mice [45]. Thanks are due to the Institute for Brain Ageing and Dementia, University of California, Irvine (Drs. C.W. Cotman and E. Head) for making the canine material available. The work was supported by FUGE, Civitan Norway (Alzheimer task force), the Polish–Norwegian Research Fund and the Research Council of Norway.

## References

[1] Haass C (2004) Take five–BACE and the γ-secretase quartet conduct Alzheimer's amyloid β-peptide generation. *EMBO J* **23**, 483-488.

[2] LaFerla FM, Green KN, Oddo S (2007) Intracellular amyloid-β in Alzheimer's disease. *Nat Rev Neurosci* **8**, 499-509.

[3] Cheng H, Vetrivel KS, Gong P, Meckler X, Parent A, Thinakaran G (2007) Mechanisms of disease: new therapeutic strategies for Alzheimer's disease–targeting APP processing in lipid rafts. *Nat Clin Pract Neurol* **3**, 374-382.

[4] Cirrito JR, Yamada KA, Finn MB, Sloviter RS, Bales KR, May PC, Schoepp DD, Paul SM, Mennerick S, Holtzman DM (2005) Synaptic activity regulates interstitial fluid amyloid-β levels *in vivo*. *Neuron* **48**, 913-922.

[5] Lazarov O, Lee M, Peterson DA, Sisodia SS (2002) Evidence that synaptically released β-amyloid accumulates as extracellular deposits in the hippocampus of transgenic mice. *J Neurosci* **22**, 9785-9793.

[6] Walsh DM, Hartley DM, Kusumoto Y, Fezoui Y, Condron MM, Lomakin A, Benedek GB, Selkoe DJ, Teplow DB (1999) Amyloid β-protein fibrillogenesis. Structure and biological activity of protofibrillar intermediates. *J Biol Chem* **274**, 25945-25952.

[7] McLaurin J, Yang D, Yip CM, Fraser PE (2000) Review: modulating factors in amyloid-β fibril formation. *J Struct Biol* **130**, 259-270.

[8] Meyer-Luehmann M, Spires-Jones TL, Prada C, Garcia-Alloza M, de Calignon A, Rozkalne A, Koenigsknecht-Talboo J, Holtzman DM, Bacskai BJ, Hyman BT (2008) Rapid appearance and local toxicity of amyloid-β plaques in a mouse model of Alzheimer's disease. *Nature* **451**, 720-724.

[9] Oddo S, Caccamo A, Shepherd JD, Murphy MP, Golde TE, Kayed R, Metherate R, Mattson MP, Akbari Y, LaFerla FM (2003) Triple-transgenic model of Alzheimer's disease with plaques and tangles: intracellular Abeta and synaptic dysfunction. *Neuron* **39**, 409-421.

[10] Torp R, Head E, Milgram NW, Hahn F, Ottersen OP, Cotman CW (2000) Ultrastructural evidence of fibrillar β-amyloid associated with neuronal membranes in behaviorally characterized aged dog brains. *Neuroscience* **96**, 495-506.

[11] Torp R, Ottersen OP, Cotman CW, Head E (2003) Identification of neuronal plasma membrane microdomains that colocalize β-amyloid and presenilin: implications for β-amyloid precursor protein processing. *Neuroscience* **120**, 291-300.

[12] Matsubara A, Laake JH, Davanger S, Usami S, Ottersen OP (1996) Organization of AMPA receptor subunits at a glutamate synapse: a quantitative immunogold analysis of hair cell synapses in the rat organ of Corti. *J Neurosci* **16**, 4457-4467.

[13] Takumi Y, Ramirez-Leon V, Laake P, Rinvik E, Ottersen OP (1999) Different modes of expression of AMPA and NMDA receptors in hippocampal synapses. *Nat Neurosci* **2**, 618-624.

[14] Cummings BJ, Cotman CW (1995) Image analysis of $\beta$-amyloid load in Alzheimer's disease and relation to dementia severity. *Lancet* **346**, 1524-1528.

[15] Gong Y, Meyer EM, Meyers CA, Klein RL, King MA, Hughes JA (2006) Memory-related deficits following selective hippocampal expression of Swedish mutation amyloid precursor protein in the rat. *Exp Neurol* **200**, 371-377.

[16] Fiala JC (2005) Reconstruct: a free editor for serial section microscopy. *J Microsc* **218**, 52-61.

[17] Harris KM, Perry E, Bourne J, Feinberg M, Ostroff L, Hurlburt J (2006) Uniform serial sectioning for transmission electron microscopy. *J Neurosci* **26**, 12101-12103.

[18] Yu WH, Cuervo AM, Kumar A, Peterhoff CM, Schmidt SD, Lee JH, Mohan PS, Mercken M, Farmery MR, Tjernberg LO, Jiang Y, Duff K, Uchiyama Y, Naslund J, Mathews PM, Cataldo AM, Nixon RA (2005) Macroautophagy–a novel $\beta$-amyloid peptide-generating pathway activated in Alzheimer's disease. *J Cell Biol* **171**, 87-98.

[19] Nixon RA (2007) Autophagy, amyloidogenesis and Alzheimer disease. *J Cell Sci* **120**, 4081-4091.

[20] Fiala JC, Feinberg M, Peters A, Barbas H (2007) Mitochondrial degeneration in dystrophic neurites of senile plaques may lead to extracellular deposition of fine filaments. *Brain Struct Funct* **212**, 195-207.

[21] Iwatsubo T, Odaka A, Suzuki N, Mizusawa H, Nukina N, Ihara Y (1994) Visualization of A$\beta_{42(43)}$ and A$\beta_{40}$ in senile plaques with end-specific A$\beta$ monoclonals: evidence that an initially deposited species is A$\beta_{42(43)}$. *Neuron* **13**, 45-53.

[22] Uchihara T, Nakamura A, Nakayama H, Arima K, Ishizuka N, Mori H, Mizushima S (2003) Triple immunofluorolabeling with two rabbit polyclonal antibodies and a mouse monoclonal antibody allowing three-dimensional analysis of cotton wool plaques in Alzheimer disease. *J Histochem Cytochem* **51**, 1201-1206.

[23] Koo EH, Squazzo SL (1994) Evidence that production and release of amyloid $\beta$-protein involves the endocytic pathway. *J Biol Chem* **269**, 17386-17389.

[24] Tienari PJ, Ida N, Ikonen E, Simons M, Weidemann A, Multhaup G, Masters CL, Dotti CG, Beyreuther K (1997) Intracellular and secreted Alzheimer $\beta$-amyloid species are generated by distinct mechanisms in cultured hippocampal neurons. *Proc Natl Acad Sci U S A* **94**, 4125-4130.

[25] Oddo S, Caccamo A, Smith IF, Green KN, LaFerla FM (2006) A dynamic relationship between intracellular and extracellular pools of A$\beta$. *Am J Pathol* **168**, 184-194.

[26] Cras P, Kawai M, Siedlak S, Mulvihill P, Gambetti P, Lowery D, Gonzalez-DeWhitt P, Greenberg B, Perry G (1990) Neuronal and microglial involvement in $\beta$-amyloid protein deposition in Alzheimer's disease. *Am J Pathol* **137**, 241-246.

[27] Palmert MR, Podlisny MB, Witker DS, Itersdorf TO, Younkin LH, Selkoe DJ, Younkin SG (1988) Antisera to an amino-terminal peptide detect the amyloid protein precursor of Alzheimer's disease and recognize senile plaques. *Biochem Biophys Res Commun* **156**, 432-437.

[28] Kawai M, Kalaria RN, Harik SI, Perry G (1990) The relationship of amyloid plaques to cerebral capillaries in Alzheimer's disease. *Am J Pathol* **137**, 1435-1446.

[29] Kawai M, Cras P, Perry G (1992) Serial reconstruction of $\beta$-protein amyloid plaques: relationship to microvessels and size distribution. *Brain Res* **592**, 278-282.

[30] Carare RO, Bernardes-Silva M, Newman TA, Page AM, Nicoll JA, Perry VH, Weller RO (2008) Solutes, but not cells, drain from the brain parenchyma along basement membranes of capillaries and arteries: significance for cerebral amyloid angiopathy and neuroimmunology. *Neuropathol Appl Neurobiol* **34**, 131-144.

[31] Weller RO, Subash M, Preston SD, Mazanti I, Carare RO (2008) Perivascular drainage of amyloid- peptides from the brain and its failure in cerebral amyloid angiopathy and Alzheimer's disease. *Brain Pathol* **18**, 253-266.

[32] Matsuzaki K (2007) Physicochemical interactions of amyloid $\beta$-peptide with lipid bilayers. *Biochim Biophys Acta* **1768**, 1935-1942.

[33] Gellermann GP, Appel TR, Tannert A, Radestock A, Hortschansky P, Schroeckh V, Leisner C, Lutkepohl T, Shtrasburg S, Rocken C, Pras M, Linke RP, Diekmann S, Fandrich M (2005) Raft lipids as common components of human extracellular amyloid fibrils. *Proc Natl Acad Sci U S A* **102**, 6297-6302.

[34] Oksman M, Iivonen H, Hogyes E, Amtul Z, Penke B, Leenders I, Broersen L, Lutjohann D, Hartmann T, Tanila H (2006) Impact of different saturated fatty acid, polyunsaturated fatty acid and cholesterol containing diets on $\beta$-amyloid accumulation in APP/PS1 transgenic mice. *Neurobiol Dis* **23**, 563-572.

[35] Li G, Larson EB, Sonnen JA, Shofer JB, Petrie EC, Schantz A, Peskind ER, Raskind MA, Breitner JC, Montine TJ (2007) Statin therapy is associated with reduced neuropathologic changes of Alzheimer disease. *Neurology* **69**, 878-885.

[36] Gellermann GP, Ullrich K, Tannert A, Unger C, Habicht G, Sauter SR, Hortschansky P, Horn U, Mollmann U, Decker M, Lehmann J, Fandrich M (2006) Alzheimer-like plaque formation by human macrophages is reduced by fibrillation inhibitors and lovastatin. *J Mol Biol* **360**, 251-257.

[37] Kusumi A, Suzuki K (2005) Toward understanding the dynamics of membrane-raft-based molecular interactions. *Biochim Biophys Acta* **1746**, 234-251.

[38] Mishra S, Joshi PG (2007) Lipid raft heterogeneity: an enigma. *J Neurochem* **103** (Suppl 1), 135-142.

[39] Hayashi H, Kimura N, Yamaguchi H, Hasegawa K, Yokoseki T, Shibata M, Yamamoto N, Michikawa M, Yoshikawa Y, Terao K, Matsuzaki K, Lemere CA, Selkoe DJ, Naiki H, Yanagisawa K (2004) A seed for Alzheimer amyloid in the brain. *J Neurosci* **24**, 4894-4902.

[40] Yamamoto N, Hirabayashi Y, Amari M, Yamaguchi H, Romanov G, Van Nostrand W E, Yanagisawaa K (2005) Assembly of hereditary amyloid $\beta$-protein variants in the presence of favorable gangliosides. *FEBS Lett* **579**, 2185-2190

[41] Zilberter Y, Harkany T, Holmgren CD (2005) Dendritic release of retrograde messengers controls synaptic transmission in local neocortical networks. *Neuroscientist* **11**, 334-344.

[42] Harkany T, Holmgren C, Hartig W, Qureshi T, Chaudhry FA, Storm-Mathisen J, Dobszay MB, Berghuis P, Schulte G, Sousa KM, Fremeau RT, Jr., Edwards RH, Mackie K, Ernfors P, Zilberter Y (2004) Endocannabinoid-independent retrograde signaling at inhibitory synapses in layer 2/3 of neocortex: involvement of vesicular glutamate transporter 3. *J Neurosci* **24**, 4978-4988.

[43] Nyilas R, Dudok B, Urban GM, Mackie K, Watanabe M, Cravatt BF, Freund TF, Katona I (2008) Enzymatic machinery for endocannabinoid biosynthesis associated with calcium stores in glutamatergic axon terminals. *J Neurosci* **28**, 1058-1063.

[44] Wilson RI, Nicoll RA (2002) Endocannabinoid signaling in the brain. *Science* **296**, 678-682.

[45] Nuntagij P, Oddo S, LaFerla FM, Kotchabhakdi N, Ottersen OP, Torp R (2009) Amyloid deposits show complexity and intimate spatial relationship with dendrosomatic plasma membranes: an electron microscopic 3D reconstruction analysis in 3xTg-AD mice and aged canines. *J Alzheimers Dis* **16**, 315-323.

# The Cholesterol-Fed Rabbit as a Model System for Cholesterol-Alzheimer's Disease Studies

R.P. Jaya Prasanthi, Gurdeep Marwarha and Othman Ghribi
*Department of Pharmacology, Physiology and Therapeutics, University of North Dakota School of Medicine and Health Sciences, Grand Forks, North Dakota, 58202*

**Abstract.** Epidemiological, animal, and cellular studies suggest that abnormalities in cholesterol metabolism are a risk factor for Alzheimer's disease (AD), potentially by increasing amyloid-β (Aβ) peptide levels. Accumulation of Aβ in the brain is suggested to play a key role in the neurodegenerative processes by triggering the hyperphosphorylation of tau and the neuronal death that develop in the course of AD. However, the mechanisms by which cholesterol increases Aβ levels are still ill-defined. Work from our laboratory using the cholesterol-fed rabbit model system indicates that hypercholesterolemia increases Aβ through multiple mechanisms that affect production, degradation and clearance of this peptide. We also have found that the oxidized cholesterol metabolite, 27-hydroxycholesterol, also increases Aβ levels in organotypic slices from rabbit hippocampus. Our results suggest that multiple signaling pathways are involved in hypercholesterolemia-induced AD pathology and suggest 27-hydroxycholesterol as the link between circulating cholesterol and AD-like pathology in the brain.

Key words: Alzheimer's disease, amyloid-β, cholesterol, 27-hydroxycholesterol, rabbit

## INTRODUCTION

Alzheimer's disease (AD) is a complex and heterogeneous disorder that presently affects more than 4 million citizens in the US and is projected to affect more than 14 million people within the next 50 years if no cure is found. Accumulation of amyloid-β (Aβ) peptide, hyperphosphorylation of tau, and oxidative stress are hallmarks of the disease. Aβ may be particularly important to the pathogenesis of AD because it accumulates in both the early onset familial and late onset sporadic AD forms. Early onset familial forms of AD are associated with mutations in the amyloid-β protein precursor (APP) and the presenilin genes. Subsequently to the identification of these genes, various cellular and animal models were designed. Transgenic animal models that express one or more of these mutations exhibit increased Aβ levels. These models have helped in understanding aspects of the pathophysiology of the early-onset familial forms of AD. However, the causes of the sporadic late-onset forms of AD, which represent the vast majority of AD cases, are still unknown. While genetic mutations are responsible for the accumulation of Aβ in familial AD in humans and transgenic animal models, the causative factors for the accumulation of Aβ in sporadic forms of AD are not known. Thus, identification of risk factors and mechanisms by which these factors increase Aβ accumulation in sporadic late onset AD may help in designing therapeutic strategies to prevent the onset of this devastating disorder.

Growing evidence suggests that abnormalities in cholesterol metabolism are important in the pathogenesis of AD (see for review [55]). However, the mechanism by which cholesterol increases the risk for AD is yet to be determined. Work by others and from our laboratory use the cholesterol-fed rabbit as a model system for Alzheimer's disease studies. Findings in this model suggest that the cholesterol-fed rabbit is a relevant model system for sporadic AD studies, as it is a unique model system that exhibits the major pathological hallmarks of AD without genetic manipulations. In this chapter, we will *first* describe factors that govern cholesterol homeostasis in normal brain and in cholesterol-fed rabbit brain. *Second*, we will provide data from the literature linking cholesterol to the pathology of AD and present results from our studies in rabbits showing the benefits of using this model for *in vivo* and *in vitro* studies of cholesterol-AD link. *Third*, we will demonstrate potential mechanisms that underlie cholesterol-induced AD-like pathology in rabbit brains and that may also be involved in humans. *Finally*, we will describe potential therapeutic strategies to protect against cholesterol-induced AD pathology.

## CHOLESTEROL HOMEOSTASIS IN THE BRAIN

*Synthesis and cellular distribution of cholesterol*

The homeostasis of cholesterol in the brain is maintained primarily through synthesis, transport, and clearance. These processes are tightly regulated in order to prevent the accumulation of cholesterol in the brain. The synthesis of cholesterol occurs in the endoplasmic reticulum and requires numerous reactions and intermediaries. The enzyme 3-hydroxy-3-methylglutaryl-CoA (HMG-CoA) reductase is the rate limiting step in this process. The brain contains approximately 25% of the body's cholesterol, essentially (>99.5%) as unesterified cholesterol [11]. Most cholesterol present in the brain is contained in oligodendrocytes that form the myelin sheaths and in the plasma membranes of astrocytes and neurons. Cholesterol homeostasis in the brain is regulated through *de novo* synthesis distinctly from the peripheral cholesterol pool. No transfer of cholesterol (or only a very poor transfer) occurs from the peripheral circulation to the brain. This is due to the impermeability of the blood brain barrier (BBB) to lipoproteins that carry cholesterol [71]. The synthesis of cholesterol in the brain is maximal during the developmental stage and declines with age. In the adulthood, recycling of cholesterol becomes the main source needed for cell functioning [11]. We have measured cholesterol levels and have found that although blood cholesterol levels are dramatically increased, brain cholesterol (esterified and unesterfied) concentrations are unchanged in hippocampus of rabbits fed a 1 % cholesterol-enriched diet for 6 months [46;47]. Our results confirm the independence of brain cholesterol from the peripheral pool and demonstrate that hypercholesterolemia does not alter cholesterol levels in the rabbit brain.

In the brain, cells either produce their own cholesterol or import the required cholesterol from neighboring cells. Oligodendrocytes produce cholesterol and use large amounts of brain-made cholesterol for the synthesis of myelin [102]. Astrocytes synthesize two to three-fold more cholesterol than do neurons (for review see [11]). During development, neurons synthesize the cholesterol required for their growth and synaptogenesis. Mature neurons, however, depend on cholesterol derived from oligodendrocytes and/or astrocytes [90;91]. We have found, using filipin as a probe for cholesterol imaging, an overaccumulation of cholesterol in hippocampal neurons from rabbits fed a cholesterol-enriched diet [47]. We have further observed that increased cholesterol content in neurons is accompanied by a dramatic decrease in cholesterol in oligodendrocytes and astrocytes in the brain of cholesterol-fed rabbits. It is reasonable to suggest that increased cholesterol in neurons results from an increased shuttling of cholesterol from oligodendrocytes and astrocytes to neurons. Collectively, our results demonstrate that cholesterol distribution, not concentration, is altered by cholesterol-enriched diets. Our data is in agreement with studies in Tg mice model for Alzheimer's disease also showing that cholesterol distribution not levels of cholesterol are important to the generation of Aβ [16].

*Cholesterol transport*

Several proteins are involved in the transport of cholesterol in the brain. These include ApoE, which binds cholesterol and acts as a ligand for LDL-receptor-related proteins (LRP-1); members of the superfamily of ATP-binding cassette, including ABCA1, ABCA2 and ABCG1; and the nuclear receptor liver X receptor (LXR).

ApoE, synthesized primarily by astrocytes and microglia, is a specific type of lipoprotein that is tightly linked to cholesterol transport. ApoE carries cholesterol from the blood into the brain [80] and shuttles cholesterol from astrocytes to neurons [72]. ApoE immunoreactivity increases in rabbit brains when fed with 2% cholesterol for 8 weeks [115]. There are three isoforms of the ApoE gene in humans: ApoE2, 3 and 4. The three ApoE alleles, E2, E3, and E4 only differ by a few amino-acid residues. While the ApoE4 genotype correlates with an increased AD risk, ApoE2 may be protective against AD. ApoE4 is also associated with high plasma cholesterol and a high risk of atherosclerosis [40;86].

On the other hand, the ApoE3 genotype, the most common form found in more than half the human population, plays an important role in delivering cholesterol to neurons.

The ATP-binding cassette transporters, ABCA1, ABCA2, and ABCG1, play an important role in shuttling cholesterol among brain cells (see for review [65]). However, the molecular mechanism by which these transporters promote cholesterol distribution is not understood. ABCA1 is primarily expressed in neurons and microglia [64], ABCA2 is primarily found in lysosomes of oligodendrocytes [103;135], and ABCG1 is associated with lysosomes, endosomes and Golgi apparatus in microglia, oligodendrocytes, neurons and astrocytes [64]. The expression of ABCA1, a key modulator of cholesterol transport across the plasma membrane, is dysregulated in the AD brain and this dysregulation is associated with increasing severity of AD [1].

ABCA1 has been reported to modulate Aβ production *in vivo* [59]. Several studies have demonstrated that transgenic mice lacking ABCA1 exhibit increased Aβ levels and plaque formation [59;66;127]. Mutations in the ABCA1 gene cause Tangier disease which is associated with cellular cholesterol accumulation, premature atherosclerosis, and peripheral neuropathy [2]. *In vitro* studies have demonstrated that ABCA1 inhibits Aβ secretion from Neuro2a cells, rat primary cortical neurons, and from a CHO cell line expressing human APP695 [15;65;67;120]. ABCA1 and ApoE are functionally linked and work in concert to regulate not only cholesterol homeostasis but also Aβ accumulation. Indeed, inactivation of ABCA1 gene leads to reduced ApoE levels and increased Aβ in a mouse model for AD [59;127]. ABCA2 is suggested to contribute to AD pathogenesis by regulating cholesterol homeostasis in oligodendrocytes [79]. Transient expression of ABCG1 in CHO cells that stably express human wild-type APP695 results in a significant reduction in Aβ production [65]. Other studies, however, have shown that the transient expression of ABCG1 in HEK cells leads to increased Aβ production [122].

LXR α and β are nuclear receptors that control cholesterol transport and metabolism in the circulation through regulation of expression levels of the cholesterol transporters ABCA1, ABCG1, ApoE, and ApoC [8]. LXRs are also expressed in the brain [128] where they regulate cholesterol trafficking among cells via ABCA1 and ApoE. LXRs can upregulate ApoE and ABCA1, causing cholesterol efflux from glia and neurons [22].

*Cholesterol clearance*

Cholesterol is catabolized in the brain through two main pathways, esterification and oxidation. Esterified cholesterol can be stored within neurons as cholesteryl esters [98]. Because alterations in cholesteryl esters in the brain are suggested to increase Aβ production and may affect AD pathology, we have measured these levels and found that they do not differ statistically between the control and cholesterol-treated groups [46;47]. The unchanged cholesteryl esters levels following hypercholesterolemia indicate that the increased cholesterol content in neurons is independent of the cholesteryl ester pool.

Oxidation of cholesterol to oxysterols is mediated by CYP46A1 and CYP27A1 [98]. CYP46A1, mainly a brain-specific enzyme, oxidizes cholesterol to 24-hydroxycholesterol, whereas CYP27A1 is expressed more broadly and oxidizes cholesterol to 27-hydroxycholesterol [76]. These oxysterols are readily transported from the brain to the circulation. Expression and distribution of these two enzymes appear to be altered in AD and are associated with amyloid plaques [15]. We have determined the extent to which feeding rabbits a 1% cholesterol diet for 6 months affects levels of CYP46A1 and CYP27A1. We have found that CYP46A1 and CYP27A1 expression levels are decreased in the hippocampus, suggesting that cholesterol turnover to oxysterols in rabbit brain is reduced by the cholesterol diet (Ghribi, unpublished data). These results are in accordance with data showing reduced expression of CYP46A1 and CYP27A1 in the brain of patients with AD [15]. The reason behind decreased protein levels of the enzymes that convert cholesterol to oxysterols may be indicative of a negative retrocontrol mechanism, due to excess entry of oxysterols from the circulation into the brain.

## CHOLESTEROL AND THE PATHOGENESIS OF AD

*Human and laboratory studies*

The first factor linked to cholesterol metabolism and the increased risk of AD is apolipoprotein E (ApoE). People who have one or two copies of the ApoE ε4 allele are at increased risk of AD [28] although the ε4 allele is not necessary or sufficient to cause AD. Of importance to the cholesterol-AD connection, carrying the ApoE ε4 is associated with high plasma cholesterol levels [40;86]. Clusterin or ApoJ is a second risk factor related to ApoE and has been recently linked to AD. Two recent genome-wide association studies have established clusterin as a new genetic risk factor for late onset AD [54;70]. What links clusterin to cholesterol is that clusterin, like ApoE, is an abundantly expressed lipoprotein in the brain. Epidemiological studies also link plasma cholesterol status to dementia. A recent epidemiological study on a large cohort of men and women has demonstrated that high cholesterol levels during mid-life are associated with a 66% increase in AD risk late in life [114]. Interestingly, even borderline levels of cholesterol are associated with increased risk of vascular dementia in this study. These findings suggest that the plasma cholesterol level at mid-age may determine our degree of risk for AD. At later age when AD progresses the correlation between AD and cholesterol levels is no longer consistent. In AD, levels of free cholesterol in neurofibrillary tangle-bearing neurons are higher than those of adjacent tangle-free neurons [35]. Early exposure to elevated cholesterol or LDL receptor dysfunction as in familial hypercholesterolemia is suggested to be a risk factor for mild cognitive impairment [134].

High cholesterol levels have been shown to increase Aβ production in mice [96;107], and in cultured cells [6;43;45;95;108]. There is also evidence that cholesterol co-localizes with fibrillar Aβ in the amyloid plaques of transgenic mice model for AD [17]. Treatment of guinea pigs with simvastatin or reduction of cholesterol in cultured cells decreases Aβ generation [41]. Importantly, cholesterol distribution, not levels of cholesterol, may be important to the generation of Aβ [16].

Lipid composition in membranes determines production, conformational changes, oligomerization and/or aggregation of Aβ peptide. The activity of β-secretase (BACE1), the enzyme that initiates the cleavage of APP to yield Aβ, has been demonstrated to be increased in lipid microdomains [29;82]. APP inside lipid raft clusters is cleaved by BACE1, whereas APP outside the lipid clusters is cleaved by α-secretase yielding non-amyloidogenic products [39]. Aβ can function as a ligand for LRP-1 to be transported from the brain to the circulation across the BBB [32]. Aβ has been recently demonstrated to promote cholesterol efflux in a similar fashion to apolipoprotein-mediated cholesterol efflux, suggesting an apolipoprotein-like function for Aβ in cholesterol homeostasis [124]. The effects of cholesterol on Aβ fibrillogeneis have been addressed recently in a recent review by Harris and Milton [55].

*Cholesterol and Alzheimer's disease-like pathology in rabbits*

The first indication of a connection between cholesterol and Aβ was reported in rabbits [116]. The rabbit demonstrates a marked response to a 12 weeks 2% cholesterol diet by exhibiting Aβ deposition in plaques. We have fed rabbits a 1 or 2% cholesterol-enriched diet for 6 or 12 weeks respectively and found that cholesterol diets increase Aβ levels in hippocampus and cortex as determined with Western blot, immunohistochemistry (Figure 1) and ELISA assays [46;47;83;92;105].

Tau hyperphosphorylation is a key component of neurofibrillary tangles (NFTs) in Alzheimer's disease. Various studies suggest that tau phosphorylation is enhanced by the increase in cholesterol levels in neurons. Free cholesterol levels were found to be higher in tangle-bearing neurons than in adjacent tangle-free neurons in AD patients [35]. Additionally, transgenic mice over-expressing human ApoE ε4, which is associated with increased plasma cholesterol levels, exhibit increased NFT load [123]. Niemann-Pick type C disease, a cholesterol storage disease, is characterized by the accumulation of NFTs [36]. The mechanisms by which cholesterol modulates NFTs formation are still to be established. Despite the large number of studies linking cholesterol to Aβ production, the effect of hypercholesterolemia on tau phosphorylation are lesser studied. Nevertheless, Woodruff-Pak and collaborators have reported increased tau levels in the cerebral cortex, cerebellum and hippocampus from rabbits fed with cholesterol-enriched diets [131]. Most studies involving rabbits have used 2% cholesterol-enriched diets. Because 2% cholesterol diets cause severe hypercholesterolemic side effects, requiring sacrifice of the cholesterol-fed rabbits at 8-12 weeks, we have anticipated that a long-term diet supplemented with a lower level of cholesterol would allow the animals to live longer, thus more closely following the course in neurodegenerative human disease. We have therefore fed rabbits with a 1% cholesterol-enriched diet for 6-7 months. We have used retired breeder female rabbits for these studies, because young female animals are protected against the cholesterol effects based on Sparks' extensive work in

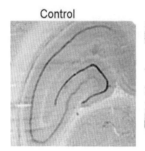 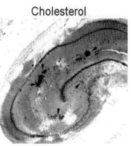

Fig. 1. Photomicrographs of hippocampus from a control rabbit and a 2% cholesterol-fed rabbit for 12 weeks immunostained for Aβ with 6E10 antibody. The hippocampus from the cholesterol-fed rabbits shows aggregated (arrows) and diffuse Aβ staining.

rabbits. This treatment has caused increased tau phosphorylation [46;47]. We have also shown that tau hyperphosphorylation is accompanied by an increase in phospho-ERK [47], suggesting that cholesterol-induced Aβ accumulation may enhance tau phosphorylation through stimulation of a MAPK pathway, which is found to be localized with the NFTs in AD brains [52]. To the best of our knowledge, our results are the first to demonstrate the involvement of a cholesterol-enriched diet in the hyperphosphorylation of tau, in addition to the accumulation of Aβ peptide in rabbit brain. The triggering of the hyperphosphorylation of tau, in addition to the accumulation of Aβ, add important features to the useful model system of rabbits fed a cholesterol-enriched diet in studies of some of the pathological hallmarks associated with AD.

Oxidative stress, in addition to Aβ and phosphorylated tau accumulation, is an important hallmark of AD, and may be the earliest change that occurs in the pathogenesis of AD [24;110-113;119]. High cholesterol levels have been reported to increase the generation of reactive oxygen species (ROS) and to cause oxidative stress [7]. Exposure of neurons to Aβ caused oxidative stress as well as cholesterol accumulation, and oxidative stress was associated with cholesterol accumulation in the brains of AD patients [30]. These results suggest that disturbances in cholesterol metabolism, oxidative stress and Aβ accumulation are intrinsically related. We have recently found that feeding rabbits a 2% cholesterol-enriched diet induces oxidative stress, as evidenced by increased levels of ROS, isoprostanes, glutathione depletion, and endoplasmic reticulum stress proteins (Ghribi, in preparation).

## Mechanisms involved in cholesterol-induced AD pathology

We have speculated that although hypercholesterolemia does not affect cholesterol levels in the brain, it might alter the cellular distribution of cholesterol in this organ. Alteration in cellular cholesterol distribution may be a mechanism by which hypercholesterolemia increases Aβ levels. Currently, the molecular mechanisms by which dietary cholesterol promotes Aβ peptide accumulation are not fully understood. Aβ levels in the brain are regulated by the production from the β-amyloid precursor protein (APP) by BACE1, the degradation by the insulin degrading enzyme (IDE), the transport from brain to blood by the low density lipoprotein receptor-related protein (LRP-1), and the transport from blood to brain by receptor for advanced glycation end products (RAGE).

Aβ is generated from APP through sequential cleavage by membrane proteases referred to as β- and γ-secretases. The β-secretase, BACE1, first cleaves APP to generate a C-terminus fragment (β-CTF), an immediate substrate for γ-secretase [18;53], which further cleaves β-CTF to yield Aβ. IDE cleaves numerous proteins that share a propensity to form β-pleated sheet–rich amyloid fibrils including insulin and APP intracellular domain (AICD) [38]. IDE is also the first protease to be implicated in the proteolytic degradation of Aβ [68]. LRP-1 is a member of the LDL receptor family which functions as a multi-functional scavenger and signaling receptor, a transporter, and a metabolizer of cholesterol and ApoE-containing lipoproteins [56]. RAGE is a multi-ligand receptor which belongs to immunoglobulin superfamily of cell surface molecules. It binds a wide range of molecules including the products of nonenzymatic glycoxidation (advanced glycation end products, or AGEs), Aβ and proinflammatory cytokine-like mediators [132]. RAGE and LRP-1 bind a wide range of ligands including ApoE, $\alpha_2$-macroglobulin, amyloid precursor protein, and Aβ [32-34]. While RAGE transports Aβ from the blood into the brain, LRP-1 transports Aβ from the brain into the blood [106]. The flux of Aβ into or out of the brain is regulated by the density and activity of RAGE and LRP-1 receptors [37].

Our published data show that the accumulation of cholesterol in neurons is associated with increased levels of APP [47;61]. BACE1 activity has been shown to be increased in a lipid-rich environment

[60;133]. Lipid rafts that are enriched in cholesterol influence cleavage of APP by BACE1 to generate Aβ; APP inside raft clusters is cleaved by BACE1. Our studies demonstrate a colocalization of cholesterol with BACE1 in neurons in brains from cholesterol-fed rabbits [47]. The increased Aβ levels may be due to increased expression levels of APP and its accelerated processing by BACE1 during hypercholesterolemia. We also demonstrate that hypercholesterolemia-induced Aβ accumulation is associated a decrease in levels of IDE, a decrease in levels of LRP-1, and an increase in levels of RAGE [61]. The changes in expression levels of these proteins may contribute to the cholesterol-induced Aβ accumulation. These results suggest that reducing Aβ accumulation in the brain may require strategies that combine reduction of generation and transport of Aβ in addition to acceleration of degradation and clearance of this peptide.

*Is 27-hydroxycholesterol the missing link between circulating cholesterol and AD in the brain?*

Due to the impermeability of the BBB, there is little if any transfer of plasma lipoproteins from the peripheral circulation to the brain [71]. The impermeability of brain to circulating cholesterol suggests that cholesterol itself does not place the brain at risk as a result of hypercholesterolemia. However, disturbances in levels of some oxysterols may place the brain at risk of injury. Of these oxysterols, 24S-hydroxycholesterol and 27-hydroxycholesterol have the ability to cross lipophilic membranes into and out of the brain [58]. While 24-hydroxycholesterol originates primarily from the brain, 27-hydroxycholesterol originates almost exclusively from the circulation [9;10;12;77]. Brain levels of 27-hydroxycholesterol have been shown to dramatically increase in AD brains [57]. Hypercholesterolemia, as it results from cholesterol-enriched diets, may lead to increased turnover of cholesterol to 27-hydroxycholesterol, thus increasing its levels in the circulation and potentially in the brain. Plasma and cerebrospinal fluid levels of 24-hydroxycholesterol have been shown to be higher in patients with AD in comparison with age-matched controls [78;87;88]. Circulating 27-hydroxycholesterol levels are 0.15-0.73 µM, and these concentrations can be in the millimolar range in some pathological situations such as atherosclerosis [14]. 27-hydroxycholesterol has recently been found to reduce extracellular Aβ levels by mechanisms that may involve its action as an liver LXR ligand [63].

We measured oxysterol levels in 2% cholesterol-fed rabbits and found that both 24-hydroxycholesterol and 27-hydroxycholesterol levels are markedly increased in the plasma. In the brain, the 27-hydroxycholesterol/24-hydroxycholesterol ratio increased from 1:4500 to 1:3000. As the concentrations of these oxysterols are tightly regulated in the brain, the increase in the 27-hydroxycholesterol/24-hydroxycholesterol ratio may have severe effects on the brain. Taken together, the above data indicate that disturbances in levels of oxysterols may represent the link by which high cholesterol levels in plasma induce pathological features in the brain suggestive of AD. In order to determine the extent to which oxysterols reproduce the effects of hypercholesterolemia on Aβ levels, we have developed two *in vitro* systems, neuronal cell cultures and hippocampal organotypically-cultured slices from adult rabbit brain (see 3.2 and 3.3 below).

We incubated organotypic slices from hippocampus of adult rabbits and human neuroblastoma SH-SY5Y cells with 27-hydroxycholesterol and 24-hydroxycholesterol and found that 27-hydroxycholesterol [83;105] not 24-hydroxycholesterol [92] increases Aβ and phosphorylated tau by mechanisms involving GSK-3αβ and the insulin degrading enzyme (IDE). While activation of GSK-3α (p-Tyr$^{279}$ GSK-3α) increases Aβ production and reduction in IDE levels leads to a decrease in Aβ degradation, activation of GSK-3β (p-Tyr$^{216}$ GSK-3β) increases phosphorylated tau levels. Interestingly, incubation of organotypic slices with cholesterol (water soluble cholesterol; Sigma) did not affect Aβ or tau levels. These results indicate that 27-hydroxycholesterol but not cholesterol or 24-hydroxycholesterol is toxic. These data further suggest that cholesterol may exert its deleterious effects through its metabolite 27-hydroxycholesterol.

## RELEVANCE OF THE RABBIT MODEL TO AD STUDIES

*In vivo model*

The cholesterol-fed rabbit was initially used as a model for experimental atherosclerosis by Nikolaj Anitschkow in 1913 and was later found by Larry Sparks [116] to exhibit Aβ plaques. We have also demonstrated that, in addition to increasing Aβ, cholesterol-enriched diets increase tau phosphorylation and oxidative stress in rabbits [46;47]. Additionally, the eye blink classical conditioning that is impaired in AD patients is also impaired in the rabbit model [131]. Furthermore, rabbits have a phylogeny closer to humans than rodents [48], and their Aβ sequence, unlike that of rodents, is similar to the Aβ sequence of the human [62]. All these findings suggest that the cholesterol-fed rabbit is a relevant model for sporadic AD studies, as it is a unique model system that exhibits the major pathological hallmarks of AD without genetic manipulations. Thus, this rabbit model provides a complementary system to the current mouse models which are formulated around the autosomal dominant forms of AD.

Various pathological hallmarks of AD are also common features of another degenerative disease, inclusion body myositis (IBM). IBM is an inflammatory myopathy, and people suffering from this most common age-related inflammatory muscle disease exhibit progressive muscle weakness and increased mortality. Similarities between some pathological findings in the AD brain and IBM-affected skeletal muscle are intriguing and include various features, such as increased oxidative stress, inflammation, Aβ accumulation, ApoE accumulation, and hyperphosphorylation of tau [5;85]. Similar to AD, the vast majority of IBM cases are sporadic, with no known genetic link or cause. An increased accumulation of free intracellular cholesterol appears to participate in the pathogenesis of sporadic IBM. Increased levels of free cholesterol increase Aβ production in non-muscle cells, and free cholesterol is colocalized with the intramuscular deposition of Aβ in sporadic IBM patients. Some of the strongest evidence for a link between cholesterol and IBM comes from our recently published studies in cholesterol-fed rabbits in which we have found features of IBM [26]. These include vacuolated muscle fibers, increased numbers of mononuclear inflammatory cells, increased intramuscular deposition of Aβ, and accumulation of hyperphosphorylated tau composed of paired helical filaments. In one-third of the rabbits fed the cholesterol-enriched diets increased numbers of muscle fibers immunopositive for ubiquitin in the skeletal muscle. These features are not detected in rabbits on control diets. Our study has demonstrated for the first time that increased ingestion of dietary cholesterol results in pathological features that closely resemble IBM, and thus may serve as an important new experimental system with which to study this common and debilitating disorder.

*Neuronal preparations from adult rabbit brain*

The use of cultured central nervous system neurons represents a valuable tool for the further characterization and investigation of intracellular mechanisms underlying hypercholesterolemia-induced AD-like pathology. In order for us to have cultures from the same species and of the same age for comparison with our *in vivo* work, we have developed a cell culture model from adult rabbit brain using a technique adapted from that described by Gregory Brewer designed to culture neurons and glial cells from adult and aged rats [13]. This technique demonstrates that hippocampal neurons can regenerate axons and dendrites if provided with adequate nutrition and if growth inhibitors are removed. The following is a brief description of the technique. Brains from adult (1-3 years of age) rabbits are promptly removed and the hippocampi dissected and processed as follows: hippocampi are incubated in 2 ml chilled Hibernate A (Brain bits) + 2% B27 (Invitrogen) + 0.5mM glutamine (Invitrogen) in 25 mm dishes. Leptomeninges are removed and tissue is transferred to sterile wet filter paper with cold HibA/B27 in a McIlwain chopper. The tissue is chopped at 0.5 mm, transferred to a tube containing 5 ml chilled HibA/B27, and placed in a shaker (150 rpm) for 5 min at 30°C. The tissue is then transferred to a 50 ml tube containing 6 ml papain solution (12mg/6ml HibA – no B27, no glutamine) and incubated for 30 min at a rotation of ~170 rpm. The papain solution is decanted and the tissue is transferred into 4 ml HibA/B27 and allowed to incubate for 5 min at room temperature before trituration. A gradient is prepared by diluting 1.32 Optiprep (Sigma) in Hibernate A 1:1 to produce a 1.15 working solution. The gradient is prepared in four 1 ml layers of 35, 25, 20, 15% solution in HibA/B27 and added carefully to a 15 ml centrifuge tube. The cell suspension is carefully applied to the gradient and centrifuged at 800g for fifteen minutes at room temperature. The upper layer contains debris and is discarded and the fractions corresponding to oligodendrocytes (second layer), neurons (third layer) and astrocytes (bottom layer) are plated. We found that hippocampal neurons and glial cells from adult rabbits (1-3 years old) live for more than 8 weeks (Figure 2). We examined

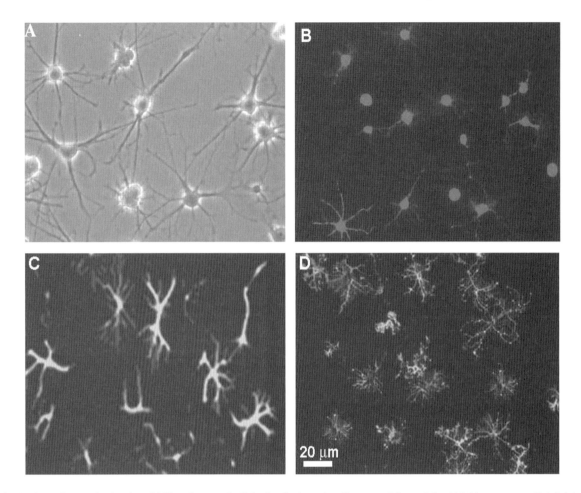

Fig. 2. Photomicrographs showing viability of neuronal, glial, oligodendrocytic cells prepared from adult rabbit hippocampus. **A)** A bright field photomicrograph. **B)** MAP-2 staining, as a marker for neurons and their processes; **C)** GFAP staining for astrocytes; **D)** Oligodendrocytes stained with anti-oligodendrocyte marker O4. Bar, 20 μm.

the effect of incubation of neuronal preparations from rabbit hippocampus with concentrations of 5, 10 and 20 μM of 27-hydroxycholesterol on Aβ levels and found that incubation of neurons with 10 and 20 μM 27-hydroxycholesterol for 48 and 72 hours has increased Aβ immunostaining.

*Organotypic slices from adult rabbit brains*

We also recently developed an organotypic hippocampal slice model from adult rabbits following method optimization of the procedure described by Stoppini in neonatal animals [117]. The procedure for preparing and maintaining organotypic slices from rabbits are similar to the procedure we recently published in adult mice [104]. The organotypic slice system has distinct advantages over other *in vitro* systems, including the maintenance of proximity and some connectivity between neurons, interneurons, and glia. It also offers several advantages over *in vivo* systems, including the simplified administration of pharmacological compounds. Organotypic slices were prepared from the hippocampi of New Zealand white rabbits (1-3 years-old) as follows. Hippocampi are dissected, trimmed of excess white matter and placed into chilled dissection media composed of hibernate A (BrainBits) containing 20% horse serum and 0.5 mM l-glutamine (Invitrogen). Isolated tissue is placed on a wetted filter paper on the Teflon stage of a MacIlwain chopper for coronal sectioning (300 μm thick). From each rabbit hippocampi, about 30 sections are cut. Sections are placed in new dissection media and allowed to rest five minutes on ice before separating and plating on membrane inserts (Millipore, Bedford, MA). Five sections are placed on each insert with a total of 6 inserts per hippocampus. Inserts are placed in 35 mm culture dishes containing 1.1 ml growth media (Neurobasal A with 20% horse serum, 0.5 mM l-glutamine, 100 U/ml penicillin, and 0.05 µM/ml streptomycin), and warmed 30 min prior to plating to ensure complete equilibration. Slices are exposed to a humidified incubator atmosphere (4.5% $CO_2$ and 35°C). Media was changed at DIV1 and at DIV4, slices are switched to a defined medium consisting of Neurobasal A, 2% B27 supplement and 0.5 mM l-glutamine. At day 10, slices are incubated with 25 µM 27-hydroxycholesterol or vehicle and then harvested 72 hr later. We have found that organotypic slices can remain alive for more than 6 weeks (Figure 3). Collectively, our *in vitro* results strongly suggest that an increase in 27-hydroxycholesterol levels is a potential pathway by which high blood cholesterol levels induce AD-like pathology in the brain.

## THERAPEUTIC STRATEGIES TARGETING CHOLESTEROL-INDUCED AD PATHOLOGY

### Statins

Members of the statin family have been reported to lower the risk of AD [41;109;129;130]. However, the mechanisms by which statins may reduce the risk of AD are still to be established. While some studies link the effects of statins on the pathogenesis of AD to decreasing Aβ levels, other studies have demonstrated that the beneficial effect of statins is not related to lower levels of Aβ peptide [23;97]. In addition to lowering cholesterol levels, statins have pleiotropic effects, including immunomodulatory, antioxidant and anti-inflammatory effects [69;89;101;118;121] that may account for their effect on AD. Currently there is no consensus as to whether statins may be beneficial to AD patients. Nonetheless, numerous studies have investigated the effects of the use of statins on AD and only very few have reported no benefits of cholesterol-lowering therapy [99;100].

### Liver X activated receptor (LXR) agonists

LXR α and β are nuclear receptors that control cholesterol transport and metabolism in the circulation

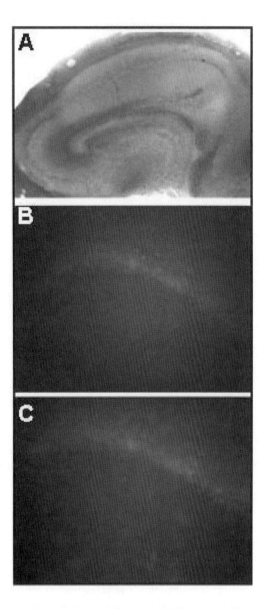

Fig. 3. Photomicrographs of organotypic hippocampal slices from a 2 year-old rabbit light photographed (A), stained with the marker for neurons MAP-2 (B), or with an oligodendrocyte marker (C).

through regulation of expression levels of the cholesterol transporters ABCA1, ABCG1, and ApoE [8]. LXRs are also expressed in the brain [128] where they regulate cholesterol trafficking among cells via ABCA1 and ApoE. Because of the involvement of both ApoE and the ABCA1 in Aβ transport and clearance, their activation with LXR agonists may influence Aβ levels in the brain. ABCA1 and ApoE are functionally linked and work in concert to regulate not only cholesterol homeostasis but also Aβ accumulation. Indeed, inactivation of ABCA1 gene leads to reduced ApoE levels and increased Aβ in a mouse model for AD [59;127]. Regulation of ABCA1 and ApoE by LXRs agonists may therefore be a useful therapeutic strategy for prevention or treatment of AD. Oxysterols, including 27-hydroxycholesterol, are endogenous activators of LXR [44]; however, little is known about the effects of the 27-hydroxycholesterol on LXR and LXR-regulated genes, such as ABCA1, ABCG1, and ApoE. Synthetic oxysterols that can regulate LXR activity in the brain may provide a novel therapeutic avenue for AD [126].

*Caffeine*

Caffeine, a multifaceted methylxanthine, has been shown to increase or decrease the production of Aβ peptide [84;93] as well as to increase and decrease Aβ-induced cell death *in vitro* [31]. In humans, the risk of developing AD has been reported to be lower in individuals with elevated plasma caffeine levels [81]. Caffeine has antioxidant properties [20] and has been shown to reduce brain levels of Aβ in transgenic mice models for early-onset familial AD [3;4;21]. Coffee intake may also have an anti-atherogenic property by increasing ABCG1 expression and enhancing HDL-mediated cholesterol efflux from the macrophages [125]. At low concentrations caffeine can sensitize ryanodine receptor (RyR), leading to increased Aβ production [84;93;94]. Caffeine can also block adenosine A1 and A2A receptors and can inhibit cAMP phosphodiesterase activity at high doses [19]. We have found that although it doesn't affect plasma and brain cholesterol levels, caffeine dramatically reduces Aβ burden and tau phosphorylation in hippocampus of cholesterol-fed rabbits (Ghribi, in preparation). Increased cholesterol levels in the blood have been shown to compromise the integrity of the BBB [25;47], thereby potentially leading to the entrance of biochemical factors that trigger deleterious effects in the brain. We have found that 3 mg caffeine, administered daily in drinking water for 12 weeks, blocks the extravasation of IgG and fibrinogen as well as Evan's blue dye in the brains of rabbits fed with 2% cholesterol-enriched diets [25]. Furthermore, caffeine has reversed the cholesterol-enriched diet-induced decrease in levels of the tight junction proteins occludin and ZO-1. Our data suggest that caffeine and related compounds can protect against the deleterious effects of hypercholesterolemia in a dose dependent manner (see for review [27]).

*Leptin*

Leptin, a 16 kDa protein primarily synthesized by the adipocytes and identified as a hormone that fights obesity, is also expressed in the brain [73]. Leptin holds promise as a novel therapeutic agent for AD as high plasma levels of leptin were associated with a lower risk of dementia in humans [74;75]. Also, leptin has been shown to promote Aβ clearance and improve learning in transgenic mice for AD [42;49-51]. Because leptin is involved in the regulation of lipid homeostasis, investigating the effects of leptin in the cholesterol-fed rabbit model is relevant to understanding the pathophysiology of AD. We found that cholesterol diets reduce leptin expression in rabbit hippocampus [83], suggesting that reduction in leptin may contribute to the increase in Aβ and phosphorylated tau levels. We examined whether oxysterols reproduce the effects of cholesterol on leptin. We found that incubation of organotypic slices with 27-hydroxycholesterol reduces leptin expression [83]; 24-hydroxycholesterol did not affect leptin levels. Interestingly, our data shows that incubation of organotypic slices with exogenous leptin reduced the 27-hydroxycholesterol-induced increase in Aβ and phosphorylated tau [83]. Leptin treatment decreases the levels of APP and BACE1 and increases the levels of sAPPα. Reduction in APP and BACE1 levels suggests that leptin reduces Aβ levels by regulating the production of Aβ peptide, through reduction of Aβ substrate, APP, and by reducing the initial cleavage by BACE1 that yields Aβ formation. Additionally, the increase in sAPPα levels indicates that leptin favors the shunting of the cleavage toward the α-secretase mediated non-amyloidogenic pathway. Collectively, our results demonstrate that leptin reduction is associated with increased Aβ and p-tau, and that leptin supplementation reduces Aβ and phosphorylated tau accumulation.

## CONCLUSION

Our published work, as well as our ongoing research, uses the cholesterol-fed rabbit model system to understand the potential linkages between cholesterol and AD pathology. Our experiments focus on

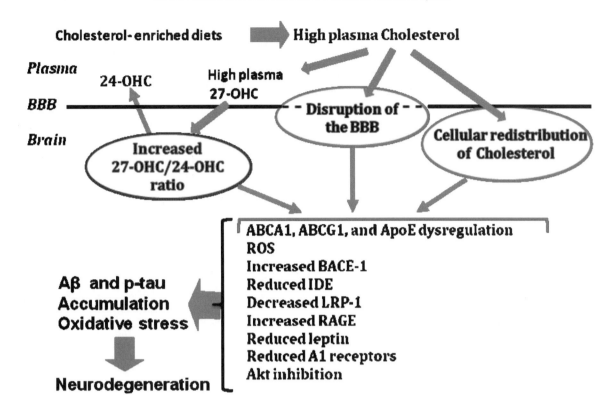

Fig. 4. A schematic summary illustrating the potential mechanisms by which cholesterol induces AD-like pathology. A high cholesterol diet increases blood cholesterol levels and leads to increased 27-hydroxycholesterol (27-OHC)/24-hydroxycholesterol (24-OHC) ratio in the brain, disruption of the BBB, and redistribution of cholesterol among brain cells. All these effects trigger a series of events involving alterations in expression of the cholesterol transporters ABCA1, ABCG1, and ApoE, generation of reactive oxygen species (ROS), increased BACE1, decreased LRP-1 and IDE, increased RAGE, reduced leptin and adenosine A1 receptors, and inhibition of Akt. These changes increase Aβ levels and tau phosphorylation and ultimately trigger neurodegeneration.

understanding the molecular and cellular mechanisms by which circulating cholesterol induces some of the pathological hallmarks of AD in the brain. We have shown that although it does not affect brain cholesterol concentrations, circulating cholesterol alters the redistribution of cholesterol in brain cells. The accumulation of cholesterol in neurons is associated with increased processing of APP by BACE1, leading to Aβ accumulation. We have further found that the cholesterol metabolite, 27-hydroxycholesterol, increases Aβ production in both organotypic hippocampal slices and neuronal preparations from adult rabbit brain. A diagram summarizing the potential mechanism by which cholesterol causes AD-like pathology is presented in Figure 4. The neuronal and organotypic slice preparations from adult rabbit hippocampus are a valuable addition to the *in vivo* model for mechanistic studies involved in cholesterol metabolism and AD. We have also extended the use of the cholesterol-fed rabbit model to studies of IBM, a disease that is pathologically linked to some features of AD. Since rabbits have a phylogeny closer to humans than rodents and their Aβ sequence, unlike rodent, is similar to that of the human, the cholesterol-treated rabbits may more closely resemble sporadic AD, thus providing a complementary system to the current mouse models which are formulated around the genetics and pathophysiology of the autosomal dominant forms of diseases.

## ACKNOWLEDGMENTS

This work was supported by Grants from the National Center for Research Resources (5P20RR017699, Centers of Biomedical Research Excellence) and the NIH (NIEHS, R01ES014826).

# References

[1] A. Akram, J. Schmeidler, P. Katsel, P.R. Hof, V. Haroutunian, Increased expression of cholesterol transporter ABCA1 is highly correlated with severity of dementia in AD hippocampus, *Brain Res.* **1318** (2010), 167-177.

[2] J.C. Antoine, M. Tommasi, S. Boucheron, P. Convers, B. Laurent, D. Michel, Pathology of roots, spinal cord and brainstem in syringomyelia-like syndrome of Tangier disease, *J. Neurol. Sci.* **106** (1991), 179-185.

[3] G.W. Arendash, T. Mori, C. Cao, M. Mamcarz, M. Runfeldt, A. Dickson, K. Rezai-Zadeh, J. Tane, B.A. Citron, X. Lin, V. Echeverria, H. Potter, Caffeine reverses cognitive impairment and decreases brain amyloid-beta levels in aged Alzheimer's disease mice, *J. Alzheimers. Dis.* **17** (2009), 661-680.

[4] G.W. Arendash, W. Schleif, K. Rezai-Zadeh, E.K. Jackson, L.C. Zacharia, J.R. Cracchiolo, D. Shippy, J. Tan, Caffeine protects Alzheimer's mice against cognitive impairment and reduces brain beta-amyloid production, *Neurosci* **142** (2006), 941-952.

[5] V. Askanas, W.K. Engel, Inclusion-body myositis: newest concepts of pathogenesis and relation to aging and Alzheimer disease, *J. Neuropathol. Exp. Neurol.* **60** (2001), 1-14.

[6] B.M. Austen, C. Sidera, C. Liu, E. Frears, The role of intracellular cholesterol on the processing of the beta-amyloid precursor protein, *J. Nutr. Health Aging* **7** (2003), 31-36.

[7] N. Aytan, T. Jung, F. Tamturk, T. Grune, N. Kartal-Ozer, Oxidative stress related changes in the brain of hypercholesterolemic rabbits, *Biofactors* **33** (2008), 225-236.

[8] S.W. Beaven, P. Tontonoz, Nuclear receptors in lipid metabolism: targeting the heart of dyslipidemia, *Annu. Rev. Med.* **57** (2006), 313-329.

[9] I. Bjorkhem, Do oxysterols control cholesterol homeostasis?, *J. Clin. Invest* **110** (2002), 725-730.

[10] I. Bjorkhem, U. Diczfalusy, Oxysterols: friends, foes, or just fellow passengers?, *Arterioscler. Thromb. Vasc. Biol.* **22** (2002), 734-742.

[11] I. Bjorkhem, S. Meaney, Brain cholesterol: long secret life behind a barrier, *Arterioscler. Thromb. Vasc. Biol.* **24** (2004), 806-815.

[12] I. Bjorkhem, S. Meaney, U. Diczfalusy, Oxysterols in human circulation: which role do they have?, *Curr. Opin. Lipidol.* **13** (2002), 247-253.

[13] G.J. Brewer, Isolation and culture of adult rat hippocampal neurons, *J. Neurosci. Methods* **71** (1997), 143-155.

[14] A.J. Brown, W. Jessup, Oxysterols and atherosclerosis, *Atherosclerosis* **142** (1999), 1-28.

[15] J. Brown, III, C. Theisler, S. Silberman, D. Magnuson, N. Gottardi-Littell, J.M. Lee, D. Yager, J. Crowley, K. Sambamurti, M.M. Rahman, A.B. Reiss, C.B. Eckman, B. Wolozin, Differential expression of cholesterol hydroxylases in Alzheimer's disease, *J. Biol. Chem.* **279** (2004), 34674-34681.

[16] M.P. Burns, U. Igbavboa, L. Wang, W.G. Wood, K. Duff, Cholesterol distribution, not total levels, correlate with altered amyloid precursor protein processing in statin-treated mice, *Neuromolecular. Med.* **8** (2006), 319-328.

[17] M.P. Burns, W.J. Noble, V. Olm, K. Gaynor, E. Casey, J. LaFrancois, L. Wang, K. Duff, Co-localization of cholesterol, apolipoprotein E and fibrillar Abeta in amyloid plaques, *Brain Res. Mol. Brain Res.* **110** (2003), 119-125.

[18] J. Busciglio, D.H. Gabuzda, P. Matsudaira, B.A. Yankner, Generation of beta-amyloid in the secretory pathway in neuronal and nonneuronal cells, *Proc. Natl. Acad. Sci. U. S. A* **90** (1993), 2092-2096.

[19] R.W. Butcher, E.W. Sutherland, Adenosine 3',5'-phosphate in biological materials. I. Purification and properties of cyclic 3',5'-nucleotide phosphodiesterase and use of this enzyme to characterize adenosine 3',5'-phosphate in human urine, *J. Biol. Chem.* **237** (1962), 1244-1250.

[20] M.M. Camouse, K.K. Hanneman, E.P. Conrad, E.D. Baron, Protective effects of tea polyphenols and caffeine, *Expert. Rev. Anticancer Ther.* **5** (2005), 1061-1068.

[21] C. Cao, J.R. Cirrito, X. Lin, L. Wang, D.K. Verges, A. Dickson, M. Mamcarz, C. Zhang, T. Mori, G.W. Arendash, D.M. Holtzman, H. Potter, Caffeine suppresses amyloid-beta levels in plasma and brain of Alzheimer's disease transgenic mice, *J. Alzheimers. Dis.* **17** (2009), 681-697.

[22] G. Cao, K.R. Bales, R.B. DeMattos, S.M. Paul, Liver X receptor-mediated gene regulation and cholesterol homeostasis in brain: relevance to Alzheimer's disease therapeutics, *Curr. Alzheimer Res.* **4** (2007), 179-184.

[23] C.M. Carlsson, C.E. Gleason, T.M. Hess, K.A. Moreland, H.M. Blazel, R.L. Koscik, N.T. Schreiber, S.C. Johnson, C.S. Atwood, L. Puglielli, B.P. Hermann, P.E. McBride, J.H. Stein, M.A. Sager, S. Asthana, Effects of simvastatin on cerebrospinal fluid biomarkers and cognition in middle-aged adults at risk for Alzheimer's disease, *J. Alzheimers. Dis.* **13** (2008), 187-197.

[24] R.J. Castellani, P.I. Moreira, G. Liu, J. Dobson, G. Perry, M.A. Smith, X. Zhu, Iron: the Redox-active center of oxidative stress in Alzheimer disease, *Neurochem. Res.* **32** (2007), 1640-1645.

[25] X. Chen, J.W. Gawryluk, J.F. Wagener, O. Ghribi, J.D. Geiger, Caffeine blocks disruption of blood brain barrier in a rabbit model of Alzheimer's disease, *J. Neuroinflammation.* **5** (2008), 12.

[26] X. Chen, O. Ghribi, J.D. Geiger, Rabbits fed cholesterol-enriched diets exhibit pathological features of inclusion body myositis, *Am. J. Physiol Regul. Integr. Comp Physiol* **294** (2008), R829-R835.

[27] X. Chen, O. Ghribi, J.D. Geiger, Caffeine Protects Against Disruptions of the Blood-Brain Barrier in Animal Models of Alzheimer's and Parkinson's Diseases, *J. Alzheimers. Dis.* (2010), in press.

[28] E.H. Corder, A.M. Saunders, W.J. Strittmatter, D.E. Schmechel, P.C. Gaskell, G.W. Small, A.D. Roses, J.L. Haines, M.A. Pericak-Vance, Gene dose of apolipoprotein E type 4 allele and the risk of Alzheimer's disease in late onset families, *Science* **261** (1993), 921-923.

[29] J.M. Cordy, I. Hussain, C. Dingwall, N.M. Hooper, A.J. Turner, Exclusively targeting beta-secretase to lipid rafts by GPI-anchor addition up-regulates beta-site processing of the amyloid precursor protein, *Proc. Natl. Acad. Sci. U. S. A* **100** (2003), 11735-11740.

[30] R.G. Cutler, J. Kelly, K. Storie, W.A. Pedersen, A. Tammara, K. Hatanpaa, J.C. Troncoso, M.P. Mattson, Involvement of oxidative stress-induced abnormalities in ceramide and cholesterol metabolism in brain aging and Alzheimer's disease, *Proc. Natl. Acad. Sci. U. S. A* **101** (2004), 2070-2075.

[31] O.P. Dall'Igna, L.O. Porciuncula, D.O. Souza, R.A. Cunha, D.R. Lara, Neuroprotection by caffeine and adeno-

sine A2A receptor blockade of beta-amyloid neurotoxicity, *Br. J. Pharmacol.* **138** (2003), 1207-1209.
[32] R. Deane, A. Sagare, B.V. Zlokovic, The role of the cell surface LRP and soluble LRP in blood-brain barrier Abeta clearance in Alzheimer's disease, *Curr. Pharm. Des* **14** (2008), 1601-1605.
[33] R. Deane, Z. Wu, A. Sagare, J. Davis, Y.S. Du, K. Hamm, F. Xu, M. Parisi, B. LaRue, H.W. Hu, P. Spijkers, H. Guo, X. Song, P.J. Lenting, W.E. Van Nostrand, B.V. Zlokovic, LRP/amyloid beta-peptide interaction mediates differential brain efflux of Abeta isoforms, *Neuron* **43** (2004), 333-344.
[34] R. Deane, Z. Wu, B.V. Zlokovic, RAGE (yin) versus LRP (yang) balance regulates alzheimer amyloid beta-peptide clearance through transport across the blood-brain barrier, *Stroke* **35** (2004), 2628-2631.
[35] R. Distl, V. Meske, T.G. Ohm, Tangle-bearing neurons contain more free cholesterol than adjacent tangle-free neurons, *Acta Neuropathol.* **101** (2001), 547-554.
[36] R. Distl, S. Treiber-Held, F. Albert, V. Meske, K. Harzer, T.G. Ohm, Cholesterol storage and tau pathology in Niemann-Pick type C disease in the brain, *J. Pathol.* **200** (2003), 104-111.
[37] J.E. Donahue, S.L. Flaherty, C.E. Johanson, J.A. Duncan, III, G.D. Silverberg, M.C. Miller, R. Tavares, W. Yang, Q. Wu, E. Sabo, V. Hovanesian, E.G. Stopa, RAGE, LRP-1, and amyloid-beta protein in Alzheimer's disease, *Acta Neuropathol.* **112** (2006), 405-415.
[38] W.C. Duckworth, R.G. Bennett, F.G. Hamel, Insulin degradation: progress and potential, *Endocr. Rev.* **19** (1998), 608-624.
[39] R. Ehehalt, P. Keller, C. Haass, C. Thiele, K. Simons, Amyloidogenic processing of the Alzheimer beta-amyloid precursor protein depends on lipid rafts, *J. Cell Biol.* **160** (2003), 113-123.
[40] M. Eto, K. Watanabe, N. Chonan, K. Ishii, Familial hypercholesterolemia and apolipoprotein E4, *Atherosclerosis* **72** (1988), 123-128.
[41] K. Fassbender, M. Simons, C. Bergmann, M. Stroick, D. Lutjohann, P. Keller, H. Runz, S. Kuhl, T. Bertsch, K. von Bergmann, M. Hennerici, K. Beyreuther, T. Hartmann, Simvastatin strongly reduces levels of Alzheimer's disease beta-amyloid peptides Abeta 42 and Abeta 40 in vitro and in vivo, *Proc. Natl. Acad. Sci. U. S. A* **98** (2001), 5856-5861.
[42] D.C. Fewlass, K. Noboa, F.X. Pi-Sunyer, J.M. Johnston, S.D. Yan, N. Tezapsidis, Obesity-related leptin regulates Alzheimer's Abeta, *FASEB J.* **18** (2004), 1870-1878.
[43] E.R. Frears, D.J. Stephens, C.E. Walters, H. Davies, B.M. Austen, The role of cholesterol in the biosynthesis of beta-amyloid, *Neuroreport* **10** (1999), 1699-1705.
[44] X. Fu, J.G. Menke, Y. Chen, G. Zhou, K.L. MacNaul, S.D. Wright, C.P. Sparrow, E.G. Lund, 27-hydroxycholesterol is an endogenous ligand for liver X receptor in cholesterol-loaded cells, *J. Biol. Chem.* **276** (2001), 38378-38387.
[45] J.L. Galbete, T.R. Martin, E. Peressini, P. Modena, R. Bianchi, G. Forloni, Cholesterol decreases secretion of the secreted form of amyloid precursor protein by interfering with glycosylation in the protein secretory pathway, *Biochem. J.* **348 (Pt 2)** (2000), 307-313.
[46] O. Ghribi, M.Y. Golovko, B. Larsen, M. Schrag, E.J. Murphy, Deposition of iron and beta-amyloid plaques is associated with cortical cellular damage in rabbits fed with long-term cholesterol-enriched diets, *J. Neurochem.* **99** (2006), 438-449.
[47] O. Ghribi, B. Larsen, M. Schrag, M.M. Herman, High cholesterol content in neurons increases BACE, beta-amyloid, and phosphorylated tau levels in rabbit hippocampus, *Exp. Neurol.* **200** (2006), 460-467.
[48] D. Graur, L. Duret, M. Gouy, Phylogenetic position of the order Lagomorpha (rabbits, hares and allies), *Nature* **379** (1996), 333-335.
[49] S.J. Greco, K.J. Bryan, S. Sarkar, X. Zhu, M.A. Smith, J.W. Ashford, J.M. Johnston, N. Tezapsidis, G. Casadesus, Chronic Leptin Supplementation Ameliorates Pathology and Improves Cognitive Performance in a Transgenic Mouse Model of Alzheimer's Disease, *J. Alzheimers. Dis.* **19** (2010), 1155-1168.
[50] S.J. Greco, S. Sarkar, G. Casadesus, X. Zhu, M.A. Smith, J.W. Ashford, J.M. Johnston, N. Tezapsidis, Leptin inhibits glycogen synthase kinase-3beta to prevent tau phosphorylation in neuronal cells, *Neurosci. Lett.* **455** (2009), 191-194.
[51] S.J. Greco, S. Sarkar, J.M. Johnston, N. Tezapsidis, Leptin regulates tau phosphorylation and amyloid through AMPK in neuronal cells, *Biochem. Biophys. Res. Commun.* **380** (2009), 98-104.
[52] S.M. Greenberg, E.H. Koo, D.J. Selkoe, W.Q. Qiu, K.S. Kosik, Secreted beta-amyloid precursor protein stimulates mitogen-activated protein kinase and enhances tau phosphorylation, *Proc. Natl. Acad. Sci. U. S. A* **91** (1994), 7104-7108.
[53] C. Haass, D.J. Selkoe, Cellular processing of beta-amyloid precursor protein and the genesis of amyloid beta-peptide, *Cell* **75** (1993), 1039-1042.
[54] D. Harold, R. Abraham, P. Hollingworth, R. Sims, A. Gerrish, M.L. Hamshere, J.S. Pahwa, V. Moskvina, K. Dowzell, A. Williams, N. Jones, C. Thomas, A. Stretton, A.R. Morgan, S. Lovestone, J. Powell, P. Proitsi, M.K. Lupton, C. Brayne, D.C. Rubinsztein, M. Gill, B. Lawlor, A. Lynch, K. Morgan, K.S. Brown, P.A. Passmore, D. Craig, B. McGuinness, S. Todd, C. Holmes, D. Mann, A.D. Smith, S. Love, P.G. Kehoe, J. Hardy, S. Mead, N. Fox, M. Rossor, J. Collinge, W. Maier, F. Jessen, B. Schurmann, B.H. van den, I. Heuser, J. Kornhuber, J. Wiltfang, M. Dichgans, L. Frolich, M. Hampel, M. Hull, D. Rujescu, A.M. Goate, J.S. Kauwe, C. Cruchaga, P. Nowotny, J.C. Morris, K. Mayo, K. Sleegers, K. Bettens, S. Engelborghs, P.P. De Deyn, B.C. Van, G. Livingston, N.J. Bass, H. Gurling, A. McQuillin, R. Gwilliam, P. Deloukas, A. Al-Chalabi, C.E. Shaw, M. Tsolaki, A.B. Singleton, R. Guerreiro, T.W. Muhleisen, M.M. Nothen, S. Moebus, K.H. Jockel, N. Klopp, H.E. Wichmann, M.M. Carrasquillo, V.S. Pankratz, S.G. Younkin, P.A. Holmans, M. O'Donovan, M.J. Owen, J. Williams, Genome-wide association study identifies variants at CLU and PICALM associated with Alzheimer's disease, *Nat. Genet.* **41** (2009), 1088-1093.
[55] J.R. Harris, N.G. Milton, Cholesterol in Alzheimer's Disease and other Amyloidogenic Disorders, *Subcell. Biochem.* **51** (2010), 47-75.
[56] J. Herz, LRP: a bright beacon at the blood-brain barrier, *J. Clin. Invest* **112** (2003), 1483-1485.
[57] M. Heverin, N. Bogdanovic, D. Lutjohann, T. Bayer, I. Pikuleva, L. Bretillon, U. Diczfalusy, B. Winblad, I. Bjorkhem, Changes in the levels of cerebral and extracerebral sterols in the brain of patients with Alzheimer's disease, *J. Lipid Res.* **45** (2004), 186-193.

[58] M. Heverin, S. Meaney, D. Lutjohann, U. Diczfalusy, J. Wahren, I. Bjorkhem, Crossing the barrier: Net flux of 27-hydroxycholesterol into the human brain, *J. Lipid Res.* **46** (2005), 1047-1052.

[59] V. Hirsch-Reinshagen, L.F. Maia, B.L. Burgess, J.F. Blain, K.E. Naus, S.A. McIsaac, P.F. Parkinson, J.Y. Chan, G.H. Tansley, M.R. Hayden, J. Poirier, N.W. Van, C.L. Wellington, The absence of ABCA1 decreases soluble ApoE levels but does not diminish amyloid deposition in two murine models of Alzheimer disease, *J. Biol. Chem.* **280** (2005), 43243-43256.

[60] R.M. Holsinger, C.A. McLean, K. Beyreuther, C.L. Masters, G. Evin, Increased expression of the amyloid precursor beta-secretase in Alzheimer's disease, *Ann. Neurol.* **51** (2002), 783-786.

[61] R.P. Jaya Prasanthi, E. Schommer, S. Thomasson, A. Thompson, G. Feist, O. Ghribi, Regulation of beta-amyloid levels in the brain of cholesterol-fed rabbit, a model system for sporadic Alzheimer's disease, *Mech. Ageing Dev.* **129** (2008), 649-655.

[62] E.M. Johnstone, M.O. Chaney, F.H. Norris, R. Pascual, S.P. Little, Conservation of the sequence of the Alzheimer's disease amyloid peptide in dog, polar bear and five other mammals by cross-species polymerase chain reaction analysis, *Brain Res. Mol. Brain Res.* **10** (1991), 299-305.

[63] W.S. Kim, S.L. Chan, A.F. Hill, G.J. Guillemin, B. Garner, Impact of 27-hydroxycholesterol on amyloid-beta peptide production and ATP-binding cassette transporter expression in primary human neurons, *J. Alzheimers. Dis.* **16** (2009), 121-131.

[64] W.S. Kim, G.J. Guillemin, E.N. Glaros, C.K. Lim, B. Garner, Quantitation of ATP-binding cassette subfamily-A transporter gene expression in primary human brain cells, *Neuroreport* **17** (2006), 891-896.

[65] W.S. Kim, C.S. Weickert, B. Garner, Role of ATP-binding cassette transporters in brain lipid transport and neurological disease, *J. Neurochem.* **104** (2008), 1145-1166.

[66] R. Koldamova, M. Staufenbiel, I. Lefterov, Lack of ABCA1 considerably decreases brain ApoE level and increases amyloid deposition in APP23 mice, *J. Biol. Chem.* **280** (2005), 43224-43235.

[67] R.P. Koldamova, I.M. Lefterov, M.D. Ikonomovic, J. Skoko, P.I. Lefterov, B.A. Isanski, S.T. DeKosky, J.S. Lazo, 22R-hydroxycholesterol and 9-cis-retinoic acid induce ATP-binding cassette transporter A1 expression and cholesterol efflux in brain cells and decrease amyloid beta secretion, *J. Biol. Chem.* **278** (2003), 13244-13256.

[68] I.V. Kurochkin, S. Goto, Alzheimer's beta-amyloid peptide specifically interacts with and is degraded by insulin degrading enzyme, *FEBS Lett.* **345** (1994), 33-37.

[69] B. Kwak, F. Mulhaupt, S. Myit, F. Mach, Statins as a newly recognized type of immunomodulator, *Nat. Med.* **6** (2000), 1399-1402.

[70] J.C. Lambert, S. Heath, G. Even, D. Campion, K. Sleegers, M. Hiltunen, O. Combarros, D. Zelenika, M.J. Bullido, B. Tavernier, L. Letenneur, K. Bettens, C. Berr, F. Pasquier, N. Fievet, P. Barberger-Gateau, S. Engelborghs, D.P. De, I. Mateo, A. Franck, S. Helisalmi, E. Porcellini, O. Hanon, M.M. de Pancorbo, C. Lendon, C. Dufouil, C. Jaillard, T. Leveillard, V. Alvarez, P. Bosco, M. Mancuso, F. Panza, B. Nacmias, P. Bossu, P. Piccardi, G. Annoni, D. Seripa, D. Galimberti, D. Hannequin, F. Licastro, H. Soininen, K. Ritchie, H. Blanche, J.F. Dartigues, C. Tzourio, I. Gut, B.C. Van, A. Alperovitch, M. Lathrop, P. Amouyel, Genome-wide association study identifies variants at CLU and CR1 associated with Alzheimer's disease, *Nat. Genet.* **41** (2009), 1094-1099.

[71] Y. Lange, J. Ye, M. Rigney, T.L. Steck, Regulation of endoplasmic reticulum cholesterol by plasma membrane cholesterol, *J. Lipid Res.* **40** (1999), 2264-2270.

[72] O. Levi, D. Lutjohann, A. Devir, B.K. von, T. Hartmann, D.M. Michaelson, Regulation of hippocampal cholesterol metabolism by apoE and environmental stimulation, *J. Neurochem.* **95** (2005), 987-997.

[73] C. Li, J.M. Friedman, Leptin receptor activation of SH2 domain containing protein tyrosine phosphatase 2 modulates Ob receptor signal transduction, *Proc. Natl. Acad. Sci. U. S. A* **96** (1999), 9677-9682.

[74] W. Lieb, A.S. Beiser, R.S. Vasan, Z.S. Tan, R. Au, T.B. Harris, R. Roubenoff, S. Auerbach, C. DeCarli, P.A. Wolf, S. Seshadri, Association of plasma leptin levels with incident Alzheimer disease and MRI measures of brain aging, *JAMA* **302** (2009), 2565-2572.

[75] W. Lieb, L.M. Sullivan, J. Aragam, T.B. Harris, R. Roubenoff, E.J. Benjamin, R.S. Vasan, Relation of serum leptin with cardiac mass and left atrial dimension in individuals >70 years of age, *Am. J. Cardiol.* **104** (2009), 602-605.

[76] E.G. Lund, J.M. Guileyardo, D.W. Russell, cDNA cloning of cholesterol 24-hydroxylase, a mediator of cholesterol homeostasis in the brain, *Proc. Natl. Acad. Sci. U. S. A* **96** (1999), 7238-7243.

[77] D. Lutjohann, O. Breuer, G. Ahlborg, I. Nennesmo, A. Siden, U. Diczfalusy, I. Bjorkhem, Cholesterol homeostasis in human brain: evidence for an age-dependent flux of 24S-hydroxycholesterol from the brain into the circulation, *Proc. Natl. Acad. Sci. U. S. A* **93** (1996), 9799-9804.

[78] D. Lutjohann, M. Marinova, B. Schneider, J. Oldenburg, B.K. von, T. Bieber, I. Bjorkhem, U. Diczfalusy, 4beta-hydroxycholesterol as a marker of CYP3A4 inhibition in vivo - effects of itraconazole in man, *Int. J. Clin. Pharmacol. Ther.* **47** (2009), 709-715.

[79] J.T. Mack, V. Beljanski, K.D. Tew, D.M. Townsend, The ATP-binding cassette transporter ABCA2 as a mediator of intracellular trafficking, *Biomed. Pharmacother.* **60** (2006), 587-592.

[80] R.W. Mahley, Apolipoprotein E: cholesterol transport protein with expanding role in cell biology, *Science* **240** (1988), 622-630.

[81] L. Maia, M.A. de, Does caffeine intake protect from Alzheimer's disease?, *Eur. J. Neurol.* **9** (2002), 377-382.

[82] L. Marlow, M. Cain, M.A. Pappolla, K. Sambamurti, Beta-secretase processing of the Alzheimer's amyloid protein precursor (APP), *J. Mol. Neurosci.* **20** (2003), 233-239.

[83] G. Marwarha, B. Dasari, R.P.J. Prasanthi, J. Schommer, O. Ghribi, Leptin reduces the accumulation of Abeta and phosphorylated tau induced by 27-hydroxycholesterol in rabbit organotypic slices, *J. Alzheimers. Dis.* **19** (2010), 1007-1019.

[84] T.J. McNulty, C.W. Taylor, Caffeine-stimulated Ca2+ release from the intracellular stores of hepatocytes is not mediated by ryanodine receptors, *Biochem. J.* **291 (Pt 3)** (1993), 799-801.

[85] P.I. Moreira, K. Honda, X. Zhu, A. Nunomura, G. Casadesus, M.A. Smith, G. Perry, Brain and brawn: parallels in oxidative strength, *Neurology* **66** (2006), S97-101.

[86] K. Murakami, M. Shimizu, N. Yamada, S. Ishibashi, H. Shimano, Y. Yazaki, Y. Akanuma, Apolipoprotein E polymorphism is associated with plasma cholesterol response in a 7-day hospitalization study for metabolic and dietary control in NIDDM, *Diabetes Care* **16** (1993), 564-569.

[87] A. Papassotiropoulos, D. Lutjohann, M. Bagli, S. Locatelli, F. Jessen, M.L. Rao, W. Maier, I. Bjorkhem, B.K. von, R. Heun, Plasma 24S-hydroxycholesterol: a peripheral indicator of neuronal degeneration and potential state marker for Alzheimer's disease, *Neuroreport* **11** (2000), 1959-1962.

[88] A. Papassotiropoulos, J.R. Streffer, M. Tsolaki, S. Schmid, D. Thal, F. Nicosia, V. Iakovidou, A. Maddalena, D. Lutjohann, E. Ghebremedhin, T. Hegi, T. Pasch, M. Traxler, A. Bruhl, L. Benussi, G. Binetti, H. Braak, R.M. Nitsch, C. Hock, Increased brain beta-amyloid load, phosphorylated tau, and risk of Alzheimer disease associated with an intronic CYP46 polymorphism, *Arch. Neurol.* **60** (2003), 29-35.

[89] D. Paris, K.P. Townsend, J. Humphrey, D.F. Obregon, K. Yokota, M. Mullan, Statins inhibit A beta-neurotoxicity in vitro and A beta-induced vasoconstriction and inflammation in rat aortae, *Atherosclerosis* **161** (2002), 293-299.

[90] F.W. Pfrieger, Cholesterol homeostasis and function in neurons of the central nervous system, *Cell Mol. Life Sci.* **60** (2003), 1158-1171.

[91] J. Poirier, A. Baccichet, D. Dea, S. Gauthier, Cholesterol synthesis and lipoprotein reuptake during synaptic remodelling in hippocampus in adult rats, *Neurosci* **55** (1993), 81-90.

[92] R.P.J. Prasanthi, A. Huls, S. Thomasson, A. Thompson, E. Schommer, O. Ghribi, Differential effects of 24-hydroxycholesterol and 27-hydroxycholesterol on beta-amyloid precursor protein levels and processing in human neuroblastoma SH-SY5Y cells, *Mol. Neurodegener.* **4** (2009), 1.

[93] H.W. Querfurth, N.J. Haughey, S.C. Greenway, P.W. Yacono, D.E. Golan, J.D. Geiger, Expression of ryanodine receptors in human embryonic kidney (HEK293) cells, *Biochem. J.* **334 (Pt 1)** (1998), 79-86.

[94] H.W. Querfurth, J. Jiang, J.D. Geiger, D.J. Selkoe, Caffeine stimulates amyloid beta-peptide release from beta-amyloid precursor protein-transfected HEK293 cells, *J. Neurochem.* **69** (1997), 1580-1591.

[95] M. Racchi, R. Baetta, N. Salvietti, P. Ianna, G. Franceschini, R. Paoletti, R. Fumagalli, S. Govoni, M. Trabucchi, M. Soma, Secretory processing of amyloid precursor protein is inhibited by increase in cellular cholesterol content, *Biochem. J.* **322 (Pt 3)** (1997), 893-898.

[96] L.M. Refolo, B. Malester, J. LaFrancois, T. Bryant-Thomas, R. Wang, G.S. Tint, K. Sambamurti, K. Duff, M.A. Pappolla, Hypercholesterolemia accelerates the Alzheimer's amyloid pathology in a transgenic mouse model, *Neurobiol. Dis.* **7** (2000), 321-331.

[97] R.G. Riekse, G. Li, E.C. Petrie, J.B. Leverenz, D. Vavrek, S. Vuletic, J.J. Albers, T.J. Montine, V.M. Lee, M. Lee, P. Seubert, D. Galasko, G.D. Schellenberg, W.R. Hazzard, E.R. Peskind, Effect of statins on Alzheimer's disease biomarkers in cerebrospinal fluid, *J. Alzheimers. Dis.* **10** (2006), 399-406.

[98] D.W. Russell, Oxysterol biosynthetic enzymes, *Biochim. Biophys. Acta* **1529** (2000), 126-135.

[99] M.N. Sabbagh, Drug development for Alzheimer's disease: where are we now and where are we headed?, *Am. J. Geriatr. Pharmacother.* **7** (2009), 167-185.

[100] M.N. Sabbagh, K. Thind, D.L. Sparks, On cholesterol levels and statins in cognitive decline and Alzheimer disease: progress and setbacks, *Alzheimer Dis. Assoc. Disord.* **23** (2009), 303-305.

[101] A. Saheki, T. Terasaki, I. Tamai, A. Tsuji, In vivo and in vitro blood-brain barrier transport of 3-hydroxy-3-methylglutaryl coenzyme A (HMG-CoA) reductase inhibitors, *Pharm. Res.* **11** (1994), 305-311.

[102] G. Saher, B. Brugger, C. Lappe-Siefke, W. Mobius, R. Tozawa, M.C. Wehr, F. Wieland, S. Ishibashi, K.A. Nave, High cholesterol level is essential for myelin membrane growth, *Nat. Neurosci.* **8** (2005), 468-475.

[103] G. Schmitz, W.E. Kaminski, ABCA2: a candidate regulator of neural transmembrane lipid transport, *Cell Mol. Life Sci.* **59** (2002), 1285-1295.

[104] M. Schrag, S. Sharma, H. Brown-Borg, O. Ghribi, Hippocampus of Ames dwarf mice is resistant to beta-amyloid-induced tau hyperphosphorylation and changes in apoptosis-regulatory protein levels, *Hippocampus* **18** (2008), 239-244.

[105] S. Sharma, R.P.J. Prasanthi, E. Schommer, G. Feist, O. Ghribi, Hypercholesterolemia-induced Abeta accumulation in rabbit brain is associated with alteration in IGF-1 signaling, *Neurobiol. Dis.* **32** (2008), 426-432.

[106] M. Shibata, S. Yamada, S.R. Kumar, M. Calero, J. Bading, J. Frangione, D.M. Holtzman, C.A. Miller, D.K. Strickland, J. Ghiso, B.V. Zlokovic, Clearance of Alzheimer's amyloid-ss(1-40) peptide from brain by LDL receptor-related protein-1 at the blood-brain barrier, *J. Clin. Invest* **106** (2000), 1489-1499.

[107] F.S. Shie, L.W. Jin, D.G. Cook, J.B. Leverenz, R.C. LeBoeuf, Diet-induced hypercholesterolemia enhances brain A beta accumulation in transgenic mice, *Neuroreport* **13** (2002), 455-459.

[108] M. Simons, P. Keller, B. De Strooper, K. Beyreuther, C.G. Dotti, K. Simons, Cholesterol depletion inhibits the generation of beta-amyloid in hippocampal neurons, *Proc. Natl. Acad. Sci. U. S. A* **95** (1998), 6460-6464.

[109] M. Sjogren, K. Gustafsson, S. Syversen, A. Olsson, A. Edman, P. Davidsson, A. Wallin, K. Blennow, Treatment with simvastatin in patients with Alzheimer's disease lowers both alpha- and beta-cleaved amyloid precursor protein, *Dement. Geriatr. Cogn Disord.* **16** (2003), 25-30.

[110] M.A. Smith, G. Perry, Free radical damage, iron, and Alzheimer's disease, *J Neurol Sci* **134 Suppl.** (1995), 92-94.

[111] M.A. Smith, G. Perry, P.L. Richey, L.M. Sayre, V.E. Anderson, M.F. Beal, N. Kowall, Oxidative damage in Alzheimer's, *Nature* **382** (1996), 120-121.

[112] M.A. Smith, C.A. Rottkamp, A. Nunomura, A.K. Raina, G. Perry, Oxidative stress in Alzheimer's disease, *Biochim. Biophys. Acta* **1502** (2000), 139-144.

[113] M.A. Smith, X. Zhu, M. Tabaton, G. Liu, D.W. McKeel, Jr., M.L. Cohen, X. Wang, S.L. Siedlak, B.E. Dwyer, T. Hayashi, M. Nakamura, A. Nunomura, G. Perry, Increased iron and free radical generation in preclinical Alzheimer disease and mild cognitive impairment, *J. Alzheimers. Dis.* **19** (2010), 363-372.

[114] A. Solomon, M. Kivipelto, B. Wolozin, J. Zhou, R.A. Whitmer, Midlife serum cholesterol and increased risk of Alzheimer's and vascular dementia three decades later, *Dement. Geriatr. Cogn Disord.* **28** (2009), 75-80.

[115] D.L. Sparks, H. Liu, D.R. Gross, S.W. Scheff, Increased density of cortical apolipoprotein E immunoreactive neurons in rabbit brain after dietary administration of cholesterol, *Neurosci. Lett.* **187** (1995), 142-144.

[116] D.L. Sparks, S.W. Scheff, J.C. Hunsaker, III, H. Liu, T. Landers, D.R. Gross, Induction of Alzheimer-like beta-amyloid immunoreactivity in the brains of rabbits with dietary cholesterol, *Exp. Neurol.* **126** (1994), 88-94.

[117] L. Stoppini, P.A. Buchs, D. Muller, A simple method for organotypic cultures of nervous tissue, *J. Neurosci. Methods* **37** (1991), 173-182.

[118] O. Stuve, S. Youssef, L. Steinman, S.S. Zamvil, Statins as potential therapeutic agents in neuroinflammatory disorders, *Curr. Opin. Neurol.* **16** (2003), 393-401.

[119] R. Sultana, D.A. Butterfield, Role of oxidative stress in the progression of Alzheimer's disease, *J. Alzheimers. Dis.* **19** (2010), 341-353.

[120] Y. Sun, J. Yao, T.W. Kim, A.R. Tall, Expression of liver X receptor target genes decreases cellular amyloid beta peptide secretion, *J. Biol. Chem.* **278** (2003), 27688-27694.

[121] Y.X. Sun, M. Crisby, S. Lindgren, S. Janciauskiene, Pravastatin inhibits pro-inflammatory effects of Alzheimer's peptide Abeta(1-42) in glioma cell culture in vitro, *Pharmacol. Res.* **47** (2003), 119-126.

[122] G.H. Tansley, B.L. Burgess, M.T. Bryan, Y. Su, V. Hirsch-Reinshagen, J. Pearce, J.Y. Chan, A. Wilkinson, J. Evans, K.E. Naus, S. McIsaac, K. Bromley, W. Song, H.C. Yang, N. Wang, R.B. DeMattos, C.L. Wellington, The cholesterol transporter ABCG1 modulates the subcellular distribution and proteolytic processing of beta-amyloid precursor protein, *J. Lipid Res.* **48** (2007), 1022-1034.

[123] I. Tesseur, D.J. van, K. Spittaels, H.C. Van den, D. Moechars, L.F. Van, Expression of human apolipoprotein E4 in neurons causes hyperphosphorylation of protein tau in the brains of transgenic mice, *Am. J. Pathol.* **156** (2000), 951-964.

[124] T. Umeda, H. Mori, H. Zheng, T. Tomiyama, Regulation of cholesterol efflux by amyloid beta secretion, *J. Neurosci. Res.* (2010), in press.

[125] H. Uto-Kondo, M. Ayaori, M. Ogura, K. Nakaya, M. Ito, A. Suzuki, S. Takiguchi, E. Yakushiji, Y. Terao, H. Ozasa, T. Hisada, M. Sasaki, F. Ohsuzu, K. Ikewaki, Coffee consumption enhances high-density lipoprotein-mediated cholesterol efflux in macrophages, *Circ. Res.* **106** (2010), 779-787.

[126] J. Vaya, H.M. Schipper, Oxysterols, cholesterol homeostasis, and Alzheimer disease, *J. Neurochem.* **102** (2007), 1727-1737.

[127] S.E. Wahrle, H. Jiang, M. Parsadanian, R.E. Hartman, K.R. Bales, S.M. Paul, D.M. Holtzman, Deletion of Abca1 increases Abeta deposition in the PDAPP transgenic mouse model of Alzheimer's disease, *J. Biol. Chem.* **280** (2005), 43236-43242.

[128] L. Wang, G.U. Schuster, K. Hultenby, Q. Zhang, S. Andersson, J.A. Gustafsson, Liver X receptors in the central nervous system: from lipid homeostasis to neuronal degeneration, *Proc. Natl. Acad. Sci. U. S. A* **99** (2002), 13878-13883.

[129] B. Wolozin, Cholesterol and the biology of Alzheimer's disease, *Neuron* **41** (2004) 7-10.

[130] B. Wolozin, J. Brown, III, C. Theisler, S. Silberman, The cellular biochemistry of cholesterol and statins: insights into the pathophysiology and therapy of Alzheimer's disease, *CNS. Drug Rev.* **10** (2004), 127-146.

[131] D.S. Woodruff-Pak, A. Agelan, V.L. Del, A rabbit model of Alzheimer's disease: valid at neuropathological, cognitive, and therapeutic levels, *J. Alzheimers. Dis.* **11** (2007), 371-383.

[132] S.D. Yan, H. Zhu, A. Zhu, A. Golabek, H. Du, A. Roher, J. Yu, C. Soto, A.M. Schmidt, D. Stern, M. Kindy, Receptor-dependent cell stress and amyloid accumulation in systemic amyloidosis, *Nat. Med.* **6** (2000), 643-651.

[133] L.B. Yang, K. Lindholm, R. Yan, M. Citron, W. Xia, X.L. Yang, T. Beach, L. Sue, P. Wong, R. Price, R. Li, Y. Shen, Elevated beta-secretase expression and enzymatic activity detected in sporadic Alzheimer disease, *Nat. Med.* **9** (2003), 3-4.

[134] D. Zambon, M. Quintana, P. Mata, R. Alonso, J. Benavent, F. Cruz-Sanchez, J. Gich, M. Pocovi, F. Civeira, S. Capurro, D. Bachman, K. Sambamurti, J. Nicholas, M.A. Pappolla, Higher incidence of mild cognitive impairment in familial hypercholesterolemia, *Am. J. Med.* **123** (2010), 267-274.

[135] C. Zhou, L. Zhao, N. Inagaki, J. Guan, S. Nakajo, T. Hirabayashi, S. Kikuyama, S. Shioda, Atp-binding cassette transporter ABC2/ABCA2 in the rat brain: a novel mammalian lysosome-associated membrane protein and a specific marker for oligodendrocytes but not for myelin sheaths, *J. Neurosci.* **21** (2001), 849-857.

# Hepatic Ceramide May Mediate Brain Insulin Resistance and Neurodegeneration in Type 2 Diabetes and Non-alcoholic Steatohepatitis

Suzanne M. de la Monte, Lascelles E. Lyn-Cook Jr.*, Margot Lawton*, Ming Tong, Elizabeth Silbermann, Lisa Longato, Ping Jiao, Princess Mark, Haiyan Xu and Jack R. Wands
*Departments of Medicine, Pathology, and Neurology, Divisions of Gastroenterology and Endocrinology, and the Liver Research Center, Rhode Island Hospital and the Warren Alpert Medical School of Brown University, Providence, RI*

Abstract: *Background*: Obesity, type 2 diabetes mellitus (T2DM), and non-alcoholic steatohepatitis (NASH) can lead to cognitive impairment and neurodegeneration. Experimental high fat diet (HFD) induced obesity with T2DM causes neurodegeneration with brain insulin resistance. *Objective*: Since ceramides are neurotoxic, cause insulin resistance, and are increased in T2DM, we investigated their potential role in neurodegeneration. *Methods*: C57BL/6 mice were pair-fed HFD or control diets for 4-20 weeks. Pro-ceramide genes and biochemical indices of neurodegeneration were measured. In vitro experiments directly examined neurodegenerative effects of ceramides. *Results*: Chronic HFD feeding gradually increased body weight, but after 16 weeks, liver weight surged ($P<0.001$) due to triglyceride accumulation ($P<0.001$), and brain weight declined ($P<0.0001$). HFD increased pro-ceramide gene expression in liver ($P<0.05$-$P<0.001$), but not brain. Temporal lobes of HFD fed mice had increased ubiquitin ($P<0.001$) and 4-hydroxynonenal ($P<0.05$ or $P<0.01$), and decreased tau, β-actin, and choline acetyltransferase levels ($P<0.05$-$P<0.001$) with development of NASH. Ceramide treatment of neuronal cultures caused cell death, oxidative stress, mitochondrial dysfunction, and insulin resistance. *Conclusions*: In obesity, T2DM, or NASH, excess hepatic production of neurotoxic ceramides that readily cross the blood-brain barrier causes cognitive impairment with brain insulin resistance via a liver-brain axis of neurodegeneration.

Keywords: Obesity; diabetes mellitus; insulin resistance; neurodegeneration; high fat diet; Alzheimer; amyloid; non-alcoholic steatohepatitis

## INTRODUCTION

The prevalence rates of Alzheimer's disease (AD), obesity, type 2 diabetes mellitus (T2DM), and non-alcoholic fatty liver disease (NAFLD)/non alcoholic steatohepatitis (NASH), which includes metabolic syndrome, have all increased dramatically over the past several decades [1-6]. The probable interrelatedness among these diseases is suggested by studies demonstrating: 1) increased risk of developing mild cognitive impairment (MCI), dementia, or AD in individuals with T2DM [7, 8] or obesity/dyslipidemic disorders [9]; 2) progressive brain insulin resistance and insulin deficiency in AD [10-13]; 3) cognitive impairment in experimental animal models of T2DM and/or obesity [14, 15]; 4) AD-type neurodegeneration and cognitive impairment in experimentally induced brain insulin resistance and

---

Research supported by AA02666, AA02169, AA11431, AA12908, and AA16126 from the National Institutes of Health, and a Child Health Research Grant from the Hood Foundation.
*These authors contributed equally to this work.
Address for correspondence: Dr. Suzanne M. de la Monte, Rhode Island Hospital, 55 Claverick Street, 4th Floor, Providence, RI 02903. Tel.: 401-444-7364; Fax: 401-444-2939; E-mail: Suzanne_DeLaMonte_MD@Brown.edu.

insulin deficiency [16-20]; 5) improved cognitive performance in experimental models of AD [21] and in human subjects with AD or MCI after treatment with insulin sensitizer agents or intranasal insulin [22-27]; and 6) similar molecular, biochemical, and mechanistic abnormalities in T2DM, NASH, and AD [7, 28-33]. Since obesity, MCI, AD, T2DM, and NASH are all associated with insulin resistance, i.e. impaired ability to respond to insulin stimulation, they may share common etiologies. On the other hand, the lack of complete overlap among these disease states suggests that specific organ systems may be differentially afflicted by the same or similar exposures, resulting in dissimilar degrees of insulin resistance with disparate long-term outcomes, e.g. neurodegeneration versus NASH.

Approximately 24 million people worldwide have dementia, and if current trends continue, it is expected that AD prevalence rates will nearly double every 20 years [34]. Moreover, since the rate of MCI progression to dementia is about 3.8/100 person-years, with 60% eventually developing AD [35], the incidence rates of AD are also likely to continue climbing. Consequently, AD and other forms of dementia are set to overwhelm our health care system. While aging is clearly the strongest risk factor for AD, emerging data suggest that T2DM and dyslipidemic states can either contribute to, or serve as co-factors in its pathogenesis [34]. This concept is supported by epidemiologic data demonstrating a significant association between T2DM and MCI or dementia, and that T2DM is a significant risk factor for developing AD [7, 36-40].

Mechanistically, increased risk of dementia in T2DM and obesity could be linked to chronic hyperglycemia, insulin resistance, oxidative stress, accumulation of advanced glycation end-products, increased production of pro-inflammatory cytokines, and/or microvascular disease [36]. However, most of these features define the core abnormalities in T2DM and NASH [41-44]. This concept led us to hypothesize that toxic/injurious agents produced in the body and associated with increased BMI, mediate similar adverse effects in different target organs and tissues. Injurious agents or toxins causing insulin resistance in adipose tissue and skeletal muscle would result in T2DM, whereas the same insult in liver would produce NASH/metabolic syndrome, and in brain, MCI or early AD-type neurodegeneration. We now propose that ceramides and related molecules, are critical agents involved in the pathogenesis of these disease states because they: 1) can be generated in liver, adipose tissue, or brain [45-48]; 2) cause insulin resistance [46]; 3) are cytotoxic [46]; 4) increase in the CNS with various dementia-associated diseases, including AD [47, 49-51]; and 5) are lipid soluble and therefore likely to readily cross the blood-brain barrier.

Ceramides comprise a family of lipids generated from fatty acid and sphingosine [46, 52, 53]. Ceramides are distributed in cell membranes, and in addition to their structural functions, they regulate intracellular signaling pathways that mediate growth, proliferation, motility, adhesion, differentiation, senescence, and apoptosis. Ceramides are generated by de novo biosynthesis through the actions of ceramide synthases and serine palmitoyltransferase [45, 48, 54]. In addition, ceramides can be generated by sphingolipid catabolism through activation of neutral or acidic sphingomyelinases [45, 53], or through degradation of complex sphingolipids and glycosphingolipids localized in late endosomes and lysosomes (Table 1) [52]. Ceramides are metabolized by ceramidases to form sphingosine, which is converted to sphingosine 1-phosphate by spingosine kinase. Ceramide, sphingosine, and sphingosine-1-phosphate have all been implicated in the pathogenesis of obesity and insulin resistance [45]. Correspondingly, inhibition of ceramide synthesis or its accumulation prevents obesity-mediated insulin resistance [55, 56].

Complex sphingolipids including gangliosides [57], and long-chain naturally occurring ceramides i.e. up to 24 carbon atoms in length, [58] positively influence cell growth and functions, whereas sphingosine-containing lipids, including shorter ceramides, have inhibitory effects on cells resulting in increased apoptosis, cytotoxicity, or impaired growth [57, 59, 60]. Sphingomyelinases are activated by pro-inflammatory cytokines, e.g. TNF-α [46], and pro-apoptotic stimuli including Fas, trophic factor withdrawal, and ionizing radiation [52, 53]. Ceramides adversely alter cellular function and cause apoptosis by modulating phosphorylation states of proteins, including those that regulate insulin signaling [61], activating enzymes such as interleukin-1β converting enzyme (ICE)-like proteases, which promote apoptosis [52], or inhibiting Akt phosphorylation and kinase activity [62] through activation of protein phosphatase 2A [63].

Potential roles for ceramides in diabetes, obesity, inflammation, NASH, and neurodegeneration have already been suggested [46, 47, 50, 64]. In obesity, adipose tissue, skeletal muscle, and liver exhibit

Table 1
Ceramide related genes and their functions

| Gene | Abbreviation/ Synonym | Molecular pathway | Effect/References |
|---|---|---|---|
| Serine palmitoyltransferase | SPTLC:<br>1. SPTLC-1: subunit 1 (non catalytic)<br>2. SPTLC-2: Subunit 2 (catalytic) | Rate limiting step in *de novo* ceramide synthesis from condensation of serine and palmitoyl-CoA | References: #44, #45, #54 |
| Sphingomyelin phosphodiesterase 1, acid lysosomal | Smpd1, aSMase, A-SMase | Ceramide generation from acid sphingomyelinase–dependent hydrolysis of sphingomyelin | References: #44, #51 |
| Sphingomyelin phosphodiesterase 3, neutral | Smpd3, nSMase2 | Ceramide generation from neutral sphingomyelinase–dependent hydrolysis of sphingomyelin | References: #44, #51 |
| Ceramide synthase 1 | LAG1 homolog, ceramide synthase 1 (Lass1) | *De novo* synthesis from N-acylation of spingosine | Synthesis of C18-ceramide #47 |
| Ceramide synthase 2 | LAG1 homolog, ceramide synthase 2 (Lass2) | *De novo* synthesis from N-acylation of spingosine | Synthesis of C20-C26 ceramide. Reference: #47 |
| Ceramide synthase 4 | LAG1 homolog, ceramide synthase 4 (Lass4) | *De novo* synthesis from N-acylation of spingosine | Synthesis of C18, C20, C24 References: #47, #53 |
| Ceramide synthase 5 | LAG1 homolog, ceramide synthase 5 (Lass5), CerS5 | *De novo* synthesis from N-acylation of spingosine | Synthesis of C16-ceramide References: #47, #44 |
| UDP-glucose ceramide glucosyltransferase | UGCG | First step in glycosphingolipid synthesis, the transfer of glucose from UDP-glucose to ceramide. | Synthesis of glucosylceramide References: #56, #57 |

major abnormalities in sphingolipid metabolism that result in increased ceramide production, inflammation, and activation of pro-inflammatory cytokines, and impairments in glucose homeostasis and insulin responsiveness [45, 46, 65, 66]. In both humans with NASH [67], and the C57BL/6 mouse model of diet-induced obesity with T2DM and NASH [68], ceramide levels in adipose tissue are elevated due to increased activation of serine palmitoyl transferase, and acidic and neutral sphingomyelinases [52]. Of note is that pro-inflammatory cytokines, such as tumor necrosis factor-alpha (TNF-α), can induce ceramide synthesis [46], and TNF-α levels are increased in both T2DM and NASH [69-72]. We investigated the potential role of ceramides as mediators of neurodegeneration and brain insulin resistance utilizing an in vivo model of chronic obesity/T2DM [73]. In addition, we extended the analysis by directly investigating the role of exogenous ceramide exposure in the pathogenesis of neurodegeneration and neuronal insulin resistance using an in vitro model.

## MATERIALS AND METHODS

### Obesity/T2DM in vivo model

Harlan adult male C57BL/6 mice (N=10 per group), starting at 4 weeks of age, were pair-fed for 4, 8, 12, 16, or 20 weeks with high fat chow diets in which 60% of the calories were derived from fat (Research Diets Inc, New Brunswick, NJ), or low fat chow diets in which 5% of the calories were from fat (Harlan, Indianapolis, IN) as reported previously [73, 74]. Mice were weighed weekly, and at the time of sacrifice, fresh liver and brain weights were obtained. Livers and brains were sectioned for snap freezing and immersion fixation [73]. Fixed tissues were embedded in paraffin, and histological sections (8 microns) were stained with Hematoxylin and eosin (H&E; liver) or Luxol fast blue, H&E (brain). Frozen tissue was stored at -80°C for biochemical and molecular studies. Our experimental protocol was approved by the Institutional Animal Care and Use Committee at the Lifespan-Rhode Island Hospital and conforms to guidelines established by the National Institutes of Health.

*In vitro ceramide exposure model*

Subconfluent cultures of human primitive neuroectodermal tumor 2 (PNET2), CNS-derived neuronal cells [75, 76] were treated with 0.25-125.8 uM of ceramide analogs, D-erythro-N-Acetyl-Sphingosine (C2 Cer), D-erythro- N-Hexanoyl-Sphingosine (C6 Cer), or D-erythro-Dihydro-N-Acetyl-Sphingosine (C2 Dihydroceramide; C2D). C2D, a structurally similar, inactive analog of C2 Cer, was used as a negative control. Vehicle-treated cells were simultaneously studied. After 48 h of treatment, cells grown in 96-well cultures were analyzed for viability and ATP content using the CyQuant (Molecular Probes, Eugene, OR) and ATP-Lite (Perkin-Elmer, Boston, MA) assays according to the manufacturers' protocols. Cells grown in 6-well cultures were used to measure gene expression by qRT-PCR or by direct binding enzyme-linked immunosorbent assays (ELISAs).

*Quantitative reverse transcriptase polymerase chain reaction (qRT-PCR) assay of gene expression*

We used qRT-PCR to measure pro-ceramide gene expression with previously described methods [16]. Briefly, total RNA isolated from liver and temporal lobe was reverse transcribed using random oligodeoxynucleotide primers. The resulting cDNA templates were used in qPCR amplification reactions with gene specific primer pairs and QuantiTect SYBR Green PCR Mix (Qiagen Inc, Valencia, CA) [16]. The amplified signals were detected and analyzed using the Mastercycler ep realplex instrument and software (Eppendorf AG, Hamburg, Germany). Relative mRNA abundance was determined from the ng ratios of specific mRNA to 18S [77, 78].

*Enzyme-linked immunosorbent assay (ELISA)*

Temporal lobes were homogenized in radioimmunoprecipitation assay buffer with protease and phosphatase inhibitors [16]. Protein concentrations were determined using the bicinchoninic acid (BCA) assay (Pierce, Rockford, IL). We used direct ELISAs to measure Tau, phospho-Tau, 4-hydroxy-2-nonenal (HNE), ubiquitin, β-actin, and choline acetyltransferase (ChAT) expression as previously described [79]. Immunoreactivity was detected with horseradish peroxidase (HRP)-conjugated secondary antibody and Amplex Red soluble fluorophore [79]. Fluorescence was measured (Ex 579/Em 595) in a SpectraMax M5 microplate reader (Molecular Devices Corp., Sunnyvale, CA). Parallel negative control assays included incubations in which the primary, secondary, or both antibodies were omitted.

*Lipid assays*

Chloroform-methanol lipid extracts of fresh frozen liver tissue were quantified using the Nile Red assay (Molecular Probes, Eugene, OR) [80-82]. Fluorescence intensity (Ex 485/ Em 572) was measured in a SpectraMax M5 microplate reader. Triglyceride levels were measured using a Serum Triglyceride Determination kit (Sigma-Aldrich Co., St. Louis, MO), and cholesterol was measured using the Amplex Red Cholesterol Assay Kit (Molecular Probes, Eugene, OR) according to the manufacturers' protocols.

*Source of reagents*

QuantiTectSYBR Green PCR Mix was obtained from (Qiagen Inc, Valencia, CA). Rabbit or goat generated monoclonal or polyclonal antibodies to ubiquitin, tau, phospho-tau, HNE, ChAT, and β-actin were purchased from Chemicon (Temecula, CA), CalBiochem (Carlsbad, CA) or Molecular Probes (Eugene, OR). Secondary antibodies were purchased from Pierce Chemical Co. (Rockford, IL). Amplex Red reagent was obtained from Molecular Probes (Eugene, OR). All other fine chemicals were purchased from either CalBiochem (Carlsbad, CA) or Sigma-Aldrich (St. Louis, MO).

*Statistical analysis*

Data depicted in the graphs represent the means ± S.E.M.'s for each group. Inter-group comparisons were made using Two-way Analysis of Variance (ANOVA) with the Bonferroni post-hoc test, or one-way ANOVA with the Dunn's multiple comparison test or the post-hoc Deming test for linear trend. Statistical analyses were performed using GraphPad Prism 5 (GraphPad Software, Inc., San Diego, CA). The computer software generated significant P-values are indicated within the graph panels.

RESULTS

*Longitudinal effects of the HFD on body, liver, and brain weights*

The chronic HFD fed mice developed obesity and T2DM associated with fasting hyperglycemia, hyperlipidemia, and increased serum pro-inflammatory cytokine levels as previously reported [73, 74]. Chronic HFD feeding caused progressive increases in mean body weight such that after 8 weeks, the HFD-fed mice were significantly heavier than the LFD control group. With increasing duration of HFD feeding, mean body weight continued to climb ($P<0.0001$ for linear trend), whereas the mean body weight of the LFD group remained relatively stable (Figure 1A), consistent with previous findings in this model [83]. During the first 12 weeks of study, HFD feeding had no significant effect on mean liver weight. Over that interval, the mean liver/body weight ratios were significantly lower in the HFD relative to LFD control group due to their progressive increases in body weight (data not shown). In contrast, at the 16 and 20 week time points, HFD-fed mice exhibited striking and significant increases in mean liver weight (Figure 1B) due to significant increases in hepatic lipid content as demonstrated with the Nile Red assay (Figure 1C). Correspondingly, the calculated mean liver-body weight ratios were also increased relative to control (data not shown). Biochemical assays detected significantly elevated mean hepatic triglyceride content after 4, 16, and 20 weeks of HFD feeding (Figure 1D), and significantly increased hepatic cholesterol after 20 weeks on the HFD (Figure 1E).

Although mean brain weights were similar in the HFD and LFD groups after 4, 8, 12, 16, or 20 weeks of paired feeding (Figure 1F), a time-dependent trend of declining mean brain weight was observed in the HFD ($R^2=0.475$; $P<0.0001$), but not the LFD group (Figure 1G, H). The combined effects of increasing body mass and declining brain weight resulted in significantly reduced mean brain/ body mass ratios in the HFD relative to control at the 8, 12, 16, and 20 week time points (Figure 1I). The time dependent trend of declining brain weight/body weight ratio was also statistically significant ($P<0.0001$). Finally, the sharpest reductions in brain/body weight ratio occurred concurrently with the striking increases in hepatic steatosis.

*Chronic HFD feeding causes non-alcoholic steatohepatitis (NASH)*

LFD livers exhibited regular cord-like architecture with minimal or no evidence of inflammation, steatosis, or apoptosis (Figure 2). In contrast, after 4, 8, or 12 weeks on the HFD, the livers had progressive increases in hepatic micro- and macrosteatosis with associated disorganization of the architecture (Figure 2B-D-F-H-J-L). After 16 weeks of HFD feeding, the livers exhibited more prominent macro- and microsteatosis with small cluster of lymphomononuclear inflammatory cells (Figure 2N-P). After 20 weeks on the HFD, the livers had established histopathological features of NASH, consisting of widespread micro- and macro-steatosis in hepatocytes (40-50% versus 5-10% at earlier time points), patchy lymphomononuclear cell inflammation, apoptosis, and necrosis (Figure 2R-T), corresponding with previous descriptions of this and closely related models [68, 84, 85].

*Neurodegeneration after chronic HFD feeding*

Histopathological studies were focused on analysis of the temporal lobes because they are major targets of AD neurodegeneration. 4, 8, or 12 weeks of HFD feeding produced no detectable effects on brain histopathology. However, 16 weeks of HFD feeding caused subtle histopathological abnormalities consisting of scattered foci of neuronal loss, apoptosis, or nuclear pyknosis, increased irregularity of neuropil spacing among neurons (consistent with cell loss), and mildly increased white matter gliosis, as previously described [73]. Immunohistochemical staining studies failed to detect neurofibrillary tangles, dystrophic neuritis, senile plaques, or APP-Aβ deposits in plaques or vessels in any of the brains, corresponding with our previous observations [73]. After 20 weeks of HFD feeding, on-going apoptosis in both the hip-

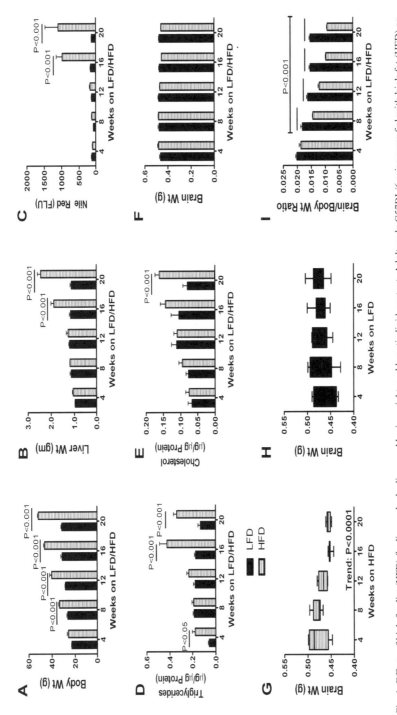

Fig. 1. Effects of high fat diet (HFD) feeding on body, liver, and brain weight, and hepatic lipid content. Adult male C57BL/6 mice were fed with high fat (HFD) or low fat (LFD) chow diets for 4, 8, 12, 16, or 20 weeks (N=10/group). Bar graphs or box plots depict mean ± range and/or S.E.M. for (A) body weight, (B) liver weight, (C) Nile Red fluorescence measurement of liver lipid content, (D) hepatic triglyceride levels, (E) hepatic cholesterol levels, (F) brain weight, (G, H) brain weight trend over time for mice fed with the (G) HFD or (H) LFD, and (I) the calculated brain/body weight ratio for each group at the time of sacrifice. Inter-group comparisons were made using Two-way ANOVA with the post-hoc Bonferroni test of significance, or within-group trend analysis. Significant P-values and trends are indicated within the panels.

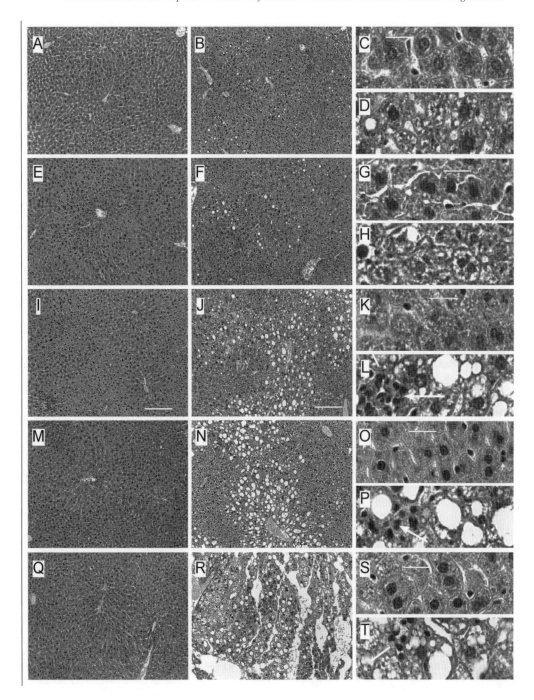

Fig. 2. Chronic HFD feeding causes NASH. Adult male C57BL/6 mice were fed with high fat (HFD) or low fat (LFD) chow for 4, 8, 12, 16, or 20 weeks (N=10/group). Liver tissue harvested at sacrifice was fixed in Histochoice, embedded in paraffin, and histological sections were stained with H&E. Photomicrographs depict representative areas of liver from mice maintained for 4 (A-D), 8 (E-H), 12 (I-L), 16 (M-P) or 20 (Q-T) weeks on the LFD (A,C,E,G,I,K,M,O,Q,S) or HFD (B,D,F,H,J,L,N,P,R,T). Over time, mice fed with the LFD exhibited no detectable change in liver histology, whereas mice fed with the HFD exhibited progressive increases in steatosis (marked by clear intracellular vacuoles) that was detectable after just 4 weeks of HFD feeding (D), and associated with inflammation (arrows) and cell loss (reduced hepatocyte nuclear density) beginning at the 12 week time point. Note extensive macro- and microsteatosis with reduced hepatocellular nuclei in Panels P and T relative to O and S. Original magnifications, A,B,E,F,I,J,M,N,Q,R= 200x; C,D,G,H,K,L,O,P,S,T=1200x. Scale bar=50 μm.

pocampal formation and temporal neocortex was more conspicuous, as evidenced by cell loss, and increased lipid peroxidation (HNE immunoreactivity), and gliosis (GFAP immunoreactivity) in these regions (Figure 3). In addition, temporal white matter in HFD fed mice exhibited increased gliosis and lipid peroxidation, reflecting fiber degeneration (Figure 3), as occurs early in the course of AD [86].

*Effects of HFD on ceramide gene expression in liver and brain*

Exploratory studies demonstrated that normal liver expresses Ceramide synthases (CER) 1, 2, 4, and 5, UDP glucose ceramide glucosyltransferase (UGCG), serine palmitoyltransferase (SPTLC) 1 and SPTLC2, and sphingomyelin phosphodiesterase (SMPD) 1 and SMPD3 (See Table 1 for definitions). QRT-PCR analysis revealed significantly higher mean levels of CER2, CER4, SPTLC1, SPTLC2, SMPD1, and SMPD3 mRNA transcripts in HFD compared with LFD livers, after 12 weeks on the diets (Figure 4). However, the pro-ceramide synthesis mRNA transcripts, i.e. CER2, CER4, SPTLC1, and SPTLC2, were increased mainly at the 12- and 16-week time points, just preceding or coinciding with the early stages of conspicuous hepatic steatohepatitis. Their expression returned to control levels (4-week time point) after 20 weeks on the HFD. UGCG mRNA levels increased in both groups between the 4- and 8-week time points, and then gradually thereafter in the control group, in which peak levels were detected at the 16- and 20-week time points. Mice fed with the HFD had earlier peaks in hepatic UGCG (at 12 or 16 weeks), and significantly lower levels than control at the 20-week time point (Figure 4E). In HFD-fed mice, SMPD1 and SMPD3 mRNA transcripts increased progressively over time, such that their mean levels were significantly higher than control at both the 16 and 20 week time points (Figure 4). In contrast to CER and SPTLC, SMPD1 and SMPD3 generate ceramide through hydrolysis of spingomyelin [45, 52, 53]. Corresponding with the qRT-PCR results, preliminary ELISA studies demonstrated higher levels of ceramide immunoreactivity in HFD relative to LFD fed mice at the 12- (21016 ± 1730 versus 16759 ± 208; P=0.025) and 16-week (15636 ± 472 versus 14151 ± 160; P=0.037) time points (Mean ± S.E.M. values in arbitrary units; Student T-tests with 4 samples per group). A more detailed analysis of HFD-induced alterations in ceramide levels and characteristics in liver and brain is currently underway and will be presented in a future report.

We also used qRT-PCR to measure expression of the same pro-ceramide genes in temporal lobe samples (Figure 5). Generally higher levels of CER2, CER5, UGCG, SPTLC2, SMPD1 and SMPD3 were detected in brain at earlier compared to later time points in the study. We interpreted this result to possibly reflect developmental or aging effects on the expression levels of these genes. At the 8 and/or 12 week time points, CER2, UGCG, SPTL C2, SMPD1 and SMPD3 were all expressed at significantly lower levels in HFD compared with LFD brains. In addition, at the 16 and/or 20 week time points, the mean levels of UGCG mRNA were also significantly lower in the HFD compared with LFD group (Figure 5). Therefore, unlike liver, temporal lobe ceramide gene expression was not increased by chronic HFD feeding.

*Effect of obesity/T2DM on molecular and biochemical indices of AD*

AD-type neurodegeneration is associated with increased levels of APP, oxidative stress, ubiquitin, and phospho-Tau, and reduced levels of Tau and choline acetyltransferase (ChAT) mRNA [11]. ELISAs performed with temporal lobe tissue demonstrated similar mean levels of tau, phospho-tau, β-actin, ubiquitin, and 4-HNE in the HFD and LFD groups at the 8 and 12 week time points (Figure 6). Only ChAT immunoreactivity was significantly reduced in the HFD group during these early time periods (Figure 6D). At the 16- and/or 20-week time points, the mean levels of tau, β-actin, and ChAT were significantly reduced, whereas the mean levels of ubiquitin and 4-HNE were significantly increased in temporal lobes of HFD fed relative to control mice (Figure 6).

*Ceramide impairs neuronal viability and energy metabolism*

PNET2 human neuronal cells were treated with vehicle, or 0.25-125.8 uM C2 ceramide, C6 ceramide or C2D inactive ceramide for 48 hours. The CyQuant assay was used to measure neuronal viability, and the ATPLite assay was used to measure ATP content as an index of energy metabolism. 7.9 uM or higher concentrations of C2 ceramide proved to be highly toxic, resulting in 50% cell loss

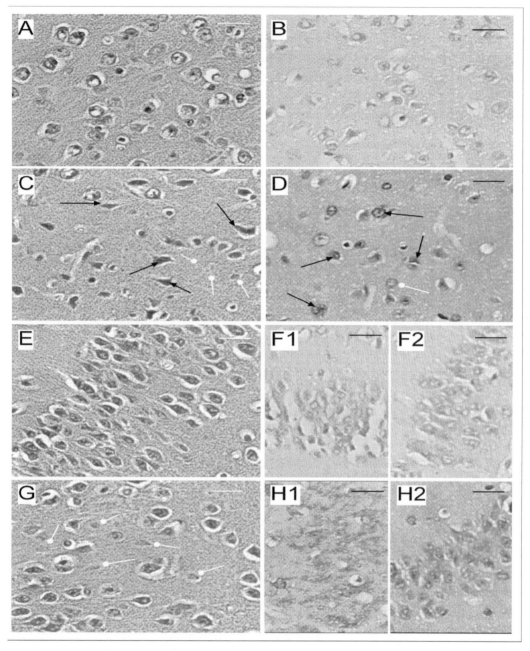

Fig. 3. Neurodegeneration in diet Induced obesity with T2DM and NASH. Adult male C57BL/6 mice were fed with low fat (LFD-A,B,E,F1,F2) or high fat (HFD-C,D,G,H1,H2) chow for 20 weeks (N=10/group). Brains harvested at sacrifice was immersion fixed in Histochoice, embedded in paraffin, and adjacent histological sections including (A-D) temporal cortex and (E-H) hippocampal formation (CA1), were either stained with (A,C,E,G) Luxol Fast Blue, H&E, or immunostained with antibodies to (B,D,F2,H2) 4-hydroxynonenol (HNE) or (F1,H1) glial fibrillary acidic protein (GFAP) to detect lipid peroxidation/oxidative stress or gliosis, respectively. Immunoreactivity was detected by the ABC method with diaminobenzidine as the chromogen. Immunostained sections were lightly counterstained with hematoxylin. HFD fed mice exhibited reduced cell density with ongoing neuronal shrinkage (black arrows) and apoptosis (white wands) in the (C) temporal cortex and (G) hippocampal formation, relative to corresponding brain regions in (A,E) LFD fed mice. Ongoing cell loss was associated with increased HNE (arrows) and GFAP immunoreactivity in the temporal cortex and/or hippocampal formation in (D,H) HFD relative to (B,F) LFD fed mice. Original magnifications, 600x for all images. Scale bar=50 μm.

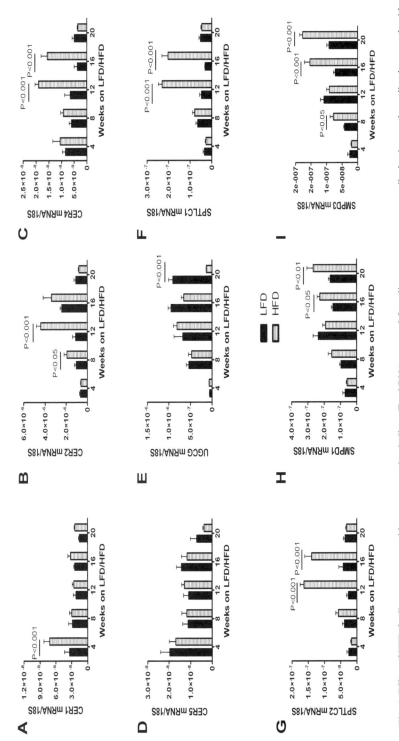

Fig. 4. Effect of HFD feeding on pro-ceramide gene expression in liver. Total RNA extracted from liver was reverse transcribed using random oligodeoxynucleotide primers, and the resulting cDNA templates were used in qRT-PCR assays to measure (A) Ceramide synthase (CER)1, (B) CER2, (C) CER4, (D) CER5, (E) UDP-glucose ceramide glycoysltransferase (UGCG), (F) Serine palmitoyltransferase 1 (SPTLC1), (G) SPTLC2, (H) sphingomyelin phosphodiesterase 1 (SMPD1), and (I) SMPD3. The mRNA levels were normalized to 18S rRNA measured in the same templates. Graphs depict the mean ± S.E.M. levels of gene expression in livers from LFD-fed or HFD-fed mice (N=6/group; see Material and Methods). Inter-group comparisons were made using Two-way ANOVA with the post-hoc Bonferroni test of significance. Significant P-values are indicated within the panels.

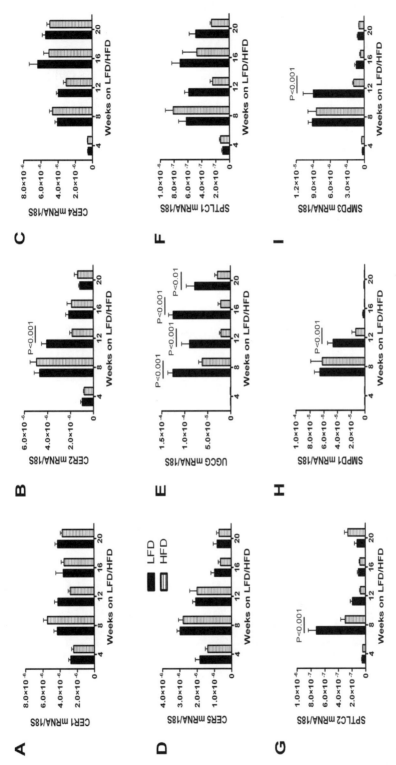

Fig. 5. Effect of HFD feeding on pro-ceramide gene expression in temporal lobe. Total RNA extracted from liver was reverse transcribed using random oligodeoxynucleotide primers, and the resulting cDNA templates were used in qRT-PCR assays to measure (A) Ceramide synthase (CER)1, (B) CER2, (C) CER4, (D) CER5, (E) UDP-glucose ceramide glycoysltransferase (UGCG), (F) Serine palmitoyltransferase 1 (SPTLC1), (G) SPTLC2, (H) sphingomyelin phosphodiesterase 1 (SMPD1), and (I) SMPD3. The mRNA levels were normalized to 18S rRNA measured in the same templates. Graphs depict the mean ± S.E.M. levels of gene expression in brains from LFD-fed or HFD-fed mice (N=6 per group; see Material and Methods). Inter-group comparisons were made using Two-way ANOVA with the post-hoc Bonferroni test of significance. Significant P-values are indicated within the panels.

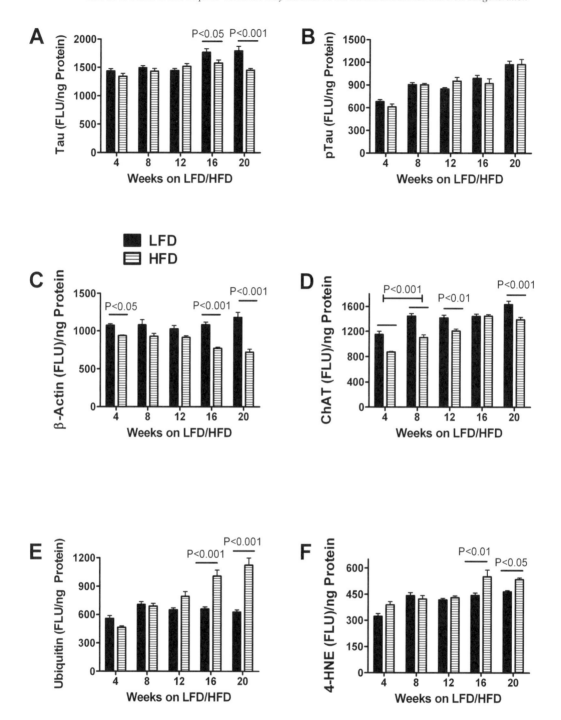

Fig. 6. Effect of HFD feeding on molecular indices of neurodegeneration. Temporal lobe protein homogenates from LFD-fed or HFD-fed mice were used to measure (A) Tau; (B) phospho (p)-Tau; (C) β-actin; (D) choline acetyltransferase (ChAT); (E) ubiquitin; and (F) 4-hydroxynonenal (4-HNE) by ELISA (see Materials and Methods). Immunoreactivity was detected with HRP-conjugated secondary antibody and Amplex Red soluble fluorophore. Fluorescence light units (FLU) were measured (Ex 579 nm/Em 595 nm) in a Spectromax M5, and results were normalized to sample protein content in the wells. Graphs depict mean ± S.E.M of results (N=8/group). Inter-group comparisons were made using Two-way ANOVA with the post-hoc Bonferroni test of significance. Significant P-values are indicated within the panels.

relative to C6 and C2D ceramide treatments [87]. In contrast, cell viability remained relative stable in cultures treated with up to 125.8 uM C6 ceramide. Mean ATP content was similar to control in cultures treated with 0.25-15.7 uM ceramide, but at higher concentrations, ATP content declined sharply in both C2 and C6 relative to C2D ceramide treated cultures [87]. Further studies were performed using 15.7 uM ceramide treatments for 48 hours.

*Ceramide impairs neuronal insulin, IGF and IRS signaling mechanisms*

C6 ceramide exposure significantly increased the mean levels of insulin, insulin receptor, IGF-1 receptor, and IRS-4, and decreased the mean levels of IGF-2 receptor and IRS-1 mRNAas demonstrated by qRT-PCR analysis (Figure 7). Although IGF-2 mRNA levels were also increased in C6 ceramide treated cells, the difference from control and C2D did not reach statistical significance (P=0.08). The effects of C2 ceramide treatment were modest relative to C6, with the major effects being reduced IRS-1 and increased IRS-2 mRNA expression relative to vehicle-treated control cells. Treatment with C2D inactive ceramide had no significant effects on the expression of insulin, IGF, or IRS genes relative to vehicle-treated controls (Figure 7).

*Ceramide induces AD-type molecular indices of neurodegeneration*

Further studies were used to determine if ceramide-induced alterations in insulin/IGF signaling mechanisms were associated with AD-type abnormalities in gene expression. QRT-PCR analysis revealed significantly increased mRNA levels of APP, and reduced levels of ChAT and GAPDH (insulin-responsive gene) in C6 ceramide treated relative to control cells (Figure 8). C2 ceramide treatment also significantly inhibited ChAT expression relative to vehicle-treated control cells, but it did not increase APP or reduce GAPDH expression. Tau mRNAwas not significantly altered by C2 or C6 ceramide, and C2D inactive ceramide had no significant effects on Tau, APP, ChAT, or GAPDH relative to vehicle-treatment.

We used ELISAs to examine the effects of C6 ceramide on neuronal indices of oxidative stress, acetylcholine homeostasis, cytoskeletal integrity, and insulin-responsive gene expression. The effects of C2 ceramide were not investigated here due to its relatively modest effects compared with C6 in all other studies. Post-hoc statistical comparisons were made with respect to the vehicle-treated control group. C6 ceramide exposure significantly increased the mean levels of 4-HNE, ubiquitin and AChE, and reduced the mean levels of GAPDH, β-Actin and ChAT [87]. In contrast, treatment with C2D inactive ceramide had minimal effects on the expression levels of these proteins, except for modestly increased levels of AChE.

## DISCUSSION

Previously, we demonstrated that chronic HFD feeding, which results in obesity and T2DM, causes relatively modest AD-type molecular and biochemical abnormalities in brain, including insulin resistance [73]. Subsequently, we re-generated the model to examine the time course of altered gene expression in brain, as well as uncover potential mechanisms by which peripheral insulin resistance causes AD-type neurodegeneration [88]. This line of investigation is important because it could lead to the discovery of biomarkers for detecting individuals at risk for developing cognitive impairment or progressing from MCI to AD in the context of obesity/T2DM. We focused our attention on the liver because previous studies showed that: 1) NASH occurs frequently with obesity/T2DM and is a feature of the model used herein; 2) NASH is associated with hepatic insulin resistance; 3) individuals with NASH can exhibit neuropsychiatric dysfunction, including anxiety and depression [89], which frequently precede cognitive impairment and dementia; 4) peripheral ceramide production is increased in adipose tissue in NASH, and ceramides, as well as long-chain fatty acids, mediate insulin resistance [46, 55, 90]; and 5) ceramide can be neurotoxic, and its levels are increased in AD as well as other injury or inflammatory disease states [47, 49-51, 62, 91].

This study demonstrated that chronic HFD feeding was sufficient to cause obesity with T2DM, including hyperglycemia, hyperinsulinemia, and hypercholesterolemia as previously reported [74, 92]. In contrast, liver weight remained relatively stable until the 16th week of HFD feeding when the liver weights increased sharply in association with increased hepatic lipid content (steatosis), inflammation, apoptosis, and necrosis, i.e. NASH. Other investigators demonstrated that NASH, in this or related models,

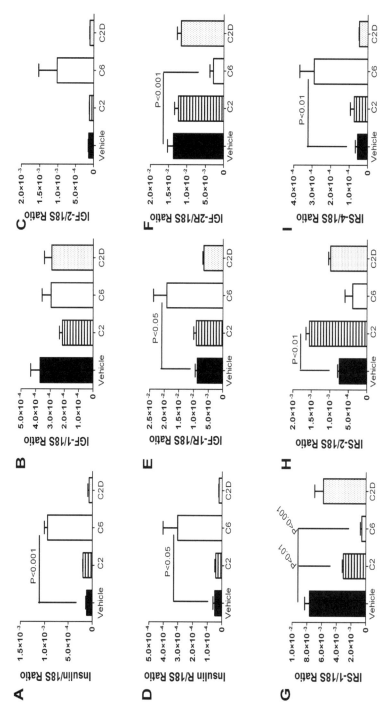

Fig. 7. Ceramide treatment alters neuronal insulin/IGF signaling mechanisms. Total RNA extracted from PNET2 cells treated with vehicle or 62.9–72.8 μM C6, C2, or C2D (inactive) ceramide for 48 h, was reverse transcribed using random oligodeoxynucleotide primers, and the cDNA templates were used in qRT-PCR assays to measure expression of (A) insulin, (B) IGF-I, (C) IGF-II, (D) insulin receptor (R), (E) IGF-I R, (F) IGF-II R, (G) IRS-1, (H) IRS-2, and (I) IRS-4, with results normalized to 18S rRNA. Graphs depict the mean ± S.E.M. levels of gene expression in each group. Inter-group comparisons were made using one-way ANOVA with post-hoc Dunn's multiple comparison test of significance. Significant P-values are indicated within the panels.

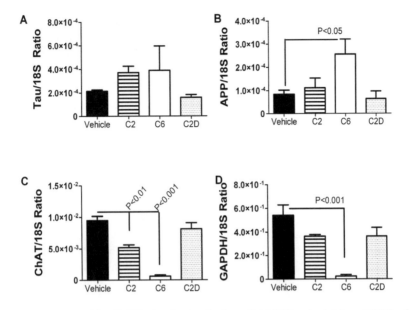

Fig. 8. Ceramide treatment alters neuronal insulin/IGF signaling mechanisms. Total RNA extracted from PNET2 cells treated with vehicle or 15.7μM C6, C2, or C2D ceramide for 48 h, was reverse transcribed using random oligodeoxynucleotide primers, and the cDNA templates were used in qRT-PCR assays to measure expression of AD-associated genes including, (A) tau, B) amyloid precursor protein-APP, C) choline acetyl-transferase-ChAT, and D) glyceraldehyde-3-phosphate dehydrogenase-GAPDH (which is insulin responsive). Results were normalized to 18S rRNA measured in parallel reactions. Graphs depict the mean ± S.E.M. levels of gene expression in each group. Inter-group comparisons were made using one-way ANOVA with post-hoc Dunn's multiple comparison test of significance relative to vehicle-treated controls. Significant P-values are indicated within the panels.

is associated with increased serum transaminase levels, reflecting hepatocellular injury [68, 85, 93]. Therefore, during the earlier phases of HFD feeding, compensatory mechanisms help sustain structural and functional integrity of the liver, but after 16 weeks of HFD feeding, a critical metabolic perturbation disrupts the homeostatic milieu, allowing hepatic steatosis to progress to NASH. One potential mediator of this response is the abrupt surge in serum tumor necrosis factor alpha (TNF-α) that occurs after 16 weeks of HFD feeding [94]. TNF-α is a potent pro-inflammatory cytokine that plays a key role in peripheral and hepatic insulin resistance [70, 72, 95, 96], and also promotes ceramide biosynthesis [46]. Although a potential source of the increased TNF-α in this model is peripheral adipose tissue with on-going adipocyte death and remodeling [94], preliminary data suggest that endogenous hepatic TNF-α expression is increased in the HFD mice (Longato, et al, unpublished). After 16 weeks of HFD feeding, the livers had histopathological features of NASH, i.e. steatosis with multiple foci of lymphomononuclear cell inflammation, necrosis, and apoptosis, consistent with previous reports [85]. Molecular studies demonstrated that throughout the period of HFD feeding, hepatic mRNA levels of various pro-ceramide genes were significantly increased relative to control, including genes responsible for generating ceramide de novo (CER 2,3,4,5), and also those involved in sphingomyelin degradation (SMPD1, and SMPD3). The gradual reductions in the expression of genes that mediate de novo ceramide synthesis could represent a compensatory response to lipid accumulation in hepatocytes. On the other hand, the persistently increased levels of SMPD3 after 16 or 20 weeks of HFD feeding, may reflect an effort to prevent further accumulation of lipids in hepatocytes through increased hydrolysis of sphingomyelin.

Consequences of increased sphingomyelinase activity include increased generation of ceramide through degradation of sphingmyelin, as well as increased production of fatty acids. Since ceramides have key roles in signaling cellular processes such as growth arrest, senescence, apoptosis, and cell death, increased expression of pro-ceramide genes could contribute to the deficits in hepatocellular repair and regenerative functions in NASH [46, 55]. Conceivably, ceramides generated by degradation of sphingolipids and glycosphingolipids localized in late endosomes and lysosomes may be more toxic and inhibitory to insulin signaling than ceramides generated via de novo synthesis pathways [46, 55, 64]. In contrast to the findings in liver, temporal lobe (brain) pro-ceramide gene expression was not significantly increased by the chronic HFD feeding. Therefore, if ceramides have a role in mediating brain insulin resistance, neurodegeneration, and cognitive impairment in obesity, T2DM, and NASH, the source is not likely to be the CNS, and instead would probably be of liver, and possibly adipocyte origin.

Corresponding with findings in our initial study, we showed that chronic HFD feeding of C57BL/6 mice causes mild neuropathological lesions with significant impairments in insulin receptor binding and reduced expression of insulin responsive genes such as ChAT and GAPDH in temporal lobe tissue [73]. Furthermore, the time course analysis revealed early HFD-associated impairments in ChAT expression, beginning after 8 weeks, and coinciding with increased expression of multiple pro-ceramide genes in liver. The facts that: 1) ceramide inhibits insulin signaling [46, 55]; 2) ChAT gene expression is insulin responsive [97]; and 3) ceramide is lipid soluble and therefore probably readily crosses the blood brain barrier, suggest that the early impairments in ChAT expression in brain may be mediated by increased hepatic ceramide production. However, it is noteworthy that ChAT expression increased sharply between 4 and 8 weeks on the LFD, whereas in mice fed with the HFD, ChAT expression increased gradually over time, such that at the 16- and 20-week time points, the levels more closely approximated those in the control group. This suggests that maturation of CNS cholinergic function may be delayed but not thoroughly impaired by brain insulin resistance states.

The other indices of neurodegeneration, including increased 4-HNE (marker of lipid peroxidation) and ubiquitin (index of abnormal protein processing and accumulation), and reduced tau (probably reflecting cytoskeletal collapse and insulin/IGF resistance) were primarily detected at the 16 and 20 week time points, coinciding with surges in steatohepatitis and serum TNF-α, and persistent elevation of hepatic SMPD3. These results provide supportive evidence that hepatic/peripheral ceramide and pro-inflammatory cytokine production play key roles in the pathogenesis of CNS oxidative stress, insulin resistance and neuronal cytoskeletal collapse, all of which occur in AD.

The in vitro experiments helped clarify the role of ceramides as mediators of neurodegeneration. Short-term exposure to C6 ceramide impaired energy metabolism and altered the expression of several genes that are critical to the insulin and IGF signaling pathways, including insulin, insulin, IGF-1, and IGF-2 receptors, IRS-1, and IRS-4. The ceramide-induced increases in insulin and IGF-1 receptors most likely reflect insulin/IGF-1 resistance. This effect could represent impaired ligand-receptor binding, as previously observed in mice chronically fed with the HFD [73]. Alternatively, the result could mark compensatory responses to decreased levels of IRS-1, which is one of the critical molecules responsible for transmitting insulin and IGF-1 stimulated signals [98, 99]. The modest increases in IGF-2 may also reflect an adaptive response to decreased levels of the corresponding receptor.

The in vitro C6 ceramide treatments also significantly increased APP, and decreased ChAT and GAPDH mRNA expression, corresponding with the findings in human AD [10, 11]. Overexpression of the APP gene results in increased APP-Aβ deposition in Down syndrome [100, 101] and transgenic mouse models [102], and similarly, the higher levels of APP expression in AD correspond with the generally increased APP-Aβ burden in AD brains [10, 11, 103]. In a previous study, C6 ceramide was demonstrated to increase APP-Aβ biogenesis by stabilizing beta-secretase BACE1 activity [104], indicating a post-translational mechanism by which increased levels of ceramide could contribute to neurodegeneration. In addition, TNF-α or APP-Aβ exposure was experimentally shown to increase CNS ceramide levels [47], suggesting the potential for establishing a reverberating loop of ever-increasing APP-Aβ and ceramide levels in brain after an initial oxidative stress insult.

The ceramide-induced reductions in ChAT and GAPDH also correspond with findings in AD, but reflect insulin and IGF resistance as both genes are modulated by insulin and IGF stimulation [21, 97]. Reduced ChAT expression reflects a neuronal cholinergic deficit, which one of the pivotal features correlating with dementia in AD as well as other neurodegenerative diseases [105, 106]. GAPDH is a key enzyme in the glycolytic pathway needed for energy metabolism, and principally regulated by insulin [107]. Reduced GAPDH expression correlated with the ceramide-mediated impairments in ATP production [87]. The ELISA and mRNA studies complemented each other in demonstrating that C6 and/or C2 ceramide exposure significantly increased neuronal level of lipid peroxidation and oxidative stress, caused cytoskeletal collapse (reduced β-actin levels), and impaired acetylcholine homeostasis, i.e.

increased AChE and decreased ChAT expression. Therefore, exogenous exposure to ceramide could lead to neuronal insulin resistance with increased oxidative stress, deficits in energy metabolism, and impairments in cholinergic function, similar to the findings in AD.

Altogether, the neurodegenerative effects of C6 ceramide treatment resembled many of the abnormalities caused by chronic HFD feeding, in that both resulted in increased insulin and IRS-4 expression, decreased ChAT and GAPDH expression, and increased 4-HNE and ubiquitin immunoreactivity [73]. However, their effects differed in that C6 ceramide treatment also increased expression of insulin, IGF-1 receptor, AChE, and APP, and decreased expression of IRS-1, whereas chronic HFD feeding did not produce these responses [73]. Overall, our results are consistent with a previous report demonstrating ceramide-inhibition of IGF-1 stimulated energy metabolism and viability in neurons, i.e. neuronal IGF-1 resistance [62], but provide new information about other adverse effects of ceramide on insulin signaling and AD-associated indices of neurodegeneration. It is noteworthy that ceramide and TNF-α both increase the levels of APP-Aβ [104], and that APP-Aβ promotes ceramide and sphingomyelin accumulation by increasing sphingomyelinase activity [47, 108]. Therefore, obesity with T2DM or NASH, has the potential to establish a reverberating loop in brain centered on ever-increasing levels of ceramide, TNF-α, and APP-Aβ, resulting in progressive insulin and IGF resistance with attendant disturbances in energy metabolism and acetylcholine homeostasis.

The present work provides additional support for the concept that extra-CNS sources of ceramide, such as liver or adipose tissue, can contribute to the pathogenesis of neurodegeneration. It is noteworthy that the effects of in vitro ceramide treatment overlapped with many features of insulin and IGF resistance observed in early and intermediate stages of AD in humans [10]. More recently, we demonstrated that in vivo administration of C2 or C6 ceramide produces sustained insulin/IGF resistance, increased oxidative stress, and molecular abnormalities that overlap with but are not identical to AD [109]. The fact that the effects of in vitro and in vivo ceramide treatment effects were not identical to the abnormalities in AD lends support to our hypothesis that T2DM, NASH, and obesity do not cause AD, but instead, they probably serve as pathogenic co-factors. This phenomenon could account for both the absence of complete overlap and the two- or three-fold increased risk of developing MCI or AD among individuals with T2DM [7-9]. Improved ability to detect increased levels of toxic ceramides and related lipids in peripheral blood and cerebrospinal fluid may help identify individuals at risk for developing cognitive impairment in the context of obesity, T2DM, and/or NASH.

Since AD shares many biochemical, molecular, and signal transduction abnormalities in common with T2DM and NASH, but the relevant pathology is fundamentally confined to the CNS in the vast majority of cases, we suggested that AD be regarded as "Type 3 diabetes" [10, 11]. It is of particular interest that, with regard to the antecedent discussion, CNS ceramide levels are elevated in both AD [50] and AD-relevant experimental animal models [47, 91]. These relatively recent observations harken back to a much earlier report showing that white matter atrophy is one of the earliest abnormalities in AD [86]. At the present time, it appears that white matter atrophy in AD could be mediated by brain insulin resistance, which is evident even in the early stages of AD [11]. Insulin resistance in white matter would result in degeneration and loss of oligodendroglia, since oligodendroglia require insulin/IGF-1 for survival signaling and myelinogenesis [110-112]. On the other hand, insulin/IGF resistance would promote oxidative stress and secondarily lead to activation of pro-inflammatory cascades in astrocytes, and attendant myelin degeneration via activation of sphingomyelinases and pro-inflammatory cytokines, including TNF-α, and attendant release of ceramide. This proposed scheme provides a mechanism for producing Type 3 diabetes without the need for hepatic/peripheral sources of ceramides or cytokines, and which would be relevant in the vast majority of sporadic AD cases. With mounting evidence pointing toward CNS insulin/IGF resistance as the mediator if not initiator of AD-type neurodegeneration, our very next goal should be to identify the agents and factors responsible for establishing this cascade.

Although a likely connection between increased CNS ceramide (and probably other toxic lipids as well) and AD has been demonstrated as discussed above, the novelty of this work is that extra-CNS sources of ceramide, such as liver or adipose tissue, can contribute to the pathogenesis of cognitive impairment and AD-type neurodegeneration. The critical variable mediating this effect may be the degree to which liver disease or adipocyte degeneration and remodeling increase ceramide production. The fact that the effects of HFD feeding were not identical to that which actually happens in AD, lends support to our hypothesis that T2DM, NASH, and obesity do not cause AD, and instead they probably serve as pathogenic co-factors. This phenomenon could account for both the absence of complete overlap and the two- or three-fold increased risk of developing MCI or AD among individuals with T2DM [7-9]. Improved ability to detect increased levels of toxic ceramides and related lipids in peripheral blood and cerebrospinal fluid may help identify individuals at risk for developing cognitive impairment in the context of obesity, T2DM, and/or NASH.

## References

[1] Rector RS, Thyfault JP, Wei Y, Ibdah JA (2008) Non-alcoholic fatty liver disease and the metabolic syndrome: an update. *World J Gastroenterol* 14, 185-192.

[2] Pradhan A (2007) Obesity, metabolic syndrome, and type 2 diabetes: inflammatory basis of glucose metabolic disorders. *Nutr Rev* 65, S152-156.

[3] Launer LJ (2007) Next steps in Alzheimer's disease research: interaction between epidemiology and basic science. *Curr Alzheimer Res* 4, 141-143.

[4] Wang XP, Ding HL (2008) Alzheimer's disease: epidemiology, genetics, and beyond. *Neurosci Bull* 24, 105-109.

[5] Delgado JS (2008) Evolving trends in nonalcoholic fatty liver disease. *Eur J Intern Med* 19, 75-82.

[6] Nugent C, Younossi ZM (2007) Evaluation and management of obesity-related nonalcoholic fatty liver disease. *Nat Clin Pract Gastroenterol Hepatol* 4, 432-441.

[7] Pasquier F, Boulogne A, Leys D, Fontaine P (2006) Diabetes mellitus and dementia. *Diabetes Metab* 32, 403-414.

[8] Verdelho A, Madureira S, Ferro JM, Basile AM, Chabriat H, Erkinjuntti T, Fazekas F, Hennerici M, O'Brien J, Pantoni L, Salvadori E, Scheltens P, Visser MC, Wahlund LO, Waldemar G, Wallin A, Inzitari D (2007) Differential impact of cerebral white matter changes, diabetes, hypertension and stroke on cognitive performance among non-disabled elderly. The LADIS study. *J Neurol Neurosurg Psychiatry* 78, 1325-1330.

[9] Martins IJ, Hone E, Foster JK, Sunram-Lea SI, Gnjec A, Fuller SJ, Nolan D, Gandy SE, Martins RN (2006) Apolipoprotein E, cholesterol metabolism, diabetes, and the convergence of risk factors for Alzheimer's disease and cardiovascular disease. *Mol Psychiatry* 11, 721-736.

[10] Rivera EJ, Goldin A, Fulmer N, Tavares R, Wands JR, de la Monte SM (2005) Insulin and insulin-like growth factor expression and function deteriorate with progression of Alzheimer's disease: link to brain reductions in acetylcholine. *J Alzheimers Dis* 8, 247-268.

[11] Steen E, Terry BM, Rivera EJ, Cannon JL, Neely TR, Tavares R, Xu XJ, Wands JR, de la Monte SM (2005) Impaired insulin and insulin-like growth factor expression and signaling mechanisms in Alzheimer's disease-is this type 3 diabetes? *J Alzheimers Dis* 7, 63-80.

[12] Craft S (2007) Insulin resistance and Alzheimer's disease pathogenesis: potential mechanisms and implications for treatment. *Curr Alzheimer Res* 4, 147-152.

[13] Craft S (2006) Insulin resistance syndrome and Alzheimer disease: pathophysiologic mechanisms and therapeutic implications. *Alzheimer Dis Assoc Disord* 20, 298-301.

[14] Winocur G, Greenwood CE, Piroli GG, Grillo CA, Reznikov LR, Reagan LP, McEwen BS (2005) Memory impairment in obese Zucker rats: an investigation of cognitive function in an animal model of insulin resistance and obesity. *Behav Neurosci* 119, 1389-1395.

[15] Winocur G, Greenwood CE (2005) Studies of the effects of high fat diets on cognitive function in a rat model. *Neurobiol Aging* 26 Suppl 1, 46-49.

[16] Lester-Coll N, Rivera EJ, Soscia SJ, Doiron K, Wands JR, de la Monte SM (2006) Intracerebral streptozotocin model of type 3 diabetes: relevance to sporadic Alzheimer's disease. *J Alzheimers Dis* 9, 13-33.

[17] Weinstock M, Shoham S (2004) Rat models of dementia based on reductions in regional glucose metabolism, cerebral blood flow and cytochrome oxidase activity. *J Neural Transm* 111, 347-366.

[18] Nitta A, Murai R, Suzuki N, Ito H, Nomoto H, Katoh G, Furukawa Y, Furukawa S (2002) Diabetic neuropathies in brain are induced by deficiency of BDNF. *Neurotoxicol Teratol* 24, 695-701.

[19] Hoyer S, Lannert H, Noldner M, Chatterjee SS (1999) Damaged neuronal energy metabolism and behavior are improved by Ginkgo biloba extract (EGb 761). *J Neural Transm* 106, 1171-1188.

[20] Biju MP, Paulose CS (1998) Brain glutamate dehydrogenase changes in streptozotocin diabetic rats as a function of age. *Biochem Mol Biol Int* 44, 1-7.

[21] de la Monte SM, Tong M, Lester-Coll N, Plater M, Jr., Wands JR (2006) Therapeutic rescue of neurodegeneration in experimental type 3 diabetes: relevance to Alzheimer's disease. *J Alzheimers Dis* 10, 89-109.

[22] Haan MN (2006) Therapy Insight: type 2 diabetes mellitus and the risk of late-onset Alzheimer's disease. *Nat Clin Pract Neurol* 2, 159-166.

[23] Reger MA, Watson GS, Green PS, Wilkinson CW, Baker LD, Cholerton B, Fishel MA, Plymate SR, Breitner JC, Degroodt W, Mehta P, Craft S (2008) Intranasal insulin improves cognition and modulates {beta}-amyloid in early AD. *Neurology* 70, 440-448.

[24] Landreth G (2007) Therapeutic use of agonists of the nuclear receptor PPARgamma in Alzheimer's disease. *Curr Alzheimer Res* 4, 159-164.

[25] Watson GS, Bernhardt T, Reger MA, Cholerton BA, Baker LD, Peskind ER, Asthana S, Plymate SR, Frolich L, Craft S (2006) Insulin effects on CSF norepinephrine and cognition in Alzheimer's disease. *Neurobiol Aging* 27, 38-41.

[26] Reger MA, Watson GS, Frey WH, 2nd, Baker LD, Cholerton B, Keeling ML, Belongia DA, Fishel MA, Plymate SR, Schellenberg GD, Cherrier MM, Craft S (2006) Effects of intranasal insulin on cognition in memory-impaired older adults: modulation by APOE genotype. *Neurobiol Aging* 27, 451-458.

[27] Pedersen WA, McMillan PJ, Kulstad JJ, Leverenz JB, Craft S, Haynatzki GR (2006) Rosiglitazone attenuates learning and memory deficits in Tg2576 Alzheimer mice. *Exp Neurol* 199, 265-273.

[28] Nicolls MR (2004) The clinical and biological relationship between Type II diabetes mellitus and Alzheimer's disease. *Curr Alzheimer Res* 1, 47-54.
[29] Yeh MM, Brunt EM (2007) Pathology of nonalcoholic fatty liver disease. *Am J Clin Pathol* **128**, 837-847.
[30] Marchesini G, Marzocchi R (2007) Metabolic syndrome and NASH. *Clin Liver Dis* **11**, 105-117, ix.
[31] Papandreou D, Rousso I, Mavromichalis I (2007) Update on non-alcoholic fatty liver disease in children. *Clin Nutr* 26, 409-415.
[32] Pessayre D (2007) Role of mitochondria in nonalcoholic fatty liver disease. *J Gastroenterol Hepatol* **22 Suppl 1**, S20-27.
[33] Wei Y, Rector RS, Thyfault JP, Ibdah JA (2008) Nonalcoholic fatty liver disease and mitochondrial dysfunction. *World J Gastroenterol* 14, 193-199.
[34] Qiu C, De Ronchi D, Fratiglioni L (2007) The epidemiology of the dementias: an update. *Curr Opin Psychiatry* 20, 380-385.
[35] Solfrizzi V, Panza F, Colacicco AM, D'Introno A, Capurso C, Torres F, Grigoletto F, Maggi S, Del Parigi A, Reiman EM, Caselli RJ, Scafato E, Farchi G, Capurso A (2004) Vascular risk factors, incidence of MCI, and rates of progression to dementia. *Neurology* 63, 1882-91.
[36] Whitmer RA (2007) Type 2 diabetes and risk of cognitive impairment and dementia. *Curr Neurol Neurosci Rep* 7, 373-380.
[37] Haan MN, Wallace R (2004) Can dementia be prevented? Brain aging in a population-based context. *Annu Rev Public Health* 25, 1 -24.
[38] Luchsinger JA, Mayeux R (2004) Cardiovascular risk factors and Alzheimer's disease. *Curr Atheroscler Rep* 6, 261-266.
[39] Luchsinger JA, Reitz C, Patel B, Tang MX, Manly JJ, Mayeux R (2007) Relation of diabetes to mild cognitive impairment. *Arch Neurol* 64, 570-575.
[40] Launer LJ (2005) Diabetes and brain aging: epidemiologic evidence. *CurrDiab Rep* 5, 59-63.
[41] Garcia-Galiano D, Sanchez-Garrido MA, Espejo I, Montero JL, Costan G, Marchal T, Membrives A, Gallardo-Valverde JM, Munoz-Castaneda JR, Arevalo E, De la Mata M, Muntane J (2007) IL-6 and IGF-1 are independent prognostic factors of liver steatosis and non-alcoholic steatohepatitis in morbidly obese patients. *Obes Surg* 17, 493-503.
[42] Gholam PM, Flancbaum L, Machan JT, Charney DA, Kotler DP (2007) Nonalcoholic fatty liver disease in severely obese subjects. *Am J Gastroenterol* **102**, 399-408.
[43] Liew PL, Lee WJ, Lee YC, Wang HH, Wang W, Lin YC (2006) Hepatic histopathology of morbid obesity: concurrence of other forms of chronic liver disease. *Obes Surg* **16**, 1584-1593.
[44] Nobili V, Manco M (2007) Measurement of advanced glycation end products may change NASH management. *J Gastroenterol Hepatol* **22**, 1354-1355.
[45] Shah C, Yang G, Lee I, Bielawski J, Hannun YA, Samad F (2008) Protection from high fat diet-induced increase in ceramide in mice lacking plasminogen activator inhibitor 1. *J Biol Chem* **283**, 13538-13548.
[46] Summers SA (2006) Ceramides in insulin resistance and lipotoxicity. *Prog Lipid Res* 45, 42-72.
[47] Alessenko AV, Bugrova AE, Dudnik LB (2004) Connection of lipid peroxide oxidation with the sphingomyelin pathway in the development of Alzheimer's disease. *Biochem Soc Trans* **32**, 144-146.
[48] Laviad EL, Albee L, Pankova-Kholmyansky I, Epstein S, Park H, Merrill AH, Jr., Futerman AH (2008) Characterization of ceramide synthase 2: tissue distribution, substrate specificity, and inhibition by sphingosine 1-phosphate. *J Biol Chem* **283**, 5677-5684.
[49] Nakane M, Kubota M, Nakagomi T, Tamura A, Hisaki H, Shimasaki H, Ueta N (2000) Lethal forebrain ischemia stimulates sphingomyelin hydrolysis and ceramide generation in the gerbil hippocampus. *Neurosci Lett* **296**, 89-92.
[50] Katsel P, Li C, Haroutunian V (2007) Gene expression alterations in the sphingolipid metabolism pathways during progression of dementia and Alzheimer's disease: a shift toward ceramide accumulation at the earliest recognizable stages of Alzheimer's disease? *Neurochem Res* **32**, 845-856.
[51] Adibhatla RM, Hatcher JF (2008) Altered Lipid Metabolism in Brain Injury and Disorders. *Subcell Biochem* 48, nihpa41041.
[52] Liu B, Obeid LM, Hannun YA (1997) Sphingomyelinases in cell regulation. *Semin Cell Dev Biol* 8, 311-322.
[53] Reynolds CP, Maurer BJ, Kolesnick RN (2004) Ceramide synthesis and metabolism as a target for cancer therapy. *Cancer Lett* 206, 169-180.
[54] Mizutani Y, Kihara A, Igarashi Y (2005) Mammalian Lass6 and its related family members regulate synthesis of specific ceramides. *Biochem J* **390**, 263-271.
[55] Holland WL, Knotts TA, Chavez JA, Wang LP, Hoehn KL, Summers SA (2007) Lipid mediators of insulin resistance. *Nutr Rev* **65**, S39-46.
[56] Chavez JA, Holland WL, BarJ, Sandhoff K, Summers SA (2005) Acid ceramidaseoverexpression prevents the inhibitory effects of saturated fatty acids on insulin signaling. *J Biol Chem* **280**, 20148-20153.
[57] Spiegel S, Merrill AH, Jr. (1996) Sphingolipid metabolism and cell growth regulation. *FASEB J* **10**, 1388-1397.
[58] Gomez-Munoz A, Frago LM, Alvarez L, Varela-Nieto I (1997) Stimulation of DNA synthesis by natural ceramide 1-phosphate. *Biochem J* **325** ( Pt 2), 435-440.
[59] Bryan L, Kordula T, Spiegel S, Milstien S (2008) Regulation and functions of sphingosine kinases in the brain. *Biochim BiophysActa* **1781**, 459-466.
[60] Van Brocklyn JR (2007) Sphingolipid signaling pathways as potential therapeutic targets in gliomas. *Mini Rev Med Chem* 7, 984-990.
[61] Silveira LR, Fiamoncini J, Hirabara SM, Procopio J, Cambiaghi TD, Pinheiro CH, Lopes LR, Curi R (2008) Updating the effects of fatty acids on skeletal muscle. *J Cell Physiol* **217**, 1-12.
[62] Arboleda G, Huang TJ, Waters C, Verkhratsky A, Fernyhough P, Gibson RM (2007) Insulin-like growth factor-1 - dependent maintenance of neuronal metabolism through the phosphatidylinositol 3-kinase-Akt pathway is inhibited by C2-ceramide in CAD cells. *Eur J Neurosci* **25**, 3030-3038.
[63] Chalfant CE, Kishikawa K, Mumby MC, Kamibayashi C, Bielawska A, Hannun YA(1999) Long chain ceramides activate protein phosphatase-1 and protein phosphatase-2A. Activation is stereospecific and regulated by phosphatidic acid. *J Biol Chem* **274**, 20313-20317.
[64] Han MS, Park SY, Shinzawa K, Kim S, Chung KW, Lee JH, Kwon CH, Lee KW, Park CK, Chung WJ, Hwang JS, Yan JJ, Song DK, Tsujimoto Y, Lee MS (2008) Lysophosphatidylcholine as a death effector in the lipoapoptosis of hepatocytes. *J Lipid Res* 49, 84-97.

[65] Delarue J, Magnan C (2007) Free fatty acids and insulin resistance. *Curr Opin Clin Nutr Metab Care* 10, 142-148.

[66] Assimacopoulos-Jeannet F (2004) Fat storage in pancreas and in insulin-sensitive tissues in pathogenesis of type 2 diabetes. *Int J Obes Relat Metab Disord* **28 Suppl 4,** S53-57.

[67] Kolak M, Westerbacka J, Velagapudi VR, Wagsater D, Yetukuri L, Makkonen J, Rissanen A, HakkinenAM, LindellM, Bergholm R, HamstenA, Eriksson P, Fisher RM, OresicM, Yki-Jarvinen H (2007) Adipose tissue inflammation and increased ceramide content characterize subjects with high liver fat content independent of obesity. *Diabetes* 56, 1960-1968.

[68] Cong WN, Tao RY, Tian JY, Liu GT, Ye F (2008) The establishment of a novel non-alcoholic steatohepatitis model accompanied with obesity and insulin resistance in mice. *Life Sci* **82,** 983-990.

[69] Sahai A, Malladi P, Pan X, Paul R, Melin-Aldana H, Green RM, Whitington PF (2004) Obese and diabetic db/db mice develop marked liver fibrosis in a model of nonalcoholic steatohepatitis: role of short-form leptin receptors and osteopontin. *Am J Physiol Gastrointest Liver Physiol* **287,** G1035-1043.

[70] Lieber CS, Leo MA, Mak KM, Xu Y, Cao Q, Ren C, Ponomarenko A, DeCarli LM (2004) Model of nonalcoholic steatohepatitis. *Am J Clin Nutr* 79, 502-509.

[71] Yalniz M, Bahcecioglu IH, Ataseven H, Ustundag B, Ilhan F, Poyrazoglu OK, Erensoy A (2006) Serum adipokine and ghrelin levels in nonalcoholic steatohepatitis. *Mediators Inflamm* **2006,** 34295.

[72] Satapathy SK, Garg S, Chauhan R, Sakhuja P, Malhotra V, Sharma BC, Sarin SK(2004) Beneficial effects of tumor necrosis factor-alpha inhibition by pentoxifylline on clinical, biochemical, and metabolic parameters of patients with nonalcoholic steatohepatitis. *Am J Gastroenterol* 99, 1946-1952.

[73] Moroz N, Tong M, Longato L, Xu H, de la Monte SM (2008) Limited Alzheimer-type neurodegeneration in experimental obesity and Type 2 diabetes mellitus. *J Alzheimers Dis* 15, 29-44.

[74] Xu H, Barnes GT, Yang Q, Tan G, Yang D, Chou CJ, Sole J, Nichols A, Ross JS, Tartaglia LA, Chen H (2003) Chronic inflammation in fat plays a crucial role in the development of obesity-related insulin resistance. *J Clin Invest* **112,** 1821-1830.

[75] Xu YY, Bhavani K, Wands JR, de la Monte SM (1995) Insulin-induced differentiation and modulation of neuronal thread protein expression in primitive neuroectodermal tumor cells is linked to phosphorylation of insulin receptor substrate-1. *J Mol Neurosci* 6, 91-108.

[76] Xu YY, Bhavani K, Wands JR, de la Monte SM (1995) Ethanol inhibits insulin receptor substrate-1 tyrosine phosphorylation and insulin-stimulated neuronal thread protein gene expression. *Biochem J* 310 ( Pt 1), 125-132.

[77] Yeon JE, Califano S, Xu J, Wands JR, De La Monte SM (2003) Potential role of PTEN phosphatase in ethanol-impaired survival signaling in the liver. *Hepatology* 38, 703-714.

[78] Xu J, Eun Yeon J, Chang H, Tison G, Jun Chen G, Wands JR, De La Monte SM (2003) Ethanol impairs insulin-stimulated neuronal survival in the developing brain: Role of PTEN phosphatase. *J Biol Chem* **278,** 26929-26937.

[79] Cohen AC, Tong M, Wands JR, de la Monte SM (2007) Insulin and insulin-like growth factor resistance with neurodegeneration in an adult chronic ethanol exposure model. *Alcohol Clin Exp Res* 31, 1558-1573.

[80] McMillian MK, Grant ER, Zhong Z, Parker JB, Li L, Zivin RA, Burczynski ME, Johnson MD (2001) Nile Red binding to HepG2 cells: an improved assay for in vitro studies of hepatosteatosis. *In Vitr Mol Toxicol* 14, 177-190.

[81] Fowler SD, Greenspan P (1985) Application of Nile red, a fluorescent hydrophobic probe, for the detection of neutral lipid deposits in tissue sections: comparison with oil red O. *J Histochem Cytochem* 33, 833-836.

[82] Greenspan P, Fowler SD (1985) Spectrofluorometric studies of the lipid probe, nile red. *J Lipid Res* 26, 781-789.

[83] Winzell MS, Magnusson C, Ahren B (2007) Temporal and dietary fat content-dependent islet adaptation to high-fat feeding-induced glucose intolerance in mice. *Metabolism* 56, 122-128.

[84] Yoshimatsu M, Terasaki Y, Sakashita N, Kiyota E, Sato H, van der Laan LJ, Takeya M (2004) Induction of macrophage scavenger receptor MARCO in nonalcoholic steatohepatitis indicates possible involvement of endotoxin in its pathogenic process. *Int J Exp Pathol* 85, 335-343.

[85] Kirsch R, Clarkson V, Verdonk RC, Marais AD, Shephard EG, Ryffel B, de la MHP (2006) Rodent nutritional model of steatohepatitis: effects of endotoxin (lipopolysaccharide) and tumor necrosis factor alpha deficiency. *J Gastroenterol Hepatol* 21, 174-182.

[86] de la Monte SM (1989) Quantitation of cerebral atrophy in preclinical and end-stage Alzheimer's disease. *Ann Neurol* 25, 450-459.

[87] Tong M, de la Monte SM (2009) Mechanisms of ceramide-mediated neurodegeneration. *J Alzheimers Dis* 16, 705-714.

[88] Lyn-Cook LE, Jr., Lawton M, Tong M, Silbermann E, Longato L, Jiao P, Mark P, Wands JR, Xu H, de la Monte SM (2009) Hepatic ceramide may mediate brain insulin resistance and neurodegeneration in type 2 diabetes and non-alcoholic steatohepatitis. *J Alzheimers Dis* 16, 715-729.

[89] Elwing JE, Lustman PJ, Wang HL, Clouse RE (2006) Depression, anxiety, and nonalcoholic steatohepatitis. *Psychosom Med* 68, 563-569.

[90] Kraegen EW, Cooney GJ (2008) Free fatty acids and skeletal muscle insulin resistance. *Curr OpinLipidol* 19, 235-241.

[91] Wang G, Silva J, Dasgupta S, Bieberich E (2008) Long-chain ceramide is elevated in presenilin 1 (PS1M146V) mouse brain and induces apoptosis in PS1 astrocytes. *Glia* 56, 449-456.

[92] Gallou-Kabani C, Vige A, Gross MS, Rabes JP, Boileau C, Larue-Achagiotis C, Tome D, Jais JP, Junien C (2007) C57BL/6J and A/J mice fed a high-fat diet delineate components of metabolic syndrome. *Obesity (Silver Spring)* 15, 1996-2005.

[93] Ito M, Suzuki J, Sasaki M, Watanabe K, Tsujioka S, Takahashi Y, Gomori A, Hirose H, Ishihara A, Iwaasa H, Kanatani A (2006) Development of nonalcoholic steatohepatitis model through combination of high-fat diet and tetracycline with morbid obesity in mice. *Hepatol Res* 34, 92-98.

[94] Strissel KJ, Stancheva Z, Miyoshi H, Perfield JW, 2nd, DeFuria J, Jick Z, Greenberg AS, Obin MS (2007) Adipocyte death, adipose tissue remodeling, and obesity com-

plications. *Diabetes* 56, 2910-2918.

[95] Diehl AM (2005) Lessons from animal models of NASH. *Hepatol Res* 33, 138-144.

[96] Solis Herruzo JA, Garcia Ruiz I, Perez Carreras M, Munoz Yague MT (2006) Non-alcoholic fatty liver disease. From insulin resistance to mitochondrial dysfunction. *RevEsp Enferm Dig* 98, 844-874.

[97] Soscia SJ, Tong M, Xu XJ, Cohen AC, Chu J, Wands JR, de la Monte SM (2006) Chronic gestational exposure to ethanol causes insulin and IGF resistance and impairs acetylcholine homeostasis in the brain. *Cell Mol Life Sci* 63, 2039-2056.

[98] Fritsche L, Weigert C, Haring HU, Lehmann R (2008) How insulin receptor substrate proteins regulate the metabolic capacity of the liver--implications for health and disease. *CurrMed Chem* 15, 1316-1329.

[99] Youngren JF (2007) Regulation of insulin receptor function. *Cell Mol Life Sci* 64, 873-891.

[100] de la Monte SM (1999) Molecular abnormalities of the brain in Down syndrome: relevance to Alzheimer's neurodegeneration. *J Neural Transm Suppl* 57, 1-19.

[101] Head E, Lott IT (2004) Down syndrome and beta-amyloid deposition. *Curr Opin Neurol* **17**, 95-100.

[102] Stein TD, Johnson JA (2002) Lack of neurodegeneration in transgenic mice overexpressing mutant amyloid precursor protein is associated with increased levels of transthyretin and the activation of cell survival pathways. *J Neurosci* **22**, 7380-7388.

[103] de la Monte SM, Wands JR (2006) Molecular indices of oxidative stress and mitochondrial dysfunction occur early and often progress with severity of Alzheimer's disease. *J Alzheimers Dis* 9, 167-181.

[104] Puglielli L, Ellis BC, Saunders AJ, Kovacs DM (2003) Ceramide stabilizes beta-site amyloid precursor protein-cleaving enzyme 1 and promotes amyloid beta-peptide biogenesis. *J Biol Chem* **278**, 19777-19783.

[105] Cuello AC, Bruno MA, Bell KF (2007) NGF-cholinergic dependency in brain aging, MCI and Alzheimer's disease. *Curr Alzheimer Res 4*, 351-358.

[106] Schaeffer EL, Gattaz WF (2008) Cholinergic and glutamatergic alterations beginning at the early stages of Alzheimer disease: participation of the phospholipaseA2 enzyme. *Psychopharmacology (Berl)* **198,** 1-27.

[107] Alexander BM, Dugast I, Ercolani L, Kong XF, Giere L, Nasrin N (1992) Multiple insulin-responsive elements regulate transcription of the GAPDH gene. *Adv Enzyme Regul* 32, 149-159.

[108] He X, Huang Y, Li B, Gong CX, Schuchman EH (2008) Deregulation of sphingolipid metabolism in Alzheimer's disease. *Neurobiol Aging*.

[109] de la Monte SM, Tong M, Ng VA, Setshedi M, Longato L, Wands JR (2010) Ceramide-mediated insulin resistance and impairment of cognitive-motor functions. *J Alzheimers Dis*, (In Press).

[110] Lopes-Cardozo M, Sykes JE, Van der Pal RH, van Golde LM (1989) Development of oligodendrocytes. Studies of rat glial cells cultured in chemically-defined medium. *J Dev Physiol* 12, 117-127.

[111] Broughton SK, Chen H, Riddle A, Kuhn SE, Nagalla S, Roberts CT, Jr., Back SA (2007) Large-scale generation of highly enriched neural stem-cell-derived oligodendroglial cultures: maturation-dependent differences in insulin-like growth factor-mediated signal transduction. *J Neurochem* 100, 628-638.

[112] Chesik D, De Keyser J, Wilczak N (2008) Insulin-like growth factor system regulates oligodendroglial cell behavior: therapeutic potential in CNS. *J Mol Neurosci* 35, 81-90.

# Animal Model of Insulin-Resistant Brain State: Intracerebroventricular Streptozotocin Injection Deteriorates Alzheimer-Like Changes in Tg2576 APP-Overexpressing Mice

Konstanze Plaschke[a,1], Jürgen Kopitz[b], Markus Siegelin[c], Reinhard Schliebs[d], Melita Salkovic-Petrisic[e], Peter Riederer[f] and Siegfried Hoyer[c]

[a]Clinic of Anesthesiology, University of Heidelberg Medical School, D-69120 Heidelberg, Germany
[b]Department of Pathology, University of Heidelberg, Heidelberg, Germany
[c]Departments of Pathology and Neuropathology, University of Heidelberg, Heidelberg, Germany
[d]Paul-Flechsig Institute for Brain Research; Neurochemistry, University of Leipzig, Germany
[e]Department of Pharmacology and Croatian Institute for Brain Research, School of Medicine, University of Zagreb, Zagreb, Croatia
[f]Clinical Neurochemistry; National Parkinson Foundation Centre of Excellence Research Laboratory, Clinic and Policlinic for Psychiatry, Psychosomatic and Psychotherapy, University Hospital of Wuerzburg, Germany

**Abstract.** Sporadic Alzheimer's disease has been proposed to start with an insulin-resistant brain state (IRBS). We investigated the effect of IRBS induced by intracerebroventricularly (icv) administered streptozotocin (STZ) on behavior, glycogen synthase kinase-3 (GSK)alpha/beta content, and the formation of AD-like morphological hallmarks Abeta-amyloid and tau protein in amyloid precursor protein (APP) tg2576 mice. Nine-month-old tg mice were investigated 6 months after a single icv injection of STZ or placebo. Spatial cognition was analyzed using the Morris water maze test. Soluble and aggregated Abeta40/42 fragments, total and phosphorylated tau protein, and GSK-3alpha/beta were determined by ELISA. Cerebral (immuno)histological analyses were performed. In tg mice, STZ treatment increased mortality, reduced spatial cognition, and increased cerebral aggregated Abeta fragments, total tau protein, and congophilic amyloid deposits. These changes were associated with decreased GSK-3alpha/beta ratio (phosphorylated/total). A linear negative correlation was detected between Abeta42 and cognition, and between GSK-3alpha/beta ratio and aggregated Abeta40+42. No marked necrotic and apoptotic changes were observed. In conclusion, IRBS may aggravate AD-like changes such as behavioral and increase the formation of pathomorphological AD hallmarks via GSK-3alpha/beta pathway in APP-overexpressing mice.

Keywords: sporadic Alzheimer disease, beta-amyloid, transgenic 2576 mice, tau, glycogen synthase kinase-3, streptozotocin, insulin

---

[1]Address for correspondence: Konstanze Plaschke, MD, PhD., Department of Anesthesiology, Im Neuenheimer Feld 110, D-69120 Heidelberg, Tel: +49-6221-566451, Fax: +49-6221-564399 E-Mail: konstanze.plaschke@med.uni-heidelberg.de

## INTRODUCTION

Among the diverse molecular defects leading to the brain lesions that have been reported in Alzheimer disease (AD) patients, the major problem of the "cause, correlation, and consequence" in this type of dementia still remains. To model this multifactorial disorder in animals, different and complex experimental approaches are needed, such as an investigation of the role of insulin resistant brain state (IRBS) in genetic mice models.

Transgenic mice of the tg2576 line overexpress human betaAPP with the Swedish mutation (bAPP695swe, K670N/M/671L) [1]. Therefore, this mouse strain is a suitable animal model for investigating the hereditary type of AD and the effects of disturbed amyloid metabolism and AD-like brain amyloidosis [2,3,4,5]. In tg2576 mice, a significant increase in soluble Abeta 1-40 and Abeta 1-42/43 peptides is found from 3 months to 9-10 months of age [1]. At ages of 1 year and older, senile amyloid plaques are formed which are associated with reactive microglia and astroglia, resembling late-onset AD pathology [3,6]. Normal spatial reference memory was obtained in 3-month-old animals, and, in a forced alternation paradigm, the spatial reference memory was impaired at 9-10 months and later in a water maze test. However, changes in behavior and cognition in tg2576 mice as compared to their age-related littermates are the subject of controversial discussion [1,7,8,9].

To our knowledge, the effect of disturbed cerebral insulin metabolism has not been investigated yet in APP transgenic mice. Indeed, it would be of interest to evaluate the effect of IRBS as an environmental factor in the generation of APP mismetabolism as this is a hallmark of AD.

Hyperinsulinemia and type II diabetes mellitus are among the age-related environmental risk factors for sporadic AD (sAD) [10,11]. High levels of insulin can negatively influence neuronal function and survival [12]. For example, the results of Neumanns' investigations showed that peripheral hyperinsulinemia correlates with an abnormal removal of Abeta and an increase in tau hyperphosphorylation as a result of augmented GSK-3 beta activity. Current review studies undertaken by our research group [13] and by Jones et al. [14] substantiate the similarities in these pathological processes and transcriptional pathways between disturbances in both diabetes mellitus type II and cerebral glucose / energy metabolism in sAD. Here, the changes in brain metabolism present as an IRBS, independently of peripheral abnormalities [13].

The intracerebroventricular (icv) administration of the diabetogenic substance streptozotocin (STZ) has been found to induce an IRBS [15]. In more detail, icv administration of subdiabetogenic doses of STZ in rats produced numerous alterations in behavior [16,17,18,19], in glucose and related pathways, and in energy metabolism [20,21,22]. Furthermore, this substance can induce some cerebral structural features [for review: 15] that resemble those ones found in human AD. Thus, although both elevated total and phosphorylated tau protein levels have been found at 1 month and 3 months after icv STZ administration [23] and beta-amyloid peptide-like aggregates have been shown to occur in rat brain capillaries [24], no diffuse or compact amyloid plaques and neurofibrillary tangles were detected in rats up to 3 months after icv STZ treatment. Thus, the icv STZ rat model is different from APP transgenic mice models because the genes involved in APP / Abeta homeostasis are not manipulated, but rather the substance targets brain insulin receptor (IR) signaling cascade function in relation to both amyloid and tau protein metabolism. A combination of the two models, however, could give new insight to the etiopathogenesis of sAD.

Several lines of evidence support possible similarities between the peripheral and cerebral mechanism of STZ action; for instance, GLUT2 has been found to be regionally distributed in the brain [25] and insulin is not only transported from the periphery into brain, but is synthesized in the particular brain regions [26]. Furthermore, peripheral treatment with low to moderate doses of STZ can cause insulin resistance by damaging insulin receptor (IR) and its tyrosine kinase function; low icv STZ doses alter brain IR and consequently the IRBS [15,26]. Brain insulin and IR are functionally linked to cognition by up-regulation of insulin mRNA in the hippocampus and increased IR accumulation in synaptic membranes [27]. One month after icv STZ injection, time-dependent changes in the phosphorylated and total glycogen synthase kinase (GSK)-3alpha/beta were found in rat brain that might be related to the formation of beta-amyloid peptide-like aggregates in brain capillaries in rats [24].

Tg2576 mice that overexpress the Swedish mutation of human APP and demonstrate age-related Abeta-amyloid accumulation and plaque formation appears to be a suitable model for investigating the consequences of disturbed cerebral insulin metabolism on APP metabolism *in vivo*. It is assumed that failures in insulin transduction represent a potential factor in triggering enhanced Abeta formation and plaque deposition. Therefore, the main goal of the present study was to demonstrate that an IRBS, induced by icv STZ injection, may aggravate or accelerate mice behavioral deficits, tau hyperphosphorylation, and the formation of human amyloid in the brain of tg2576 APP-overexpressing mice. Our study provides the first evidence in tg2576 mice that the disturbed insulin signaling pathway may represent a metabolic abnormality which severely affects the APP cleavage associated with permanent learning and memory deficits.

## MATERIALS AND METHODS

### Animals

The experimental protocol was approved by the appropriate review committee of the Medical Faculties of the Universities of Leipzig and Heidelberg, Germany, and complied with the guidelines of the responsible government agencies and with international standards.

For animal experiments, tg2576 transgenic mice [1] expressing the Swedish mutant of APP (tg, n=25) were used. Mice with a primary body weight of 22.3 ± 3.6 g were aged 3 months at the start of the experiments. For STZ analysis, mice of both sexes were randomly divided into two treatment groups: icv injections of placebo in 12 (plac tg group, 6 male, 6 female) and of streptozotocin in 13 (STZ tg group, 6 male, 7 female) tg mice. Five mice with different signs on their ears were placed together in one cage, and 1 week before the experimental period, animals were transferred to a temperature-controlled room at 22 ± 0.5°C with a reversed day - night cycle (12 h:12 h, light on at 7 p.m.), and tested during the dark period. Free access to food (Altromin, standard no. 1320, Lage, Germany) and water was allowed throughout the experimental period of 6 months. The experimental design is shown in figure 1.

### Streptozotocin (STZ) injections

Intracerebroventricular (icv) injections of streptozotocin [STZ, (2-deoxy-2-(3-(methyl-3-nitrosoureido)-D-glucopyranose))] were given at a single dosos under anesthesia with 1.5 vol% halothane and nitrous oxide / oxygen (70:30) using a mouse-adapted mask. The STZ-treated mice received a single icv injection into both lateral brain ventricles using a Hamilton syringe (Fisher Scientific Ca, Pittsburgh, Pennsylvania, USA) 3 days before the first psychometric experiments were performed while secured in a stereotactic apparatus (UHL, Wetzlar, Germany). The coordinates according to bregma were: -0.9 mm lateral, -0.1 mm posterior, and - 3.1 mm below [28]. For this purpose, STZ (Sigma-Aldrich, Germany) was dissolved in artificial cerebrospinal fluid (CSF = placebo) with the following composition: 120 mM NaCl, 3 mM KCl, 1.15 mM $CaCl_2$, 0.8 mM $MgCl_2$, 0.33 mM $NaH_2PO_4$, 27 mM $NaHCO_3$, pH 7.2 adjusted by $CO_2$ insufflations). STZ solution was produced freshly before it was injected at doses of 1.25 mg/kg body weight [22] dissolved in a final volume of 2 μl (one μl in each lateral ventricle) artificial CSF. As placebo, two μl of artificial CSF was given in the same way as described for STZ.

### Psychometric investigation

Morris Water Maze Testing was performed according to Plaschke et al. [29]. In detail, the Morris water maze consisted of a white-colored round basin (EthoVision, Noldus Information Technology, Freiburg, Germany, 150 cm in diameter; 70 cm in depth) which was filled with water (22 ± 1°C) up to 15 cm below the rim. The pool was divided into four quadrants of equal area, arbitrarily called north, south, east, and west. A square transparent platform (10 x 10 cm) made of Perspex® was submerged to 1 cm below the water surface in the middle of the target quadrant. The basin was surrounded by a white wall (~52 cm away) to the north, east, and west. The experimenter and rat cages were positioned to the south. During trials, the experimenter and rat cages were in separate rooms with a halfway-closed door hidden from the swimming mouse. Visual cues were positioned around the basin, 60 to 90 cm from its rim. Two 250 W quartz halogen lamps were positioned on the bottom of the laboratory aimed at the ceiling to indirectly illuminate the water surface. A closed-circuit television camera (Philips AG, Hamburg, Germany) was mounted onto the ceiling directly above the center of the pool to convey subject swimming trajectories and parameters to an electronic image analyzer (EthoVision®, Noldus Information Technology, Freiburg, Germany) at an image frequency of 4.2 per sec. For offline analysis, raw data were transferred to Excel and SPSS.

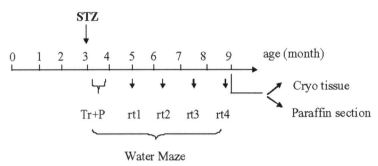

Fig. 1. Experimental design
Tr: training, P: probe trial, rt: retest, STZ: streptozotocin (bilateral intracerebroventricular injections)

For the Morris water maze procedure, each mouse was placed into the water, immediately facing the perimeter, for exactly 90 seconds. The mice started each trial from a different cardinal compass point. The three release points were predetermined to be quasi-random and nonsequential; each mouse was released an equal number of times from each point. Finding the platform was defined as staying on it for at least 4 seconds before the time frame of 90 seconds ended. Mice that crossed the platform without stopping (jumping immediately into the water) were left to swim. After staying on the platform for 10 seconds to memorize the location of the platform, the mouse was gently picked up, returned to its home cage, and allowed to warm up. If the mouse failed to find the platform in the allotted time, it was placed onto the platform for 10 seconds and assigned a latency of 90 seconds. Animals had three training trials per day with a 30- to 40-min intertrial interval. The mean values from three trials were calculated.

The following parameters were calculated: (1) escape performance during training by time to find the platform (escape latency in seconds), and (2) escape path length (distance moved into the basin in cm) during the 90-second trial.

*Visible platform (cued trial) and hidden platform (unchanged position) testing*

The cued trial was performed to assess both the mice motivation to escape from the water and its sensorimotor integrity. The platform was placed in the north quadrant and had a black and a visible cue (yellow striped flag at the top, 13 cm high, 3 cm diameter). Place learning, search time (escape latency), and path length were scored for each mouse with a visible platform from three equidistant points (situated left, opposite, and right relative to the platform quadrant) and averaged for three trials.

For the following days (1-12), the platform was also located in the north quadrant, where it was now hidden from view. Behavioral parameters were scored for each mouse from three equidistant points (situated left, opposite, and right relative to the platform quadrant).

*Probe trial*

A probe trial was administered after the last retest for a hidden platform. Each animal was placed into the water diagonally from the target quadrant and for 90 seconds was allowed to search for the platform, which had been removed from the water. The time (percentage of total time spent in each quadrant) spent searching for the platform in the former platform quadrant (north) and the other three quadrants was measured for each mouse.

*Retest*

Every 6 weeks after STZ injections all mice were retested. Therefore, mice from both groups had to swim to a constant hidden platform as described above three times from a starting point opposite to the target quadrant. The means from three trials were calculated.

*Biochemical determinations of tau protein, APP derivatives, and GSK-3alpha/beta*

*Tissue preparation for ELISA*

Mice (n=7 for placebo tg and n=5 for STZ tg) were decapitated under anesthesia (2.5 vol% halothane, oxygen/nitrous oxide (30/70). Brains were quickly removed and dissected into the right and left hemisphere. The two hemispheres were randomly dissected in gray brain matter containing parieto-temporal cerebral cortex or hippocampus. Brain tissues were quickly transferred to ice-cold isopentane in a randomly defined order and were stored at -80° C until tau and Abeta 40 / 42 analyses and GSK-3 alpha/beta determination.

For biochemical analysis, respective brain samples were homogenized by Potter homogenization in ice-cold lysis buffer, pH 7.4, comprised of 10 mmol/l Tris-HCl, 150 mmol/l NaCl, 1 mmol/l EDTA, and 1 mmol/l EGTA containing protease inhibitors. The lysates were centrifuged at 20,000 x $g$ for 10 min at 2°C to remove insoluble debris. Protein concentration in the supernatant was determined in triplicate using the Bradford protein assay [15].

*Sandwich ELISA technique*

The content of total and phosphorylated glycogen-synthase kinase *(GSK)*-3alpha/beta was measured using the DuoSet IC ELISA (R&D Systems GmbH Wiesbaden, Germany) according to the principle that an immobilized capture antibody specific for GSK-3alpha and GSK-3beta binds both phosphorylated and unphosphorylated protein. After washing away unbound material, a biotinylated detection antibody specific for GSK-3alpha phosphorylated at S21, and GSK-3beta phosphorylated at S9, was used to detect only phosphorylated protein, utilizing a standard streptavidin-HRP format.

To detect *tau* protein (total and phosphorylated) in mice brain, commercial tau ELISA kits were used according to the manufacturer's instructions (SIGMA-Aldrich, Saint Louis, Missouri, USA). These kits are two mouse, solid-phase, sandwich enzyme-linked immuno-sorbent assays (ELISA) for quantitatively determining *in vitro* mouse tau in mouse brain, recognizing both natural and recombinant tau.

The soluble and aggregated *β40 and β42 APP* fragments were determined in hippocampus in accordance with the manufacturer's instructions [WAKO Pure Chemical Industries, Osaka, Japan]. The kits are constructed as a sandwich ELISA format kit with two kinds of antibodies. The monoclonal antibody is coated on 96-well surfaces of a separate microplate and acts as a capture antibody for Abeta (x-40) or Abeta (x-42) to the two ELISA test kits, respectively. Captured antibody is recognized by another antibody, which specifically detects the C-terminal portion of A40 or A42 labeled with HRP. After adding TMB solution, positive samples developed a blue color. The reaction was terminated by adding stop solution, which produces a yellow color. The absorbance is then measured at 450 run (Tecan Genios, MTX Lab Systems Inc. Virginia, USA).

*Histological analysis*

Two 9-month-old mice from each group were killed 6 months after icv injection of STZ or placebo under anesthesia with 2.5 vol% halothane and nitrous oxide / oxygen (70:30) and transcardial infusion of saline and 4% paraformaldehyde (PFA) in 0.1 mol /1 PBS. Thereafter, mice brains were post-fixed in the same fixative overnight. After paraffin embedding, 3-μm-thick sections were cut using a microtome (Leica, Bensheim, Germany), mounted on poly-1-lysine-coated glass slides, and used for histological investigations. Modified versions of hematoxylin-eosin (H&E), Bielschowski silver, and Congo red stainings according to Puchtler and Higmann were performed [30]. The sections were coverslipped and observed under a polarization microscope (OLYMPUS, Hamburg. Germany).

*Immunohistochemistry*

Immunohistochemical investigations were performed for hyperphosporylated tau protein and Abeta. To unmask the antigens, deparaffinized and hydrated sections were incubated in trypsin-EDTA (DakoCytomation, Hamburg, Germany) at 70°C for 25 min to enhance the immunoreactivity, and thereafter incubated in 1% normal goat serum in PBS at 4°C for 1 h. The specimens were blocked for 15 min with 3% $H_2O_2$ in distilled water at room temperature. For tau detection, sections were immunostained overnight at 4°C using primary monoclonal antibody AT8, which recognizes human tau phosphorylated at Ser202 and Thr205 (Thermo Scientific, MN 1020, Lot: JL 126853, 1:100). The primary antiserum was omitted in a subset of slides, which resulted in no immunostaining (negative control). Samples were then incubated in streptavidin / HRP (horse radish peroxidase, DakoCytomation, Hamburg, Germany, 1:400) for 30 min, washed again, and subsequently stained with DAB (3,3'Diaminobenzidine, BIOMOL, Hamburg, Germany), followed by counterstaining with hematoxylin & eosin (H&E, SIGMA-Aldrich, Munich, Germany) before embedding in Elvanol.

For Abeta determination, human monoclonal antibody against beta amyloid (DAKO, Clone 6F, 3D monoclonal antibody, diluted 1:40) was used for staining in accordance with the manufacturer's data and employing an automatic analyzing system (BENCHMARK, VENTANA Medical system INC., Illkirch, CEDEX, France). Thereafter, slides were counterstained with H&E staining before embedding in Elvanol.

Qualitative microscopic investigations of tau- and Abeta-positive cells were performed using light microscopy at various magnifications (Zeiss Axioskop, 10x-- 400x) in predefined mouse brain sections: subventricular zone, the basal ganglia, hippocampus, and cerebral cortex. Cells were counted by an examiner blinded to the experimental conditions in a defined area of 50 mm$^2$ that was determined by using digital imaging software Spot-Basic™ (Sterling Heights, MI, USA).

Respective cell count differences were determined using two-tailed, non-paired, Student's t-test. P-values<0.05 were considered statistically significant.

To determine apoptotic changes at 6 months after icv STZ injections, 3-um-thick paraffin slices were deparaffinized and hydrated and analyzed for apoptotic changes using a TUNEL assay (Neuro-Tacs II, R&D Systems GmbH, Wiesbaden, Germany). Testing was conducted according to the manufacturer's instructions.

*Statistics*

All measurements were performed by an independent investigator blinded to the experimental conditions. Results are expressed in figures as mean ± standard error of means (SEM). Differences within or between the normally distributed data were analyzed by analysis of variance (ANOVA) using SPSS 16.0 version followed by *post-hoc* Tukey test. In the case of non-normally distributed data for biochemical analyses, the significance of between-group differences was tested by Wilcoxon or Mann-Whitney U test. Statistical significance was set to the $p \leq 0.05$ level. Correlation analyses were done according to Pearson.

Variables from the Morris Water Maze test were quantified with the EthoVision 3.1 software (Noldus Information Technology, Freiburg, Germany), and mean values for each trial were calculated for each subject. The following parameters were calculated: (1.) escape performance during training by time to find the platform (escape latency in seconds), and (2.) escape path length (distance moved into the basin in cm) during a 90-second trial. For statistical analysis, the effect of treatment (STZ tg versus placebo tg) on the acquisition of the water escape task was assessed by using ANOVA with the repeated measure factor sessions (number of days). For probe trial analysis in the water maze, within-group differences in the time spent in each quadrant were analyzed by means of one-factor ANOVA with repeated measure on the within-subject factor time spent in quadrant (four levels). When ANOVA detected a significant strain treatment or time effects, pair-wise differences between means for a given variable were evaluated using Bonferroni post hoc tests for multiple pair-wise comparisons, with significance set to $p < 0.05$.

## RESULTS

*Baseline studies on wild-type mice*

We compared blood glucose levels and body weights in tg2576 mice and wild-type (wt) control mice (n=16) in a control study; we did not detect any differences *(p > 0.05)* between 9-month-old tg mice and strain-, age-, and gender-matched wild-type mice (data not shown).

*Survival was decreased after STZ*

In STZ tg mice, the mortality until 9 months of age was significantly about 1.8-fold (6 out of 13, 46%) higher than in placebo tg-injected mice (3 out of 12, 25%, figure 2).

*STZ reduced spatial cognitive performance in Morris water maze*

To test whether a single icv STZ injection induces spatial memory deficits, we assessed performance in the Morris water maze task, a hippocampus-dependent spatial learning task in which mice are required to learn to locate an escape platform in a pool of water.

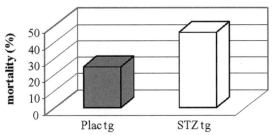

Fig. 2. Mortality. During a 6-month observation period, streptozotocin (STZ) induced a nearly 2-fold increased mortality rate as compared to placebo (plac) mice (ANOVA, p < 0.05).

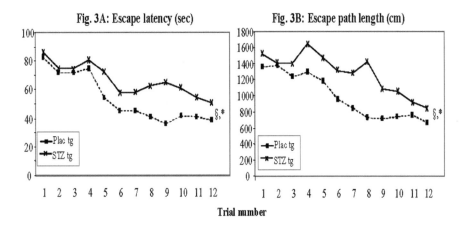

Fig. 3. Learning effects in Morris Water maze after training Means during training from trial 1 until trial 12 are represented. Plac: placebo (n=9): STZ: streptozotocin (n=7), tg: transgenic. ANOVA, p < 0.05. §: Time effect in both groups: trial 1 versus trial 12, *: significant differences between the groups.

In a cued trial with a visible platform, all mice found the platform and no significant changes between the placebo tg and STZ tg groups were observed (21.8 ± 6.5 sec versus 22.3 ± 8.9 sec, respectively), suggesting that no motivational or sensory motor deficit differences exist. Furthermore, no significant difference was observed in swimming speed between placebo tg (20.8 ± 5.5 cm / sec) and STZ tg (21.0 ± 3.6 cm / sec) mice.

In the hidden platform trial, a marked training effect between trial 1 and 12 (§: intragroup factor time, $p < 0.05$) was observed in both groups for escape latency (Fig. 3A) and escape path length (Fig. 3B). Placebo tg mice as compared to the STZ tg mice were able to locate the platform in less time [$F(1,4) = 33.7$, *: $p < 0.05$] and in a shorter distance [$F(1,4) = 28.2$, *: $p < 0.05$].

## STZ-treated mice have spatial memory deficits

To determine whether mice were using a spatial memory strategy to locate the platform, animals were subjected to a probe trial conducted after the training procedure. The results showed (Fig. 4) that STZ tg mice spent less time searching in the target quadrant where the platform had been located during training than in each of the other quadrants (*: $p < 0.05$), indicating that spatial learning had deteriorated in STZ tg as compared to placebo tg mice. Only placebo tg mice showed a significant (§: $p < 0.05$) effect if compared the time searching in the target quadrant as compared to the other quadrants.

## STZ reduced memory capacities during aging

To determine the age-dependency after STZ on Morris water maze parameters, all mice were subjected to retesting every 6 weeks. In retest settings 3 (mice age: 7.5 months) and 4 (mice age: 9 months), placebo tg animals showed better cognitive spatial performance than STZ tg animals (Fig. 5A and B). Thus, with increasing mice age, a cognitive decline was obtained in STZ tg mice in escape la-

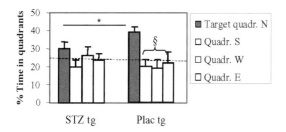

Fig. 4. Probe trial test in Morris water maze. Streptozotocin (STZ)-injected mice spent less time in target quadrant as compared to placebo (plac)-treated mice (p< 0.05). Bars represent mean (+ SEM) time spent in the target quadrant and in the other 3 maze quadrants. Horizontal dashed line indicates a chance level of 25%. ANOVA, p < 0.05 *: significant effect between the groups, §: significant (p < 0.05) effects within the different quadrants (quadr.). N: nord, W: west, E: east, S: south

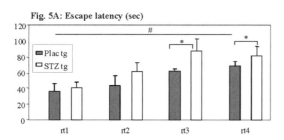

Fig. 5. Effect of age-related retesting in Morris water maze Means + SEM during mice' aging, Plac: placebo: STZ: streptozotocin, tg: transgenic, rt: retesting (age of mice: rt 1= 4.5 month, rt2 = 6 month, rt 3 = 7.5 month, rt 4 = 9 month). n=7 and n=9. ANOVA. *: Group effect with hidden platform: streptozotocin (STZ) versus placebo (plac), p < 0.05. #: Time effect in both groups with hidden platform: mice' age of 9 month versus 3-month, ANOVA, p < 0.05.

tencies (times to platform) and in escape path length (distance to platform).

In addition to the STZ effect, an age-dependent deterioration was found over the 6-month observation period, which was significant for the escape latency (#: $p < 0.05$).

## STZ increases APP fragments and diffuse amyloid plaques

Amyloid load was significantly increased in tg mice after STZ (Fig. 6A). Six months following a single STZ treatment, the content of all measured soluble and aggregated A40beta and A42beta fragments was higher in STZ tg mice than in placebo tg animals. Significant differences were observed for the aggregated forms of Abeta 40 and 42 (Fig. 6B). Interestingly, a significant linear correlation was found between Abeta 42 fragment and escape latency ($r = 0.726$, $p = 0.00019$) and Abeta 42 and escape path length ($r = 0.634$, $p = 0.0043$) regardless of treatment. Abeta40 fragment did not show significant correlation to the investigated water maze parameters.

Histologically, Bielschowski staining did not detect *compact* amyloid plaques. Congophilic brain capillaries as has been detected by Congo red staining according to Puchtler and Higman were detected in both mice groups (Fig. 6C, above).

*Diffuse* plaques, however, were significantly increased after STZ as has been detected by Congo red staining (Fig. 6C, below) and by immunohistochemical investigation (data not shown).

## STZ did not induce apoptosis

Using the TUNEL technique, apoptosis staining did not detect any marked cell death. Thus, no significant apoptotic structural deterioration was observed until mice reached the age of 9 months (placebo mice) and after STZ injections.

## STZ increased total tau protein

The results of total and phosphorylated tau proteins ELISA analyses are illustrated in figure 7. Accordingly, STZ tg mice had more total tau protein than placebo animals at 6 months (Fig. 7A, $p < 0.05$). In relation to phosphorylated tau protein, differences of means failed to reach statistical significance in ELISA analysis (Fig. 7B). Immunohistochemical detection only revealed sporadic tau-positive cells in STZ tg mice brain as compared to placebo tg, two representative figures are shown as inserted pictures in figure 7B. No marked differences between both groups were obtained.

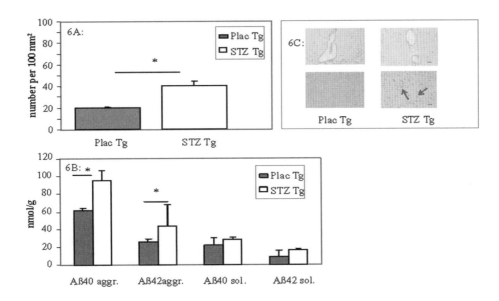

Fig. 6. Amyloid burden (A) and beta fragment formation (B) after STZ Effect of streptozotocin (STZ) is represented as means + SEM in transgenic (tg) mice on amyloid burden (fig. 6A, n=2 per group) and on aggregated (aggr.) and soluble (sol) Abeta40 and Abeta42 fragments (fig. 6B) in comparison to placebo-treated (plac) mice. Means from n=5 (STZ) and n=7 (plac) mice (+ SD) per group are represented for ELISA APP fragment analysis. *: STZ vs. placebo, ANOVA, $p < 0.05$. Amyloid burden = number of Congo red diffuse plaques and the number of congophilic brain capillaries in mouse cerebral tissue which was counted under a 12.5 fold magnification in four brain sections. The number of Congo red cells was related to an area of 100 mm$^2$ brain tissue. Figures 6C represent representative photographies of congophilic cerebral capillares (left) in both groups and diffuse amlyoid plaques (arrow on the right picture) in a 9-month-old STZ treated tg mouse brain (Congo red staining).

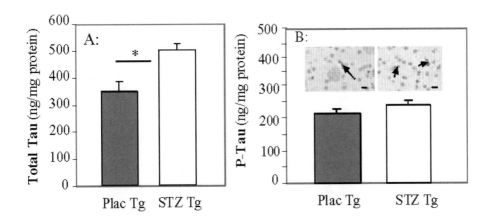

Fig. 7. Effect of STZ on tau protein. Tau-ELISA data are represented as means+SEM. As compared to placebo (plac) treatment (n=7), streptozotocin (STZ, n=5) induces an increase in total tau protein concentration (ANOVA, $p < 0.05$, fig. 7A). Phosphorylated tau (p-tau, fig. 7B) was not changed markedly in transgenic (tg) animals after STZ versus placebo. AT-8 immunostaining showed only isolated tau-positive cells (arrows) in both mice groups. Two representative photographies are shown as inserted pictures in Fig. 7B (magnif.: x400, scale bar: 20 μm).

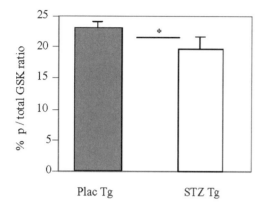

Fig. 8. Effect of STZ on GSK-3 alpha /beta ratio. Ratio of phosphorylated (p) and total GSK-3 alpha /beta was calculated. Both forms of GSK-3 alpha /beta was measured by ELISA technique and are represented as means+SEM. As compared to placebo (plac) treatment (n=7), streptozotocin (STZ, n=5) induced a 22% decrease in the ratio (ANOVA, *: p = 0.05).

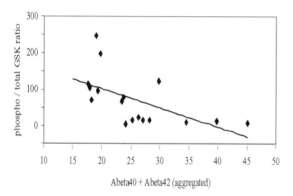

Fig. 9. Correlation analysis Figure 9 demonstrates that there was a significant linear correlation (p = 0.015) between GSK-3alpha/beta ratio and aggregated amyloid fragments (Abeta 40+42) with a negative correlation coefficient of rho = -0.58.

*STZ treatment increased total GSK-3'alpha/beta and decreased phosphorylated /total GSKalpha/beta ratio*

The results of the present study showed that in STZ tg mice the content of total GSK-3 alpha/beta increased to 1352.2 ± 326.3 pg / mg (mean ± SD) as compared to placebo tg (1008.6 ± 360.6 pg / mg, $p = 0.049$). Significant changes in phosphorylated GSK-3 alpha/beta were not observed between placebo tg-treated mice (223.6 ± 41.8 pg / mg) and STZ tg mice (255.6 ± 66.7 pg / mg).

The ratio between phosphorylated and total GSK-3 alpha/beta was decreased by about 22% in STZ tg mice as compared to placebo tg mice *(p = 0.05, figure 8)*.

Figure 9 demonstrates that there was a significant linear correlation *(p = 0.015)* between GSK-3 alpha/beta ratio and aggregated amyloid fragments (Abeta 40+42) with a negative correlation coefficient of rho = -0.58.

## DISCUSSION

The metabolic pathway of the genetically modified APP based on the amyloid cascade hypothesis [31] has been well documented. Hereditary Alzheimer's disease (AD) has been considered to be caused by this pathology. In contrast, no such definite genetic abnormality has been found for sporadic AD. For the latter, insulin-resistant brain state (IRBS) has been proposed to represent the core abnormality [15,13,32]. It is as yet unclear whether the effect of an IRBS might be a more powerful and detrimental abnormality than mutated APP to generate AD-like abnormalities and what the effect of these two disturbances would be if present concomitantly.

For the first time, we now present evidence that AD-like functional and structural changes in tg2576 mice are aggravated by an IRBS due to icv injection of STZ. This procedure increased mice mortality, increased formation of aggregated Abeta fragments and congophilic amyloid deposits, and increased total tau protein content and decreased phosporylated/total GSK-3alpha/beta ratio in tg mice. Furthermore, icv STZ administration reduced spatial cognitive functions in tg mice.

In both AD patients and animal models, there is increasing evidence that cerebral dysfunction occurs before B-amyloid-containing compact plaques accumulate and neurodegeneration ensues [33,34]. Thus, memory deficits occurred despite the lack of significant amyloid plaque or NFT pathology. Instead, changes in brain energy state [35] or neurotransmitters [36], stress hormones [37] etc. may be responsible for cognitive dysfunction, as has been shown in many other studies too. The present results showed an age-dependent decrease in mice spatial cognitive abilities starting at the age of 6 - 9 months without compact plaque formation. It is well documented that in the tg2576 mice line – in which the human betaAPP with the Swedish mutation is overexpressed – Abeta40 and Abeta42/43 peptides were increased at 3 to 9-10 months of age [1]. However, amyloid plaque depositions resembling late-onset AD pathology were found later, not earlier than 10 - 12 months of age [7,8,9]. This raised the question of whether any age-related metabolic factors or processes trigger or induce beta-amyloid plaque deposition and / or amyloidogenesis, such as alterations in cholinergic transmission, in oxidative stress, in glucose / energy metabolism, and in cerebral insulin transduction [38,39,40]. Indeed, after icv STZ administration markedly more aggregated Abeta fragments were detected in mice aged 9 months (Fig. 6). Furthermore, a significant negative linear correlation between aggregated Abeta42 and cognitive abilities was found in these STZ-tg mice, similar to that found in tg2576 mice as also reported by Hsiao et al. [1] and Westerman et al. [8], emphasizing the principal significance of Abeta42 fragments in AD-like cognitive dysfunction. However, despite the increased amyloid burden in tg mice brain (Fig. 6A), no compact plaques could be detected histologically after STZ using Bielschowsky staining in mice aged 9 months. Thus, further studies are necessary in mice 12 months of age and older to investigate the time dependency of histological changes including synaptic degeneration and the APP fragmentation process dysfunction following icv STZ administration.

Tau protein - one of the main microtubule-associated proteins - is hyperphosphorylated and loses its ability to bind microtubules when the homeostasis of phosphorylation and dephosphorylation in neurons is disturbed [41,42]. Accumulated Abeta and hyperphosphorylated tau are thought to be co-existent [43]. In accordance with other investigations [3], we did not detect a significant increase in neurofibrillary tangles, or overt neuronal loss, including apoptosis, in tg2576 mice. However, total tau protein was increased in the brains of tg2576 mice 6 months after icv STZ administration (Fig. 7). This is in line with our previous work on rats that showed increased tau protein in the hippocampus of STZ-treated animals after 1 month in comparison to the levels in control animals (placebo) [24]. Furthermore, in studies on rats dealing with the peripheral administration of STZ, a mild hyperphosphorylation of tau could be detected 10, 20, and 30 days after STZ injection, which preceded a massive hyperphosphorylation of tau observed after 40 days. However, neither mild nor massive tau phosphorylation induced tau aggregation [44]. To explain the differences between our results and the literature data, the following facts may need to be taken into account: i) the long-term effect on tau hyperphosphorylation 6 months after icv STZ administration has not been published yet and therefore the different post-STZ administration periods could account for the inconsistent results; ii) insulin dysfunction may induce abnormal tau hyperphosphorylation through two distinct mechanisms: one is related to hypothermia [44] (we did not measure body temperature changes in the present study); the other is temperature-independent, inherent to insulin depletion, and probably caused by inhibition of the protein phosphatase 2A activity [44]; and iii) in AD not only phosphorylation, but also the modifications such as glycosylation processes play a role in abnormal tau formation [45].

Different mechanisms of STZ action on the insulin cascade involved in amyloid and NFT generation are currently being discussed, including alterations on the insulin receptor and its tyrosine kinases, the protein phosphatase 2A, the Akt/PKB, and the insulin-degrading enzyme (IDE), and the GSK-3 enzymes [27,46,47]. Among these is the GSK-3 alpha/beta, a serine/threonine kinase, which has been reported to play a key role in relation to hallmarks of AD, such as amyloid plaques [47] and hyperphosphorylated tau [48,44,49]. Interestingly, GSK-3 alpha/beta was found to accumulate in neurons probably prior to developing tangles [50,51,52]. Using the rat IRBS model, it has been shown that, in the hippocampus of icv STZ-treated rats, alterations in Akt/PKB-GSK-3 enzymes have been found downstream of the PI3 kinase pathway of the IR signaling cascade [24]. In detail, the level of pGSK-3 alpha/beta was significantly increased 1 month after icv STZ treatment and then decreased to below the control values by 3 months, which was accompanied by a decreased p/total GSK-3 alpha/beta ratio; the decrement in p/total GSK alpha/beta ratio indicates increased GSK-3 alpha/beta activity. Inconsistent results, however, have been reported with respect to the GSK-3 protein levels measured in brain in humans with AD [53,54,55].

The results of the present study showed a decreased ratio of phosphorylated to total GSK-3 at 6 months after STZ (Fig. 8), corresponding to the findings in human AD [55, 56]. This effect is mainly mediated by an increase in the active (nonphosphorylated) total GSK-3alpha/beta form with concurrently unchanged phosphorylated GSK-3 and might favor the formation of beta-amyloid peptides. Indeed, an increase in aggregated Abeta fragments after STZ administration was not only associated with cognitive impairments, as discussed above, but there was also a linear correlation between aggregated Abeta and GSK-3 alpha/beta ratio (Fig. 9).

This study includes some limitations: (1) the results of wt control mice were not available for comparison of the effects in this longitudinal study, and (2) additional experiments are necessary to establish the mechanisms-of-action in relation to STZ and APP. These main points will be addressed in further studies.

We conclude from the present study, that (1) early, metabolic changes in the form of IRBS in sAD pathomechanism are responsible for alterations in cognitive function rather than compact plaque formation or tau-hyperphosphorylation; and (2) these early changes seem to be caused by altered GSK-3 pathway downstream of the insulin receptor signaling cascade as their progression correlates with p/total GSK-3 ratio. However, further studies on insulin transduction pathway are needed to support this assumption.

In summary, this study provides the first evidence that, via disturbed GSK-3alpha/beta function, metabolic conditions such as IRBS may aggravate AD-like functional and structural alterations in the life-time of tg2576 mice, a model of hereditary AD.

## ACKNOWLEDGEMENTS

The authors would like to thank Mr. Roland Galmbacher for his support in the animal surgery, Mr. Klaus Stefan for his help with the histological investigations, and Ms. Sigrun Himmelsbach for her help with the biochemical analysis. Furthermore, we would like to thank Julia Tuchy for her support in performing the water maze tests and Prof. Dr. Clemens Sommer; University of Mainz, Germany, for discussions of the histological results. The authors like to express their gratitude to Dr. Karen Hsiao Ashe, Department of Neurology, University of Minnesota, USA, for kindly providing three tg2576 founder mice to R.S. that have been used for breeding of an own animal stock.

## References

[1] Hsiao K, Chapman P, Nilsen S, Eckman C, Harigaya Y, Younkin S, Yang F, Cole G (1996) Correlative memory deficits, Abeta elevation, and amyloid plaques in transgenic mice. *Science* **274**, 99-102.

[2] Carlson GA, Borchelt DR, Dake A, Turner S, Danielson V, Coffin JD, Eckman C, Meiners J, Nilsen SP, Younkin SG, Hsiao KK (1997) Genetic modification of the phenotypes produced by amyloid precursor protein overexpression in transgenic mice. *Hum Mol Genet* 6, 1951-1959.

[3] Irizarry MC, McNamara M, Fedorchak K, Hsiao K, Hyman BT (1997) APPSw transgenic mice develop age-related A beta deposits and neuropil abnormalities, but no neuronal loss in CA1. *J Neuropathol Exp Neurol* 56, 965-973.

[4] Chapman PF, White GL, Jones MW, Cooper-Blacketer D, Marshall VJ, Irizarry M, Younkin L, Good MA, Bliss TV, Hyman BT, Younkin SG, Hsiao KK (1999) Impaired synaptic plasticity and learning in aged amyloid precursor protein transgenic mice. *Nat Neurosci* 2, 271-276.

[5] Kawarabayashi T, Younkin LH, Saido TC, Shoji M, Ashe KH, Younkin SG (2001) Age-dependent changes in brain, CSF, and plasma amyloid (beta) protein in the Tg2576 transgenic mouse model of Alzheimer's disease. *J Neurosci* 21, 372-381.

[6] Frautschy SA, Yang F, Irrizarry M, Hyman B, Saido TC, Hsiao K, Cole GM (1998) Microglial response to amyloid plaques in APPsw transgenic mice. *Am J Pathol* **152**, 307-17.

[7] Ashe KH (2001) Learning and memory in transgenic mice modeling Alzheimer's disease. *Learn Mem* 8, 301-308.

[8] Westerman MA, Cooper-Blacketer D, Mariash A, Kotilinek L, Kawarabayashi T, Younkin LH, Carlson GA, Younkin SG, Ashe KH (2002) The relationship between Abeta and memory in the Tg2576 mouse model of Alzheimer's disease. *J Neurosci* **22**, 1858-1867.

[9] Kotilinek LA, Bacskai B, Westerman M, Kawarabayashi T, Younking L, Hyman BT, Younkin S, Ashe KH (2002) Reversible memory loss in a mouse transgenic model of Alzheimer's disease. *J Neurosci* **22**, 6331-6335.

[10] Leibson CL, Rocca WA, Hanson VA, Cha R, Kokmen E, O'Brien PC, Palumbo PJ (1997) Risk of dementia among persons with diabetes mellitus: a population-based cohort study. *Am J Epidemiol* **145**, 301-308.

[11] Stolk RP, Breteler MM, Ott A, Pols HA, Lamberts SW, Grobbee DE, Hofman A (1997) Insulin and cognitive function in an elderly population. The Rotterdam Study. *Diabetes Care* 20, 792-795.

[12] Neumann KF, Rojo L, Navarrete LP, Farias G, Reyes P, Maccioni RB (2008) Insulin resistance and Alzheimer's disease: molecular links & clinical implications. *Curr Alzheimer Res* 5: 438-447.

[13] Salkovic-Petrisic M, Osmanovic J, Grünblatt E, Riederer P, Hoyer S (2010) Modeling sporadic Alzheimer disease: the insulin resistant brain state generates multiple long-term morphobiological abnormalities inclusive hyperphosphorylated tau protein and amyloid. A synthesis. *J Alzheimer's Disease* (in press)

[14] Jones A, Kulozik P, Ostertag A, Herzig S (2009) Common pathological processes and transcriptional pathways in Alzheimer's disease and type 2 diabetes. *J Alzheimers Dis* 16, 787-808.

[15] Salkovic-Petrisic M, Hoyer S (2007) Central insulin resistance as a trigger for sporadic Alzheimer-like pathology: an experimental approach. *J Neural Transm Suppl* 72, 217-233.
[16] Blokland A, Jolles J (1993) Spatial learning deficit and reduced hippocampal ChAT activity in rats after an ICV injection of streptozotocin. *Pharmacol Biochem Behav* 44, 491-494.
[17] Lannert H, Hoyer S (1998) Intracerebroventricular administration of streptozotocin causes long-term diminutions in learning and memory abilities and in cerebral energy metabolism in adult rats. *Behav Neurosci* 112, 1199-1208.
[18] Sharma M, Gupta YK (2001) Intracerebroventricular injection of streptozotocin in rats produces both oxidative stress in the brain and cognitive impairment. *Life Sci* 68, 1021-1029.
[19] Shoham S, Bejar C, Kovalev E, Weinstock M (2003) Intracerebroventricular injection of streptozotocin causes neurotoxicity to myelin that contributes to spatial memory deficits in rats. *Exp Neurol* 184, 1043-1052.
[20] Nitsch R, Hoyer S (1991) Local action of the diabetogenic drug, streptozotocin, on glucose and energy metabolism in rat brain cortex. *Neurosci Lett* 128, 199-202.
[21] Hellweg R, Nitsch R, Hock C, Jaksch M, Hoyer S (1992) Nerve growth factor and choline acetyltransferase activity levels in the rat brain following experimental impairment of cerebral glucose and energy metabolism. *JNeurosci Res* 31, 479-486.
[22] Plaschke K, Hoyer S (1993) Action of the diabetogenic drug streptozotocin on glycolytic and glycogenolytic metabolism in adult rat brain cortex and hippocampus. *Int JDevNeurosci* 11, 477-483.
[23] Griinblatt E, Salkovic-Petrisic M, Osmanovic J, Riederer P, Hoyer S (2007) Brain insulin system dysfunction in streptozotocin intracerebroventricularly treated rats generates hyperphosphorylated tau protein. *JNeurochem* 101: 757-770.
[24] Salkovic-Petrisic M, Tribl F, Schmidt M, Hoyer S, Riederer P (2006) Alzheimer-like changes in protein kinase B and glycogen synthase kinase-3 in rat frontal cortex and hippocampus after damage to the insulin signalling pathway. *J Neurochem* 96, 1005-1015.
[25] Leloup C, Arluison M, Lepetit N, Cartier N, Marfaing-Jallat P, Ferre P, Penicaud L (1994) Glucose transporter 2 (GLUT 2): expression in specific brain nuclei. *Brain Res* 638, 221-226.
[26] Havrankova J, Schmechel D, Roth J, Brownstein M (1978) Identification of insulin in rat brain. *ProcNatlAcadSci USA* 75, 5737-5741.
[27] Zhao L, Teter B, Morihara T, Lim GP, Ambegaokar SS, Ubeda OJ, Frautschy SA, Cole GM. (2004) Insulin-degrading enzyme as a downstream target of insulin receptor signaling cascade: implications for Alzheimer's disease intervention. *J Neurosci* 24, 11120-11126.
[28] Franklin KBJ, Paxinos G (1997) *The mouse brain stereotaxic coordinates*, Academic Press, San Diego.
[29] Plaschke K, Staub J, Ernst E, Marti HH (2008) VEGF overexpression improves mice cognitive abilities after unilateral common carotid artery occlusion. *Exp Neurol* 214, 285-292.
[30] Schenk EA, Churukian C, Willis C, Stotz E (1983) Staining problems with eosin Y: a note from the Biological Stain Commission. *Stain Technol* 58, 377-379.
[31] Hardy J, Selkoe DJ (2002) The amyloid hypothesis of Alzheimer's disease: progress and problems on the road to therapeutics. *Science* 297, 353-356.
[32] Hoyer S (1998) Risk factors for Alzheimer's disease during aging. Impacts of glucose/energy metabolism. *J Neural Transm Suppl* 54, 187-194.
[33] Jacobsen JS, Wu CC, Redwine JM, Comery TA, Arias R, Bowlby M, Martone R, Morrison JH, Pangalos MN, Reinhart PH, Bloom FE (2006) Early-onset behavioral and synaptic deficits in a mouse model of Alzheimer's disease. *ProcNatlAcadSci USA* 103, 5161-5166.
[34] Middei S, Daniele S, Caprioli A, Ghirardi O, Ammassari-Teule M (2006) Progressive cognitive decline in a transgenic mouse model of Alzheimer's disease overexpressing mutant hAPPswe. *Genes Brain Behav* 5: 249-256.
[35] Plaschke K, Yun SW, Martin E, Hoyer S, Bardenheuer HJ (2000) Linear relation between cerebral phosphocreatine concentration and memory capacities during permanent brain vessel occlusions in rats. *Ann N YAcad Sci* 903, 299-306.
[36] Benfenati F (2007) Synaptic plasticity and the neurobiology of learning and memory. *Ada Biomed* 78, Suppl 1, 58-66.
[37] Lupien SJ, McEwen BS, Gunnar MR, Heim C (2009) Effects of stress throughout the lifespan on the brain, behaviour and cognition. *Nat Rev Neurosci* 10, 434-445.
[38] Bigl M, Apelt J, Eschrich K, Schliebs R (2003) Cortical glucose metabolism is altered in aged transgenic Tg2576 mice that demonstrate Alzheimer plaque pathology. *J Neural Transm* 110, 77-94.
[39] Apelt J, Bigl M, Wunderlich P, Schliebs R (2004) Aging-related increase in oxidative stress correlates with developmental pattern of beta-secretase activity and beta-amyloid plaque formation in transgenic Tg2576 mice with Alzheimer-like pathology. *Int JDev Neurosci* 22, 475-484.
[40] Schliebs R (2005) Basal forebrain cholinergic dysfunction in Alzheimer's disease—interrelationship with beta-amyloid, inflammation and neurotrophin signaling. *Neurochem Res* 30, 895-908.
[41] Braak H, Braak E (1997) Diagnostic criteria for neuropathologic assessment of Alzheimer's disease. *Neurobiol Aging* 18(4 Suppl), S85-S88.
[42] Markesbery WR (1997) Neuropathological criteria for the diagnosis of Alzheimer's disease. *Neurobiol Aging* 18(Suppl), S13-S19.
[43] Noda-Saita K, Terai K, Iwai A, Tsukamoto M, Shitaka Y, Kawabata S, Okada M, Yamaguchi T (2004) Exclusive association and simultaneous appearance of congophilic plaques and AT8-positive dystrophic neurites in Tg2576 mice suggest a mechanism of senile plaque formation and progression of neuritic dystrophy in Alzheimer's disease. *Acta Neuropathol* 108, 435-442.
[44] Planel E, Tatebayashi Y, Miyasaka T, Liu L, Wang L, Herman M, Yu WH, Luchsinger JA, Wadzinski B, Duff KE, Takashima A (2007) Insulin dysfunction induces in vivo tau hyperphosphorylation through distinct mechanisms. *J Neurosci* 27, 13635-13648.
[45] Alonso AC, Li B, Grundke-Iqbal I, Iqbal K (2008) Mechanism of tau-induced neurodegeneration in Alzheimer disease and related tauopathies. *Curr Alzheimer Reserach* 5: 375-384.
[46] Myers A, Holmans P, Marshall H, Kwon J, Meyer D, Ramie D, Shears S, Booth J, DeVrieze FW, Crook R, Hamshere M, Abraham R, Tunstall N, Rice F, Carty S, Lillystone S, Kehoe P, Rudrasingham V, Jones L, Lovestone S, Perez-Tur J, Williams J, Owen MJ, Hardy J, Goate AM (2000) Susceptibility locus for Alzheimer's disease on chromosome 10. *Science* 290, 2304-2305.

[47] Phiel CJ, Wilson CA, Lee VM, Klein PS (2003) GSK-3alpha regulates production of Alzheimer's disease amyloid-beta peptides. *Nature* **423,** 435-439.

[48] Rankin CA, Sun Q, Gamblin TC (2007) Tau phosphorylation by GSK-3beta promotes tangle-like filament morphology. *Mol Neurodegener* 2:12.

[49] Resende R, Ferreiro E, Pereira C, Oliveira CR (2008) ER stress is involved in Abeta-induced GSK-3beta activation and tau phosphorylation. *J Neurosci Res* 86, 2091-2099.

[50] Leroy K, Boutajangout A, Authelet M, Woodgett JR, Anderton BH, Brion JP (2002) The active form of glycogen synthase kinase-3beta is associated with granulovacuolar degeneration in neurons in Alzheimer's disease. *Acta Neuropathol* **103,** 91-99.

[51] Pei JJ, Khatoon S, An WL, Nordlinder M, Tanaka T, Braak H, Tsujio I, Takeda M, Alafuzoff I, Winblad B, Cowburn RF, Grundke-Iqbal I, Iqbal K (2003) Role of protein kinase B in Alzheimer's neurofibrillary pathology. *Ada Neuropathol* **105,** 381-392.

[52] Griffin RJ, Moloney A, Kelliher M, Johnston JA, Ravid R, Dockery P, O'Connor R, O'Neill C (2005) Activation of Akt/PKB, increased phosphorylation of Akt substrates and loss and altered distribution of Akt and PTEN are features of Alzheimer's disease pathology. *J Neurochem* 93, 105-117.

[53] Pei JJ, Tanaka T, Tung YC, Braak E, Iqbal K, Grundke-Iqbal I (1997) Distribution, levels, and activity of glycogen synthase kinase-3 in the Alzheimer disease brain. *J Neuropathol Exp Neurol* 56, 70-78.

[54] Steen E, Terry MB, Riviera JE, Cannon LJ, Neely RT, Tavares R, Xu JX, Wandas RJ, de la Monte MS (2005) Impaired insulin and insulin-like growth factor expresion and signaling mechanisms in Alzheimer's disease-is this type 3 diabetes? *J Alzheimer's Disease* 7, 63-80.

[55] Preece P, Virley DJ, Costandi M, Coombes R, Moss SJ, Mudge AW, Jazin E, Cairns NJ (2003) Beta-secretase (BACE) and GSK-3 mRNA levels in Alzheimer's disease. *Brain Res Mol Brain Res* **116,** 155-158.

[56] Hye A, Kerr F, Archer N, Foy C, Poppe M, Brown R, Hamilton G, Powell J, Anderton B, Lovestone S (2005) Glycogen synthase kinase-3 is increased in white cells early in Alzheimer's disease. *Neurosci Lett* **373,** 1-4.

# Hippocampal Alterations in Rats Submitted to Streptozotocin-Induced Dementia Model: Neuroprotection with Aminoguanidine

Letícia Rodrigues[1], Regina Biasibetti[2], Alessandra Swarowsky[2], Marina C. Leite[2], André Quincozes-Santos[2], Matilde Achaval[1], Carlos-Alberto Gonçalves[1,2]
[1]*Programa de Pós-Graduação em Neurociências, Instituto de Ciências Básicas da Saúde, Universidade Federal do Rio Grande do Sul, Porto Alegre, Brazil*
[2]*Departamento de Bioquímica, Instituto de Ciências Básicas da Saúde, Universidade Federal do Rio Grande do Sul, Porto Alegre, Brazil*

**Abstract**. Although the exact cause of Alzheimer's disease remains elusive, many possible risk factors and pathological alterations have been used in the elaboration of *in vitro* and *in vivo* models of this disease in rodents, including intracerebral (ICV) infusion of streptozotocin (STZ). Using this model, we evaluated spatial cognitive deficit and neurochemical hippocampal alterations, particularly astroglial protein markers such as glial fibrillary acidic protein (GFAP) and S100B, glutathione content, NO production and cerebrospinal fluid (CSF) S100B; in addition, prevention of these alterations by aminoguanidine administration was evaluated. Results confirm a spatial cognitive deficit and nitrative stress in this dementia model as well as specific astroglial alterations, particularly S100B accumulation in the hippocampus and decreased CSF S100B. The hippocampal astroglial activation occurred independently of the significant alteration in GFAP content. Moreover, all these alterations were completely prevented by aminoguanidine administration, confirming the neuroprotective potential of this compound, but suggesting that nitrative stress and/or glycation may be underlying these alterations. Findings contribute to the understanding of diseases accompanied by cognitive deficits and the STZ-model of dementia.

Keywords: aminoguanidine, astrocyte, hippocampus, nitrosative stress, S100B, streptozotocin

## INTRODUCTION

Dementia is a serious and growing public heath problem that is pathologically characterized by a progressive decline in cognitive function that affects cortical and/or subcortical brain structures beyond what might be expected from normal aging. The condition affects about 5% of the elderly population of over 65 years old and 25% of those over 80 years old, where Alzheimer's disease (AD) represents more than 50% of cases [1, 2].

Although the exact cause of AD remains elusive, many possible risk factors and pathological alterations have been used in the elaboration of *in vitro* and *in vivo* models of this disease. Non-transgenic animal models include chronic cerebral hypoperfusion, intracerebroventricular (ICV) infusion of beta-amyloid peptide [3] or streptozotocin (STZ) [4] and lesion of nucleus basalis magnocellularis [5]. The usefully of these models is indicated by the presence of beta-amyloid aggregates, tau hyperphosphorylation, and glial abnormalities, particularly in the cerebral cortex and hippocampus. Alterations of the insulin-mediated signaling in brain tissue resembling those in sporadic AD have been found in the ICV infusion of STZ rat model and are associated with increase of tau protein phosphorylation and beta-amyloid production [6, 7].

In fact, ICV infusion of STZ provides a relevant animal model of chronic brain dysfunction that is characterized by long-term and progressive deficits in learning, memory, and cognitive behavior, indicated by decreases in working and reference memory, along with a permanent and ongoing cerebral

---

Corresponding author: Carlos-Alberto Gonçalves, Departamento de Bioquímica, ICBS, Universidade Federal do Rio Grande do Sul, Ramiro Barcelos, 2600-Anexo, 90035-003, Porto Alegre, RS, Brazil. Email: casg@ufrgs.br

energy deficit [6, 8]. This compound is a methylnitrourea linked to D-glucose, and biochemical alterations generated in brain tissue include oxidative stress [9, 10], increase in acetylcholinesterase (AChEase) activity [11-13], impairment of insulin/insulin-like growth factor signaling and decrease in glucose uptake [14-18].

Astrocytes are intimately associated with neurons and the importance of these glial cells in brain disorders, including AD, has been strongly suggested [19]. Astrocytes, beyond energetic support [20], are involved in the brain antioxidant defense and secretion of neurotrophic factors. Glial activation in response to injury stimuli commonly involves changes in glial fibrillary acidic protein (GFAP), S100B protein, synthesis and release of glutathione and glutamate uptake and metabolism. Therefore, the evaluation of glial activation in STZ-induced and other models of dementia is extremely useful to understand the role of astrocytes in these diseases, as well as to identify possible molecular therapeutic targets.

GFAP is a specific astrocyte marker and its increase is taken as a sign of astrogliosis, observed in many conditions of brain injury (see [21] for a review). A cortical astrogliosis, evaluated by immunohistochemistry for GFAP, has been observed in a STZ-induced model of dementia [22]. S100B is a calcium-binding protein found in brain tissue, predominantly in astrocytes, where it regulates protein phosphorylation of cytoskeleton components and transcriptional factors [23]. Moreover, S100B is secreted and, extracellularly, it plays a trophic role in neuronal and glial cells. Cerebrospinal fluid (CSF) and serum S100B levels have been used as marker of brain insult [24]. Although many cognitive tasks dependent on hippocampal integrity, only a few studies have broached glial activation in this brain region.

Many putatively neuroprotective compounds have been evaluated in the STZ-induced model of dementia, including 17-beta-esteroid [25], antioxidants [9, 11, 14, 26-29], calcium channels blockers [30], AChEase inhibitors [12] and drugs acting on glucose uptake [15]. However, no reports are available on drugs that reduce protein glycation (e.g. aminoguanidine) due to the lower glucose utilization observed in cerebral cortex and all subfields of hippocampus [18]. Aminoguanidine has been proposed as a neuroprotective agent in many conditions of brain injury (e.g. [31]) and is apparently effective against many systemic alterations observed in STZ-induced diabetes (e.g. [32]). The complete activity of this compound is not known, but its inhibition of inducible NO synthetase and inhibition of glycation helps to explain many actions [33].

In this study, our aim was to evaluate spatial cognitive deficit and hippocampal alterations in rats submitted to the STZ-induced dementia model, particularly with regard to astroglial protein markers (GFAP and S100B), glutathione content, NO production and CSF S100B. In parallel, we investigated the ability of aminoguanidine to prevent these alterations.

## MATERIAL AND METHODS

### Chemicals

Sodium carbonate, albumin, Tween-20, streptozotocin, aminoguanidine hemisulfate, glutamylhydroxamate, sodium nitrate, nitrate reductase, 3,3-diaminobenzidine (DAB), o-phenylenediamine (OPD) and monoclonal anti-S100B antibody were purchased from Sigma. Anti-S100 antibody conjugated with peroxidase and anti-GFAP antibody were from Dako. Peroxidase- and fluorescein-conjugated secondary antibodies were from Amersham and Calbiochem, respectively.

### Animals

Forty-seven male Wistar rats (90–days old, weighing 250-320 g) were obtained from our breeding colony (at the Department of Biochemistry, Universidade Federal do Rio Grande do Sul), and were maintained under controlled light and environmental conditions (12 hour light/12 hour dark cycle at a constant temperature of 22 ± 1°C) with free access to food and water. All animal experiments were carried out in accordance with the National Institute of Health Guide for the Care and Use of Laboratory Animals (NIH Publications No. 80-23) revised 1996, and following the regulations of the local animal house authorities.

Rats were divided into 4 groups: sham (N=11), sham-aminoguanidine (N=12), STZ (N=12) and STZ-aminoguanidine (N=12). Two set of experiments were carried out, using 5 or 7 rats of each group in each set, respectively. After behavioral tasks, rats were anaesthetized, as subsequently described, for CSF puncture, immunohistochemistry and hematoxylin-eosin staining (first set of experiments) or for brain slice preparation (second set of experiments), aiming to evaluate S100B and GFAP contents, S100B secretion, NO and glutathione contents.

## Surgical procedure

There are many protocols for the ICV administration of STZ for induction of cognitive deficit in rodents, varying from 0.5 to 3 mg/Kg, in a single or two doses. We used 3 mg/Kg in a single dose as described elsewhere [34]. Briefly, on the day of the surgery, the animals were anesthetized with ketamine/xylazine (75 and 10 mg/Kg, respectively, i.p.) and placed in a stereotaxic apparatus. A midline sagittal incision was made in the scalp. Burr holes were drilled in the skull on both the sides over the lateral ventricles. The lateral ventricles were accessed using the following coordinates [35]: 0.9 mm posterior to bregma; 1.5 mm lateral to sagittal suture; 3.6 mm beneath the surface of brain. Rats received a single bilateral infusion of 5 µL STZ (3 mg/Kg) or vehicle (Hank's balanced salt solution - HBSS - containing in mM: 137 NaCl; 0.63 $Na_2HPO_4$; 4.17 $NaHCO_3$; 5.36 KCl; 0.44 $KH_2PO_4$; 1.26 $CaCl_2$; 0.41 $MgSO_4$; 0.49 $MgCl_2$ and 10 glucose, in pH 7.4) using a 5 µL Hamilton syringe. After the surgical procedure, rats were placed on a heating pad to maintain body temperature at 37.5 ± 0.5 °C, and were kept there until recovery from anesthesia. The animals were submitted to behavioral tasks and biochemical analysis 2-3 weeks after SZT-injection.

The STZ/aminoguanidine and aminoguanidine groups also received aminoguanidine hemisulfate (100 mg/Kg i.p.) dissolved in HBSS on the 2$^{nd}$ and 4$^{th}$ day after ICV-infusion of STZ. There are different schedules for aminoguanidine administration. Our protocol was based on (i) short-term administration, avoiding long-term procedures, which are stressful "per se" and putatively, could affect the behavioral tasks; (ii) administration after (not together with) surgical procedure and STZ injection, aiming to avoid a potential interference; (iii) a potentially neuroprotective dose (e.g. [31]).

## Cognitive evaluation

Two weeks after surgery, rats were submitted to training in the Morris Water Maze [36, 37]. The apparatus consisted of a circular pool (180 cm diameter, 60 cm high), filled with water (depth 30 cm; 24 ± 1 °C), placed in a room with consistently located spatial cues. An escape platform (10 cm diameter) was placed in the middle of one of the quadrants, 1.5 cm below the water surface, equidistant from the sidewall and the middle of the pool. The platform provided the only escape from the water and was located in the same quadrant every trial. Four different starting positions were equally spaced around the perimeter of the pool. On each training day, all four start positions were used once in a random sequence, i.e., four training trials per day. A trial began by placing the animal in the water facing the wall of the pool at one of the starting points. If the animal failed to escape within 60 s it was gently conducted to the platform by the experimenter. The rat was allowed to stay there for 20 s. The inter-trial interval was 10 min. After each trial, the rats were dried, and returned to their cages at the end of the session. Animals were trained for 5 days. Twenty-four hours after the last training session, the rats were submitted to a test session (three weeks after surgery). Before this session, the submerged platform was removed. The retention test consisted of placing the animals in the water for 1 min. The number of crossings over the original position of the platform and time spent in the target quadrant compared to the opposite quadrant were measured.

## Obtaining CSF and hippocampal samples

Animals were anesthetized as described above and then positioned in a stereotaxic holder and cerebrospinal fluid was obtained by cisterna magna puncture using an insulin syringe (27 gauge x 1/2" length). CSF was frozen (-20°C) until further analysis [38]. The animals were killed by decapitation, the brains were removed and placed in cold saline medium with the following composition (in mM): 120 NaCl; 2 KCl; 1 $CaCl_2$; 1 $MgSO_4$; 25 HEPES; 1 $KH_2PO_4$ and 10 glucose, adjusted to pH 7.4 and previously aerated with $O_2$. The hippocampi were dissected and transverse slices of 0.3 mm were obtained using a McIlwain Tissue Chopper. Slices were then frozen (-20°C) (for measurement of GFAP and S100B) or transferred immediately to 24-well culture plates, each well containing 0.3 mL of saline medium for measuring S100B secretion, as described before [39].

*Quantification of S100B and GFAP*

S100B content in the hippocampus and CSF was measured by ELISA [40]. Briefly, 50 μL of sample plus 50 μL of Tris buffer were incubated for 2 h on a microtiter plate previously coated with monoclonal anti-S100B (SH-B1). Polyclonal anti-S100B was incubated for 30 min and then peroxidase-conjugated anti-rabbit antibody was added for a further 30 min. A colorimetric reaction with o-phenylenediamine was measured at 492 nm. The standard S100B curve ranged from 0.025 to 2.5 ng/mL. ELISA for GFAP [41] was carried out by coating the microtiter plate with 100 μL samples containing 30 μg of protein for 48 h at 4°C. Incubation with a rabbit polyclonal anti-GFAP for 2 h was followed by incubation with a secondary antibody conjugated with peroxidase for 1h, at room temperature; the standard GFAP curve ranged from 0.1 to 10 ng/mL.

*Immunohistochemistry for GFAP*

Rats were anesthetized using ketamine/xylazine and were perfused through the left cardiac ventricle with 200 mL of saline solution, followed by 200 mL of 4% paraformaldehyde in 0.1 M phosphate buffer, pH 7.4. The brains were removed and left for post-fixation in the same fixative solution at 4°C for 2 h. After this, the material was cryoprotected by immersing the brain in 30% sucrose in phosphate buffer at 4°C [5]. The brains were sectioned (50 μm) on a cryostat (Leitz). The sections were then preincubated in 2% bovine serum albumin (BSA) in phosphate-buffered saline (PBS) containing 0.3% Triton X-100 for 30 min and incubated with polyclonal anti-GFAP from rabbit, diluted 1:200 in 2% BSA in PBS-Triton X-100, for 48 h at room temperature. After washing several times, tissue sections were incubated in a fluoresceine-conjugated anti-rabbit IgG, diluted 1:200 in PBS, at room temperature for 2 h. Afterwards, the sections were mounted on slides with Fluor Save® and covered with coverslips. Images were viewed with a Nikon microscope and images transferred to a computer with a digital camera. Slices from the same brains used for GPAP immunohistochemistry were also stained with hematoxylin-eosin.

*Glutathione and NO contents*

The glutathione content was determined as described before [42]. Briefly, hippocampal slices were homogenized in sodium phosphate buffer (0.1 M, pH 8.0) containing 5 mM EDTA and protein was precipitated with 1.7% meta-phosphoric acid. Supernatant was assayed with o-phthaldialdehyde (1 mg/mL of methanol) at room temperature for 15 min. Fluorescence was measured using excitation and emission wavelengths of 350 and 420 nm, respectively. A calibration curve was performed with standard glutahione solutions (0-500 μM). Glutahione concentrations were expressed as nmol/mg protein. NO metabolites, $NO_3^-$ (nitrate) and $NO_2^-$ (nitrite), were determined according to [43]. Briefly, homogenates from hippocampal slices were mixed with 25% trichloroacetic and centrifuged at 1800 $g$ for 10 min. The supernatant was immediately neutralized with 2 M potassium bicarbonate. $NO_3^-$ was reduced $NO_2^-$ by nitrate reductase. The total $NO_2^-$ in the incubation was measured by a colorimetric assay at 540 nm, based on the Griess reaction. A standard curve was performed using sodium nitrate (0–80 μM). Results were expressed as μM of nitrite.

*Statistical analysis*

Parametric data from the experiments are presented as means ± standard error and statistically evaluated by two-way analysis of variance, followed by the Tuckey's test, assuming $p < 0.05$. The escape latency parameter in the water maze task was evaluated by repeated measures analysis of variance, assuming $p < 0.05$.

## RESULTS

*Behavioral effects*

The Morris water maze was used to evaluate reference memory in the four groups: Sham, sham/aminoguanidine, STZ and STZ/aminoguanidine. There was a decline in the average time to find the platform (escape latency) from day 2 onwards in the sham group (Fig 1A) ($F (3, 23) = 4.429$, $p = 0.035$). In addition, STZ rats spent less time in the target quadrant, as compared to the sham

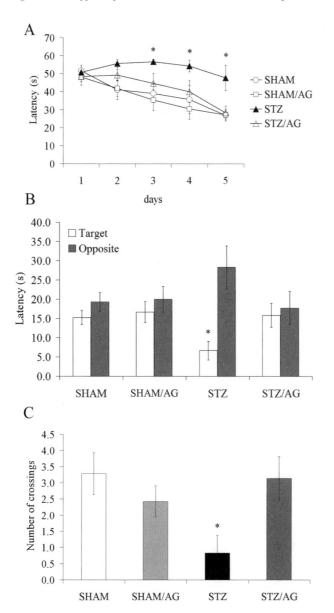

Fig. 1. Cognitive performance of rats submitted to ICV-STZ injection, evaluated in the water maze. (A) Performance in the reference memory protocol, based on escape latency. Each line represents the mean ± standard error. * Significant differences were detected from day 2 onwards when compared to the control group (N = 7, repeated measures analysis of variance, $p < 0.05$); (B) Memory in the probe trial of reference memory, as measured by time spent (in s) in the target quadrant. Values are mean ± standard error. * Significantly different from the control group (N = 7, two-way ANOVA followed by Tukey's test, $p < 0.05$); **(C)** Number of crossings over the platform position. Values are mean ± standard error. * Significantly different from control group (N = 7, two-way ANOVA followed by Tukey's test, $p < 0.05$)

group (Fig 1B) ($F_{(3, 26)} = 3.999$, $p = 0.049$). The performance of STZ/aminoguanidine was not different from the sham group, based on the average time to find the platform ($p = 0.896$), and time spent in the target quadrant (Fig 1B; $F_{(3, 26)} = 0.718$, $p = 0.668$). The number of crossings over the platform location was significantly lower in the STZ group, compared to the other experimental groups ($F_{(3, 26)} = 4.500$, $p = 0.023$) (Fig 1C). Notice that aminoguanidine "per se" had no effect on analyzed behavior.

## Changes in S100B and GFAP contents

A significant increase in S100B immunocontent ($F_{(3, 23)} = 4.942$, $p = 0.019$) of the hippocampus was observed in STZ-treated rats (Fig 2A) and this increase was not found in the STZ/AG group ($p = 0.899$); this effect in STZ group was not found in cerebral cortex and aminoguanidine by itself did not change S100B and GFAP contents in both brain regions (data not shown). Hippocampal GFAP content was apparently higher in the STZ group, however this increase was not statistically significant ($F_{(3, 23)} = 0.444$, $p = 0.724$; Fig 2B) nor was it significantly higher in the cerebral cortex (data not shown).

## Immunohistochemistry for GFAP

In order to evaluate hippocampal astroglial changes observed in rats submitted to ICV-injection of STZ, we carried out a GFAP immunohistochemistry study. The photomicrographs of astrocyte GFAP immunoreactive (ir) did not indicate signs of astrogliosis in CA1 (Fig 3) or CA3 and DG (data not shown). We also observed neuron nuclei in the pyramidal layer of the hippocampus using hematoxylin and eosin staining. There is a clear impairment of marked nucleus density (asterisk) in the STZ group, when compared to the sham group, indicating neuronal loss provoked by ICV-injection of STZ.

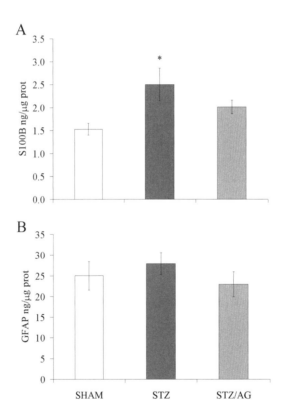

Fig. 2. GFAP and S100B content in hippocampus of rats submitted to ICV- STZ injection. Adult rats were submitted to ICV injection of STZ. Hippocampi and cerebral cortex were dissected out and the contents of S100B (panel A) and GFAP (panel B) were measured by ELISA in the third week. Values are mean ± standard error of 7 rats in each group. * Significantly different from control (two-way ANOVA, $p < 0.05$).

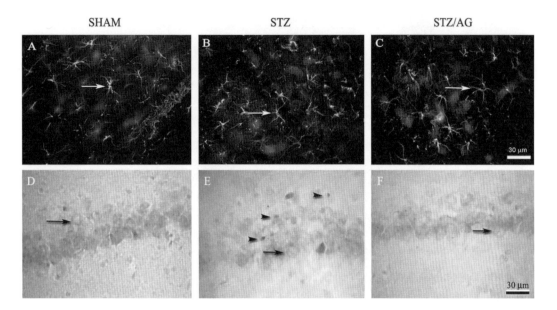

Fig. 3. Immunohistochemistry for GFAP and histological analysis of the hippocampi from rats submitted to ICV-STZ injection. Photomicrographs showing GFAP immunoreactive cells (ir) in the hippocampus of sham (panel A), STZ (panel B) and STZ/ aminoguanidine (AG) (panel C) groups. No differences were observed in GFAP-ir in the radiatum layer of CA1 region (white arrows) between groups. Hippocampal sections show a higher density of hematoxylin-eosin stained nuclei of pyramidal layer (indicated by an asterisk) in sham group (panel D) than in STZ group (panel E). In panel E, black arrow and arrow-heads indicate normal and pyknotic nuclei, respectively. Scale bars are indicated in panels C and F.

*Alterations in CSF S100B*

Interestingly, a significant decrease in CSF S100B was observed in the STZ treated group, as compared to the sham group and this difference was prevented by aminoguanidine administration (Fig 4A) ($F (3, 15) = 3.804$, $p = 0.040$). CSF S100B in sham group was not different to that of the sham-aminoguanidine group (data not shown). Hippocampal slice preparations from the 4 groups were used to evaluate *in vitro* S100B secretion (Fig 4B). No significant changes in basal S100B secretion were found at 1 h ($F (3, 11) = 0.242$, $p = 0.742$).

*Oxidative stress in the hippocampus*

Glutathione content and NO production (based on nitrite content) were used as parameters to evaluate a possible hippocampal oxidative stress (Fig 5). Glutathione content was lower in STZ-treated rats and aminoguanidine prevented this decrease (Fig 5A) ($F (3, 29) = 19.680$, $p = 0.006$). Interestingly, aminoguanidine "per se" induced an increase in glutathione content ($p = 0.004$). Conversely, in the sham-aminoguanidine group, no changes in hippocampal NO content were observed (Fig 5B) ($p = 0.229$). NO content was higher in the STZ group than in the control group and STZ-treated rats that received aminoguanidine exhibited a lower NO content than the other two groups ($F (3, 17) = 22.510$, $p = 0.003$).

## DISCUSSION

Based on oxidative stress [44] and glucose metabolism abnormalities observed in sporadic AD [45], the ICV-injected STZ model of dementia has been widely used. This model presents long-term and progressive deficits in learning, memory, and cognitive behavior, along with a permanent and ongoing cerebral energy deficit [8]. Neuroprotective strategies, including several compounds such as 17-beta-estradiol, curcumin, melatonin, resveratrol, tacrine and lercanidipine, and treadmill running, have been evaluated in this model [10, 12, 25-27, 30, 46]. Our data suggest a neuroprotective effect of aminoguanidine in the STZ-induced model of dementia. All parameters evaluated in the water-maze task, which is widely recognized to evaluate hippocampal integrity, were impaired in the STZ group

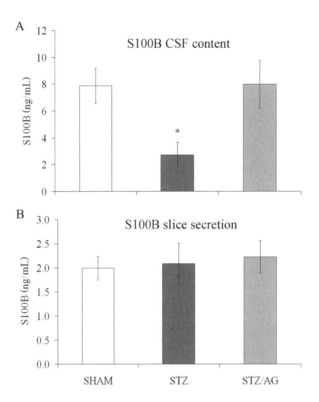

Fig. 4. S100B levels in the cerebrospinal fluid and secreted by hippocampal slices of rats submitted to ICV-STZ injection. Seven adult rats were submitted to ICV-injection of STZ. (A) Three weeks later, cerebrospinal fluid (CSF) was collected by cisterna magna puncture. S100B content was measured by ELISA.(B) Hippocampi were dissected out and chopped into 0.3 mm slices for measurement of S100B secretion. Values are mean ± standard error of 5 rats in each group. * Significantly different from control (two-way ANOVA, $p < 0.05$).

and prevented in the STZ/AG group. Notice that aminoguanidine by itself did not alter cognitive behavior, S100B and GFAP contents, nor NO content. However, it was able to induce an increase in glutathione content.

Although the mechanism of brain damage and cognitive deficit caused by STZ is not well understood, oxidative stress is associated with such damage and could be, at least in part, underlying the injury [45]. STZ is selectively toxic to insulin-producing pancreas cells, where it possibly enters cells by GLUT2 and is able to induce DNA methylation, subsequently activating poly (ADP-ribose) polymerase (PARP) and leading to consumption of $NAD^+$ and ATP (see [47] for a review). NO release and protein glycosylation are potentially additional involved mechanisms [47, 48]. In brain tissue, sub-diabetogenic amounts of ICV STZ (about 100 times lower than that used to induce diabetes) impair glucose uptake and metabolism [18]. Please, see Fig 1. Other biochemical alterations, including oxidative stress and impairment of insulin/ insulin-like growth factor signaling, confirm the neurotoxicity of this compound [14, 16]. The presence of GLUT2 in astrocytes and neurons, particularly in synaptic contacts, suggests that this transporter may allow the entrance of STZ [49, 50] . However, similar changes in the insulin-signaling pathway and cognitive

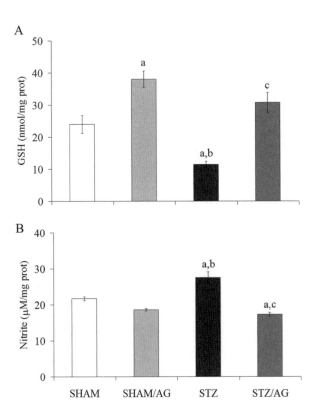

Fig. 5. Glutathione and NO levels in the hippocampus of rats submitted to ICV- STZ injection. Adult rats were submitted to ICV injection of STZ. Three weeks later, hippocampi were dissected out and homogenized for measurement of glutathione (in panel A) or NO (in panel B, measuring nitrate content). Values are mean ± standard error of 7 rats in each group. Two-way ANOVA, followed by Tukey's test, $p < 0.05$: [a], differs from sham group; [b], differs from sham/aminoguanidine group (AG); [c], differs from STZ group.

deficit induced by STZ have been observed with ICV 5-thio-D-glucose, an inhibitor of GLUT2 [51]. This indicates that GLUT2 plays a key role in this model and, possibly, in Alzheimer's disease, but it is unclear whether the effect of STZ in brain tissue depends upon GLUT2 and whether STZ directly affects GLUT2. It is important to emphasize that NO, released directly by STZ and STZ, could affect protein phosphorylation signaling, via peroxynitrite nitration in Tyr residues [52] or by inhibition of the O-GlcNAc-selective glucosaminidase, respectively [48, 53]. Deregulated protein phosphorylation at the level of the insulin receptor [6, 45, 54], protein kinase B pathway and/or glycogen synthase kinase-3 targets [51], could be involved in the alterations observed in AD and STZ models of dementia: insulin resistance, beta-amyloid production [55] and tau hyperphosphorylation in neurons [53, 54, 56]. Astrocytes are strongly involved in the brain antioxidant defense, particularly producing and secreting glutahione [20]. Many studies have demonstrated the decrease in glutathione in ICV-injected STZ rats (e.g.[27]). This decrease is reverted by antioxidants such as resveratrol [26] and coenzyme Q [28] and some drugs used in diabetes mellitus treatment such as pioglitazone [15] and pitavastatin [57]. Aminoguanidine was also able to prevent the glutathione decrease in hippocampus, induced by STZ, and was also able to prevent the increase in NO. This increase in NO is agreement with a recent observation of nitrative stress, observed in ICV-injected STZ rats [22].

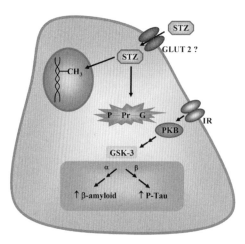

Fig. 6. Schematic representation of the mechanism of STZ in neurons and astrocytes. *STZ* may be enters cells by GLUT2. STZ induces DNA alkylation (indicated *in gray* nucleus) and inhibits O-GlcNAc-selective glucosaminidase, altering the crosstalk between phosphorylation (P) and glycosylation (G) in nucleocytoplasmic proteins (Pr). Modified cell signaling involves alterations in insulin receptor (IR), protein kinase B (PKB) and alpha and beta subunits of glycogen synthase kinase -3 (GSK-3), which could result in increased beta-amyloid production and tau hyperphosphorylation, respectively. Rectangle refers to specific alterations in neurons.

There are few studies regarding GFAP hippocampal alterations in STZ-treated rodents. In diabetes mellitus models, both increases [58, 59] and decreases in GFAP [60] have been described in hippocampus. Recently, scattered reactive astrocytes were reported in the CA1 hippocampal region in ICV-injected STZ rats [22]. We did not, herein, find any signs of hippocampal astrogliosis, as indicated by GFAP content (measured by ELISA) and confirmed by immunohistochemistry in different hippocampal regions. This discrepancy could be due to methodological differences.

Interestingly, we found a significant increment in hippocampal S100B in the STZ group that was prevented by AG administration. This finding reinforces the concept of astrogliosis, not necessarily accompanied by changes in GFAP content. S100B is a calcium-binding protein predominantly expressed and secreted by astrocytes in vertebrate brain [61, 62]. Intracellularly, S100B binds to many protein targets, possibly modulating cytoskeleton plasticity, cell proliferation and astrocyte energy metabolism [see [23] for a review]. S100B alterations could contribute to the down-regulation of the glycolytic pathway in ICV-injected STZ of rats [45], particularly since some glycolytic enzymes are putative targets of S100B [23]. In agreement, high levels of brain tissue S100B have been observed in neurodegenerative disorders, including Alzheimer's disease [63]. It is important to mention that S100B is able to stimulate the protein phosphatase calcineurin [64], which appears to be involved in the inflammatory activation of astrocytes in transgenic Alzheimer's models [65].

On the other hand, the present study showed that CSF S100B is lower in the STZ group and AG administration prevented this decrease. Assuming that extracellular S100B has neurotrophic activity [23, 62], this reduction could indicate impairment in astroglial function in some brain regions in STZ-treated rats. In agreement, rats exposed to chronic cerebral hypoperfusion also exhibited lower levels of CSF S100B [66]. No change was observed in the *ex-vivo* basal secretion of S100B secretion in hippocampal slices of ICV-injected STZ rats, contrasting with the idea that elevated intracellular concentrations of S100B result in increased S100B secretion or, conversely, that augmented intracellular S100B occurs as a consequence of the lower S100B secretion. This reinforces the idea of two independent events: intracellular S100B accumulation and S100B secretion in astrocytes [62]; however, further studies involving S100B-stimulated secretion in different brain regions are necessary to clarify this hypothesis.

In the present study, aminoguanidine was able to prevent, predictably, NO production, and interestingly, glutathione decrease, intracellular S100B accumulation and CSF S100B decrease in the hippocampus of ICV-injected STZ rats. However, no direct effect of this compound was analyzed on neuronal death, microglia activation or beta-amyloid deposition described in this model [51, 67], which, in further studies, should be investigated to evaluate the neuroprotective effect of aminoguanidine (and other AGE-inhibitors).

Moreover, it remains unclear as to how this compound "per se" increased hippocampal glutathione content. Long-term administration of aminoguanidine prevented apoptosis in retina glial cells of diabetic rats, but did not alter intracellular glutathione content [68]. On the other hand, acute administration of aminoguanidine was able to increase gastric GSH content of rats submitted to brain ischemia [69]. In astrocyte cultures, aminoguanidine prevented cell death induced by 1-methyl-4-phenylpyridinium, apparently involving changes in glutathione content, but a direct effect of aminoguanidine on GSH levels was not reported [70]. Therefore, at this moment, it is difficult to know whether aminoguanidine prevention of the glutathione decrease in STZ-treated rats, under our conditions, is only due to an opposite and independent effect.

In summary, our results in ICV-injected STZ rats confirm the spatial cognitive deficit and nitrative stress in this model and demonstrate some astroglial alterations, particularly S100B accumulation in the hippocampus and the decrease in CSF S100B. Such alterations were prevented by aminoguanidine administration, confirming the potential neuroprotective of this compound. Findings contribute to understanding diseases accompanied by cognitive deficits and the STZ model of dementia.

## ACKNOWLEDGEMENTS

Brazilian funds from "Conselho Nacional de Desenvolvimento Científico e Tecnológico" (CNPq), "Coordenação de Aperfeiçoamento de Pessoal do Ensino Superior" (CAPES), FINEP/Rede IBN 01.06.0842-00 and INCT-National Institute of Science and Technology for Excitotoxicity and Neuroprotection.

## References

[1] Ritchie K, Lovestone S (2002) The dementias. *Lancet* **360**, 1759-1766.
[2] Blennow K, de Leon MJ, Zetterberg H (2006) Alzheimer's disease. *Lancet* **368**, 387-403.
[3] Craft JM, Watterson DM, Hirsch E, Van Eldik LJ (2005) Interleukin 1 receptor antagonist knockout mice show enhanced microglial activation and neuronal damage induced by intracerebroventricular infusion of human beta-amyloid. *J Neuroinflammation* **2**, 15.
[4] Weinstock M, Shoham S (2004) Rat models of dementia based on reductions in regional glucose metabolism, cerebral blood flow and cytochrome oxidase activity. *J Neural Transm* **111**, 347-366.
[5] Swarowsky A, Rodrigues L, Biasibetti R, Leite MC, de Oliveira LF, de Almeida LM, Gottfried C, Quillfeldt JA, Achaval M, Goncalves CA (2008) Glial alterations in the hippocampus of rats submitted to ibotenic-induced lesion of the nucleus basalis magnocellularis. *Behav Brain Res* **190**, 206-211.
[6] de la Monte SM, Wands JR (2005) Review of insulin and insulin-like growth factor expression, signaling, and malfunction in the central nervous system: relevance to Alzheimer's disease. *J Alzheimers Dis* **7**, 45-61.
[7] Salkovic-Petrisic M, Hoyer S (2007) Central insulin resistance as a trigger for sporadic Alzheimer-like pathology: an experimental approach. *J Neural Transm Suppl*, 217-233.
[8] Lannert H, Hoyer S (1998) Intracerebroventricular administration of streptozotocin causes long-term diminutions in learning and memory abilities and in cerebral energy metabolism in adult rats. *Behav Neurosci* **112**, 1199-1208.
[9] Tiwari V, Kuhad A, Bishnoi M, Chopra K (2009) Chronic treatment with tocotrienol, an isoform of vitamin E, prevents intracerebroventricular streptozotocin-induced cognitive impairment and oxidative-nitrosative stress in rats. *Pharmacol Biochem Behav* **93**, 183-189.
[10] Ishrat T, Hoda MN, Khan MB, Yousuf S, Ahmad M, Khan MM, Ahmad A, Islam F (2009) Amelioration of cognitive deficits and neurodegeneration by curcumin in rat model of sporadic dementia of Alzheimer's type (SDAT). *Eur Neuropsychopharmacol* **19**, 636-647.
[11] Pinton S, da Rocha JT, Zeni G, Nogueira CW Organoselenium improves memory decline in mice: involvement of acetylcholinesterase activity. *Neurosci Lett* **472**, 56-60.
[12] Saxena G, Singh SP, Agrawal R, Nath C (2008) Effect of donepezil and tacrine on oxidative stress in intracerebral streptozotocin-induced model of dementia in mice. *Eur J Pharmacol* **581**, 283-289.
[13] Blokland A, Bothmer J, Honig W, Jolles J (1993) Behavioural and biochemical effects of acute central metabolic inhibition: effects of acetyl-l-carnitine. *Eur J Pharmacol* **235**, 275-281.
[14] Agrawal R, Mishra B, Tyagi E, Nath C, Shukla R Effect of curcumin on brain insulin receptors and memory functions in STZ (ICV) induced dementia model of rat. *Pharmacol Res* **61**, 247-252.
[15] Pathan AR, Viswanad B, Sonkusare SK, Ramarao P (2006) Chronic administration of pioglitazone attenuates intracerebroventricular streptozotocin induced-memory impairment in rats. *Life Sci* **79**, 2209-2216.
[16] Isik AT, Celik T, Ulusoy G, Ongoru O, Elibol B, Doruk H, Bozoglu E, Kayir H, Mas MR, Akman S (2009) Curcumin ameliorates impaired insulin/IGF signalling and memory deficit in a streptozotocin-treated rat model. *Age (Dordr)* **31**, 39-49.
[17] Grunblatt E, Hoyer S, Riederer P (2004) Gene expression profile in streptozotocin rat model for sporadic Alzheimer's disease. *J Neural Transm* **111**, 367-386.
[18] Duelli R, Schrock H, Kuschinsky W, Hoyer S (1994) Intracerebroventricular injection of streptozotocin induces discrete local changes in cerebral glucose utilization in rats. *Int J Dev Neurosci* **12**, 737-743.

[19] Halassa MM, Fellin T, Haydon PG (2007) The tripartite synapse: roles for gliotransmission in health and disease. *Trends Mol Med* **13**, 54-63.

[20] Takuma K, Baba A, Matsuda T (2004) Astrocyte apoptosis: implications for neuroprotection. *Prog Neurobiol* **72**, 111-127.

[21] Eng LF, Ghirnikar RS, Lee YL (2000) Glial fibrillary acidic protein: GFAP-thirty-one years (1969-2000). *Neurochem Res* **25**, 1439-1451.

[22] Shoham S, Bejar C, Kovalev E, Schorer-Apelbaum D, Weinstock M (2007) Ladostigil prevents gliosis, oxidative-nitrative stress and memory deficits induced by intracerebroventricular injection of streptozotocin in rats. *Neuropharmacology* **52**, 836-843.

[23] Donato R (2001) S100: a multigenic family of calcium-modulated proteins of the EF-hand type with intracellular and extracellular functional roles. *Int J Biochem Cell Biol* **33**, 637-668.

[24] Rothermundt M, Peters M, Prehn JH, Arolt V (2003) S100B in brain damage and neurodegeneration. *Microsc Res Tech* **60**, 614-632.

[25] Lannert H, Wirtz P, Schuhmann V, Galmbacher R (1998) Effects of Estradiol (-17beta) on learning, memory and cerebral energy metabolism in male rats after intracerebroventricular administration of streptozotocin. *J Neural Transm* **105**, 1045-1063.

[26] Sharma M, Gupta YK (2002) Chronic treatment with trans resveratrol prevents intracerebroventricular streptozotocin induced cognitive impairment and oxidative stress in rats. *Life Sci* **71**, 2489-2498.

[27] Sharma M, Gupta YK (2001) Effect of chronic treatment of melatonin on learning, memory and oxidative deficiencies induced by intracerebroventricular streptozotocin in rats. *Pharmacol Biochem Behav* **70**, 325-331.

[28] Ishrat T, Khan MB, Hoda MN, Yousuf S, Ahmad M, Ansari MA, Ahmad AS, Islam F (2006) Coenzyme Q10 modulates cognitive impairment against intracerebroventricular injection of streptozotocin in rats. *Behav Brain Res* **171**, 9-16.

[29] Baluchnejadmojarad T, Roghani M (2006) Effect of naringenin on intracerebroventricular streptozotocin-induced cognitive deficits in rat: a behavioral analysis. *Pharmacology* **78**, 193-197.

[30] Sonkusare S, Srinivasan K, Kaul C, Ramarao P (2005) Effect of donepezil and lercanidipine on memory impairment induced by intracerebroventricular streptozotocin in rats. *Life Sci* **77**, 1-14.

[31] Cockroft KM, Meistrell M, 3rd, Zimmerman GA, Risucci D, Bloom O, Cerami A, Tracey KJ (1996) Cerebroprotective effects of aminoguanidine in a rodent model of stroke. *Stroke* **27**, 1393-1398.

[32] Wilkinson-Berka JL, Kelly DJ, Koerner SM, Jaworski K, Davis B, Thallas V, Cooper ME (2002) ALT-946 and aminoguanidine, inhibitors of advanced glycation, improve severe nephropathy in the diabetic transgenic (mREN-2)27 rat. *Diabetes* **51**, 3283-3289.

[33] Schimchowitsch S, Cassel JC (2006) Polyamine and aminoguanidine treatments to promote structural and functional recovery in the adult mammalian brain after injury: a brief literature review and preliminary data about their combined administration. *J Physiol Paris* **99**, 221-231.

[34] Rodrigues L, Biasibetti R, Swarowsky A, Leite MC, Quincozes-Santos A, Quilfeldt JA, Achaval M, Goncalves CA (2009) Hippocampal alterations in rats submitted to streptozotocin-induced dementia model are prevented by aminoguanidine. *J Alzheimers Dis* **17**, 193-202.

[35] Paxinos G (1997) *The Rat Nervous System*, Academic Press, San Diego.

[36] Morris R (1984) Developments of a water-maze procedure for studying spatial learning in the rat. *J Neurosci Methods* **11**, 47-60.

[37] Silva MC, Rocha J, Pires CS, Ribeiro LC, Brolese G, Leite MC, Almeida LM, Tramontina F, Ziegler DR, Goncalves CA (2005) Transitory gliosis in the CA3 hippocampal region in rats fed on a ketogenic diet. *Nutr Neurosci* **8**, 259-264.

[38] Vicente E, Tramontina F, Leite MC, Nardin P, Silva M, Karkow AR, Adolf R, Lucion AB, Netto CA, Gottfried C, Goncalves CA (2007) S100B levels in the cerebrospinal fluid of rats are sex and anaesthetic dependent. *Clin Exp Pharmacol Physiol* **34**, 1126-1130.

[39] Tramontina AC, Tramontina F, Bobermin LD, Zanotto C, Souza DF, Leite MC, Nardin P, Gottfried C, Goncalves CA (2008) Secretion of S100B, an astrocyte-derived neurotrophic protein, is stimulated by fluoxetine via a mechanism independent of serotonin. *Prog Neuropsychopharmacol Biol Psychiatry*.

[40] Leite MC, Galland F, Brolese G, Guerra MC, Bortolotto JW, Freitas R, Almeida LM, Gottfried C, Goncalves CA (2008) A simple, sensitive and widely applicable ELISA for S100B: Methodological features of the measurement of this glial protein. *J Neurosci Methods* **169**, 93-99.

[41] Tramontina F, Leite MC, Cereser K, de Souza DF, Tramontina AC, Nardin P, Andreazza AC, Gottfried C, Kapczinski F, Goncalves CA (2007) Immunoassay for glial fibrillary acidic protein: antigen recognition is affected by its phosphorylation state. *J Neurosci Methods* **162**, 282-286.

[42] Browne RW, Armstrong D (1998) Reduced glutathione and glutathione disulfide. *Methods Mol Biol* **108**, 347-352.

[43] Hevel JM, Marletta MA (1994) Nitric-oxide synthase assays. *Methods Enzymol* **233**, 250-258.

[44] Petersen RB, Nunomura A, Lee HG, Casadesus G, Perry G, Smith MA, Zhu X (2007) Signal transduction cascades associated with oxidative stress in Alzheimer's disease. *J Alzheimers Dis* **11**, 143-152.

[45] Hoyer S, Lannert H (2007) Long-term abnormalities in brain glucose/energy metabolism after inhibition of the neuronal insulin receptor: implication of tau-protein. *J Neural Transm Suppl*, 195-202.

[46] Jee YS, Ko IG, Sung YH, Lee JW, Kim YS, Kim SE, Kim BK, Seo JH, Shin MS, Lee HH, Cho HJ, Kim CJ (2008) Effects of treadmill exercise on memory and c-Fos expression in the hippocampus of the rats with intracerebroventricular injection of streptozotocin. *Neurosci Lett* **443**, 188-192.

[47] Lenzen S (2008) The mechanisms of alloxan- and streptozotocin-induced diabetes. *Diabetologia* **51**, 216-226.

[48] Konrad RJ, Kudlow JE (2002) The role of O-linked protein glycosylation in beta-cell dysfunction. *Int J Mol Med* **10**, 535-539.

[49] Arluison M, Quignon M, Nguyen P, Thorens B, Leloup C, Penicaud L (2004) Distribution and anatomical localization of the glucose transporter 2 (GLUT2) in the adult rat brain--an immunohistochemical study. *J Chem Neuroanat* **28**, 117-136.

[50] Arluison M, Quignon M, Thorens B, Leloup C, Penicaud L (2004) Immunocytochemical localization of the glucose transporter 2 (GLUT2) in the adult rat brain. II. Electron microscopic study. *J Chem Neuroanat* **28**, 137-146.

[51] Salkovic-Petrisic M, Tribl F, Schmidt M, Hoyer S, Riederer P (2006) Alzheimer-like changes in protein kinase B and glycogen synthase kinase-3 in rat frontal cortex and hippocampus after damage to the insulin signalling pathway. *J Neurochem* **96**, 1005-1015.

[52] Reynolds MR, Berry RW, Binder LI (2007) Nitration in neurodegeneration: deciphering the "Hows" "nYs". *Biochemistry* **46**, 7325-7336.

[53] Gong CX, Liu F, Grundke-Iqbal I, Iqbal K (2005) Post-translational modifications of tau protein in Alzheimer's disease. *J Neural Transm* **112**, 813-838.

[54] Grunblatt E, Salkovic-Petrisic M, Osmanovic J, Riederer P, Hoyer S (2007) Brain insulin system dysfunction in streptozotocin intracerebroventricularly treated rats generates hyperphosphorylated tau protein. *J Neurochem* **101**, 757-770.

[55] Phiel CJ, Wilson CA, Lee VM, Klein PS (2003) GSK-3alpha regulates production of Alzheimer's disease amyloid-beta peptides. *Nature* **423**, 435-439.

[56] Arioka M, Tsukamoto M, Ishiguro K, Kato R, Sato K, Imahori K, Uchida T (1993) Tau protein kinase II is involved in the regulation of the normal phosphorylation state of tau protein. *J Neurochem* **60**, 461-468.

[57] Sharma B, Singh N, Singh M (2008) Modulation of celecoxib- and streptozotocin-induced experimental dementia of Alzheimer's disease by pitavastatin and donepezil. *J Psychopharmacol* **22**, 162-171.

[58] Saravia FE, Revsin Y, Gonzalez Deniselle MC, Gonzalez SL, Roig P, Lima A, Homo-Delarche F, De Nicola AF (2002) Increased astrocyte reactivity in the hippocampus of murine models of type 1 diabetes: the nonobese diabetic (NOD) and streptozotocin-treated mice. *Brain Res* **957**, 345-353.

[59] Baydas G, Reiter RJ, Yasar A, Tuzcu M, Akdemir I, Nedzvetskii VS (2003) Melatonin reduces glial reactivity in the hippocampus, cortex, and cerebellum of streptozotocin-induced diabetic rats. *Free Radic Biol Med* **35**, 797-804.

[60] Coleman E, Judd R, Hoe L, Dennis J, Posner P (2004) Effects of diabetes mellitus on astrocyte GFAP and glutamate transporters in the CNS. *Glia* **48**, 166-178.

[61] Marenholz I, Heizmann CW, Fritz G (2004) S100 proteins in mouse and man: from evolution to function and pathology (including an update of the nomenclature). *Biochem Biophys Res Commun* **322**, 1111-1122.

[62] Goncalves CA, Concli Leite M, Nardin P (2008) Biological and methodological features of the measurement of S100B, a putative marker of brain injury. *Clin Biochem* **41**, 755-763.

[63] Griffin WS, Sheng JG, Royston MC, Gentleman SM, McKenzie JE, Graham DI, Roberts GW, Mrak RE (1998) Glial-neuronal interactions in Alzheimer's disease: the potential role of a 'cytokine cycle' in disease progression. *Brain Pathol* **8**, 65-72.

[64] Leal RB, Frizzo JK, Tramontina F, Fieuw-Makaroff S, Bobrovskaya L, Dunkley PR, Goncalves CA (2004) S100B protein stimulates calcineurin activity. *Neuroreport* **15**, 317-320.

[65] Norris CM, Kadish I, Blalock EM, Chen KC, Thibault V, Porter NM, Landfield PW, Kraner SD (2005) Calcineurin triggers reactive/inflammatory processes in astrocytes and is upregulated in aging and Alzheimer's models. *J Neurosci* **25**, 4649-4658.

[66] Vicente E, Degerone D, Bohn L, Scornavaca F, Pimentel A, Leite MC, Swarowsky A, Rodrigues L, Nardin P, de Almeida LM, Gottfried C, Souza DO, Netto CA, Goncalves CA Astroglial and cognitive effects of chronic cerebral hypoperfusion in the rat. *Brain Res, in press*.

[67] Tahirovic I, Sofic E, Sapcanin A, Gavrankapetanovic I, Bach-Rojecky L, Salkovic-Petrisic M, Lackovic Z, Hoyer S, Riederer P (2007) Brain antioxidant capacity in rat models of betacytotoxic-induced experimental sporadic Alzheimer's disease and diabetes mellitus. *J Neural Transm Suppl*, 235-240.

[68] Giardino I, Fard AK, Hatchell DL, Brownlee M (1998) Aminoguanidine inhibits reactive oxygen species formation, lipid peroxidation, and oxidant-induced apoptosis. *Diabetes* **47**, 1114-1120.

[69] Hung CR (2006) Role of gastric oxidative stress and nitric oxide in formation of hemorrhagic erosion in rats with ischemic brain. *World J Gastroenterol* **12**, 574-581.

[70] McNaught KS, Jenner P (1999) Altered glial function causes neuronal death and increases neuronal susceptibility to 1-methyl-4-phenylpyridinium- and 6-hydroxydopamine-induced toxicity in astrocytic/ventral mesencephalic co-cultures. *J Neurochem* **73**, 2469-2476.

# Section 3
# Alzheimer's Disease Therapies Using Animal Models

# Developing Immunotherapies for Alzheimer's Disease Using a Cholesterol-Fed Rabbit Model in the Context of Th Cell Differentiation

Richard Coico[1] and Diana Woodruff-Pak[2]
[1]Department of Microbiology and Immunology, Temple University School of Medicine,
[2]Department of Psychology, College of Liberal Arts, Temple University, Philadelphia, PA

**Abstract.** The success of strategies that therapeutically exploit immune responses to self antigens such as those expressed on melanoma cells to treat patients suffering from melanoma is well established. In transgenic Alzheimer's disease (AD) animal models, elimination of β-amyloid (Aβ) pathology using Aβ immunization galvanized the AD research community a decade ago. Using an animal model of AD with closer parallels to human AD, the cholesterolemic rabbit model, we tested immunotherapeutic strategies that could maximize humoral immune responses while minimizing proinflammatory responses. The finding that some patients with AD enrolled in a clinical trial to inoculate against A experienced a misdirected polarization of Th cells reminds us that our knowledge of T cell biology, immune regulation, and the precise functional properties of adjuvants is incomplete. In this article, we review this knowledge and consider the advantages of the rabbit for immunological studies. The langomorph species is proximate to primates on the phylogenetic scale, its amino acid sequence of A is identical to the human A sequence, and the rabbit model system is extensively characterized on a form of associative learning with parallels in normal aging in rabbits and humans that is severely impaired in human AD. Cholesterol-fed rabbits treated with Aβ immunotherapy generate high titer anti-Aβ responses. The cholesterol-fed rabbit model of AD with its close parallels to human genetics and physiology, along with its validity from molecular to cognitive levels as a model of human AD, provides a promising vehicle for development of immunotherapies.

Translational research refers to the "bench-to-bedside" enterprise of harnessing knowledge from basic sciences to produce new drugs, devices, and treatment options for patients. Reducing or preventing the morbidity and mortality associated with many common diseases is the ultimate goal of translational science. The list of successful translational research endeavors that have produced promising new treatments of disease continues to expand. In some cases, these discoveries have been brought to market enabling their broad application as cutting edge arsenals against disease. Within this arena, immunotherapeutic strategies to treat or prevent a variety of diseases including cancer and autoimmunity, to name a few, stand out because of their high impact potential and versatility. As discussed below, however, occasionally the animal models that provide us with valuable scientific insights regarding the feasibility and predicted efficacy of immunotherapeutic strategies to treat disease in humans convince us to translate discoveries prematurely. The clinical trials carried out using patients with Alzheimer's disease (AD) immunized with β-amyloid (Aβ) were justifiably terminated when an unexpected immune-based lethal outcome occurred in some patients. One lesson learned from this clinical study is that we should exhaustively explore the potential outcomes of immunotherapy using a variety of animal models before we convince ourselves that it is scientifically appropriate to move

---

Corresponding Author: Richard Coico, Ph.D., Professor of Microbiology and Immunology, Temple University School of Medicine, 3500 North Broad Street, 1111J, Philadelphia, PA 19140. Tel.:215.707.4605; Fax: 215.707.2718; E-mail: rcoico@temple.edu

a discovery from bench-to-bedside. Pre-clinical studies are essential and wherever possible, a diverse set of animal models should be used as experimental tools to generate a preponderance of evidence that translation to clinical applications is appropriate. To accomplish this objective, we need to critically examine how we should orient and optimally align the interface between basic science and clinical medicine to advance translational discoveries that will manifest as more effective therapies for the treatment and, ideally, the prevention of AD. This is clearly an imperative given recent public health estimates indicating that if no scientific advances alter the incidence and progression of AD, we can expect to see a fourfold increase in the number of people in the United States (~12 million currently to 48 million) suffering from the illness by the year 2050 [54].

Alzheimer's disease is a progressive neurodegenerative disorder that is characterized by A$\beta$ deposition as one of the earliest events in the pathologic process. Indeed, there is growing evidence that aggregation and accumulation of A$\beta$ proteins are the central pathologic processes associated with neurodegeneration in AD [37]. Transgenic mouse models for AD have clearly demonstrated that active and passive anti-A$\beta$ immunotherapies reduce AD-like pathology and restores cognitive deficits in treated animals. These studies prompted the rapid implementation of translational studies using A$\beta$ immunotherapy in humans. A multi-center, placebo-controlled clinical trial involving active vaccination of patients with mild-to-moderate AD was carried out using an aggregated A$\beta_{1-42}$ peptide (AN1792) as the immunogen [42]. AN1792 was administered together with a surface active saponin as adjuvant (QS21) and a preservative component (polysorbate-80). The Phase I trial demonstrated a trend toward slower cognitive decline in the A$\beta$-immunized patients as compared with controls. However, cognitive performance was not restored to normal levels as had been shown in the transgenic mouse AD model [22]. The strategy was advanced to a Phase II trial that was halted when 6% of the immunized patients developed aseptic meningoencephalitis. Autopsy results from the brains of four individuals treated with AN1792 revealed absent or sparse plaques in the neocortex. Neuropathological analysis also revealed infiltration of pro-inflammatory T lymphocytes and abnormalities of cerebral white matter, including excessive macrophage infiltration and a reduced density of myelinated fibers [42].

Analysis of anti-A$\beta$ serum titers revealed no correlation between the signs and symptoms observed and meningoencephalitis.

The misdirected polarization of Th cells responding to A$\beta$ in a cohort of patients enrolled in the AN1792 study reminds us that our knowledge of T cell biology, immune regulation, and the precise functional properties of adjuvants is far from complete. Before we embark upon new clinical trials using active A$\beta$ immunotherapy strategies, we need to: (1) use animal models, such as the rabbit model described in this review, that closely mimic the pathological and cognitive indices of AD in humans; and, (2) develop a better understanding of the impact of adjuvant-based active A$\beta$ vaccine formulations on immune responses in humans. When autoantibody responses are the desired outcome of an immunotherapeutic regimen, as is the case with A$\beta$ vaccination, it is essential that we leave no pre-clinical experimental stone unturned in order to do no harm to the patients receiving these vaccinations.

Before discussing the utility of the cholesterol-fed rabbit AD model in helping us better understand how A$\beta$ immunotherapeutic strategies can and should be applied to humans suffering from or predisposed to AD, we begin with a summary of what we know about the functional diversity of T cells and how T cells regulate adaptive immune responses. T cells are the master orchestrators of virtually all adaptive immune responses. Our ability to optimally harness their immunoregulatory properties is essential to the design of safe and efficacious immunotherapeutic protocols for human subjects.

*Functionally heterogeneous T cell populations*

Although an extensive summary of our current understanding of how T cells regulate immune responses is beyond the scope of this review, several salient features of T cell differentiation and functional diversity are particularly noteworthy. Optimal antibody responses to A$\beta$ vaccines depend upon T cell help and, more importantly, the right T cell help. Over the past two decades, our knowledge of the diversity of effector T cell populations has grown considerably. The T helper type 1 (Th1)-Th2 paradigm provided a framework for understanding T cell biology and the interplay of innate and adaptive immunity [41]. The Th1-Th2 hypothesis postulated that subsets of CD4 T cells produce reciprocal patterns of immunity through their production of distinct profiles of cytokine secretion—either pro-inflammatory cell-mediated immunity (Th1) or anti-inflammatory humoral immunity (Th2). Furthermore, each subset promotes its own development and inhibits the development of the other subset, also via their secreted cytokines such that the induction of one type of response suppresses the induction of the other [18, 50]. This hypothesis ultimately provided us with a more enlightened understanding of immune regulation by CD4 T cells and led to the discovery that effector CD4 T cells are functionally heterogeneous. Adaptive immune responses are deterministically classified into humoral or cell-mediated depending on the pattern of Th cell polarization into Th1 or Th2. But the story of how T cells orchestrate adaptive immune responses and, alarmingly, how they can contribute to the development of autoimmune diseases, does not end here.

In the same twenty-year period in which the Th1-Th2 paradigm arose, two other effector T cell populations were identified, namely regulatory T (Treg) cells and Th17 cells. Treg cells play a pivotal role in tolerance to self-antigens and tissue grafts, and suppression of autoimmune reactions [34, 45]. They do this by modulating the intensity and quality of immune reactions through attenuation of the cytolytic activities of reactive immune cells. Treg cells operate primarily at the site of inflammation, in close spatial proximity to and through direct interaction with the effector cells. They modulate the immune reaction through three major mechanisms: a) direct killing of cytotoxic cells through cell-to-cell contact, b) inhibition of cytokine production by cytotoxic cells, in particular interleukin-2, c) direct secretion of immunomodulatory cytokines, in particular TGF-β and interleukin-10 (IL-10). Evidence is growing to support the notion that multiple Treg cell functions act either directly or indirectly at the site of antigen presentation to create a regulatory milieu that promotes bystander suppression and infectious tolerance [67]. Thus, it appears that Treg cells exert their complex and diverse suppressive activities both in an antigen-specific and antigen-nonspecific fashion [5]. Much attention has been given to the immunotherapeutic roles of Th1 and Th2 cells but have we given enough attention to how Treg cells may be called into play? What role(s) do Treg cells have in giving self-tolerant cells "permission" to respond to Aβ? Is this permission granted on a permanent basis or can it be reversed by signals capable of regenerating antigen-specific Treg function? Currently, there are no answers to these important questions although we now have many of the experimental tools needed to address them with scientific acumen.

The discovery of yet another T cell lineage (Th17 cells) as a separate arm of adaptive immunity provided us with a unifying model that helped to explain many heretofore confusing aspects of immune regulation, immune pathogenesis, and host defense [33]. The spectrum of functional activities of Th17 cells in host defense against pathogens and their potentially pathogenic role in many autoimmune diseases is only beginning to emerge. Th17 cells have become notorious for their involvement in a range of autoimmune diseases, but an exclusive role as mediators of pathology is unlikely to be their primary function. Evidence suggests that an important physiologic role of Th17 cells and the family of cytokines they produce (the IL-17 family) is their ability to generate host defense mechanisms against extracellular bacteria [8, 26]. Questions analogous to those mentioned above in the context of Treg cells may also be asked with regard to Aβ immunotherapy. For example, what role, if any, did Th17 cells play in mediating the neuropathology observed in AD patients with mengioencephalitis following Aβ inoculation? Fortunately, we can retrospectively investigate this question using archived patient tissue samples collected at autopsy. So far, no studies have been reported.

*Differentiation of naïve T cell: cytokine regulation of cell signaling*

Our current understanding of the cytokines and signaling cascades that polarize CD4 cell differentiation into a growing number of functionally distinct effector T cell populations (reviewed in 86). Figure 1 illustrates differentiation of naïve T cells into four of the eight currently recognized subsets, namely Th1, Th2, Treg, and Th17 cells. Depending on the type of antigen, site, and cells involved in antigen presentation, different soluble factors and signaling molecules trigger different genes driving the cell fate and functions of different subsets [69]. Naïve CD4 T cells differentiate towards Th1 phenotype in the presence of IL-12, which upregulates interferon-γ (IFNγ) via signal transducer and activator of transcription 4 (Stat4). This leads to IFNγ–mediated Stat1 activation and induction of the Th1 lineage determining transcription factor Tbet. Th2, by contrast,

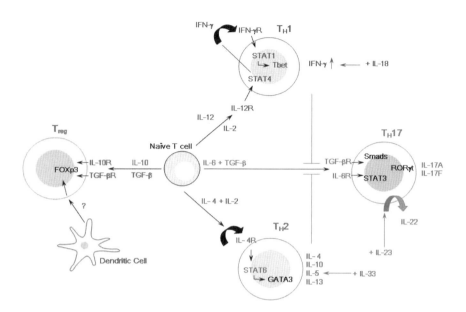

Fig. 1. Differentiation of peripheral CD4 T-cell subsets into four distinct effector populations: Th1, Th2, Th17, and Treg cells. Each population is generated from naïve CD4 T cells following their exposure to pathway-biased activation signals (e.g. cytokines produced by DCs and other antigen-presenting cells) resulting in expression of lineage determining transcription factors.

differentiates in response to IL-4, which activates Stat6, resulting in induction of GATA3. The Th17 T-cell subset develops in response to IL-6 and TGFβ, and this differentiation step is strongly inhibited by Th1 or Th2 cytokines. Signaling via IL-6 activates Stat3 and the lineage-determining transcription factor RORβt. Signaling through TGFβ receptor is also essential for Th17 development, as T cells defective in TGFβRII signaling cannot differentiate to Th17. Additional cytokines modify the response of the three T-cell subsets. IL-18 strongly increases IFNβ secretion by Th1, IL33 upregulates IL-5 and IL-13 in Th2, and IL-23 induces IL-22 secretion in Th17.

Treg cells differentiate from naïve T cells both within the thymus during T cell development (so-called natural Treg cells) and in the periphery (induced Treg cells). Peripheral naïve T cells are believed to differentiate into Treg cells following exposure to IL-10 and TGF$^β$. A subpopulation of DCs is also believed to help promote differentiation towards these Foxp3+ Treg cells [9]. The DC-elaborated signals that drive this set of events have not been identified although the elaborate cross-talk between Treg cells and DCs lead to DC silencing and expansion of the Treg cell repertoire, both of which help to actively establish and reinforce immune quiescence and self-tolerance. It is noteworthy that like mice, humans with mutations in *FOXP3* develop multi-organ autoimmune diseases with fatal consequences [20, 49, 85].

Natural Treg cells are believed to be stimulated intrathymically by high-avidity TCR–self peptide–MHC interactions augmented by CD28 signaling and an additional, unidentified, perhaps limiting, signal that induces the transcription factor, Foxp3 (not shown in Figure 1). Because intrathymic TCR signals are of an increased strength or duration, this favors the development of MHC class II–restricted CD4+ T cells. Thus, Foxp3-expressing Treg cells are highly enriched in but are not restricted to this T cell population. Thus, intrathymic differentiation of naïve T cell into CD8 Treg cells also occurs, although to a much lower frequency as compared with CD4 Treg cells. Phenotypically, in addition to the high constitutive expression of the high-affinity IL-2 receptor β chain (CD25), Treg cells are positive for cytotoxic T lymphocyte-associated antigen (CTLA)–4 and glucocorticoid-induced tumor necrosis factor receptor.

*Pro- versus anti-inflammatory immunotherapies: lessons from host defense mechanisms*

The Th1/Th2 paradigm hypothesized that the polarity of T cell differentiation following antigen stimulation was not entirely stochastic. If this is true, and by extension, if the same applies to the differentiation of Treg and Th17 cells, why would this functional diversity be necessary? Moreover, what is the nature of the extrinsic pathway-biased signals that initiate differentiation of naive T cells? The teleological answer to the first question is that evolutionary pressures imposed by exposure to a plethora of diverse pathogens selected host defense systems that can tailor adaptive immune responses to effectively protect the host thereby ensuring survival. If we focus only on the adaptive responses in which Th1 or Th2 cells are involved for the purpose of this review, the evidence to support this is, in fact, derived from the predictable microbial suspects. Host defense against bacterial pathogens requires adaptive immune responses that generate antibody responses. Intrinsic bacteria-associated adjuvants (e.g. LPS) predominantly trigger Th2 responses that promote the generation of protective antibody responses. In contrast, intrinsic virus-associated adjuvants (e.g. dsRNA) predominantly trigger Th1 responses that promote cell-mediated responses including the generation of cytotoxic T cells capable of killing virally-infected cells. What happens when we extrinsically challenge self-tolerant naive T cells (and B cells) with an autoantigen under conditions that enable these cells to "escape" central tolerance? The simple answer is that we induce an autoimmune response. Central tolerance can be "broken" using certain adjuvants. Many examples of animal models for autoimmune diseases demonstrate this potential effect of adjuvants (e.g. collagen-induced arthritis in rats inoculated with collagen together with complete Freund's adjuvant). The subject of this review underscores the therapeutic potential of anti-self immune responses but, as pointed out earlier, extrinsic adjuvant-induced manipulation of T cell help to ensure optimal immunotherapeutic responses to A$\beta$ to treat AD needs to be exquisitely tailored to achieve the desired immune response. Unfortunately, our ability to unidirectionally polarize T cell responses in humans to exclusively support antibody responses (Th2 polarization) or cell-mediated responses (Th1 polarization) remains an elusive challenge. In experimental animal models, a variety of adjuvants that effectively polarize T cell responses have been successfully employed to elicit optimal immunotherapeutic responses. This brings us to the subject of the next section of this review where we discuss adjuvants which have often been called the *dirty little secrets* of vaccines as much about how adjuvants work is a mystery.

*Adjuvants: Intrinsic versus artificial*

Long before we had any understanding of intrinsic pathogen-derived adjuvants such as bacterial lipopolysaccharide (LPS) or viral dsRNA we unknowingly benefited from their immunomodulatory properties as components of protective vaccines. For example, vaccines that contain attenuated live or heat-killed viruses or bacteria include components that can engage Toll-like receptors (TLRs) expressed on innate immune cells, receptors which recognize molecular patterns on microbes that are often described as pathogen-associated signatures. To date, thirteen distinct TLRs have been identified in humans [1]. Microbial recognition of TLRs facilitates their dimerization. TLRs are believed to homodimerize triggering activation of signaling pathways that originate from a cytoplasmic TLR domain. In the signaling pathways downstream of this domain, a domain-containing adaptor, MyD88, has been shown to be essential for induction of inflammatory cytokines such as TNF-$\alpha$ and IL-12 through all TLRs except TLR3. However, subsequent studies have demonstrated that individual TLR signaling pathways are divergent, and there are MyD88-dependent and MyD88-independent pathways [1].

Ligation of TLRs by microbial molecular signatures serving as intrinsic adjuvants stimulates a program of synergistic immune events that bridge the antigen non-specific innate arm of the immune system with the antigen-specific effector components [38]. Exploiting these intrinsic microbial adjuvants to stimulate immune responses to poorly immunogenic vaccines, such as peptides, is a work-in-progress for vaccine development. Indeed, given the increasing importance of extrinsic adjuvants to modern vaccines, the relative under-development of the science of adjuvants, compared with the rapidly advancing knowledge of immune responses to vaccines, and, in particular, peptide-based vaccines such as A$\beta$, is compromising our ability to optimally harness the power of immunotherapy. It is extraordinary that the exact mechanism of action remains unknown for even the oldest known artificial adjuvants including alum. These adjuvants can act in several non-mutually exclusive ways to augment adaptive immune responses and to generate effective immunological memory. Many of their immune augmenting effects seem to be on antigen-presenting cells such as dendritic cells (DCs). Thus, artificial adjuvants can affect the migration, maturation, antigen presentation, and expression of costimulatory molecules by DCs. These events, in turn, can directly enhance T and B cell responses to antigens. Artificial adjuvants, via DCs, can also affect the nature of CD4 T cell responses, with some adjuvants promoting pro-inflammatory Th1 responses and others preferentially inducing anti-inflammatory Th2-biased effects.

While experimental vaccine adjuvants comprise a diverse group of molecules and formulations, for human vaccines, aluminum salts (alum) remain the most commonly used adjuvants and the only ones currently approved for use in humans by the FDA. However, alum suffers from a number of downsides, including its "inability" to induce cytotoxic T-lymphocyte (CTL) responses critical in many cases for viral protection and clearance [12]. In the case of Aβ vaccination where the goal is to stimulate anti-Aβ antibody responses, this property of alum would, in fact, make it a good adjuvant candidate. Alum activates innate responses *in vivo* and promotes anti-inflammatory Th2-biased responses. Exactly how alum mediates its Th2-biased adjuvant effects is not well understood. One hypothesis is that alum, because it adsorbs antigens, serves as a depot, releasing the antigen slowly into the body, thereby allowing antigen-specific lymphocytes to be exposed to antigen for a longer time [25]. A more contemporary view of alum's adjuvant properties is that it induces T cells with a Th2 phenotype because of its effects on DCs as mentioned above. DCs exposed directly to alum do not fully upregulate costimulatory molecules required for cognate cell-cell interactions and they do not produce Th1-driving cytokines, the canonical changes to DC induced by TLR signals intrinsically provided by many pathogens [55]. Rather, alum enhances the secretion of IL-1 and IL-18 by DCs by activating caspase-1 in a myeloid differentiation factor 88 (MyD88)-independent manner [35].

Despite the known Th2-polarizing activity of alum, the AN1792 Aβ clinical trials with aggregated Aβ$_{1-42}$ peptide employed a saponin-based adjuvant (QS21). Saponin is a soluble extract from the bark of the *Quillaja saponaria* tree. While saponin has been shown to promote Th2 responses that help manifest IgG antibody responses to antigens in animals and to experimental vaccines in humans, it has also been shown to promote Th1 responses against antigens [7, 51, 71]. Interestingly, when QS-21 is combined with alum, T cell responses to antigens are predominantly of the Th2 type thereby facilitating B cell activation and IgG antibody responses [2]. In light of the mixed Th1/Th2-promoting adjuvant properties of saponin as compared with the predominant Th2 skewing of T cell responses using alum – alone or together with saponin, it is reasonable to retrospectively question the wisdom of using QS21 in the AN1792 clinical trials.

*Rabbits as models for human learning, aging, and chronic disease*

It is commonly (but mistakenly) held that rabbits are large members of the order Rodentia. Rabbits are in the order Lagomorpha. Graur et al. [24] pointed out that molecular data support the monophyletic status of Lagomorphia. Rabbits are distinct organisms from rodents and have a separate ancestry. To assess the evolutionary relationships of Lagomorphia to other taxa, Graur and colleagues used orthologous protein sequences [24]. A total of 91 protein sequences were analyzed, and reconstructions revealed that primates are phylogenetically closer to lagomorphs than are Rodentia. These results raised the question of whether primates are the closet relatives of the lagomorphs or whether other orders are closer to Lagomorphia than are primates. Results indicated that four taxa other than primates are significantly closer phylogenetically to lagomorphia than are the rodents. From a phylogenetic standpoint, rabbits are closer than rodents to human primate ancestry.

From the standpoint of the Aβ proteins that are the central pathologic processes associated with neurodegeneration in AD, rabbit models have a definite advantage. The rabbit's closer phylogenetic relation to primates is expressed in the amino acid sequence of Aβ, which is identical to the human sequence [31]. Other species in addition to rabbit in which the sequence of Aβ is close to identical to that in humans are dog, polar bear, cow, pig, and guinea pig. These animals may form Aβ plaques. Species such as mouse and rat have a different amino acid sequence for Aβ and may be an evolutionarily distinct group from those with a conserved Aβ amino acid sequence.

As a species for cognitive assessment, rabbits have another advantage. On no other species is there such a large body of parametric data on the classically conditioned eyeblink response as on the rabbit [82]. Much of the general literature on classical conditioning is based on data collected in the rabbit. From the early work of Hilgard in the 1930s with human eyeblink conditioning, Gormezano in the 1960s with rabbit eyeblink conditioning, and subsequent work of their colleagues and students, the parallels between rabbit and human eyeblink classical conditioning have been demonstrated empirically for decades. If salivation to a bell is the archetype of classical conditioning, blinking to a tone is the prototype.

It was the initial knowledge that eyeblink conditioning was impaired in normal aging in rabbits and humans that made this well-characterized model attractive for the investigation of neurobiological substrates of age-related memory impairment. The model system of eyeblink classical conditioning might be "the Rosetta stone for brain substrates of age-related deficits in learning and memory," [68]. Both rabbits and humans show age-associated deficits in conditioning, and these can be easily dissociated from age-associated changes in sensory systems (i.e., differences in sensory thresholds for detection of the conditioned stimulus) or motor systems (differences in reflexive eye blink amplitude). Rabbits and humans also show parallel age-associated changes in the neural substrates critical for eyeblink classical conditioning, the cerebellum and the hippocampus. Age-associated deficits can be artificially induced in both young rabbit and young human subjects with drugs that affect cholinergic neurotransmission, and age-associated deficits can be reversed in rabbits with drugs approved to treat cognitive impairment in AD [83]. The fact that the same classical conditioning paradigm can be tested in both rabbits and humans makes eyeblink conditioning a valuable preclinical test of cognition-enhancing drugs.

One of the strengths of using eyeblink conditioning as a model of age-associated deficits in learning and memory is that both humans and rabbits show similar deficits as they age. Humans begin to show age-associated deficits in eyeblink conditioning at about 40-50 years of age. Rabbits begin to show age-associated deficits at approximately two years of age. Based on declines in reproductive capacity, a two-year-old rabbit is equivalent to a 40-45-year-old human, which suggests similar onset of age-associated declines in eyeblink conditioning in both humans and rabbits. Furthermore, the results found with rabbits appear to generalize to other non-human species, such as mice, rats, and cats.

The combined advantages of rabbits as a research species are numerous. Rabbits have a closer phylogenetic proximity to primates than do rodents, rabbits have an extensively characterized profile on a measure of learning and memory that closely parallels human performance, and rabbits and humans have a documented profile of age-related parallels in learning impairment. The model system of rabbit eyeblink conditioning has been used extensively to elucidate mechanisms of learning and memory in normal aging. Age is currently the best predictor of AD in humans. As a model species for investigations of immunotherapy in AD, rabbits are advantageous as they have a close to identical amino acid sequence of Aβ to humans. The creation and extensive investigation of a rabbit model of AD prepared the way for our research.

*The cholesterol-fed rabbit as a model of Alzheimer's disease*

Recognition of parallels between risk factors for Alzheimer's and cardiovascular disease has provided insights about potential causes and mechanisms in AD. Long recognized as a risk factor for cardiovascular disease, the overrepresentation of APOE $E4$ allele is associated with an increased risk for developing sporadic AD [28, 39, 46, 47, 52]. By the mid-1990s the $E4$ association with AD had been confirmed in 100 laboratories [48]. Recent data suggest that APOE plays a role in facilitating the proteolytic clearance of soluble Aβ from the brain [30].

Vascular risk factors such as hypertension, type 2 diabetes, abnormal lipid levels, and obesity are other risk factors common to Alzheimer's and cardiovascular disease, well recognized in the twenty-first century [3, 32]. However, even late in the twentieth century, investigators attempted to segregate out patients with vascular signs in an attempt to identify patients with "pure" AD [75]. In this climate, Larry Sparks, working in forensics in the 1980s, had a difficult time publishing his observations that Aβ containing senile plaques occurred almost exclusively in the brains of individuals dying of critical coronary artery disease [61, 62] and/or diagnosed with hypertension [64].

Observations of Aβ containing senile plaques in the brains of human patients with cardiovascular disease led Sparks to examine the heretofore unexplored brains of a long-standing animal model of heart disease, the cholesterol-fed rabbit. As an aside, it is noteworthy that at about the same time that Sparks and colleagues were exploring the use of rabbits as an animal model for AD in humans, Graur et al.'s [24] comparative phylogenetic analyses of 91 orthologous protein sequences, mentioned previously, revealed that the order to which rabbits belong is significantly more closely related to primates than to rodents. While mouse models have greatly contributed to our understanding of the pathophysiology of AD, the rabbit model may, in fact, be superior given the closer phylogenetic proximity of this species to humans as compared with mice.

An extensive elaboration of this rabbit model of AD has been presented by Sparks and colleagues [60-65]. His group and investigators in other laboratories have demonstrated that the cholesterol-fed rabbit is a valid model of AD on a number of levels. At a molecular level, the brain of this animal model has numerous (>12) features similar to the pathology observed in the AD brain including elevated Aβ concentration, neuronal accumulation of Aβ immunoreactivity, extracellular Aβ plaques, reduced levels of acetylcholine, elevated brain cholesterol, apolipoprotein E immunoreactivity, breaches in the blood brain barrier, microgliosis, and neuronal loss. Feeding rabbits over a long time period with cholesterol causes an increase in the cholesterol content in neurons, and this high cholesterol content in neurons is accompanied by an increase in BACE1 levels [21]. The accumulation of β-amyloid in the brain can be associated with BACE1, the enzyme that initially cleaves amyloid precursor protein (APP) to generate Aβ and cause the accumulation of $A_{1-42}$ peptide. The accumulation of Aβ is associated with the phosphorylation of tau [e.g., 66]. Data demonstrating excessive cholesterol content in neurons following a seven-month diet adding 1% cholesterol to rabbit chow triggers a cascade of neuropathological changes. The increase in cholesterol content in neurons may underlie the increase in BACE1 and Aβ levels that in turn trigger the phosphorylation of tau [21].

Sparks also discovered that increased levels of Aβ in the brains of cholesterol-fed rabbits was dependent on the quality of the water the animals were administered. Animals given tap water accumulated considerably more Aβ in the brain than animals administered distilled water [63]. It was in a collaborative attempt to demonstrate cognitive validity of cholesterol-fed rabbits that Sparks discovered that it was trace amounts of copper in the drinking water that were responsible for the higher levels of Aβ in the brain in comparison to the brains of rabbits administered the cholesterol diet and distilled water. In this issue, he elaborates his position that cholesterol in the diet causes accumulation of Aβ, whereas copper introduced in the drinking water impairs elimination of Aβ.

To examine the impact of the cholesterol diet on cognition, Sparks collaborated with Schreurs at the University of West Virginia. To the investigators' surprise, cholesterol-fed rabbits produced in West Virginia and tested on eyeblink classical conditioning actually performed better than control rabbits [53]. However, when Sparks investigated the neuropathology in West Virginia cholesterol-fed rabbits, it was quite limited. Evaluation of trace metals in the water in West Virginia indicated that copper was the one trace metal that was absent. Subsequent studies adding a trace amount of copper to the water of cholesterol-fed rabbits in Schreurs' laboratory in West Virginia and Sparks' laboratory in Arizona demonstrated that neuropathology formation was accelerated [60]. Rabbits fed a diet of 2% cholesterol and trace amounts of copper in distilled drinking water were impaired in eyeblink classical conditioning [65].

*Eyeblink classical conditioning as an assessment of cognitive impairment in AD*

Severe memory loss is the most prominent clinical symptom of AD, and this memory impairment has long been associated with impairment in acetylcholine neurotransmission. Disrupted cholinergic neurotransmission is not the single cause of memory loss in AD, but this deficit characteristic of AD clearly impairs memory [6]. Disruption of the brain cholinergic system in AD links this dementing disease to the model system of eyeblink conditioning in mammals, including humans [74]. Eyeblink conditioning impairment in AD may reflect medial-temporal lobe atrophy and associated central nervous system cholinergic dysfunction that occurs early in the disease progression.

In a sample of 40 elderly adults, half of whom were diagnosed with probable AD, there were very significant differences in eyeblink conditioning between the patients and age-matched, non-demented control subjects [75]. Normal older adults are impaired in eyeblink classical conditioning compared to young adults, but all age groups of normal, non-demented adults, including adults ≥80years of age, show clear evidence of associative learning [77]. In probable AD, there is very limited eyeblink conditioning in the first session of testing. However, when given a sufficient number of training trials (e.g., four or five days of 90-trial presentations), patients diagnosed with probable AD acquire conditioned eyeblink responses (CRs) [56, 81]. This slowing of the rate of acquisition occurs in the animal model when antagonists to acetylcholine neurotransmission are introduced [40, 76]. There is strong evidence that the site of interference of cholinergic antagonists in rabbits is the hippocampus [58]. The facts that patients with AD are slow to acquire CRs and that AD patients do eventually acquire CRs are parallel to results in rabbits injected with antagonists to acetylcholine. Disruption of the hippocampal cholinergic system prolongs acquisition of CRs, but organisms eventually condition.

Solomon et al. [57] reported disruption of eyeblink conditioning in the 400 ms delay paradigm in a sample of probable AD patients in their early 70s – a decade younger than the Woodruff-Pak et al. [75] sample. Woodruff-Pak et. al. [79] carried out an additional replication in the 400 ms delay procedure testing 28 patients with probable AD and 28 healthy age-matched control subjects. Patients diagnosed with cerebrovascular dementia were also tested in this study, and eyeblink conditioning was shown to differentiate some cerebrovascular dementia patients from patients with probable AD. Eyeblink conditioning in adults over the age of 35 years with Down's syndrome and presumably AD was similar to conditioning in probable AD patients [44, 80]. Conditioning in patients with neurodegenerative disease that does not affect the critical cerebellar eyeblink conditioning circuitry or disrupt the hippocampal cholinergic system is relatively normal. For example, patients with Huntington's disease [78] and Parkson's disease [10, 59] have intact eyeblink conditioning.

We combined the data over several studies from 61 probable AD patients and 100 normal age-matched older adults that we had tested in the 400 ms delay eyeblink conditioning procedure. To these data we applied a bivariate logistic regression analysis using percentage of CRs to predict diagnosis [15]. With percentage of CRs as the only predictor, the area under the receiver operating characteristic curve (AROC) was 73%. Prediction was improved when age and gender were added to the model. In a multivariable logistic regression model including age, gender, and percentage of CRs, the model accurately distinguished patients with and without probable AD with AROC of 79.8% (Figure 2). Both the bivariate and the multivariable logistic models were statistically significant ($p < 0.0001$) overall and both models passed the Hosmer Lemeshow goodness of fit test ($p > 0.7$).

Longitudinal results suggest that eyeblink conditioning has utility in the early detection of dementia. A 3-year longitudinal study of non-demented adults tested on eyeblink conditioning revealed that three of eight normal subjects testing in the AD range (producing less than 25% CRs) at Time 1 became demented within two or three years [17]. A fourth "normal" subject who scored just above criterion (26% CRs) also developed dementia within two years of the initial testing. A fifth subject in this group died within a year of the initial testing. Thus, of eight non-demented subjects age-matched to probable AD patients who scored on eyeblink conditioning in the AD range, only three were cognitively normal at the end of a three-year period and 63% became demented within 2-3 years.. Age-matched non-demented subjects scoring above criterion remained cognitively intact during the period of the longitudinal investigation. A second longitudinal study followed 20 cognitively normal elderly participants (half good-conditioners; half poor-conditioners) over a 2-year period [13]. A neuropsychological test battery administered two years after the initial eyeblink conditioning test revealed significantly worse performance in poor conditioners on visuospatial abilities, semantic memory, and language: abilities showing early decline in AD. Over a 2-year period, 2 of the poor conditioners (and none of the good conditioners) failed significantly, and one was diagnosed with probable AD.

*Eyeblink classical conditioning in transgenic mouse models of AD*

Research on AD has made significant advances with the development of the promising models using transgenic mice. Mini-gene copies of human APP and *Presenelin 1* and *2* (*PS1* and *PS2*) genes implicated in familial AD have been used to create

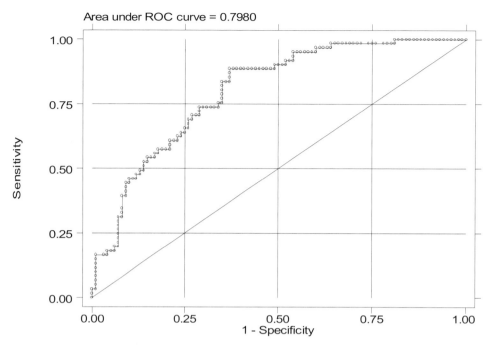

Fig. 2. Area under the receiver operating characteristic curve (ROC) distinguishing 61 patients diagnosed with probable AD and 100 normal age-matched adults. Independent variables were percentage of conditioned responses, gender, and age, and using a multivariable logistic regression model including these three variables, patients with and without probable AD were accurately distinguished 79.8% of the time, which was statistically significant ($p < 0.0001$) (From [15])

transgenic mouse models of AD. Mice doubly transgenic for *APP* and *PS1* mutations (*APP* + *PS1*) show rapid progressive development of compact plaques beginning at 3 months of age and reaching a plateau at 12 months [27]. *APP* + *PS1* mice are impaired in spatial learning tasks by the age of 15 months [4, 23]. Given the validity of eyeblink classical conditioning in detecting AD in humans, we tested 12-month-old *APP* + *PS1* mice on eyeblink classical conditioning [16]. Mice were also tested on a motor learning task (rotorod) and on sensory tasks (prepulse inhibition [PPI] and acoustic startle). *APP* + *PS1* mice performed similarly to controls on both 500 ms delay and 500 ms trace eyeblink conditioning as well as on PPI and acoustic startle, however, transgenic mice had impaired motor performance on rotorod. Within the group of transgenic mice tested in trace conditioning, cortical amyloid burden correlated significantly with decreased conditioning. Among the 14 transgenic mice, cortical amyloid burden and activation of microglia in hippocampus correlated significantly with decreased PPI. Whereas there were no differences in learning and sensory performance between transgenic and control mice, there was an association between compact plaques and microglia activation and impaired behavioral performance within the group of *APP* + *PS1* mice. The *APP* + *PS1* mouse model is a model for amyloidosis, and there is little neuronal loss or neurofibrillary tangle formation. This mouse model does not have the full compliment of AD neuropathology.

In most AD mouse models, other AD neuropathologies such as neurofibrillary tangles are not produced. An exception is the triple-transgenic mouse model of AD that has both plaque and tangle neuropathology [43]. This model with closer parallels in human AD neuropathology is impaired in eyeblink conditioning and responds to therapeutic interventions.

*Cholesterol-fed rabbits, eyeblink conditioning and therapeutic interventions*

Sparks and Schreurs' [65] demonstration that eyeblink classical conditioning was impaired in cholesterol-fed rabbits, in the light of our own work on eyeblink conditioning in human AD and classical conditioning in normal aging rabbits and humans, was a stimulating discovery that led us to initiate work with AD model rabbits. Our aim was to replicate AD neuropathology in the cholesterol-fed rabbit model, test eyeblink conditioning in this model, and determine if galantamine (Razadyne™) would ameliorate impaired conditioning. Rabbit chow with 2% cholesterol and drinking water with 0.12 PPM copper sulfate were administered for 10 weeks. Control rabbits received normal food and distilled water. Rabbit brains were probed for neuropathology. AD model rabbits had significant neuronal loss in frontal cortex, hippocampus and cerebellum. Changes in neurons in the hippocampus were consistent with neurofibrillary degeneration and cytoplasmic immunoreactivity for -amyloid and tau. In a second experiment, cholesterol-fed model rabbits were injected daily with vehicle or 3.0 mg/kg galantamine and tested on 750 ms trace and delay eyeblink conditioning. Galantamine improved eyeblink conditioning significantly over vehicle. We concluded that this animal model of AD has validity from neuropathological to cognitive levels and offers a promising addition to the available animal models of AD. Galantamine ameliorated impaired eyeblink conditioning, extending the validity of the model to treatment modalities [72].

One of the results highlighted in our report was the evident loss of Purkinje neurons and staining for Aβ and tau in the cerebellum of cholesterol-fed rabbits (confirmed by Sparks, unpublished observations). We examined the cerebellum in cholesterol-fed rabbits primarily because the cerebellum is the essential substrate for eyeblink classical conditioning, a behavior on which they were impaired. Although the cerebellum has traditionally been viewed as one of the last structures to be impacted by classical AD pathology in humans, evidence has mounted to indicate that the cerebellum is indeed affected by AD. Antibodies to Aβ revealed the presence of diffuse plaques in the molecular layer of the AD and aged-Down's syndrome cerebellum [29, 36, 70]. Ultrastructural studies have revealed only rare amyloid fibrils in diffuse plaques, whereas the dystrophic neuritis in these plaques contain membranous and vesicular dense bodies but no paired helical filaments [11, 84]. Purkinje cell loss and reactive astrocytosis of the cerebellum in familial AD was found to be more severe than in sporadic AD in a Japanese sample, but the Aβ deposition in the cerebellum in both familial and sporadic AD were similar [19]. In a British sample, there was lysosomal abnormality in Purkinje cells in AD. These differences may be associated with the absence of senile plaques and the presence of "diffuse" amyloid plaques in the cerebellum in AD [14]. Abnormalities in AD model rabbits in cerebellar Purkinje cells are consistent with findings in the cerebellum of humans with AD supporting the validity of the AD model rabbit.

Sparks and Schreurs [65] had tested trace eyeblink conditioning in cholesterol-fed rabbits because medial temporal lobe structures affected in AD are essential for trace conditioning. In their study, rabbits tested in a short-delay paradigm were not impaired. We used the same interval between the tone conditioned stimulus and airpuff unconditioned stimulus in the delay and trace paradigms and found both to be impaired. However, both Sparks and Schreurs [65] and our study tested the trace paradigm first. We followed up with a study of cholesterol-fed rabbits to determine if delay eyeblink conditioning would be impaired when tested before trace conditioning [73]. Cholesterol-fed rabbits were significantly impaired in the 750 ms delay eyeblink conditioning paradigm when that paradigm was tested first, after 10 weeks of the 2% cholesterol and trace copper diet. Just as human patients diagnosed with AD are significantly impaired in delay eyeblink conditioning, so are cholesterol-fed rabbits.

We demonstrated therapeutic efficacy with one drug approved to treat cognitive impairment in AD, and Sparks reports the effects of cholinergic altering drugs on the responses of cholesterol-fed rabbits. Whereas these drugs can ameliorate cognitive impairment in AD, their efficacy is modest, postponing dementia onset by six months to a year in some patients who tolerate the drug. More dramatic therapeutic effects in transgenic mouse models of AD have been achieved via modulation of the immune system. Immunotherapy may have the potential to prevent the development of AD in later life as well as to treat and reverse some of its symptoms in AD patients.

As mentioned previously, lagomorphs are more closely related to primates than are rodents [24] and have an amino acid sequence of Aβ that is almost identical to that of humans, whereas the rodent amino acid sequence of Aβ is not [31]. The cholesterol-fed rabbit model of AD with its close parallels to human genetics and physiology, along with its validity as a model of human AD from molecular to cognitive levels, provides an interesting vehicle for testing immunotherapies for AD. Our initial work has confirmed that cholesterol-fed rabbits treated with Aβ immunotherapy generate high titer anti-Aβ responses similar to those seen in control rabbits

(Coico, Manns, and Woodruff-Pak, in preparation). The widespread replication of AD neuropathology in sites comparable to those in human AD brains coupled with cognitive impairment on a task similar to the impairment observed in AD makes this model attractive for future experimentation with immunotherapies.

## ACKNOWLEDGEMENTS

We wish to thank Joanne Manns, Ph.D., for her assistance with the preparation of this manuscript and her expert help with ongoing Aβ immunotherapy experiments using the cholesterol-fed rabbit model described. Portions of the research with the cholesterol-fed rabbit model were supported by National Institute on Aging grant AG021925 to DSW-P.

## References

[1] S. Akira and K. Takeda, K., Toll-like receptor signaling. *Nat. Rev. Immunol.* **4** (2004), 499-511.
[2] J. Alexander, M.K. del Guercio, A. Maewal, L. Qiao, J. Fikes, R.W. Chestnut, J. Paulson, D.R. Bundle, S. DeFrees, and A. Sette, Linear PADRE T helper epitope and carbohydrate B cell epitope conjugates induce specific high titer IgG antibody Responses, *J. Immunol.* **164** (2002), 1625–1633.
[3] K.J. Anstey, D.M. Lipnicki, and L.F. Low, Cholesterol as a risk factor for dementia and cognitive decline: a systematic review of prospective studies with meta-analysis, *Am J Geriatr Psychiatry* **16** (2008), 343-354.
[4] G.W. Arendash, D.L. King, M.N. Gordon, D. Morgan, J.M. Hatcher, C.E. Hope and D.M. Diamond, Progressive, age-related behavioral impairments in transgenic mice carrying both mutant amyloid precursor protein and presenilin-1 transgenes, *Brain Res.* **891** (2001), 42-53.
[5] N. Askenasy, A. Kaminitz, and S. Yarkonu, Mechanisms of T regulatory cell function, *Autoimmunity Rev.* **7** (2008) 370-375.
[6] R.T. Bartus, On neurodegenerative diseases, models, and treatment strategies: Lessons learned and lessons forgotten a generation following the cholinergic Hypothesis, *Expt Neurol* **163** (2000), 495-529.
[7] S.R. Chavali and J.B. Campbell, Adjuvant effects of orally administered saponins on humoral and cellular immune responses in mice, *Immunobiology* **174** (1987) 347-359.
[8] D.R Chung, D.L Kasper, R.J. Panzo, T. Chitnis, M.J. Grusby, M.H. Sayegh, and A.O. Tzianabos, CD4+ T cells mediate abscess formation in intra-abdominal sepsis by an IL-17-dependent mechanism, *J. Immunol.* **170** (2003), 1958-1963.
[9] J.L. Coombes, K.R. Siddiqui, C.V. Arancibia-Cárcamo, J. Hall, C.M., Y. Belkaid , and F. Powrie, A functionally specialized population of mucosal CD103+ DCs induces Foxp3+ regulatory T cells via a TGF-β and retinoic acid-dependent mechanism, *J. Exp. Med.* **204** (2007), 1757–1764.
[10] I. Daum, M.M. Schugens, C. Breitenstein, H. Topka and S. Spieker, Classical eyeblink conditioning in Parkinson's disease, *Mov. Disord.* **11** (1996), 639-646.
[11] D.W. Dickson, A. Wertkin, L.A. Mattiace, E. Fier, Y. Kress, P. Davies and S-H. Yen, Ubiquitin immunoelectron microscopy of dystrophic neurites in cerebellar senile plaques of Alzheimer's disease, *Acta. Neuropathol.* **79** (1990), 486-493.
[12] P.C. Doherty, S.J. Turner, R.G. Webby, and P.G. Thomas, Influenza and the challenge for immunology, *Nature Immunol.* **7** (2006), 449–455.
[13] M.M. Downey-Lamb and D.S. Woodruff-Pak, Early detection of cognitive deficits using eyeblink classical conditioning. *Alzheimer's Reports* **2** *(1999)*, 37-44.
[14] J.H. Dowson, C.Q. Mountjoy, M.R. Cairns, H. Wilton-Cox and W. Bondareff, Lipopigment changes in Purkinje cells in Alzheimer's disease, *J. Alzheimers Dis.* **1** (1998), 71-79.
[15] M. Ewers, L.E. Braitman, and D.S. Woodruff-Pak, Eyeblink conditioning distinguished Alzheimer's disease in older adults, *Society for Neuroscience Abstracts* **27**, (2001), 282.
[16] M. Ewers, D. Morgan, M. Gordon and D.S. Woodruff-Pak, Associative and motor learning in 12-month-old transgenic APP + PS1 mice, *Neurobiol. Aging* **27** (2006), 1118-1128.
[17] L.S. Ferrante, and D.S. Woodruff-Pak, Longitudinal investigation of eyeblink classical conditioning in elderly human subjects, *J Gerontol: Psy Sci* **50B** (1995) P42-50.
[18] J. D. Fontenot and A. Y. Rudensky, A well adapted regulatory contrivance: regulatory T cell development and the forkhead family transcription factor Foxp3, *Nature Immunology* **6** (2005), 331–337.
[19] Y. Fukutani, N.J. Cairns, M.N. Rossor and P.L. Lantos, Cerebellar pathology in sporadic and familial Alzheimer's disease including APP 717 (Val→Ile) mutation cases: A morphometric investigation, *J. Neurol. Sci.* **149** (1997), 177-184.
[20] E. Gambineri, T.R. Torgerson, and H.D. Ochs, Immune dysregulation, polyendocrinopathy, enteropathy, and X-linked inheritance (IPEX), a syndrome of systemic autoimmunity caused by mutations of FOXP3, a critical regulator of T-cell homeostasis, *Curr. Opin. Rheumatol.* **15** (2003), 430–435.
[21] O. Ghribi, B. Larsen, M. Schrag and M.M. Herman, High cholesterol content in neurons increases BACE, - amyloid, phosphorylated tau levels in rabbit hippocampus, *Exp. Neurol.* **200** (2006), 460-467.
[22] S. Gilman, M. Koller, R.S. Black, L. Jenkins, S.G. Griffith, N.C Fox, L. Eisner, L. Kirby, M.B. Rovira, F. Ferette, J.M. Orgogoza, AN179s (QS-21)-Study Team, Clinical effects of Abeta immunization (AN1792) in patients with AD in an interrupted trial, *Neurology* **64** (2005), 1553-1562.
[23] M.N. Gordon, D.L. King, D.M. Diamond, P.T. Jantzen, K.V. Boyett, C.E. Hope, J.M. Hatcher, G. DiCarlo, W.P. Gottschall, D. Morgan and G.W. Arendash, Correlation between cognitive deficits and Abeta deposits in trans-

[24] genic APP+PS1 mice, *Neurobiol. Aging* **22** (2001), 377-385.
[24] D. Graur, L. Duret, and M. Gouy, M. Phylogenetic position of the order Lagomorpha (rabbits, hares and allies), *Nature* **379** (1996), 333-335.
[25] R.K. Gupta, Aluminum compounds as vaccine adjuvants, *Adv. Drug Deliv. Rev.* **32** (1998),155–172.
[26] K.I. Happel, P.J. Dubin, M. Zheng, N. Ghilardi, C. Lockhart, L.J. Quinton, A.R. Odden, J.E. Shellito, G.J. BagbyJ, Nelson S, et al., Divergent roles of IL-23 and IL-12 in host defense against Klebsiella pneumoniae. *J. Exp. Med.* **202** (2005) 761-769.
[27] L. Holcomb, M.N. Gordon, E. McGowan, X. Yu, S. Benkovic, P. Jantzen, K. Wright, I. Saad, R. Mueller, D. Morgan, S. Sanders, C. Zehr, K. O'Campo, J. Hardy, C.M. Prada, C. Eckman, S. Younkin, K. Hsiao and K. Duff, Accelerated Alzheimer-type phenotype in transgenic mice carrying both mutant amyloid precursor protein and presenilin 1 transgenes, *Nat. Med.* **4** (1998), 97-100.
[28] C. Hulette, B. Crain, D. Goldgaber, and A.D. Roses, Association of apolipoprotein E allele *E*4 with late-onset familial and sporadic Alzheimer's disease, *Neurology* **43** (1993), 1467-1472.
[29] S.I. Ikeda, D. Allsop and G.G. Glenner, The morphology and distribution of plaque and related deposits in the brains of Alzheimer's disease and control cases: An immunohistochemical study using amyloid beta-protein antibody, *Lab. Invest.* **60** (1989), 113-122.
[30] Q. Jiang, C.Y.D. Lee, S. Mandrekar, B. Wilkinson, P. Cramer, N. Zelcer, K. Mann, B. Lamb, t.M. Willson, J.L. Collina, J.C. Richardson, J.D. Smith, T.A. Comery, D. Riddell, D.M. Holzman, P. Tontonoz, and G.E. Landreth. ApoE promotes the proteolytic degradation of A$\beta$, *Neuron* **58** (2008), 681-693.
[31] E.M. Johnstone, M.O. Chaney, F.H. Norris, R. Pascual and S.P. Little, Conservation of the sequence of the Alzheimer's disease amyloid peptide in dog, polar bear and five other mammals by cross-species polymerase chain reaction analysis, *Brain. Res. Mol. Brain. Res.* **10** (1991), 299-305.
[32] R.P. Kloppenborg, E. van den Berg, L.J. Kappelle, and G.J. Biessels, Diabetes and other vascular risk factors for dementia: Which factor matters most? A systematic review, *Eur J Pharmacol.* **585** (2008), 97-108.
[33] J.K. Kolls and A. Linden, Interleukin-17 family members and inflammation, *Immunity* **21** ( 2004), 467-476.
[34] Y. Kuniyasu, T. Takahashi, M. Itoh, J. Shimizu, G. Toda, S. Sakaguchi, Naturally anergic and suppressive CD25+CD4+ T cells as a functionally and phenotypically distinct immunoregulatory T cell subpopulation, *Int. Immunol.* **12** (2000), 1145–55.
[35] H. Li, S. Nookala, and F. Re, Aluminum hydroxide adjuvants activate caspase-1 and induce IL-1beta and IL-18 release, *J. Immunol.* **178** (2007), 5271–5276.
[36] Y-T. Li, D.S. Woodruff-Pak and J.Q. Trojanowski, Amyloid plaques in cerebellar cortex and the integrity of Purkinje cell dendrites, *Neurobiol. Aging* **15** (1994) 1-9.
[37] J.R. Lopez, A. Lyckman, S. Oddo, F.M. Laferia, H.W. Querfurth, and A. Shtifman, Increased intraneuronal resting [Ca2+] in adult Alzheimer's disease mice, *J. Neurochem.* **105** (2008), 262-271.
[38] V. Mata-Haro, C. Cekic, M. Martin, P.M. Chilton, C.R. Casella, and T.C. Mitchell, T.C., The vaccine adjuvant monophosphoryl lipid A as a TRIF-biased agonist of TLR4, *Science* **316** (2007), 1628–1632.
[39] R. Mayeux, Y. Stern, R. Ottman, T.K. Taternichi, M.-X. Tang, G. Maestre, C. Ngai, B. Tycko, and H. Ginsber. The apolipoprotein *E*4 allele in patients with Alzheimer's disease, *Ann of Neurol* **34** (1993), 752-754.
[40] J.W. Moore, N.A. Goodell and P.R. Solomon, Central cholinergic blockade by scopolamine and habituation, classical conditioning, and latent inhibition of the rabbit's nictitating membrane response, *Physiol. Psychol.* **4** (1976), 395-399.
[41] T.R. Mosmann TR and R.L. Coffman, Two types of mouse helper T-cell clone, *Immunol. Today* **8** (2003), 223-27.
[42] J.A. Nicoll, D. Wilkinson, C. Holmes, P. Steart, H. Markham, and R.O Weller, Alzheimer disease after immunization with amyloid-beta peptide: a case report, *Nat. Med.* **9** (2003), 448-452.
[43] S. Oddo, A. Caccamo, J.D. Shepherd, M.P. Murphy, T.E. Golde, R. Kayed, R. Metherate, M.P. Mattson, Y. Akbari and F.M. LaFerla, Triple-transgenic model of Alzheimer's disease with plaques and tangles: intracellular Abeta and synaptic dysfunction, *Neuron* **39** (2003), 409-421.
[44] M. Papka, E.W. Simon, and D.S. Woodruff-Pak, A one-year longitudinal investigation of eyeblink classical conditioning and cognitive and behavioral tests in adults with Down's syndrome, *Aging and Cognition* **1** (1994), 89-104.
[45] C.A. Piccirillo, J.J. Letterio, A.M. Thornton, et al. CD4 + CD25 + regulatory T cells can mediate suppressor function in the absence of transforming growth factor ß1 production and responsiveness, *J. Exp. Med.* **196** (2002), 237–46.
[46] J. Poirier, J. Davignon, D. Bouthillier, S. Kogan, P. Bertrand, and S. Gauthier, Apolipoprotein E polymorphism and Alzheimer's disease, *Lancet* **342** (1993), 697-699.
[47] G.W. Rebeck, J.S. Reiter, D.K. Strickland, and B.T. Hyman, Apolipoprotein E in sporadic Alzheimer's disease: Allelic variation and receptor interactions, *Neuron* **11** (1993), 575-580.
[48] A.D. Roses, Apolipoprotein E and Alzheimer's disease: A rapidly expanding field with medical and epidemiological consequences. In *Apolipoprotein E genotyping in Alzheimer's disease*, N.R. Relkin, Z. Khachaturian, and S. Gandy (eds.), New York: New York Academy of Sciences, 1993, pp. 50-57.
[49] A.Y. Rudensky, M. Gavin, and Y. Zheng, FOXP3 and NFAT: Partners in Tolerance, *Cell* **126** (2006), 253-256.
[50] S. Sakaguchi, Naturally arising Foxp3-expressing CD25$^+$ CD4$^+$ regulatory T cells in immunological tolerance to self and non-self, *Nature Immunology* **6** (2005), 345–352.
[51] F.N. Santos, G.P. Borja-Cabrera, L.M. Miyashiro, J. Grechi, A.B. Reis, M.A.B. Moreira, O.A. Martins Filho, M.C.R. Luvizotto, I. Menz, L.M. Pessoa, P.R. Goncalves, M. Palatnik, C.B. Palatnik-de-Sousa, Immunotherapy against experimental canine visceral leishmaniasis with the saponin enriched-Leishmune vaccine, *Vaccine* **25** (2007), 6176–6190.
[52] A.M. Saunders, W.J. Strittmatter, D. Schmechel, P.H. St. George-Hyslop, M.A. Pericak-Vance, S.H. Joo, B.L. Rosi, J.F. Gusella, D.R. Crapper-MacLachlan, M.J. Alberts, Hulette, B. Crain, D. Goldgaber, and A.D. Roses,

Association of apolipoprotein E allele *E*4 with late-onset familial and sporadic Alzheimer's disease. *Neurology*, **43** (1993) 1467-1472.

[53] B.G. Schreurs, C.A. Smith-Bell, J. Lochhead, and D.L. Sparks. Cholesterol modifies classical conditioning of the rabbit (Oryctolagus cuniculus) nictitating membrane response, *Behav Neurosci* **117** (2003), 1220-1232.

[54] P.D. Sloane, S. Zimmerman, C. Suchindran, P. Reed, L. Wang, M. Boustani, and S. Sudha, The public health impact of Alzheimer's disease, 2000-2050: Potential implication of treatment advances, *Ann. Rev. Public Health* **23** (2002) 213-231.

[55] A. Sokolovska, S.L. Hem, and H. HogenEsch, Activation of dendritic cells and induction of CD4(+) T cell differentiation by aluminum-containing adjuvants, *Vaccine* **25** (2007), 4575–4585.

[56] P.R. Solomon M. Brett, M. Groccia-Ellison, C. Oyler, M. Tomasi, and W.W Pendlebury, Classical conditioning in patients with Alzheimer's disease: A multiday study, *Psychol Aging* **10** (1995), 248-254.

[57] P.R. Solomon, E. Levine, T. Bein and W.W. Pendlebury, Disruption of classical conditioning in patients with Alzheimer's disease, *Neurobiol. Aging* **12** (1991), 283-287.

[58] P.R. Solomon, S.D. Solomon, E. Vander Schaaf and H.E. Perry, Altered activity in the hippocampus is more detrimental to classical conditioning than removing the structure, *Science* **220** (1983), 329-331.

[59] M. Sommer, J. Grafman, K. Clark and M. Hallett, Learning in Parkinson's disease: eyeblink conditioning, declarative learning, and procedural learning, *J. Neurol. Neurosurg. Psychiatry* **67** (1999), 27-34.

[60] D.L. Sparks, R. Friedland, S. Petanceska, B.G. Schreurs, J. Shi, G. Perry, M.A. Smith, A. Sharma, S. Derosa, C. Ziolkowski, G. Stankovic. Trace copper levels in the drinking water, but not zinc or aluminum influence CNS Alzheimer-like pathology, *J Nutr Health Aging* **10** (2006), 247-254.

[61] D.L. Sparks, J. Hunsaker, III, S. Scheff, R. Kryscio, J. Henson and W. Markesbery, Cortical senile plaques in coronary artery disease, aging and Alzheimer's disease, *Neurobiol Aging* **11** (1990), 601-607.

[62] D.L. Sparks, H. Liu, S.W. Scheff, C.M. Coyne and J.C. Hunsaker III, Temporal sequence of plaque formation in the cerebral cortex of non-demented individuals, *J Neuropath Exp Neurol* **52** (1993), 135-142.

[63] D.L. Sparks, J. Lochhead, D. Horstman, T. Wagoner and T. Martin, Water quality has a pronounced effect on cholesterol-induced accumulation of Alzheimer Amyloid β (Aβ) in rabbit brain, *JAD* **4** (2002), 523-529.)

[64] D.L. Sparks, S.W. Scheff, H. Liu, T. Landers, C.M. Coyne and J.C. Hunsaker, Increased density of neurofibrillary tangles (NFT) in non-demented individuals with hypertension, *J Neurological Sci* **131** (1995), 162-169.

[65] D.L. Sparks and B.G. Schreurs, Trace amounts of copper in water induce beta-amyloid plaques and learning deficits in a rabbit model of Alzheimer's disease, *Proc Natl Acad Sci U S A.* **100** (2003), 11065-11069.

[66] A. Takashima, T. Honda, K. Yasutake, G. Michel, O. Murayama, M. Murayama, K. Ishiguro and H. Yamaguchi, Activation of tau protein kinase I/glycogen synthase kinase-3beta by amyloid beta peptide (25-35) enhances phosphorylation of tau in hippocampal neurons, *Neurosci. Res.* **31** (1998), 317-323

[67] Q. Tang and J.A. Bluestone, The Foxp3+ regulatory T cell: a jack of all trades, master of regulation, *Nature Immunology* **9** (2008), 239-244.

[68] R. F. Thompson, Classical conditioning: The Rosetta stone for brain substrates of age-related deficits in learning and memory, *Neurobiol Aging* **9** (1988), 547-548.

[69] C.T. Weaver. R.D. Hatton, P.R. Mangan, and L.E. Harrington, IL-17 Family Cytokines and the Expanding Diversity of Effector T Cell Lineages, *Ann. Rev. Immunol.* **25** (2007), 821–52.

[70] H.M. Wisniewski, C. Bancher, M. Barcikowska, G.Y. Wen and J. Currie, Spectrum of morphological appearance of amyloid deposits in Alzheimer's disease, *Acta. Neuropath.* **78** (1989), 337-347.

[71] C.P. Wong, C.Y. Okada, and R. Levy, TCR vaccines against T cell lymphoma: QS-21 and IL-12 adjuvants induce a protective CD8+ T cell response, *J. Immunol.* **162** (1999), 2251–2258.

[72] D.S. Woodruff-Pak, A. Agelan and L. Del Valle, A rabbit model of AD: Valid at neuropathological, cognitive, and therapeutic levels, *JAD* **11** (2007), 371-383.

[73] D.S. Woodruff-Pak, A. Agalan, L. Del Valle, L., and M. Achary, An animal model of Alzheimer's disease highlighting targets for computational modeling. In: *Cognitive Neurodynamics*, R. Wang & G. Fanji, eds. Berlin: Springer-Verlag, 2008.

[74] D.S. Woodruff-Pak, R.G. Finkbiner, and I.R. Katz, A model system demonstrating parallels in animal and human aging: Extension to Alzheimer's disease. In: *Novel approaches to the treatment of Alzheimer's disease* E.M. Meyer, J.W. Simpkins, and J. Yamamoto (eds.), NY: Plenum, 1989, pp. 355-371.

[75] D.S. Woodruff-Pak, R.G. Finkbiner, and D.K. Sasse, Eyeblink conditioning discriminates Alzheimer's patients from non-demented aged, *NeuroReport*, **1** (1990), 45-48.

[76] D.S. Woodruff-Pak and R.M. Hinchliffe, Scopolamine- or mecamylamine-induced learning impairment: Reversed by nefiracetam. *Psychopharmacol* **131** (1997), 130-139.

[77] D.S. Woodruff-Pak, and M.E. Jaeger, Predictors of eyeblink classical conditioning over the adult age span, *Psychol Aging* **13** (1998), 193-205.

[78] D.S. Woodruff-Pak and M. Papka, Huntington's disease and eyeblink classical conditioning: Normal learning but abnormal timing, *J. Int. Neuropsychol. Soc.* **2** (1996), 323-334.

[79] D.S. Woodruff-Pak, M. Papka, S. Romano and Y-T. Li, Eyeblink classical conditioning in Alzheimer's disease and cerebrovascular dementia, *Neurobiol. Aging* **17** (1996), 505-512.

[80] D.S. Woodruff-Pak, M. Papka, and E.W. Simon, Eyeblink classical conditioning in Down's syndrome, Fragile X Syndrome, and normal adults over and under age 35, *Neuropsychol.* **8** (1994), 14-24.

[81] D.S. Woodruff-Pak, S. Romano, and M. Papka, Training to criterion in eyeblink classical conditioning in Alzheimer's disease, Down's syndrome with Alzheimer's disease, and healthy elderly, *Behav Neurosci* **110** (1996), 22-29.

[82] D.S. Woodruff-Pak and J.E. Steinmetz (Eds.), *Eyeblink classical conditioning: Volume II: Animal models.* Boston: Kluwer Academic Publishers, 2000.

[83] D.S. Woodruff-Pak, R.W. Vogel III and G.L.Wenk, Galantamine: Effect on nicotinic receptor binding, ace-

tylcholinesterase inhibition, and learning, *PNAS* **98** (2001), 2089-2094.

[84] T. Yamazaki, H. Yamaguchi, Y. Nakazato, K. Ishiguro, T. Kawarabayashi and S. Hirai, Ultrastructural characterization of cerebellar diffuse plaques in Alzheimer's disease, *J. Neuropathol. Exp. Neurol.* **51** (1992), 281-286.

[85] S.G. Zheng, J.H. Wang, W. Stohl, K.S. Kim, J.D. Gray, and D.A. Horwitz, TGF- Requires CTLA-4 Early after T Cell Activation to Induce FoxP3 and Generate Adaptive CD4+ CD25+ Regulatory Cells, *J. Immunol.* 176 (2006), 3321-3329.

[86] J. Zhum H. Yamane, and W.E. Paul, Differentiation of effector CD4 T cell populations, Ann. Rev. Immunol 28 (2010), 445-489.

# Anti-Amyloid-$\beta$ Immunotherapy in Alzheimer's Disease: Relevance of Transgenic Mouse Studies to Clinical Trials

Donna M. Wilcock* and Carol A. Colton
*Duke University Medical Center, Department of Medicine Division of Neurology, Durham, NC, USA*

**Abstract.** Therapeutic approaches to the treatment of Alzheimer's disease are focused primarily on the amyloid-$\beta$ peptide which aggregates to form amyloid deposits in the brain. The amyloid hypothesis states that amyloid is the precipitating factor that results in the other pathologies of Alzheimer's disease. One such therapy that has attracted significant attention is anti-amyloid-$\beta$ immunotherapy. First described in 1999, immunotherapy uses anti-amyloid-$\beta$ antibodies to lower brain amyloid levels. Active and passive immunization were shown to lower brain amyloid levels and improve cognition in multiple transgenic mouse models. Mechanisms of action were studied in these mice and revealed a complex set of mechanisms that depended on the type of antibody used. When active immunization advanced to clinical trials a subset of patients developed meningoencephalitis, an event not predicted in mouse studies. It was suspected that a T-cell response due to the type of adjuvant used was the cause. Passive immunization has also advanced to Phase III clinical trials on the basis of successful transgenic mouse studies. Reports from the active immunization clinical trial indicated that, similarly to effects observed in mouse studies, amyloid levels in brain were reduced.

Keywords: Amyloid, antibody, cerebral amyloid angiopathy, clinical trial, immunization, inflammation

## ALZHEIMER'S DISEASE AND THE AMYLOID HYPOTHESIS

Alzheimer's disease (AD) is a neurodegenerative disorder leading to a dementia with progressive loss of brain function. The primary risk factor for AD is age, with onset typically occurring between 70–90 years of age. The mean life expectancy is anywhere from 7 to 15 years after the initial diagnosis, however, rates of progression vary significantly between patients. While diagnosis of AD may be made through a battery of cognitive tests, a definite diagnosis can only be made at autopsy by microscopic examination of the brain tissue. According to the NIA-Reagan criteria a diagnosis of AD requires the presence of amyloid deposits, neurofibrillary tangles and neurodegeneration as well as dementia [52]. Amyloid plaques are insoluble, extracellular accumulations of amyloid-$\beta$ (A$\beta$) peptides. Neurofibrillary tangles are intraneuronal accumulations of hyperphosphorylated, aggregated tau protein (a microtubule binding protein) that redistributes to the neuronal soma. There are many accompanying pathologies in AD including cerebral amyloid angiopathy (CAA; accumulation of amyloid in the cerebrovasculature) and neuroinflammation (microglial and astrocytic reactivity to the abnormal proteins in the AD brain). These likely play a significant role in the disease progression.

The amyloid hypothesis of AD is based upon the pathologic characteristics and genetics of the disease. Early onset-familial Alzheimer's disease (FAD) is a rare, genetic form of the disease. To date, all genes known to cause FAD are involved in the production of A$\beta$, and therefore amyloid. These genes are the amyloid-$\beta$ protein precursor (A$\beta$PP) gene, and the pre-

---

*Corresponding author: Donna M. Wilcock, Duke University Medical Center, Department of Medicine Division of Neurology, Bryan Research Bldg, Box 2900, Durham, NC 27710, USA. Tel: +1 919 668 3998; E-mail: donna.wilcock@duke.edu.

senilin 1 (PS1) and presenilin 2 (PS2) genes. A$\beta$PP is a single membrane-spanning protein whose exact physiological function is unknown. However, data suggest that A$\beta$PP may be involved in synapse formation and stability, cell adhesion, memory and even possibly may act as a G-protein coupled receptor (reviewed by [72]). A$\beta$PP can be cleaved by three enzymes; $\alpha$-, $\beta$- and $\gamma$-secretase. Cleavage by $\beta$ and $\gamma$ produces the A$\beta$ peptide, the length of which is determined by the $\gamma$-secretase cleavage. Under normal conditions, an $\alpha$ cleavage is the dominant cleavage, which produces non-amyloidogenic fragments (reviewed by [55]). The presenilins are highly conserved proteins with 8 transmembrane domains and are now known to be part of the $\gamma$-secretase complex. Both PS1 and PS2 are physiologically cleaved forming 2 polypeptides that may function in the control of apoptosis. It is also known that genetic deletion of presenilins is lethal due to alteration of Notch processing and signaling (reviewed by [36]).

Very simply, the amyloid cascade hypothesis states that deposition of A$\beta$ in the brain is the precipitating factor that then results in tau hyperphosphorylation, aggregation and, ultimately, neurofibrillary tangles. Amyloid deposition and tau pathology are then thought to both contribute to neuronal degeneration, which results in the cognitive decline in AD [29]. In support of the amyloid hypothesis, all FAD mutations either increase total A$\beta$ production (via A$\beta$PP mutations) or shift A$\beta$ production to the more fibrillogenic A$\beta_{1-42}$ species (via PS mutations) (reviewed by [71]). Also supporting this hypothesis is the pathology of Down's syndrome. Down's syndrome is caused by a triplication of chromosome 21. This chromosome carries the A$\beta$PP gene, therefore, A$\beta$PP is triplicated along with a number of other important genes. It is well known that Down's syndrome patients develop AD. By 40 years of age, 25% of Down's syndrome patients develop clinical AD and, by 60 years of age, 65% develop AD. At autopsy, all Down's syndrome patients have significant amyloid deposition in their brains [45]. Similarly, there are families carrying duplication of the A$\beta$PP locus which results in autosomal dominant early-onset AD with CAA [63,64,68].

## OVERVIEW OF TRANSGENIC MICE

Mouse models of AD are primarily focused around the familial mutations in A$\beta$PP or the PS1 and PS2 genes. Table 1 summarizes the transgenic mouse models that will be discussed in this review. The PDAPP transgenic mouse was the first reported A$\beta$PP transgenic mouse to develop amyloid deposits similar to those found in the brains of AD patients. This mouse carries the V717F A$\beta$PP mutation that in humans is associated with early onset AD. An 18-fold overexpression of mutated human A$\beta$PP is observed, and amyloid deposits are first detected at approximately six months of age. The PDAPP mice also develop mild CAA by 18 months [28]. A similar mouse, the Tg2576 mouse carries the Swedish mutation in A$\beta$PP, the K670M/ M671L. The overexpression of mutant human A$\beta$PP in the Tg2576 mouse is 5-fold, and the rate of amyloid deposition is moderate with amyloid deposits first detected at six months of age. CAA is sparse and is most commonly observed by 18 months of age [35]. The majority of studies examining immunotherapy have used either the PDAPP or Tg2576 mice.

Other commonly used A$\beta$PP transgenic mice include the APP23 mouse which carries the K670N, M671N mutations in human A$\beta$PP, and has higher level of expression than the Tg2576 mouse [70] and the TgCRND8 mouse, which carries both the V717F and the K670M, M671N mutations in A$\beta$PP [9]. Because of the mutations that affect both $\beta$-secretase and $\gamma$-secretase activities, the TgCRND8 mouse has a very aggressive rate of amyloid deposition. Finally, the APP:V717I mouse carries an A$\beta$PP mutation at the same site as the PDAPP mouse, and it is pathology closely resembles that of the PDAPP [48]. However, it should be noted that all of these mouse models are most useful as models of amyloid deposition, as opposed to models of AD, since they do not develop significant tau pathology or neuron loss (the APP23 mouse has been shown to have some cortical neuron loss [7]). A model of vascular amyloid pathology is the APPSw-DI mouse, which carries the K670M/N671L, E693Q and the D694N APP mutations, corresponding to the Swedish, Dutch and Iowa mutations. This mouse has a rapid rate of amyloid deposition with a high percentage of the amyloid associated with the vasculature [16]. This mouse was later shown to have a deficient clearance of A$\beta$ across the blood-brain barrier [17]. The mutations in presenilin alone do not result in amyloid deposition [21]. However, when the mouse carrying the M146L mutation in PS1 [33] was crossed with the Tg2576 transgenic mouse [35], it was found that the rate of amyloid deposition was significantly accelerated, and A$\beta$ production favored the more fibrillogenic A$\beta_{1-42}$ species [33]. The A$\beta$PP/PS1 transgenic mouse

Table 1
Summary of transgenic mice discussed in this review. + scale indicates severity of given pathology

| Transgenic mouse model | Mutation(s) | Expression level | Rate of amyloid deposition | Tau Pathology | Neuron loss | CAA severity | Primary Ref |
|---|---|---|---|---|---|---|---|
| **PDAPP** | Human AβPP V717F | 18-fold | Moderate | +/- | - | + | (Games et al., 1995 Nature 373, 523-527) |
| **Tg2576** | Human AβPP K670M/ M671L | 5-fold | Moderate | +/- | - | + | (Hsiao et al., 1996 Science 274, 99-102) |
| **APP23** | Human AβPP K670N/ M671L | 7-fold | Moderate | +/- | + | ++ | (Sturchler-Pierrat et al., 1997 Proc Natl Acad Sci USA 94, 13287-13292) |
| **TgCRND8** | Human AβPP V717F and K670M/M671L | 5-fold | Rapid | +/- | - | ++ | (Chishti et al., 2001 J Biol Chem 276, 21562-21570) |
| **APP:V717I** | Human AβPP V717I | 10-fold | Moderate | +/- | - | ++ | (Moechars et al., 1999 J Biol Chem 274, 6483-6492) |
| **APPSwDI** | Human AβPP K670N/ M671L, E693Q, D694N | 0.5-fold | Rapid | +/- | - | +++ | (Davis et al., 2004 J Biol Chem 279, 20296-20306) |
| **APP+PS1 Tg2576 X M146L** | Human AβPP KM670/671NL; PS1 M146L | 5-fold (AβPP) | Rapid | +/- | - | + | (Holcomb et al., 1998 Nat Med 4, 97-100) |
| **3XTg** | Human AβPP KM670/671NL; PS1 M146V. tau P301L | 4-fold (AβPP) | Moderate | ++ | - | (unknown) | (Oddo et al., 2003 Neuron 39, 409-421) |

also does not develop significant tau pathology or neuron loss.

More recent advances in AD mouse models involve the addition of tau pathology. Mutations in the tau protein have been found in humans diagnosed with frontotemporal dementia (FTD). The first transgenic mouse model incorporating a FTD human tau mutation was the JNPL3 mouse, which carries the human tau P301L mutation. The P301L mouse shows hyperphosphorylated tau at disease associated sites, redistribution of tau to the dendrites and soma of neurons, and formation of neurofibrillary tangle-like structures within neurons [44]. This mouse has provided useful information on the role of mutated human tau in neurofibrillary tangle formation and in neuronal pathology. In 2003, Oddo and colleagues presented a triple transgenic (3XTg) mouse that carried the Swedish mutation in human AβPP (same AβPP mutations as Tg2576), the M146L mutation in human PS1, and the P301L mutation in human tau. This mouse develops significant intra-neuronal Aβ and parenchymal amyloid deposits, as well as intraneuronal hyperphosphorylated tau aggregates [56]. Neuron counts from this 3XTg mouse by stereological methods show no significant loss of neurons, even in brain regions with high levels of tau and amyloid pathology [50].

Most recently, several novel approaches to the generation of transgenic mice modeling AD have been significantly more successful in demonstrating progression of disease beyond amyloid deposition. Capsoni and colleagues (2000) showed that a mouse expressing recombinant antibodies neutralizing nerve growth factor, develops both amyloid plaques and tau pathology [8]. This represents a significant advance in transgenic mouse modeling as the amyloid and tau pathologies occur with normal mouse proteins. Also, we have recently shown in two different AβPP transgenic mice (the Tg2576; APPSw and the APPSwDI mice) that genetic deletion of nitric oxide synthase 2 results in progression of amyloid pathology to include normal mouse tau pathology and significant neuron loss [11, 82] (nitric oxide synthase 2 deletion is reviewed in this issue [12]).

## THE BEGINNING OF ANTI-Aβ IMMUNOTHERAPY

In 1999, an astounding publication by Schenk and colleagues of Elan Pharmaceuticals suggested that immunization against Aβ peptide could be used as a potential therapeutic for AD. In this study, PDAPP mice were either immunized with fibrillar $A\beta_{1-42}$ in an immune adjuvant prior to the onset of pathology or at an age when significant amyloid pathology was present. Mice that were immunized prior to the onset of pathology and continued to be immunized monthly had low levels of detectable amyloid. Furthermore, in those mice that were immunized at an age when significant amyloid pathology was already present, impressive reductions in amyloid deposition were noted [65]. Following this publication, two additional reports demonstrated that $A\beta$ vaccination in the APP + PS1 [49] or TgCRND8 [37] transgenic mice improved performance in learning and memory tasks. The same $A\beta$ immunization protocol used by Schenk et al. was utilized in both studies, and behavioral deficits in the radial-arm water maze task [49] and the Morris water maze task [37] were significantly reduced by vaccination. Together, these initial data indicated that reduction of amyloid deposition could be achieved by immunization and was sufficient alone to improve learning and memory in mice.

The early vaccination protocols in mice used an active immunization approach. Active immunization uses an immunogen, in this case $A\beta$, combined with an adjuvant to stimulate an immune response, in this case Freund's adjuvant. The animal / patient's immune system generates anti-$A\beta$ antibodies that are then thought to result in reductions in amyloid deposition (Fig. 1A). Another approach of anti-$A\beta$ immunotherapy was suggested that would allow greater control of dose, and also allow a mechanism to withdraw treatment should any adverse events become apparent. This approach was passive immunization. Passive immunization involves the *in vitro* production of monoclonal anti-$A\beta$ antibodies, and then direct infusion of these antibodies into the patient / mouse (Fig. 1B).

## ACTIVE IMMUNIZATION

Administration of an antigen plus an adjuvant is the classical approach to an active immunization. An adjuvant is a solution that is designed to stimulate an immune response in order to generate antibodies to the immunogen. Many times, an adjuvant consists of bacteria fragments. For example, the adjuvant used in many of the mouse studies was Freund's adjuvant. Freund's consists of inactivated and dried mycobacteria, usually *mycobacterium tuberculosis*. The goal of this process is to generate a significant antibody response against the antigen via activation of B cell antibody production and increased T cell activation. However, unregulated T cell responses, or the wrong type of T cell response (type 1 vs. type 2 for example), is detrimental and can result in an inappropriate immune response [66]. Active immunization against $A\beta$ advanced to clinical trials in patients with AD based on the initial transgenic mouse data showing reduced amyloid deposition and cognitive decline. The trial was called AN1792 and involved up to five immunizations over a 36 week period. The clinical trial was suspended due to an occurrence of subacute meningoencephalitis (inflammation of the brain and meninges) in approximately 6% of patients [57]. It is important to note that meningoencephalitis had not been observed in mouse studies using the amyloid deposition models available at that time. Following this disrupted clinical trial, subsequent active immunization studies have focused on: 1) testing various adjuvants to overcome an inappropriate t-cell response; and 2) alternate immunization protocols (summarized in Table 2).

Speculation on the reasons for failure of the initial clinical trial on active vaccination focused on the type of adjuvant used to promote the antibody response. The adjuvant used in the AN1792 trial was QS21, which is a Th1 type adjuvant. This type of adjuvant can initiate the production of pro-inflammatory cytokines such as interferon-$\gamma$ (IFN$\gamma$) [13]. QS21 likely stimulated a T-cell reaction leading to development of meningoencephalitis. Multiple studies using transgenic mouse models explored alternatives to the QS21 adjuvant that would stimulate a Th2 response, as opposed to a Th1 response. For example, when alum, an aluminum salt based Th2-based adjuvant, was used to stimulate a B cell humoral response to $A\beta_{1-42}$ in Tg2576 mice, significant reductions in amyloid levels were observed. The effect on amyloid levels, however, was only seen in mice treated from 11- 24 months of age. Tg2576 mice immunized with $A\beta_{1-42}$ from 19 to 24 months of age showed no significant changes [1]. Another elegant study compared the humoral and cellular immune responses produced by two different adjuvants used in subcutaneous $A\beta$ vaccination. In this study, monophosphoryl lipid A (MPL)/trehalose dicorynomycolate (TDM) was compared to Escherichia coli heat-labile enterotoxin LT (R192G). MPL/TDM generated a much greater antibody titer than LT(R192G), and was accompanied by a moderate splenocyte proliferation and IFN$\gamma$ production indicating a cellular response [46].

A critical study separately tested complete Freund's adjuvant (CFA), alum, TiterMax Gold (TMG) and

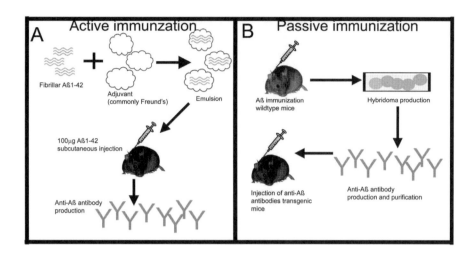

Fig. 1. The two types of immunotherapy for Alzheimer's disease. Panel A depicts active immunization. Fibrillar A$\beta_{1-42}$ is combined with an adjuvant by emulsification. This mixture is then injected into the mouse. The mouse produces anti-A$\beta$ antibodies in response to the vaccination. Panel B depicts passive immunization. In this case, mice are immunized with A$\beta$ as with active immunization. Hybridomas are then produced and selected for the correct antibody. This antibody is then harvested, purified and administered to another mouse for treatment.

QS21 as adjuvants for A$\beta_{1-42}$ vaccination. QS21 and CFA induced significantly high antibody titers. Titers were intermediate with alum and lowest with TMG. Alum primarily produced a Th2 biased response (IgG1 antibody production) whereas the other adjuvants produced a Th1 biased response (IgG2a production). Because Th1-type responses have been implicated in autoimmune disease, adjuvants that induce Th-1 responses are likely to be less useful in A$\beta$ vaccination protocols. Also, both QS21 and CFA stimulated a significant T-cell response as indicated by IFN$\gamma$ production [13]. Cribbs and colleagues have further mapped the T-cell epitope of A$\beta$ to the 6–28 sequence, while the dominant B cell epitope was found within the 1–15 region [13]. Consequently, vaccines can be designed that preferentially use this identified B-cell epitope over those regions of the A$\beta$ peptide that initiate adverse T-cell responses. Together, these data provide valuable information for the design of future clinical trials, including improved antigen and adjuvant approaches. It is still uncertain whether, indeed, the adjuvant in the AN1792 trial did cause meningoencephalitis. If a T-cell response was the cause of the meningoencephalitis, these mouse studies provide valuable information regarding the importance of adjuvant selection for design of future clinical trials.

Another approach to increase the safety of active immunization is to examine alternate approaches to administration of the vaccination as well as more sophisticated methods of stimulating anti-A$\beta$ antibody production. Lemere and colleagues from Harvard University were the first to suggest the use of intranasal administration of A$\beta$ vaccination as an alternative to the subcutaneous route. They showed that weekly intranasal immunizations using A$\beta_{1-40}$ alone could stimulate antibody responses in PDAPP transgenic mice sufficient for reduction in brain A$\beta$ levels (50–60% after 7 months of treatment) [41]. Importantly, it was shown that there was no detectable T-cell response by measuring IFN$\gamma$ levels. It was later shown that addition of the LT (R192G) adjuvant produced 16-fold higher antibody titers than with no adjuvant at all [42].

Solomon and colleagues of Tel Aviv University have studied the use of filamentous phage displaying four immunogenic residues of A$\beta$. Filamentous phage stimulate a humoral immune response for antibody production. The phage can therefore be used to display short immunogenic determinants fused to their surface resulting in antibody production against these short fragments, in this case the EFRH residues of A$\beta$. Solomon et al. showed in 1997 that N-terminal anti-A$\beta$ antibodies bound to preformed amyloid fibrils and caused a disaggregation and neutralization of their neurotoxicity [69]. The EFRH residues of A$\beta$, corresponding to positions 3–6 of A$\beta$, were found to be the epitope of the anti-aggregating properties of A$\beta$ [24,25]. A filamentous phage displaying the EFRH peptide, the sequence required for disaggregation of amyloid, was used as antigen for immunization of wildtype [26] and APP:V717I transgenic mice [27]. In wildtype mice,

Table 2
Summary of alternate approaches to active vaccination protocols

| Component | Modification | Effect |
| --- | --- | --- |
| Adjuvant | IL-4 and GM-CSF | Aß + IL-4 + GM-CSF in TgCRND8 mice generated significant antibody titers and reduces brain Aß. DaSilva et al, Neurobiol Dis 2006 23:433-444. |
| | Alum | Induces a beneficial Th1-type response and significant antibody titers. Cribbs et al, Int Immunol 2003 15:505-514. |
| | LT(R192G) | No evidence of Th2 response. Maier et al, Vaccine 2005 23:5149-5159. Significant antibody titers and reduces brain Aß in PDAPP transgenic mice. Lemere et al, Neurobiol Aging 2002 23:991-100 |
| Immunogen | Dendrimeric Aß1-15 | Significant antibody titers and reduces brain Aß in J20 transgenic mice. Seabrook et al, J Neuroinflamm 2006 3:14. |
| | Aß1-11 | Significant antibody titers, improved cognition and reduces brain Aß. Movsesyan N et al, PLoS ONE 2008 3:e2124. |
| | Aß3-6 | Significant antibody titers and reduces brain Aß in APPV:717F mice. Frenkel et al, Vaccine 2003 21:1060-1065. |
| Mode of Administration | Intranasal | Useful for boosting, significant antibody titers and reduces brain Aß. Lemere et al, Ann NY Acad Sci 2000 920:328-331. |
| Mode | Peptide | Dendrimeric Aß1-15 (see above), Aß1-42 produces significant antibody titer and brain Aß reductions. Schenk et al, Nature 1999 400:173-177. |
| | DNA | Plasmid containing 3 X Aß1-11, PADRE epitope and CCL22 chemokine improve cognition and reduce brain Aß in 3XTg mice. Movseyan et al, PLoS ONE 2008 3:e2124. Amplicon vector containing Aß and IL-4 packaged in HSV amplicon produces significant antibody titers and lowers brain Aß. Frazer et al, Mol Ther 2008:845-853. |
| | Phage | Filamentous phage displaying Aß3-6 produce significant antibody titers and reduces brain Aß. Frenkel et al, Vaccine 2003 21:1060-1065. |

the EFRH phage was found to generate significant antibody titers. In APP:V717I transgenic mice the EFRH phage was shown to reduce amyloid plaques levels by approximately 50% [27]. A similar approach to direct antibody production to the N-terminal of $A\beta$ is to use dendrimeric $A\beta_{1-15}$ (therefore excluding the T cell epitope of $A\beta$ in the $A\beta_{15-42}$ region). $A\beta_{1-15}$ alone was insufficient for antibody production [43]. However, when 16 copies of $A\beta_{1-15}$ were assembled on a branching tree ($dA\beta_{1-15}$) and used as the immunogen with LT (R192G) as the adjuvant, significant antibody, titers were produced. In J20 transgenic mice (carrying the same mutations as the PDAPP mouse), a single $A\beta_{40/42}$ immunization followed by $dA\beta_{1-15}$ boosts produced robust antibody titers, and significantly lowered amyloid plaque levels in the brain by approximately 60% [67]. Because the B cell epitope is located in the N-terminal of $A\beta$ and the T cell epitope is located in the middle of $A\beta$, generation of N-terminal antibodies may avoid an inappropriate T cell response resulting in adverse events.

Approaches to active immunization have recently focused on avoiding the use of classical adjuvants and instead employ techniques that direct the stimulated immune response to antibody production. Both IL-4 and GM-CSF are cytokines that drive dendritic cell differentiation and direct a Th2-type response. DaSilva and colleagues (2006) used these cytokines in combination to elicit a Th2-type response as a replacement for a traditional adjuvant in TgCRND8 mice. This combination elicited high antibody titers and significantly reduced amyloid deposition [15]. Bowers et al. [6] directly stimulated an immune response by using herpes simplex virus (HSV)-amplicon mediated $A\beta$ vaccination. HSV amplicon vectors have been shown to elicit a vigorous transgene product-specific immune response *in vivo* [83]. In this case, an amplicon vector coding for $A\beta$, with and without a tetanus toxin Fragment C (TxFC) adjuvant, was packaged in an HSV amplicon. The HSV amplicon was then used in a monthly vaccination protocol for three months in Tg2576 mice. Antibody titers were elevated and no significant difference was seen with the TxFC adjuvant alone. HSV-$A\beta$ resulted in a significant T-cell response and some T-cell infiltration into the brain parenchyma. However, only moderate amyloid reductions were observed with this protocol [6]. More recently, the authors improved this approach by including IL-4 with $A\beta$ in the expression vector. The rationale for this approach was that IL-4 is a Th2 cytokine that promotes a Th2-biased immune response. When administered to the 3Xtg mice HSV-$A\beta$-IL4 initiated a Th2 response, produced antibody titers, and reduced $A\beta$ levels to undetectable levels [23].

The most recent approach to active vaccination has been to engineer a molecule that will stimulate an irrelevant T-cell response (i.e., not directed to A$\beta$) along with A$\beta_{1-11}$. A non-self promiscuous epitope termed PADRE (pan HLA DR-binding epitope) was combined with A$\beta_{1-11}$ using a DNA vaccine approach. This vaccination study utilized a plasmid containing three copies of the gene encoding the B cell epitope A$\beta_{1-11}$, and the gene for the synthetic T cell peptide, PADRE. Also, expression of the macrophage derived chemokine (MDC/CCL22) was included, which has been shown to both stimulate Th2-type responses and suppress Th1-type responses. The construct produced pMDC-3A$\beta_{1-11}$-PADRE fusion protein in CHO cells *in vitro*. *In vivo*, a significant humoral response was detected with a highly polarized Th2-type response. Critically, cognitive performance of 3XTg mice was improved and amyloid deposition was significantly reduced using the DNA vaccine protocols [51].

## PASSIVE IMMUNIZATION

Passive immunization describes the direct injection of antibodies, bypassing the requirement for an immune response to generate the antibodies. The benefits of this immunization technique include targeting of specific epitopes of the A$\beta$ molecule, controlling the amount of antibody administered, and the ability to rapidly withdraw treatment if adverse events are discovered. The major disadvantage of this method is the expense required to produce monoclonal antibodies. Bard and colleagues were the first group to describe the use of passive immunization in mice [4]. Their study used several different monoclonal anti-A$\beta$ antibodies including 3D6 (IgG2b A$\beta_{1-15}$), 10D5 (IgG1 A$\beta_{3-7}$), or 16C11 (IgG1 A$\beta_{33-42}$) that targeted various A$\beta$ epitopes, and represented different IgG isotypes. The antibodies were injected intraperitoneally in 11-12 month old PDAPP mice weekly for 6 months. Amyloid deposits were significantly reduced in mice treated with 3D6 (IgG2b A$\beta_{1-15}$), or 10D5(IgG1 A$\beta_{3-7}$), whereas 16C11 (IgG1 A$\beta_{33-42}$) had no effect. The authors later mapped isotype and epitope specificities in PDAPP mice using antibodies that represented various epitopes and plaque binding abilities, as well as different IgG isotypes. The conclusion from this series of studies was that IgG2a antibodies recognizing the N-terminal of A$\beta$ (3D6 -IgG2b A$\beta_{1-15}$ or 10D5 -IgG1 A$\beta_{3-7}$), were most effective at reducing brain amyloid [5]. IgG2a has the highest affinity for the Fc$\gamma$ receptor in mouse, and therefore, the authors suggest that Fc$\gamma$ receptor-mediated clearance by microglia is a requirement for amyloid removal. The use of N-terminal antibodies, however, has been associated with adverse events. In 2002, Pfeifer et al. showed that passive immunization of aged APP23 mice with an IgG1 N-terminal A$\beta$ antibody caused a significant increase in the occurrence of CAA-associated microhemorrhage as well as acute hematomas [59].

Epitope requirements for A$\beta$ remain controversial. For example, an IgG1 antibody directed toward the mid-domain of A$\beta$ (A$\beta_{13-28}$) was shown by DeMattos and coworkers [18] to significantly reduce brain amyloid deposits in 4 month old PDAPP mice treated every other week for five months. Interestingly, this antibody did not bind *in vivo* plaques but does bind soluble, monomeric A$\beta$ with high affinity. Serum A$\beta$ levels increased rapidly, and significantly, following a single intravenous injection of the antibody (called m266). The authors, therefore, suggested that the m266 antibody worked to reduce brain amyloid via a peripheral clearance as opposed to a central phagocytic mechanism [18]. Later studies by this group showed that passive immunization the m266 mid-domain antibody did not cause microhemorrhage. This is in contrast to an N-terminal antibody that caused a significant increase in CAA-associated microhemorrhage occurrence in PDAPP transgenic mice [60].

C-terminal antibodies were also examined in aged Tg2576 transgenic mice. The 2286 antibody is an IgG1 that recognizes A$\beta_{28-40}$. This antibody was administered intraperitoneally to Tg2576 mice with a staggered start such that mice received weekly injections for one, two, or three months and were all sacrificed at 22 months of age. Unlike the C-terminal antibody included in the 2003 Bard et al. study [5], this antibody was found to bind amyloid plaques *in vivo* after systemic administration. A transient but significant microglial reaction was observed but resolved by the three month treatment timepoint. Significant amyloid reductions were observed following two months of treatment and cognitive improvement was also observed [79]. This same C-terminal antibody was later shown to significantly increase CAA levels and CAA-associated microhemorrhage [80]. An IgG2b C-terminal antibody (2H6) was shown to have the same effects as the 2286 antibody. A deglycosylated version of 2H6 was also examined and showed significant reductions in brain amyloid deposits, despite the absence of significant microglial reactivity. Normally, IgG molecules are heavily glycosylated on their Fc portion. This glycosylation is important for high affinity binding by Fcγ receptors, so therefore, deglycosylation of anti-Aβ antibodies should significantly impair microglial Fcγ receptor-mediated amyloid clearance. Importantly, this deglycosylated antibody attenuated the increase in CAA and CAA-associated microhemorrhage observed with the intact 2H6 antibody [81].

Recently, antibody therapy has been focused on oligomeric forms of Aβ, which are small, soluble assemblies of Aβ (dimers, trimers, tetramers etc.) that have been shown, both *in vitro* and *in vivo*, to be neurotoxic at relatively low concentrations (reviewed by [74]). NAB61 is a monoclonal antibody that recognizes a complex conformational epitope in the N-terminal of Aβ oligomers. Passive immunization of Tg2576 mice with NAB61 did not alter brain amyloid deposition or AβPP processing, but did, in fact, improve learning and memory [40]. Also, another antibody (NU-6) has been developed that is specific for Aβ oligomers and binds amyloid plaques in human AD tissue. This antibody has been shown to neutralize Aβ oligomers *in vitro* and future studies will examine this antibody as a potential passive immunotherapeutic [39].

## UNCOVERING THE MECHANISMS OF ACTION IN TRANSGENIC MOUSE MODELS

There have been three proposed mechanisms of action of immunotherapy. These are summarized in Fig. 2 and Table 3.

*Mechanism 1: Microglial mediated removal*

Microglial mediated phagocytosis was first suggested by Schenk et al. [65] in their original description of Aβ vaccination. The authors suggested that microglial Fc-receptor phagocytosis may be responsible for removal of existing plaque. Microglia have been shown to phagocytose Aβ both *in vitro* and *in vivo* through several different mechanisms involving opsonization through the complement cascade [62], or the scavenger receptor [58]. Indeed, across a series of active immunization studies it was shown that the reduction of compact, Congophilic amyloid deposits correlated with the degree of microglial activation [76]. Bard's study using passive immunization methods demonstrated that only antibodies of an isotype with high affinity for Fcγ receptors (IgG2a) [61] were able to lower brain amyloid deposits [4]. These data further supported the hypothesis that microglia are responsible for the clearance of Aβ. Direct imaging of amyloid deposits in living mice using multiphoton microscopy [2] was used to examine the effects of antibody application to the brains of PDAPP mice. Three days following direct application of anti-Aβ antibodies to the brain, there was significant removal of amyloid deposits accompanied by activation of microglia surrounding the remaining deposits. Direct injection of anti-Aβ antibodies into the brains of Tg2576 mice resulted in rapid removal of diffuse amyloid deposits (by 24 hr), followed by removal of compact deposits (between two and three days) accompanied by a transient microglial reaction (peaked at three days) [77].

Conflicting data suggest that effective clearance of Aβ by anti-Aβ antibodies can be obtained in the absence of Fc receptors. Das and coworkers (2003) showed that active immunization of Tg2576 transgenic mice crossed with Fc receptor knockout mice demonstrated the same amount of Aβ reductions as immunized, age-matched AβPP transgenic mice [14]. As shown by live imaging using multiphoton microscopy, F(ab')$_2$ fragments are capable of reducing amyloid deposition as effectively as the complete IgG molecule when applied directly to the brain [3]. F(ab')$_2$ fragments made from an anti-A$\beta_{28-40}$ IgG also lowered amyloid deposits when injected intracranially into Tg2576 mice [78]. Inhibition of microglial activation after intracranial injection of antibody using dexamethasone had no effect on removal of diffuse amyloid deposits. However, there was no apparent reduction in compact, thioflavine-S positive, amyloid deposits indicating that microglial activation facilitates the removal of compact amyloid deposits [78].

*Mechanism 2: Catalytic disaggregation*

Catalytic disaggregation describes the interaction between an antibody and an amyloid deposit whereby the binding of the antibody disrupts the tertiary structure of the plaque resulting in disaggregation. Solomon et al. (1997) showed that monoclonal anti-Aβ antibodies were capable of inhibiting amyloid plaque formation *in vitro* [69]. Later, Frenkel and colleagues showed that anti-Aβ antibodies are capable of disaggregating Aβ plaques and neutralizing their neurotoxicity [24]. This indicated that antibody binding to Aβ causes the Aβ aggregates to dissolve forming monomeric Aβ. Studies in which F(ab')$_2$ fragments were applied directly to the brain [3,78] and reduce Aβ indicated a non-Fc mediated clearance mechanism, possibly via a direct

Table 3
Mechanisms of action for immunotherapy and brief description of some of the studies providing evidence for the given mechanism

| Mechanism | Method | Findings | Ref. |
|---|---|---|---|
| Microglial mediated phagocytosis. | Human microglia cultures incubated with antibody-opsonized Aß. | Enhanced uptake of Aß by microglial cells when antibody was present. | (Lue & Walker, 2002, J Neurosci Res 70,599-610) |
| | Intracranial injection of anti-Aß antibodies or F(ab')$_2$ fragments in Tg2576 mice. | Inhibition of microglial activation due to IgG also inhibited removal of compact plaques. F(ab')$_2$ fragments did not result in significant compact plaque reductions. | (Wilcock et al., 2003 J Neurosci 23,3745-3751) |
| | Passive immunization and of PDAPP mice and ex vivo analysis of phagocytosis. | Those antibodies that initiated with most robust Aß reductions in PDAPP mice were also the antibodies that demonstrated the most robust phagocytosis on brain slices. | (Bard et al., 2003 Proc Natl Acad Sci USA 100,2023-2028) |
| Catalytic disaggregation. | Addition of N-terminal anti-Aß antibodies to in vitro fibrillar Aß. | Disaggregation of Aß fibrils after incubation with N-terminal antibodies that also reduced their toxicity in PC12 cells. | (Solomon et al., 1997 Proc Natl Acad Sci USA 94,4109-4112) |
| | Intracranial administration of F(ab')' fragments of antibodies. | F(ab') fragments injected intracranially reduce amyloid deposition. | (Bacskai et al., 2002 J Neurosci 22,7873-7878) |
| | Inhibition of microglial activation by dexamethasone following intracranial anti-Aß antibody injection. | Diffuse (non-fibrillar) amyloid deposits were cleared despite complete inhibition of microglial activation. | (Wilcock et al, 2004, Neurobiol Dis 15,11-20) |
| Peripheral sink | Systemic injection of anti-Aß antibodies. | Systemic injection of anti-Aß antibodies into PDAPP mice resulted in a significant increase in plasma Aß levels and decrease in brain Aß. | (DeMattos et al., 2001 Proc Natl Acad Sci USA 98,8850-8855) |
| | Administration of anti-Aß antibodies to mice with deficient blood brain barrier Aß clearance (APPSwDI) | Systemic injection of anti-Aß antibodies into a mouse with impaired blood brain barrier Aß transport APPSwDI mice does not cause increased plasma Aß or decreased brain Aß. | (Vasilevko et al., 2007 J Neurosci 27,13376-13383) |

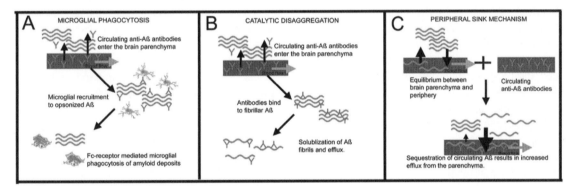

Fig. 2. The three major proposed mechanisms of action for immunotherapy in amyloid reduction. Panel A shows the mechanism of microglial phagocytosis. In this case, amyloid fibers (shown in blue) are opsonized by antibodies (shown in green) entering the brain from the bloodstream. Microglia then recognize the opsonized Aβ and phagocytose the amyloid via the Fcγ receptor. Panel B shows the mechanism of catalytic disaggregation. In this case amyloid fibers are bound by antibodies which then disrupt the tertiary structure of the amyloid deposit. This results in solublization of the Aβ and efflux out of the brain. Panel C shows the peripheral sink mechanism. In this case, monomeric soluble Aβ circulating in the bloodstream is bound by the circulating antibodies. This sequestration of circulating Aβ produces a shift in the concentration gradient of Aβ between the brain and the blood causing an efflux of Aβ out of the brain.

interaction with amyloid deposits. Also, the reduction in diffuse amyloid deposits by intracranially administered anti-Aβ antibodies despite complete inhibition of microglial activation suggested a direct interaction between the antibody and amyloid [78].

*Mechanism 3: Peripheral sink*

Studies involving passive immunization have suggested that the primary mechanism for Aβ clearance is peripheral and is not due to the antibodies entering the CNS. Instead, the Aβ antibodies act to reduce circulating Aβ levels. The peripheral sink mechanism is derived from studies using anti-Aβ antibodies that were specifically designed to not bind to amyloid plaques in the brain (the m266 antibody). When these antibodies were administered by intraperitoneal injection in the PDAPP mouse, a rapid 1,000-fold increase in circulating plasma Aβ levels was observed. These data suggested that circulating A$\beta$ antibodies bind to plasma A$\beta$ and consequently transiently reduce the circulating levels of soluble A$\beta$. In turn, this reduction promotes the removal of soluble A$\beta$ from the brain by mass action transfer across the blood brain barrier to the vasculature, hence the term peripheral sink [18]. Dodart et al. further demonstrated that treatment of PDAPP mice with m266 antibody reversed memory deficits one day after injection, without a reduction in amyloid burden in the brain [20]. The authors suggested that this rapid reversal of cognitive deficits was due to removal of soluble A$\beta$ from the CNS as opposed to reducing brain amyloid plaque burden. Cognitive improvement following passive immunization was also been shown in the Tg2576 mouse with an antibody recognizing A$\beta_{1-12}$, which also did not reduce brain A$\beta$ levels but did reverse memory deficits [38]. In a time-course study of weekly systemic anti-A$\beta$ antibody injection in Tg2576 mice, circulating A$\beta$ levels in the serum were increased 100-fold following 1 month of treatment, and remained significantly elevated following two and three months of treatment [79].

Another study which supports the peripheral sink mechanism as being a primary mechanism of A$\beta$ removal used the APPSwDI transgenic mouse, which has been shown to have deficient clearance of A$\beta$ across the blood-brain barrier. Active vaccination of these mice produced significantly high antibody titers, however, no increase in plasma A$\beta$ was observed, and no change in brain A$\beta$ levels were apparent. The authors confirmed that there were equal amounts of IgG entering the brain by performing an IgG western blot on brain homogenates comparing immunized Tg2576 mice (that are known to have antibody entering the brain) with immunized APPSwDI mice [73].

In summary, evidence for each of the proposed mechanisms for amyloid removal is strong. It is likely that all proposed mechanisms occur as a result of immunotherapy. However, the dominant mechanism may be determined by several factors including blood brain barrier integrity, antibody epitope and isotype and individual disease characteristics.

## CLINICAL TRIALS

Based on the impressive data from transgenic mouse studies, active A$\beta$ immunization progressed rapidly into a clinical trial called AN1792. Although the early response of the individuals was promising, the trial was suspended when 6% of patients in Phase II of the study developed subacute meningoencephalitis, a debilitating, life threatening inflammation of the brain and meninges [57]. Many of these patients made at least a partial recovery following treatment with corticosteroids. Over the course of several years after cessation of the trial, several autopsy reports were published on patients from the trial. These reports indicated that vaccination may have lowered amyloid levels in the brain and, therefore, interest in immunotherapy for treatment of AD has remained intense.

Comparisons between data from mouse studies and data from the human clinical trials are provided in Table 4. The initial reports following the human trial focused on the patients' ability to generate anti-A$\beta$ antibodies in response to the vaccination. Indeed, some (but not all) patients generated significant antibody titers. Characterization of these antibodies showed that they cross reacted with amyloid deposits in brain (including CAA and diffuse amyloid), but did not cross react with full length A$\beta$PP [31]. Cognition was followed in a subset of patients from the AN1792 clinical trial in Zurich, Switzerland after suspension of the trial. Patients who generated anti-A$\beta$ antibodies that would cross-react with amyloid plaques in tissue from AD patients showed a slowed cognitive decline when compared to those patients receiving placebo control [32]. These findings are reminiscent of the multiple reports of behavioral improvement following vaccination in various transgenic mouse models.

The first postmortem pathology report from a trial patient was published in 2003 [53]. This patient received five vaccinations (4 A$\beta$+ QS21 and 1 A$\beta$+QS21 +polysorbate-80) and six weeks following the final injection developed meningoencephalitis. The patient died from a pulmonary embolism 12 months after the last vaccination. Amyloid plaques were described as "patchy" in contrast to reasonably uniform plaques in an untreated AD control matched for Braak & Braak staging (V-VI). Quantification of amyloid plaque load in the patient compared to seven untreated AD controls revealed 60–70% less amyloid plaque throughout the neocortex compared to a group of disease stage-matched untreated AD tissue. Some regions that typically showed high levels of amyloid, such as the inferior, middle and superior temporal gyri, were almost completely devoid of plaques. Other features of AD such as CAA, neurofibrillary tangles and neuropil threads were described as unchanged compared to untreated AD controls. Significant T-lymphocyte infiltration was noted, mostly CD4+ with few CD8+ T-cells.

Table 4
A table summarizing the effects of vaccination in the human clinical trial and in transgenic mice. Bold font indicates where human and mouse showed the same effects

| Vaccination effects reported in human clinical trial | Vaccination effects reported in AβPP transgenic mice |
|---|---|
| Meningoencephalitis. 6% of patients in the phase II AN1792 active immunization trial (Orgogozo et al, Neurology 61:46-54). | No sickness behavior has been reported. |
| **Reduced Aβ plaque density (Nicoll et al, J Neuropathol Exp Neurol 65:1040-1048; Masliah et al, Neurology 64:129-131).** | **Reduced Aβ plaque density (Schenk et al, Nature 400:173-178; Das et al, J Neurosci 25:8532-8538; Wilcock et al, J Neurosci 24:6144-2151).** |
| Neurofibrillary tangles and tau pathology unaffected (Ferrer et al, Brain Pathol 14:11-20; Nicoll et al, Nat Med 9:448-452, Masliah et al, Neurology 64:129-131). | Some early tau species reduced (Billings et al, Neuron 45:675-688). |
| **Microhemorrhage in association with CAA (Ferrer et al, Brain Pathol 14:11-20).** | **Microhemorrhage in association with CAA (Pfeifer et al, Science 298:1379; Wilcock et al, J Neuroinflamm 1:24; Racke et al, J Neurosci 25:629-636; Wilcock et al Neuroscience144:950-960).** |
| **Persistence of CAA (Ferrer et al, Brain Pathol 14:11-20; Nicoll et al, Nat Med 9:448-452, Masliah et al, Neurology 64:129-131).** | **Persistence and increase in CAA (Pfeifer et al, Science 298:1379; Wilcock et al, J Neuroinflamm 1:24; Racke et al, J Neurosci 25:629-636; Wilcock et al Neuroscience144:950-960).** |
| Loss of myelin in cerebral white matter (Ferrer et al, Brain Pathol 14:11-20) – possibly complicated by encephalitis. | Not reported. |
| **Slowed cognitive decline (Hock et al, Neuron 38:547-554).** | **Improved cognition (Morgan et al, Nature 408:982-985).** |

A second pathology report in a trial patient that developed meningoencephalitis and later died echoed many of the same findings as Nicoll and colleagues [22]. The patient received only two immunizations and nine months following the final injection showed symptoms of aseptic meningoencephalitis with both CD4+ and CD8+ t-cell infiltrates. The patient was treated but the encephalitis reactivated resulting in death. The brain showed extensive T cell infiltration consistent with the meningoencephalitis. CAA and cerebral microhemorrhages were also noted. Plaques were described as "collapsed" and were surrounded by activated microglial cells. Many areas showed significantly fewer amyloid plaques than untreated AD brains of the same Braak & Braak staging. Examination of tau pathology showed reduction / disappearance of neuritic tau pathology, however, there was still extensive tau pathology present. These reductions were in comparison to disease stage-matched AD tissue.

The first pathology report from a trial patient that did not develop encephalitis was published in 2005 [47]. The patient received three immunizations, developed significant antibody titers and did not develop any obvious adverse reaction. It should be noted, despite the absence of encephailits, B and T cell infiltrates were observed in the brain. The patient died one year after the final injection and cause of death was described as "failure to thrive". The brain showed significant reductions in amyloid load, as detected by immunohistochemistry and biochemistry. Tangle score was generally lower than the average for untreated AD tissue. CAA score was the same as untreated controls. Astrogliosis was slightly lower and microgliosis was unchanged. Despite low antibody titers, another autopsy report also revealed significantly fewer amyloid deposits in the brain [54]. It was recently reported at the New York Academy of Sciences conference that, taken together, autopsy data indeed indicates that CAA levels are increased as well as an increased occurrence of microhemorrhage. However, four years after vaccination some patients showed resolution of even the CAA. Disappointingly, however, the patients failed to show a long term cognitive benefit of the vaccination [34].

Data from the AN1792 clinical trial suggest that immunotherapy is, indeed, effective in the removal (or prevention of accumulation) of amyloid from the human AD brain, however, tau pathology appeared to be relatively unaffected. It should be mentioned, however, that there remains the possibility that the non-specific inflammatory response to the adjuvant resulted in $A\beta$ reduction. In mice, stimulation of an inflammatory response by LPS certainly resulted in $A\beta$ reduction [19,30]. It is also unclear whether there was active clearance of amyloid or, simply, prevention of further amyloid accumulation. Further studies from the other patients involved in this trial will determine the exact effect of vaccination on other AD pathologies. Most recently, another setback for active immunization occurred when the trial ACC-001 failed. This trial used a 7 amino acid fragment of $A\beta$ from the N-terminal conjugated to a mutated diphtheria toxin protein called CRM 197. The vaccination was designed to avoid the T-cell response seen in the first trial. ACC-001 caused a vasculitis (inflammation of the blood vessels) resulting in skin lesions (see story at http://www.alzforum.org/new/detail.asp?id=1807). The cause is currently unknown.

Several trials examining passive immunization are currently ongoing. While the antibodies used in mouse studies are somewhat equivalent to those in clinical trial, the antibodies must be humanized to avoid an immune response by the patient. Elan Pharmaceuticals are in Phase IIb continuing to Phase III with their antibody called Bapineuzumab in trial AAB-001. While there has been no data published from this trial to date, the advancement to Phase IIb-III suggest that the antibody is well tolerated. Eli Lilly Pharmaceuticals are in Phase II with their monoclonal antibody 266 in trial LY2062430. Patients enrolled in the Phase I trial received one or three doses of the antibody and showed increased plasma and levels of A$\beta$ in cerebrospinal fluid as predicted from the mouse studies. The key in these studies will be cognitive performance and avoidance of adverse effects predicted in mice, i.e., elevated CAA levels and subsequent microhemorrhage.

## CONCLUSION

Discovery and development of immunotherapy as a treatment for AD would not have been possible without the development of transgenic mouse models. The extensive mouse model data has supported the clinical trials in immunotherapy. Further, studies examining the mechanisms of action have allowed for advancement in development of types of immunotherapy. Adverse events have been predicted by mouse model studies (CAA and microhemorrhage effects, for example), while some adverse events were not apparent (meningoencephalitis) in mice. Also, all immunotherapy studies in mice have shown cognitive benefit, where tested. However, the AN1792 trial failed to show any cognitive benefit over the longer term. It is important to note here that while transgenic mouse models are valuable to the study of AD, they are generally incomplete. There are some significant differences between mouse and human, and one example is the immune response. It has been known for some time that human macrophages *in vitro* produce significantly less NO when stimulated under conditions that produce high NO levels in mouse macrophages *in vitro* [10,75]. The lack of neuron loss in A$\beta$PP transgenic mice suggests more subtle and reversible pathways leading to the cognitive deficits in these mice. It is likely that in human AD, neuron loss is the main contributor to the clinical dementia and, therefore, is an irreversible process once initiated. This would suggest that treatments such as immunotherapy may be most valuable in the early stages of disease to halt any further progression. Improvements on mouse models continue, and the generation of models demonstrating amyloid plaques, normal tau pathology and accompanying neuron loss will be critical in the future testing of potential therapeutics to assess not only the efficacy pathologically, but also the point at which therapeutic benefit is likely to occur. In conclusion, development of better transgenic mouse models will result in improved understanding of the disease process and therefore improved development of therapeutic interventions.

## References

[1] A.A. Asuni, A. Boutajangout, H. Scholtzova, E. Knudsen, Y.S. Li, D. Quartermain, B. Frangione, T. Wisniewski and E.M. Sigurdsson, Vaccination of Alzheimer's model mice with Abeta derivative in alum adjuvant reduces Abeta burden without microhemorrhages, *Eur J Neurosci* **24** (2006), 2530-2542.

[2] B.J. Bacskai, S.T. Kajdasz, R.H. Christie, C. Carter, D. Games, P. Seubert, D. Schenk and B.T. Hyman, Imaging of amyloid-beta deposits in brains of living mice permits direct observation of clearance of plaques with immunotherapy, *Nat Med* **7** (2001), 369-372.

[3] B.J. Bacskai, S.T. Kajdasz, M.E. McLellan, D. Games, P. Seubert, D. Schenk and B.T. Hyman, Non-Fc-mediated mechanisms are involved in clearance of amyloid-beta in vivo by immunotherapy, *J Neurosci* **22** (2002), 7873-7878.

[4] F. Bard, C. Cannon, R. Barbour, R.L. Burke, D. Games, H. Grajeda, T. Guido, K. Hu, J. Huang, K. Johnson-Wood, K. Khan, D. Kholodenko, M. Lee, I. Lieberburg, R. Motter, M. Nguyen, F. Soriano, N. Vasquez, K. Weiss, B. Welch, P. Seubert, D. Schenk and T. Yednock, Peripherally administered antibodies against amyloid beta-peptide enter the central nervous system and reduce pathology in a mouse model of Alzheimer disease, *Nat Med* **6** (2000), 916-919.

[5] F. Bard, R. Barbour, C. Cannon, R. Carretto, M. Fox, D. Games, T. Guido, K. Hoenow, K. Hu, K. Johnson-Wood, K. Khan, D. Kholodenko, C. Lee, M. Lee, R. Motter, M. Nguyen, A. Reed, D. Schenk, P. Tang, N. Vasquez, P. Seubert and T. Yednock, Epitope and isotype specificities of antibodies to beta -amyloid peptide for protection against Alzheimer's disease-like neuropathology, *Proc Natl Acad Sci U S A* **100** (2003), 2023-2028.

[6] W.J. Bowers, M.A. Mastrangelo, H.A. Stanley, A.E. Casey, L.J. Milo, Jr. and H.J. Federoff, HSV amplicon-mediated Abeta vaccination in Tg2576 mice: differential antigen-specific immune responses, *Neurobiol Aging* **26** (2005), 393-407.

[7] M.E. Calhoun, K.H. Wiederhold, D. Abramowski, A.L. Phinney, A. Probst, C. Sturchler-Pierrat, M. Staufenbiel, B. Sommer and M. Jucker, Neuron loss in APP transgenic mice, *Nature* **395** (1998), 755-756.

[8] S. Capsoni, G. Ugolini, A. Comparini, F. Ruberti, N. Berardi and A. Cattaneo, Alzheimer-like neurodegeneration in aged antinerve growth factor transgenic mice, *Proc Natl Acad Sci U S A* **97** (2000), 6826-6831.

[9] M.A. Chishti, D.S. Yang, C. Janus, A.L. Phinney, P. Horne, J. Pearson, R. Strome, N. Zuker, J. Loukides, J. French, S. Turner, G. Lozza, M. Grilli, S. Kunicki, C. Morissette, J. Paquette, F. Gervais, C. Bergeron, P.E. Fraser, G.A. Carlson, P.S. George-Hyslop and D. Westaway, Early-onset amyloid deposition and cognitive deficits in transgenic mice expressing a double mutant form of amyloid precursor protein 695, *J Biol Chem* **276** (2001), 21562-21570.

[10] C. Colton, S. Wilt, D. Gilbert, O. Chernyshev, J. Snell and M. Dubois-Dalcq, Species differences in the generation of reactive oxygen species by microglia, *Mol Chem Neuropathol* **28** (1996), 15-20.

[11] C.A. Colton, M.P. Vitek, D.A. Wink, Q. Xu, V. Cantillana, M.L. Previti, W.E. Van Nostrand, J.B. Weinberg and H. Dawson, NO synthase 2 (NOS2) deletion promotes multiple pathologies in a mouse model of Alzheimer's disease, *Proc Natl Acad Sci U S A* **103** (2006), 12867-12872.

[12] C.A. Colton, D.M. Wilcock, D.A. Wink, J. Davis, W.E. Van Nostrand and M.P. Vitek, The Effects of NOS2 Gene Deletion on Mice Expressing Mutated Human AβPP, *J Alzheimers Dis* **15** (2008), 571–587.

[13] D.H. Cribbs, A. Ghochikyan, V. Vasilevko, M. Tran, I. Petrushina, N. Sadzikava, D. Babikyan, P. Kesslak, T. Kieber-Emmons, C.W. Cotman and M.G. Agadjanyan, Adjuvant-dependent modulation of Th1 and Th2 responses to immunization with beta-amyloid, *Int Immunol* **15** (2003), 505-514.

[14] P. Das, V. Howard, N. Loosbrock, D. Dickson, M.P. Murphy and T.E. Golde, Amyloid-beta immunization effectively reduces amyloid deposition in FcRgamma-/- knock-out mice, *J Neurosci* **23** (2003), 8532-8538.

[15] K. DaSilva, M.E. Brown, D. Westaway and J. McLaurin, Immunization with amyloid-beta using GM-CSF and IL-4 reduces amyloid burden and alters plaque morphology, *Neurobiol Dis* **23** (2006), 433-444.

[16] J. Davis, F. Xu, R. Deane, G. Romanov, M.L. Previti, K. Zeigler, B.V. Zlokovic and W.E. Van Nostrand, Early-onset and robust cerebral microvascular accumulation of amyloid beta-protein in transgenic mice expressing low levels of a vasculotropic Dutch/Iowa mutant form of amyloid beta-protein precursor, *J Biol Chem* **279** (2004), 20296-20306.

[17] J. Davis, F. Xu, J. Miao, M.L. Previti, G. Romanov, K. Ziegler and W.E. Van Nostrand, Deficient cerebral clearance of vasculotropic mutant Dutch/Iowa Double A beta in human A betaPP transgenic mice, *Neurobiol Aging* **27** (2006), 946-954.

[18] R.B. DeMattos, K.R. Bales, D.J. Cummins, J.C. Dodart, S.M. Paul and D.M. Holtzman, Peripheral anti-A beta antibody alters CNS and plasma A beta clearance and decreases brain A beta burden in a mouse model of Alzheimer's disease, *Proc Natl Acad Sci U S A* **98** (2001), 8850-8855.

[19] G. DiCarlo, D. Wilcock, D. Henderson, M. Gordon and D. Morgan, Intrahippocampal LPS injections reduce Abeta load in APP+PS1 transgenic mice, *Neurobiol Aging* **22** (2001), 1007-1012.

[20] J.C. Dodart, K.R. Bales, K.S. Gannon, S.J. Greene, R.B. DeMattos, C. Mathis, C.A. DeLong, S. Wu, X. Wu, D.M. Holtzman and S.M. Paul, Immunization reverses memory deficits without reducing brain Abeta burden in Alzheimer's disease model, *Nat Neurosci* **5** (2002), 452-457.

[21] K. Duff, C. Eckman, C. Zehr, X. Yu, C.M. Prada, J. Pereztur, M. Hutton, L. Buee, Y. Harigaya, D. Yager, D. Morgan, M.N. Gordon, L. Holcomb, L. Refolo, B. Zenk, J. Hardy and S. Younkin, Increased amyloid-beta42(43) in brains of mice expressing mutant presenilin 1, *Nature* **383** (1996), 710-713.

[22] I. Ferrer, M. Boada Rovira, M.L. Sanchez Guerra, M.J. Rey and F. Costa-Jussa, Neuropathology and pathogenesis of encephalitis following amyloid-beta immunization in Alzheimer's disease, *Brain Pathol* **14** (2004), 11-20.

[23] M.E. Frazer, J.E. Hughes, M.A. Mastrangelo, J.L. Tibbens, H.J. Federoff and W.J. Bowers, Reduced pathology and improved behavioral performance in Alzheimer's disease mice vaccinated with HSV amplicons expressing amyloid-beta and interleukin-4, *Mol Ther* **16** (2008), 845-854.

[24] D. Frenkel, M. Balass and B. Solomon, N-terminal EFRH sequence of Alzheimer's beta-amyloid peptide represents the epitope of its anti-aggregating antibodies, *J Neuroimmunol* **88** (1998), 85-90.

[25] D. Frenkel, M. Balass, E. Katchalski-Katzir and B. Solomon, High affinity binding of monoclonal antibodies to the sequential epitope EFRH of beta-amyloid peptide is essential for modulation of fibrillar aggregation, *J Neuroimmunol* **95** (1999), 136-142.

[26] D. Frenkel, O. Katz and B. Solomon, Immunization against Alzheimer's beta-amyloid plaques via EFRH phage administration, *Proc Natl Acad Sci U S A* **97** (2000), 11455-11459.

[27] D. Frenkel, I. Dewachter, F. Van Leuven and B. Solomon, Reduction of beta-amyloid plaques in brain of transgenic mouse model of Alzheimer's disease by EFRH-phage immunization, *Vaccine* **21** (2003), 1060-1065.

[28] D. Games, D. Adams, R. Alessandrini, R. Barbour, P. Berthelette, C. Blackwell, T. Carr, J. Clemens, T. Donaldson, F. Gillespie et al., Alzheimer-type neuropathology in transgenic mice overexpressing V717F beta-amyloid precursor protein, *Nature* **373** (1995), 523-527.

[29] J. Hardy and D.J. Selkoe, The amyloid hypothesis of Alzheimer's disease: progress and problems on the road to therapeutics, *Science* **297** (2002), 353-356.

[30] D.L. Herber, L.M. Roth, D. Wilson, N. Wilson, J.E. Mason, D. Morgan and M.N. Gordon, Time-dependent reduction in Abeta levels after intracranial LPS administration in APP transgenic mice, *Exp Neurol* **190** (2004), 245-253.

[31] C. Hock, U. Konietzko, A. Papassotiropoulos, A. Wollmer, J. Streffer, R.C. von Rotz, G. Davey, E. Moritz and R.M. Nitsch, Generation of antibodies specific for beta-amyloid by vaccination of patients with Alzheimer disease, *Nat Med* **8** (2002), 1270-1275.

[32] C. Hock, U. Konietzko, J.R. Streffer, J. Tracy, A. Signorell, B. Muller-Tillmanns, U. Lemke, K. Henke, E. Moritz, E. Garcia, M.A. Wollmer, D. Umbricht, D.J. de Quervain, M. Hofmann, A. Maddalena, A. Papassotiropoulos and R.M. Nitsch, Antibodies against beta-amyloid slow cognitive decline in Alzheimer's disease, *Neuron* **38** (2003), 547-554.

[33] L. Holcomb, M.N. Gordon, E. McGowan, X. Yu, S. Benkovic, P. Jantzen, K. Wright, I. Saad, R. Mueller, D. Morgan, S. Sanders, C. Zehr, K. O'Campo, J. Hardy, C.M. Prada, C. Eckman, S. Younkin, K. Hsiao and K. Duff, Accelerated Alzheimer-type phenotype in transgenic mice carrying both mutant amyloid precursor protein and presenilin 1 transgenes, *Nat Med* **4** (1998), 97-100.

[34] C. Holmes, D. Boche, D. Wilkinson, G. Yadegarfar, V. Hopkins, A. Bayer, R.W. Jones, R. Bullock, S. Love, J.W. Neal, E. Zotova and J.A. Nicoll, Long-term effects of Abeta42 immunisation in Alzheimer's disease: follow-up of a randomised, placebo-controlled phase I trial, *Lancet* **372** (2008), 216-223.

[35] K. Hsiao, P. Chapman, S. Nilsen, C. Eckman, Y. Harigaya, S. Younkin, F. Yang and G. Cole, Correlative memory deficits, Abeta elevation, and amyloid plaques in transgenic mice, *Science* **274** (1996), 99-102.

[36] M. Hutton and J. Hardy, The presenilins and Alzheimer's disease, *Hum Mol Genet* **6** (1997), 1639-1646.

[37] C. Janus, J. Pearson, J. McLaurin, P.M. Mathews, Y. Jiang, S.D. Schmidt, M.A. Chishti, P. Horne, D. Heslin, J. French, H.T. Mount, R.A. Nixon, M. Mercken, C. Bergeron, P.E. Fraser, P. St George-Hyslop and D. Westaway, A beta peptide immunization reduces behavioural impairment and plaques in a model of Alzheimer's disease, *Nature* **408** (2000), 979-982.

[38] L.A. Kotilinek, B. Bacskai, M. Westerman, T. Kawarabayashi, L. Younkin, B.T. Hyman, S. Younkin and K.H. Ashe, Reversible memory loss in a mouse transgenic model of Alzheimer's disease, *J Neurosci* **22** (2002), 6331-6335.

[39] M.P. Lambert, P.T. Velasco, L. Chang, K.L. Viola, S. Fernandez, P.N. Lacor, D. Khuon, Y. Gong, E.H. Bigio, P. Shaw, F.G. De Felice, G.A. Krafft and W.L. Klein, Monoclonal antibodies that target pathological assemblies of Abeta, *J Neurochem* **100** (2007), 23-35.

[40] E.B. Lee, L.Z. Leng, B. Zhang, L. Kwong, J.Q. Trojanowski, T. Abel and V.M. Lee, Targeting amyloid-beta peptide (Abeta) oligomers by passive immunization with a conformation-selective monoclonal antibody improves learning and memory in Abeta precursor protein (APP) transgenic mice, *J Biol Chem* **281** (2006), 4292-4299.

[41] C.A. Lemere, R. Maron, E.T. Spooner, T.J. Grenfell, C. Mori, R. Desai, W.W. Hancock, H.L. Weiner and D.J. Selkoe, Nasal A beta treatment induces anti-A beta antibody production and decreases cerebral amyloid burden in PD-APP mice, *Ann N Y Acad Sci* **920** (2000), 328-331.

[42] C.A. Lemere, E.T. Spooner, J.F. Leverone, C. Mori and J.D. Clements, Intranasal immunotherapy for the treatment of Alzheimer's disease: Escherichia coli LT and LT(R192G) as mucosal adjuvants, *Neurobiol Aging* **23** (2002), 991-1000.

[43] J.F. Leverone, E.T. Spooner, H.K. Lehman, J.D. Clements and C.A. Lemere, Abeta1-15 is less immunogenic than Abeta1-40/42 for intranasal immunization of wild-type mice but may be effective for "boosting", *Vaccine* **21** (2003), 2197-2206.

[44] J. Lewis, E. McGowan, J. Rockwood, H. Melrose, P. Nacharaju, M. Van Slegtenhorst, K. Gwinn-Hardy, M. Paul Murphy, M. Baker, X. Yu, K. Duff, J. Hardy, A. Corral, W.L. Lin, S.H. Yen, D.W. Dickson, P. Davies and M. Hutton, Neurofibrillary tangles, amyotrophy and progressive motor disturbance in mice expressing mutant (P301L) tau protein, *Nat Genet* **25** (2000), 402-405.

[45] I.T. Lott and E. Head, Alzheimer disease and Down syndrome: factors in pathogenesis, *Neurobiol Aging* **26** (2005), 383-389.

[46] M. Maier, T.J. Seabrook and C.A. Lemere, Modulation of the humoral and cellular immune response in Abeta immunotherapy by the adjuvants monophosphoryl lipid A (MPL), cholera toxin B subunit (CTB) and E. coli enterotoxin LT(R192G), *Vaccine* **23** (2005), 5149-5159.

[47] E. Masliah, L. Hansen, A. Adame, L. Crews, F. Bard, C. Lee, P. Seubert, D. Games, L. Kirby and D. Schenk, Abeta vaccination effects on plaque pathology in the absence of encephalitis in Alzheimer disease, *Neurology* **64** (2005), 129-131.

[48] D. Moechars, I. Dewachter, K. Lorent, D. Reverse, V. Baekelandt, A. Naidu, I. Tesseur, K. Spittaels, C.V. Haute, F. Checler, E. Godaux, B. Cordell and F. Van Leuven, Early phenotypic changes in transgenic mice that overexpress different mutants of amyloid precursor protein in brain, *J Biol Chem* **274** (1999), 6483-6492.

[49] D. Morgan, D.M. Diamond, P.E. Gottschall, K.E. Ugen, C. Dickey, J. Hardy, K. Duff, P. Jantzen, G. DiCarlo, D. Wilcock, K. Connor, J. Hatcher, C. Hope, M. Gordon and G.W. Arendash, A beta peptide vaccination prevents memory loss in an animal model of Alzheimer's disease, *Nature* **408** (2000), 982-985.

[50] D.A. Morrissette and F.M. LaFerla, Stereological quantification of neuronal cells in 3xTg-AD mice, *Soc Neurosci Abstr* (2007), Program No. 888.8.

[51] N. Movsesyan, A. Ghochikyan, M. Mkrtichyan, I. Petrushina, H. Davtyan, P.B. Olkhanud, E. Head, A. Biragyn, D.H. Cribbs and M.G. Agadjanyan, Reducing AD-like pathology in 3xTg-AD mouse model by DNA epitope vaccine - a novel immunotherapeutic strategy, *PLoS ONE* **3** (2008), e2124.

[52] K.L. Newell, B.T. Hyman, J.H. Growdon and E.T. Hedley-Whyte, Application of the National Institute on Aging (NIA)-Reagan Institute criteria for the neuropathological diagnosis of Alzheimer disease, *J Neuropathol Exp Neurol* **58** (1999), 1147-1155.

[53] J.A. Nicoll, D. Wilkinson, C. Holmes, P. Steart, H. Markham and R.O. Weller, Neuropathology of human Alzheimer disease after immunization with amyloid-beta peptide: a case report, *Nat Med* **9** (2003), 448-452.

[54] J.A. Nicoll, E. Barton, D. Boche, J.W. Neal, I. Ferrer, P. Thompson, C. Vlachouli, D. Wilkinson, A. Bayer, D. Games, P. Seubert, D. Schenk and C. Holmes, Abeta species removal after abeta42 immunization, *J Neuropathol Exp Neurol* **65** (2006), 1040-1048.

[55] J. Nunan and D.H. Small, Regulation of APP cleavage by alpha-, beta- and gamma-secretases, *FEBS Lett* **483** (2000), 6-10.

[56] S. Oddo, A. Caccamo, J.D. Shepherd, M.P. Murphy, T.E. Golde, R. Kayed, R. Metherate, M.P. Mattson, Y. Akbari and F.M. LaFerla, Triple-transgenic model of Alzheimer's disease with plaques and tangles: intracellular Abeta and synaptic dysfunction, *Neuron* **39** (2003), 409-421.

[57] J.M. Orgogozo, S. Gilman, J.F. Dartigues, B. Laurent, M. Puel, L.C. Kirby, P. Jouanny, B. Dubois, L. Eisner, S. Flitman, B.F. Michel, M. Boada, A. Frank and C. Hock, Subacute meningoencephalitis in a subset of patients with AD after Abeta42 immunization, *Neurology* **61** (2003), 46-54.

[58] D.M. Paresce, R.N. Ghosh and F.R. Maxfield, Microglial cells internalize aggregates of the Alzheimer's disease amyloid beta-protein via a scavenger receptor, *Neuron* **17** (1996), 553-565.

[59] M. Pfeifer, S. Boncristiano, L. Bondolfi, A. Stalder, T. Deller, M. Staufenbiel, P.M. Mathews and M. Jucker, Cerebral hemorrhage after passive anti-Abeta immunotherapy, *Science* **298** (2002), 1379.

[60] M.M. Racke, L.I. Boone, D.L. Hepburn, M. Parsadainian, M.T. Bryan, D.K. Ness, K.S. Piroozi, W.H. Jordan, D.D. Brown, W.P. Hoffman, D.M. Holtzman, K.R. Bales, B.D. Gitter, P.C. May, S.M. Paul and R.B. DeMattos, Exacerbation of cerebral amyloid angiopathy-associated microhemorrhage in amyloid precursor protein transgenic mice by immunotherapy is dependent on antibody recognition of deposited forms of amyloid beta, *J Neurosci* **25** (2005), 629-636.

[61] S. Radaev and P.D. Sun, Recognition of IgG by Fcgamma receptor. The role of Fc glycosylation and the binding of peptide inhibitors, *J Biol Chem* **276** (2001), 16478-16483.

[62] J. Rogers, L.F. Lue, D.G. Walker, S.D. Yan, D. Stern, R. Strohmeyer and C.J. Kovelowski, Elucidating molecular mechanisms of Alzheimer's disease in microglial cultures, *Ernst Schering Res Found Workshop* (2002), 25-44.

[63] A. Rovelet-Lecrux, D. Hannequin, G. Raux, N. Le Meur, A. Laquerriere, A. Vital, C. Dumanchin, S. Feuillette, A. Brice, M. Vercelletto, F. Dubas, T. Frebourg and D. Campion, APP locus duplication causes autosomal dominant early-onset Alzheimer disease with cerebral amyloid angiopathy, *Nat Genet* **38** (2006), 24-26.

[64] A. Rovelet-Lecrux, T. Frebourg, H. Tuominen, K. Majamaa, D. Campion and A.M. Remes, APP locus duplication in a Finnish family with dementia and intracerebral haemorrhage, *J Neurol Neurosurg Psychiatry* **78** (2007), 1158-1159.

[65] D. Schenk, R. Barbour, W. Dunn, G. Gordon, H. Grajeda, T. Guido, K. Hu, J. Huang, K. Johnson-Wood, K. Khan, D. Kholodenko, M. Lee, Z. Liao, I. Lieberburg, R. Motter, L. Mutter, F. Soriano, G. Shopp, N. Vasquez, C. Vandevert, S. Walker, M. Wogulis, T. Yednock, D. Games and P. Seubert, Immunization with amyloid-beta attenuates Alzheimer-disease-like pathology in the PDAPP mouse, *Nature* **400** (1999), 173-177.

[66] V.E. Schijns, Immunological concepts of vaccine adjuvant activity, *Curr Opin Immunol* **12** (2000), 456-463.

[67] T.J. Seabrook, L. Jiang, K. Thomas and C.A. Lemere, Boosting with intranasal dendrimeric Abeta1-15 but not Abeta1-15 peptide leads to an effective immune response following a single injection of Abeta1-40/42 in APP-tg mice, *J Neuroinflammation* **3** (2006), 14.

[68] K. Sleegers, N. Brouwers, I. Gijselinck, J. Theuns, D. Goossens, J. Wauters, J. Del-Favero, M. Cruts, C.M. van Duijn and C. Van Broeckhoven, APP duplication is sufficient to cause early onset Alzheimer's dementia with cerebral amyloid angiopathy, *Brain* **129** (2006), 2977-2983.

[69] B. Solomon, R. Koppel, D. Frankel and E. Hanan-Aharon, Disaggregation of Alzheimer beta-amyloid by site-directed mAb, *Proc Natl Acad Sci U S A* **94** (1997), 4109-4112.

[70] C. Sturchler-Pierrat, D. Abramowski, M. Duke, K.H. Wiederhold, C. Mistl, S. Rothacher, B. Ledermann, K. Burki, P. Frey, P.A. Paganetti, C. Waridel, M.E. Calhoun, M. Jucker, A. Probst, M. Staufenbiel and B. Sommer, Two amyloid precursor protein transgenic mouse models with Alzheimer disease-like pathology, *Proc Natl Acad Sci U S A* **94** (1997), 13287-13292.

[71] R.E. Tanzi, D.M. Kovacs, T.W. Kim, R.D. Moir, S.Y. Guenette and W. Wasco, The gene defects responsible for familial Alzheimer's disease, *Neurobiol Dis* **3** (1996), 159-168.

[72] P.R. Turner, K. O'Connor, W.P. Tate and W.C. Abraham, Roles of amyloid precursor protein and its fragments in regulating neural activity, plasticity and memory, *Prog Neurobiol* **70** (2003), 1-32.

[73] V. Vasilevko, F. Xu, M.L. Previti, W.E. Van Nostrand and D.H. Cribbs, Experimental investigation of antibody-mediated clearance mechanisms of amyloid-beta in CNS of Tg-SwDI transgenic mice, *J Neurosci* **27** (2007), 13376-13383.

[74] D.M. Walsh and D.J. Selkoe, A beta oligomers - a decade of discovery, *J Neurochem* **101** (2007), 1172-1184.

[75] J.B. Weinberg, M.A. Misukonis, P.J. Shami, S.N. Mason, D.L. Sauls, W.A. Dittman, E.R. Wood, G.K. Smith, B. McDonald, K.E. Bachus and et al., Human mononuclear phagocyte inducible nitric oxide synthase (iNOS): analysis of iNOS mRNA, iNOS protein, biopterin, and nitric oxide production by blood monocytes and peritoneal macrophages, *Blood* **86** (1995), 1184-1195.

[76] D.M. Wilcock, M.N. Gordon, K.E. Ugen, P.E. Gottschall, G. DiCarlo, C. Dickey, K.W. Boyett, P.T. Jantzen, K.E. Connor, J. Melachrino, J. Hardy and D. Morgan, Number of Abeta inoculations in APP+PS1 transgenic mice influences antibody titers, microglial activation, and congophilic plaque levels, *DNA Cell Biol* **20** (2001), 731-736.

[77] D.M. Wilcock, G. DiCarlo, D. Henderson, J. Jackson, K. Clarke, K.E. Ugen, M.N. Gordon and D. Morgan, Intracranially administered anti-Abeta antibodies reduce beta-amyloid deposition by mechanisms both independent of and associated with microglial activation, *J Neurosci* **23** (2003), 3745-3751.

[78] D.M. Wilcock, S.K. Munireddy, A. Rosenthal, K.E. Ugen, M.N. Gordon and D. Morgan, Microglial activation facilitates Abeta plaque removal following intracranial anti-Abeta antibody administration, *Neurobiol Dis* **15** (2004), 11-20.

[79] D.M. Wilcock, A. Rojiani, A. Rosenthal, G. Levkowitz, S. Subbarao, J. Alamed, D. Wilson, N. Wilson, M.J. Freeman, M.N. Gordon and D. Morgan, Passive amyloid immunotherapy clears amyloid and transiently activates microglia in a transgenic mouse model of amyloid deposition, *J Neurosci* **24** (2004), 6144-6151.

[80] D.M. Wilcock, A. Rojiani, A. Rosenthal, S. Subbarao, M.J. Freeman, M.N. Gordon and D. Morgan, Passive immunotherapy against Abeta in aged APP-transgenic mice reverses cognitive deficits and depletes parenchymal amyloid deposits in spite of increased vascular amyloid and microhemorrhage, *J Neuroinflammation* **1** (2004), 24.

[81] D.M. Wilcock, J. Alamed, P.E. Gottschall, J. Grimm, A. Rosenthal, J. Pons, V. Ronan, K. Symmonds, M.N. Gordon and D. Morgan, Deglycosylated anti-amyloid-beta antibodies eliminate cognitive deficits and reduce parenchymal amyloid with minimal vascular consequences in aged amyloid precursor protein transgenic mice, *J Neurosci* **26** (2006), 5340-5346.

[82] D.M. Wilcock, M.R. Lewis, W.E. Van Nostrand, J. Davis, M.L. Previti, N. Gharkholonarehe, M.P. Vitek and C.A. Colton, Progression of amyloid pathology to Alzheimer's disease pathology in an amyloid precursor protein transgenic mouse model by removal of nitric oxide synthase 2, *J Neurosci* **28** (2008), 1537-1545.

[83] R.A. Willis, W.J. Bowers, M.J. Turner, T.L. Fisher, C.S. Abdul-Alim, D.F. Howard, H.J. Federoff, E.M. Lord and J.G. Frelinger, Dendritic cells transduced with HSV-1 amplicons expressing prostate-specific antigen generate antitumor immunity in mice, *Hum Gene Ther* **12** (2001), 1867-1879.

# Heterogeneity in Red Wine Polyphenolic Contents Differentially Influences Alzheimer's Disease-Type Neuropathology and Cognitive Deterioration

Lap Ho[a,b,1,*], Ling Hong Chen[a,1], Jun Wang[a], Wei Zhao[a], Stephen T. Talcott[c], Kenjiro Ono[d], David Teplow[d], Nelson Humala[a], Alice Cheng[a], Susan S. Percival[e], Mario Ferruzzi[f], Elsa Janle[f], Dara L. Dickstein[g], Giulio Maria Pasinetti[a,b,g]
[a]*Department of Neurology, Mount Sinai School of Medicine, New York, NY, USA*
[b]*Geriatric Research and Clinical Center, James J. Peters VA Medical Center, Bronx, NY, USA*
[c]*Department of Nutrition and Food Science, Texas A&M University, College Stations, Texas, USA*
[d]*Department of Neurology, David Geffen School of Medicine at UCLA, Los Angeles, CA, USA*
[e]*Food Science and Human Nutrition Department, University of Florida, Gainesville, FL, USA*
[f]*Foods and Nutrition Department, Purdue University, West Lafayette, IN, USA*
[g]*Department of Neuroscience, Mount Sinai School of Medicine, New York, NY, USA*

**Abstract.** We recently found that moderate consumption of two un-related red wines generate from different grape species, a Cabernet Sauvignon and a muscadine wine that are characterized by distinct component composition of polyphenolic compounds, significantly attenuated the development of Alzheimer's disease (AD)-type brain pathology and memory deterioration in a transgenic AD mouse model. Interestingly, our evidence suggests that the two red wines attenuated AD phenotypes through independent mechanisms. In particular, we previously found that treatment with Cabernet Sauvignon reduced the generation of AD-type beta-amyloid (Aβ) peptides. In contrast, evidence from our present study suggests that muscadine treatment attenuates Aβ neuropathology and Aβ-related cognitive deterioration in Tg2576 mice by interfering with the oligomerization of Aβ molecules to soluble high-molecular-weight Aβ oligomer species that are responsible for initiating a cascade of cellular events resulting in cognitive decline. Collectively, our observations suggest that distinct polyphenolic compounds from red wines may be bioavailable at the organism level and beneficially modulate AD phenotypes through multiple Aβ-related mechanisms. Results from these studies suggest the possibility of developing a "combination" of dietary polyphenolic compounds for AD prevention and/or therapy by modulating multiple Aβ-related mechanisms.

Keywords: Alzheimer's disease dementia, amyloid-β protein precursor (AβPP), high-molecular weight amyloid-β oligomer species, polyphenols

## INTRODUCTION

Alzheimer's disease (AD) is a growing public health concern with potentially devastating effects.

There is presently no cure for or means of preventing AD. There is increasing consensus that the production and accumulation of β-amyloid (Aβ) peptides is central to the pathogenesis of AD. Deposition of Aβ peptides into insoluble fibrous aggregates known as amyloid plaques in the brain is a major hallmark of AD neuropathology. Moreover, increasing evidence suggests that cognitive decline in AD may be directly caused by accumulation of soluble high molecular weight (HMW) oligomeric Aβ spe-

---

[1]Both authors contributed equally to this work.
*Corresponding author: Lap Ho, Ph.D., Department of Neurology, Mount Sinai School of Medicine, New York, NY 10029, USA. Tel. +1 212 241 8017; E-mail: Lap.Ho@mssm.edu

cies in the brain that are generated by aggregation of Aβ peptides [1,2,3]. Genetic factors are important risk factors for AD, especially for early-onset AD cases. Despite a relatively lower penetrance of genetic factors among late-onset sporadic AD cases, genetic factors remained to be highly relevant for the vast majority of late-onset sporadic AD cases, which is the most common form of AD [4]. Non-genetic factors, including modifiable life-style dietary regimens such as moderate consumption of certain alcoholic beverages, are receiving increasing attention in AD research, especially in light of recent epidemiological studies indicating that moderate wine consumption may reduce the relative risk for AD clinical dementia [5].

Consistent with the hypothesis that red wines might provide beneficial disease-modifying activities in AD, we recently demonstrated that moderate consumption of a red Cabernet Sauvignon wine generated from *Vitis Vinifera L.* grapes attenuated the onset of cognitive deterioration in the Tg2576 transgenic AD mouse model by reducing the accumulation of amyloid pathology in the brain [6]. Moreover, our evidence suggests that the polyphenolic components from Cabernet Sauvignon may attenuate the development of AD-phenotypes, in part, by promoting α-secretase activity in the brain that is responsible for "non-amyloidogenic" processing of the amyloid precursor protein (APP) that precludes the generation of amyloidogenic Aβ peptides [6].

To gather insights on the specific polyphenolic component(s) in red wines that might exert beneficial AD-modifying activities *in vivo*, we recently assessed for potential AD-disease modifying activity in a red muscadine wine, which is generated from *Vitis rotundifolia* grapes and is characterized by a distinctly different polyphenol composition compared to the Cabernet Sauvignon used in our previous study.

We report that moderate consumption of the muscadine wine significantly reduced AD-type Aβ neuropathology and attenuated spatial memory decline in the Tg2576 AD mouse model. However, in contrast to our previous observation with Cabernet Sauvignon [6], we found that muscadine treatment did not modulate α-secretase activity in the brain. Instead, we demonstrated that the beneficial impact of muscadine treatment is associated with reducing the aggregation of Aβ peptides into soluble HMW oligomeric Aβ species in the brain. Results from these studies suggest the possibility of developing a "combination" of dietary polyphenolic compounds for AD prevention and/or therapy by modulating multiple Aβ-related mechanisms.

## MATERIALS AND METHODS

### Muscadine wine

Muscadine was generated from *Vitis rotundifolia* at the University of Florida as previously described [7]. Red muscadine grapes (cv. Noble) were obtained from a local vineyard. Grapes were crushed, de-stemmed and allowed to ferment on the skins for 7 days at 13°C. The soluble solids of the Noble grapes were adjusted to 21% before fermentation. The must was pressed and allowed to finish fermenting to dryness (< 0.05% reducing sugar) at 13°C. The muscadine wine was then treated with 100 mg/L of potassium metabisulfite, cold-stabilized at 3°C for 2 months, filtered and stored at 13°C for approximately 3.5 years. The muscadine wine contains approximately 12% alcohol – as determined by ebulliometry [8], a titratable acidity (as tartaric acid) of 6.9 g/L and a pH of 3.00. The phenolic composition (as gallic acid and measured by the Folin-Coicalteau method [8] was 1,731 mg/L.

### Chemical analysis of wine polyphenolic component composition

Chemical compositions of muscadine (and Cabernet Sauvignon) were assessed by reverse phase chromatography using HPLC and an Octadecyl silane column (4.6 X 250 mm) and the solvent conditions of Talcott and Lee [9]. Briefly, 2 ml of each wine was freeze dried. Dried extracts were re-dissolved in 0.1M citric acid buffer at pH 3.0 and injected into a Waters Alliance 2695 HPLC system that linked to a Waters 996 Photodioarray detector. Compounds were separated on a Dionex Acclaim 120 C18 column (4.6 x 250 mm) using a gradient mobile phase that consisted of Phase A (100% water adjusted to pH 2.4 with o-phosphoric acid) and Phase B (60:40 methanol:water, also at pH 2.4 with o-phosphoric acid). Phase A changed from 100% to 50% in 30 min, to 20% in 10 min, and to 0% in 10 min and held isocratic for 10 min. Compounds were detected at 280 nm (phenolic acids), 370 nm (flavonoids), and 520 nm (anthocyanins), with select compounds identified based on spectroscopic interpretations from 200-600 nm [9,10].

## Tg2576 mice and wine treatment

In this study, 4-month-old female Tg2576 mice (Taconic, Germantown Inc.) were randomly assigned to muscadine-treatment or control, non-treatment groups. Muscadine was delivered to mice by diluting the wine into the drinking water - to a final ethanol concentration of 6%. Animals had free access to the liquid and standard rodent chow. Drinking solutions were changed every three days. Liquid consumption, food intake and animal body weight were monitored weekly throughout the study. Mice were treated chronically with the muscadine wine for 10 months. In parallel control studies, wild-type (WT) mice matching with Tg2576 mice in respect to strain, gender and age were similarly treated with the muscadine wine for 10 months; control, non-treated WT mice received unadulterated drinking water. At 14 months of age, spatial memory functions of muscadine-treated and non-treated Tg2576 and WT mice were assessed by the Morris water maze test (see below). Thereafter, all mice were anesthetized with the general inhalation anesthetic 1-chloro-2,2,2-trifluoroethyl difluoromethyl ether (Baxter Healthcare, Deerfield, IL) and sacrificed by decapitation. Brains were harvested and hemidissected. One hemisphere was fixed in 4% paraformaldehyde for 24 hours for morphological studies; hippocampus, and cingulate and parietal neocortex were dissected from the opposite hemisphere, rapidly frozen, pulverized in liquid nitrogen and stored at -80°C for biochemical studies. All studies involving animal subjects are conducted in accord with the guide of the Mount Sinai School of Medicine Institutional Animal Care and Use Committee.

## Assessment of wine toxicity

Blood was collected after sacrifice by cardio puncture. Serum was collected by clotting blood specimens for 10 minutes at room temperature followed by centrifugation at 2,500g for 20 minutes at 4°C. Serum was immediately subjected to total bilirubin, alanine aminotransferase (ALT) and aspartate aminotransferase (AST) analysis using commercial enzyme assays according to the manufacturer's instructions (Stanbio laboratory, Texas). Serum alcohol content was measured using a commercially available kit (BioAssay Systems, CA).

## Behavioral assessment of cognitive functions by the Morris Water Maze (MWM) test

Spatial learning memory was assessed at 14 months of age by the Morris water maze behavioral test, as previously described [11-12]. Following 10-months of treatment with the muscadine wine, approximately 14 months old mice were tested in a 1.25 m circular pool filled with water mixed with non-toxic white paint (Dick Blick Art Materials, Galesburg, IL). The temperature of the water was maintained at 27 °C. Mice were trained in absence of visual cue to mount a hidden/submerged (1.5 cm below-water surface) escape platform (14x14 cm) in a restricted region of the pool. Thereafter, learning trials (four 60 second trials per day with a 10-min inter-trial, for 7 consecutive days) were conduct to test spatial memory by evaluating the latency to locate the hidden escape platform in an environment enriched in spatial cues as a function of learning trials. At the end of the learning trials, animals were rested for 24-hours before they were subjected to a 45 second probe trial in the water maze in the presence of the identical spatial cues used in the learning trials, but in the absence of the hidden escape platform. The pool was further divided into four hypothetical quadrants of identical surface area. The percentage of time spent in the target quadrant where the former escape platform was located was recorded to reflect the capability of the animals to the use spatial cues to identify the appropriate location in the pool where the escape platform ought to be found.

In control studies, a visible platform test was used to determine whether differences in visual and/or swimming capability might confound interpretations of the behavioral performance data gathered from the MWM tests. Visual capability was assessed by the latency of the animals to escape onto the visible escape platform as a function of trial days. Motor functions were reflected by swimming speed (cm/second) as a function of trial days during the visible platform test.

All water maze activities were monitored with the San Diego Instrument Poly-Track video tracking system (San Diego, CA). All behavior analyses were conducted during the last 4 hours of the day portion of the light cycle in an environment with minimal stimuli (e.g., noise, movement, or changes in light or temperature).

*Assessment of AD-type amyloid neuropathology in Tg2576 mice*

AD-type amyloid plaque density in the brain at 14 months of age was assessed using stereological procedures. Freshly harvested mouse brain hemispheres were immersion fixed overnight in 4% paraformaldehyde. They were then sectioned in the coronal plane on a Vibratome at a nominal thickness of 50 μm. Amyloid plaques were visualized by thioflavin-S staining. Every 12th section was selected from a random start position and processed for thioflavin-S staining, as previously described [13,14]. All stereological analysis was performed using a Zeiss Axiophoto photomicroscope equipped with a Zeiss motorized stage and MSP65 stage controller, a high-resolution MicroFire digital camera, and a Dell computer running the custom-designed software Stereo Investigate (MBF Bioscience). Amyloid plaque burden was estimated using the Cavalieri principle with a small-size grid (25 × 25 μm) for point counting, as previously described [6,13].

*Assessments of α-, β-, γ-secretase activity in the brain*

α-, β-, and γ- secretase activities in brain specimen from 14 months old Tg2576 mice were assessed using commercially available kits (R&D Systems) [6,12-13]. Brain samples were homogenized in supplied buffers. Homogenate was then added to secretase-specific APP peptide conjugated to the reporter molecules EDANS and DABCYL. In the uncleaved form, fluorescent emissions from EDANS were quenched by the physical proximity of the DABCYL moiety. Cleavage of APP peptide by secretase physically separates the EDANS and DABCYL reporter molecules, allowing for the release of a fluorescent signal. The level of secretase enzymatic activity is proportional to the fluorometric reaction in the homogenate (R&D Systems). In parallel control studies, holo-APP expression was examined by Western blot analysis with the C8 antibody (raised against aa 676–695 of human APP cytoplasmic domain; gift of Dennis Selkoe, Brigham and Women's Hospital, Boston, Massachusetts, USA).

*Assessment of HMW soluble oligomeric Aβ oligomerization in the brain by dot blot and western blot assays*

Soluble proteins were extracted from brain samples (cerebral cortex) of 14 months old Tg2576 mice with phosphate-buffered saline in the presence of protease inhibitors followed by centrifugation at 78,500 g for 1 hour at 4°C [15]. In dot-blot assays, four micrograms of extracellular soluble protein isolated from the cerebral cortex or from the hippocampal formation of muscadine-treated or control, non-treated Tg2576 mice was directly applied to a nitrocellulose membrane, air dried, and blocked with 5% nonfat milk, as previously described [16]. The membrane was probed with either the anti-oligomer A11 antibody (1:1,000; Biosource) or the 6E10 antibody (1:1,000; Signet) [16]. In western blot assays, seventy-five μg of soluble proteins were resolved by electrophoresis through a 10-20% Tris-Tricine gel, transferred to a nitrocellulose membrane and blocked with 10% non-fat milk for 1 hr. The membrane was then incubated with the 6E10 antibody (Signet). For both dot-blot and western blot analysis, immunoreactive signals were visualized by using enhanced chemiluminescence detection and quantified densitometrically (Quantity One; Bio-Rad).

*Polyphenolic extraction for in vitro studies*

For *in vitro* Aβ peptide oligomerization studies, a muscadine polyphenol extract was prepared by solid phase extraction using C18 cartridges (Waters). Muscadine was diluted 1:4 in acidified water (water buffered to pH 2.4 with o-phosphoric acid) to optimize preservation of the polyphenol compounds in the wine. Diluted wine was loaded onto C18 cartridges followed by washing the columns with 3-volumes of acidified water. Bound organic compounds were eluted in 12% methanol diluted in acidified water. The eluants were dispensed into small aliquots and dried in a speedi-vac. For anti-oligomerization assays, dried polyphenol extracts were dissolved into phosphate-buffered saline immediately before use.

*Gel electrophoresis Aβ peptide oligomerization assay in vitro*

Lyophilized $A\beta_{1-42}$ peptide was dissolved in 1,1,1,3,3,3,-hexafluoro-2-propanol (HFIP; Sigma-Aldrich), incubated at room temperature for 60 minutes, aliquoted, vacuum dried, and stored at –80°C. Aβ peptide was dissolved in DMSO and diluted into ddH$_2$O to a final concentration of 100 μg/ml. 5 μl of the peptide was then mixed with 1-2 μl of muscadine wine, the final volume adjusted to 10 μl and the reaction was incubated at 37°C for 3 hrs [17]. Following incubation, samples were mixed with 2×

SDS sample buffer and separated on a 10%–20% Tris-Tricine gradient SDS gel (Invitrogen). The separated peptides were subjected to Western blotting using 6E10 antibody (1:1,000; Signet). Immunoreactive signals were visualized using ECL detection (Amersham) and quantified densitometrically (Quantity One; Bio-Rad).

*Photo-induced cross-linking of unmodified proteins (PICUP) assay*

Freshly isolated low molecular weight $A\beta_{1-40}$ (30-40 µM) peptide was mixed with 1µl of 1 mM tris(2,2'-bipyridyl)dichlororuthenium (II) (Ru(bpy)) and 1µl of 20 mM ammonium persulfate in 10mM phosphate, pH 7.4, either in the presence or absence of muscadine wine or a muscadine polyphenolic extract. The mixture was irradiated for 1 second, and quenched immediately with 10 µl of Tricine sample buffer (Invitrogen) containing 5% β-mercaptoethanol [18]. The reaction was subjected to SDS-PAGE and visualized by silver staining (SilverXpress, Invitrogen).

*Statistical analysis*

All values are expressed as mean and standard error of the mean (SEM). Differences between means were analyzed using either 2-way repeated measures ANOVA or 2-tailed Student $t$ test. In all analyses, the null hypothesis was rejected at the 0.05 level. All statistical analyses were performed using the Prism Stat program (GraphPad Software, Inc., San Diego CA).

**RESULTS**

*Chemical analysis of muscadine and Cabernet Sauvignon wines*

Chemical compositions of the muscadine wine used in this study and the red Cabernet Sauvignon wine we used in our previous study [6] were analyzed using HPLC C18 reverse phase chromatography. We confirmed that the two wines are characterized by distinct component compositions of phenolic acid, flavonoid and anthocyanin polyphenolic compounds (Fig. 1A-F).

*Moderate consumption of muscadine is well-tolerated in Tg2576 Mice*

Based on our observation that each mouse consumed an approximately 4 ml of wine-adulterated water per day, we calculated that daily wine consumption by muscadine-treated Tg2576 mice in this study is equivalent to moderate wine consumption in the human as defined by the United States Department of Agriculture and Health and Human Services [19]. Moreover, we found that moderate consumption of muscadine for 10 months is compatible with general good health in Tg2576 mice, as reflected by no detectable changes in food/water intakes, body weights, and indexes of liver functions in 14-months old muscadine-treated, compared to age- gender-matched control, non-treated Tg2576 mice. This observation is consistent with our recent report that moderate red wine consumption is well-tolerated by Tg2576 mice [6]. Details on translating daily wine intake in muscadine-treated Tg2576 mice to equivalent wine consumption in the human, and assessments of food/liquid consumptions, body weights, and liver functions are provided.

*Muscadine treatment attenuates AD-type cognitive deterioration in Tg2576 mice*

In this study, we explored the potential impact of muscadine treatment on cognitive behavioral functions and AD-type amyloid neuropathology in 14-months old Tg2576 mice, when these animals typically exhibit moderate-to-severe amyloid neuropathology and cognitive impairment [20].

Using the MWM test, we found that muscadine treatment for 10 months significantly attenuated spatial memory activity decline in approximately 14-month-old Tg2576 mice relative to age- and gender-matched control, non-treated Tg2576 mice, as reflected by a shorter escape latency as a function of learning trials during the learning phase [2-way ANOVA analysis of Tg-muscadine vs. Tg-control groups for muscadine treatment ($p<0.05$, F=4.24, DFn=1, DFd=84) and for training days ($p<0.05$, F=6.43, DFn=6, DFd=84)]. (Fig. 2A). That muscadine-treatment significantly attenuated cognitive impairment in Tg2576 mice was confirmed by analysis of spatial memory retention in a probe trial conducted following removal of the escape platform,

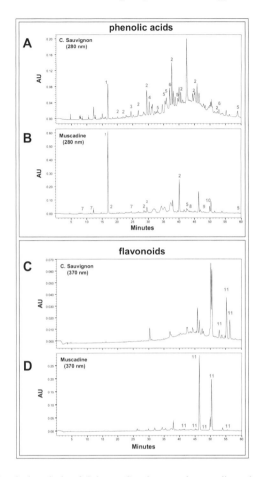
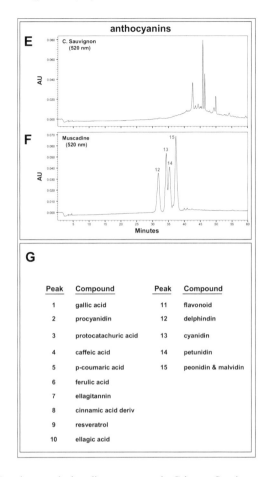

Fig. 1. Chemical analysis of Cabernet Sauvignon and muscadine wines. Constituent polyphenolic components in Cabernet Sauvignon (**A,C,E**) and muscadine (**B,D,F**) wines were analyzed by reverse phase HPLC using a C18 column. (**A,B**) Detection of phenolic acid compounds at 280 nm. (**C,D**) Detection of flavonoids at 370 nm. (**E,F**) Detection of anthocyanins at 520 nm. (**G**) Identification of polyphenols corresponding to peaks detected in panel (**A-F**) based on spectroscopic interpretations.

which showed that muscadine-treated mice spent significantly more time in the target quadrant area relative to non-treated Tg2576 mice (p<0.05, 2-tailed Student *t* test). (Fig. 2B).
Moreover, we confirmed that improved Morris water maze behavioral performance in muscadine-treated mice is not due to non-specific promotion in either visual or motor functions as both muscadine-treated and non-treated Tg2576 mice exhibited comparabe capability to locate a visible escape platform (Fig. 2C) and demonstrated similar swim speeds (Fig. 2D). In parallel control studies, we assess for MWM performances of 14-months old WT mice that are strain-, gender-, and age-matched with Tg2576 mice used in behavioral testing. We found that 10-months of muscadine treatment did not lead to any detectable change in MWM performances in muscadine-treated compared to non-treated WT mice (data not shown). This suggests that muscadine treatment attenuated cognitive impairments in Tg2576 mice through mechanisms associated with expression of the APP transgene in this AD mouse model.

*Muscadine treatment attenuates the development of AD-type Aβ neuropathology in Tg2576 mice*

The initiatising age in Tg2576 mice is due to age-dependent proon and progression of cognitive decline with increagressive accumulations of AD-type Aβ neuropathology in the brain [20]. We used stereological histological methodologies to quantitatively explore the accumulation of AD-type amyloid neuritic plaques in the brain of 14 months old Tg2576 mice in response to 10-months of muscadine treatment. We found significantly lower amyloid

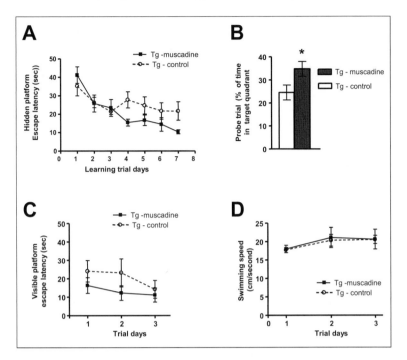

Fig. 2. Muscadine treatment improves spatial memory function in Tg2576 mice. Assessments of spatial memory behavioral functions of 14-months old muscadine-treated (Tg-muscadine) and control, gender- and age-matched non-treated (Tg-control) Tg2576 mice using the Morris water maze (MWM) protocol. (**A**) Learning trial hidden-platform acquisition curves. Tg-muscadine group performed significantly better than the control, non-treated group (Tg-control) [2-way ANOVA analysis of Tg-muscadine vs. Tg-control groups for muscadine treatment ($p<0.05$, F=4.24, DFn=1, DFd=84) and for training days ($p<0.05$, F=6.43, DFn=6, Dfd=84)]. (**B**) Probe trial conducted 24 hours after completion of hidden-platform training. Muscadine-treated Tg2576 mice exhibited a significantly higher preference for the target platform compared to control, non-treated Tg2576 mice ($p<0.05$, 2-tailed Student $t$ test). (**C**) Visible-platform learning curves. There is no significant difference in visible platform performance among Tg-muscadine compared to Tg-control mice. (**D**) Average swimming speed. There is no significant difference in swimming ability of muscadine-treated animals compared to control, non-treated animals. In (**A-D**) Values represent group mean (+SEM); n=7-9 mice per group.

neuritic plaque density in both the cortical and the hippocampal formation regions of the brain of 14-months old muscadine-treated compared to age- and gender-matched control, non-treated Tg2576 mice (Fig. 3A-B). Thus, coincidental to attenuating the development of AD-type spatial memory functions, muscadine treatment also reduced the accumulation of AD-type Aβ neuropathology in the Tg2576 AD mouse model. Ongoing studies exploring the impact of muscadine treatment on amyloid neuropathology as a function of age in the Tg2576 mouse AD model will clarify whether the red wine might benefit AD by delaying the onset of amyloid plaque deposition, and/or by attenuating the continue progression of existing amyloid neuropathology.

The efficacy of muscadine treatment to attenuate AD-type Aβ neuropathology in Tg2576 mice is not due to ethanol exposure from the wine. In control studies, we found that long-term exposure of Tg2576 mice to a drinking water solution containing 6% ethanol (the same ethanol content in the muscadine-adulterated drinking water used in this study) did not lead to any detectable changes Aβ neuropathology in this AD mouse model [6].

*Muscadine treatment is not associated with detectable changes in APP processing in the brain of Tg2576 Mice*

Multiple enzyme activities are known to regulate the generation of Aβ peptides from the ubiquitous amyloid precursor protein (APP) [21]. Aβ species with different carboxy termini are generated from APP through sequential proteolysis by β- and γ-secretase. APP can also be processed by a non-amyloidogenic pathway in which α-secretase leaves APP within the Aβ domain, thereby precluding the generation (and deposition) of intact amyloidogenic Aβ peptides in the brain [22]. In an initial step to explore the

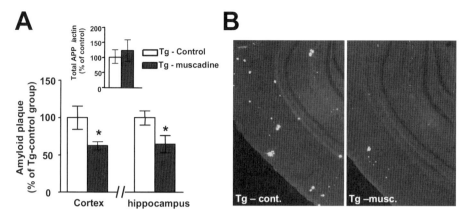

Fig. 3. Muscadine treatment significantly reduced Aβ neuropathology in Tg2576 mice. Tg2576 mice were assessed for indexes of AD-type amyloid burden in the brain in response to muscadine treatment using stereological technologies; the same 14-months old Tg2576 mice used for behavioral assessments in Fig. 2 were used in this study. (A) Assessments of amyloid neuritic plaque density in cerebral cortex and in the hippocampal formation of brain specimens from muscadine-treated and control, non-treated Tg2576 mice. (A, inset) Control studies confirming muscadine treatment did not modulate expression of the holo-APP in the brain (cerebral cortex). Bar graphs represent means ± SEM., n = 6-8 per group; * $P < 0.05$ vs. non-treated control Tg2576 group (2-tailed Student's $t$ test). (B) Representative micrograph of brain specimen stained for amyloid neuritic plaques in muscadine-treated (Tg-musc.) or in control, non-treated (Tg-cont.) Tg2576 mice.

mechanism by which the muscadine treatment might have attenuate AD-type Aβ neuropathology, we examined the potential impact of muscadine treatment on the regulation of α-, β- and γ-secretase activities in the brain of the same 14 months-old muscadine-treated and control Tg2576 animals we have used for behavioral and Aβ neuropathology studies. We found no detectable change in α-, β-, or γ-secretase activity in the brain of the muscadine-treated in comparison to control, non-treated Tg2576 mice (Fig. 4). Thus, it is unlikely that muscadine treatment might have attenuated the development of AD-type Aβ neuropathology and cognitive deterioration in the Tg2576 AD mouse model by modulating secretase activities in the brain that play key roles in the generation of Aβ peptides from APP.

*Attenuation of Aβ neuropathology in response to muscadine treatment is associated with reduced accumulation of soluble HMW oligomeric Aβ species in the brain*

Recent evidence indicates that accumulations of soluble HMW oligomeric Aβ species in the brain, rather than deposition of amyloid plaques per se, may be specifically related to spatial memory reference deficits in mouse models of AD [1,2]. Based on this, we continue to explore whether muscadine treatment in this study might have attenuated the accumulation of soluble HMW oligomeric Aβ species in the brain.

In a dot-blot assay using the A11 antibody that detects oligomeric species [16], we found significantly reduced contents of A11-immunoreactive oligomeric Aβ species in the cerebral cortex (Fig. 5A, left panel) and in the hippocampal formation (Fig. 5A, right panel) of brain specimens from muscadine-treated relative to control, non-treated Tg2576 mice. We also assess the content of Aβ oligomers in the brain of Tg2576 mice using an independent western blot analysis that identifies Aβ oligomers based on their characteristic molecular size. Consistent with our observation from the dot-blot assay, we found significantly lower contents of HMW oligomeric Aβ peptides in the cerebral cortex of muscadine-treated compared to control Tg2576 mice (Fig. 5B). Thus based on two independent assays, we found that muscadine treatment might have attenuated AD pathologic and cognitive phenotypes in Tg2576 mice, in part, by reducing the accumulation of HMW oligomeric Aβ peptides in the brain.

*Muscadine treatment might reduce the accumulation of HMW oligomeric Aβ species by interfering with aggregation of Aβ peptides*

We explored whether muscadine wine or bioactive component from the wine may mechanistically

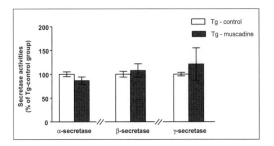

Fig. 4. No detectable changes in α-, β- and γ-secretase activities in the brain of Tg2576 mice in response to muscadine treatment. α-, β-, and γ-secretase enzymatic activities were assessed in cerebral cortex brain specimens from the same 14-months old Tg2576 animals used in the behavioral studies in Fig. 2 and neuropathology assessments in Fig. 3. (A) α-secretase activity, (B) β-secretase activity, and (C) γ-secretase activity in Tg-muscadine and Tg-control mice. (A-C).Bar graphs represent group mean ($\pm$ SEM), n = 6-8 per group.

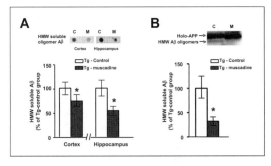

Fig. 5. Muscadine treatment significantly attenuated the accumulation soluble HMW Aβ species in the brain of Tg2576 mice. The contents of HMW Aβ oligomeric species in cerebral cortex brain specimens from muscadine-treated or control, non-treated 14-months old Tg2576 mice were assessed by independent dot-blot (A) and western blot studies (B). (A) Immunodetection of HMW Aβ oligomeric species in the cerebral cortex and in the hippocampal formation using the oligomer-specific antibody A11 antibody in an dot blot assay. (A, inset) Representative A11-immunoreactive dot-blot analysis of cortical and hippocampal formation brain specimens. Abbreviations: C: control, non-treated Tg2576 mice; M: muscadine-treated Tg2576 mice. (B) Western blot analysis in which HMW Aβ oligomeric species are resolved by gel electrophoresis followed by immunodetection using a pan-Aβ 6E10 antibody. (B, Inset) Representative western blot analysis of muscadine-treated and control non-treated Tg2576 mice. Immunoreactive holo-APP and relatively less abundant HMW Aβ species are identified. In (A, B), bar graphs represent means ± SEM., n = 6-8 per group; * $P < 0.05$ vs. non-treated control Tg2576 group (2-tailed Student's $t$ test).

reduce the accumulation of HMW Aβ oligomers in the brain by interfering with aggregations of Aβ peptides that is necessary for the generation of HMW Aβ oligomers. In initial *in vitro* studies using a gel electrophoresis protocol, we found that muscadine wine prevented aggregations of synthetic Aβ1-42 peptides into SDS-stable HMW oligomers (Fig. 6A). In parallel control studies, we found that the application of ethanol does not have any detectable influence on aggregations of synthetic Aβ1-42 peptides into HMW oligomeric species (Fig. 6A), suggesting that the alcohol contents in the muscadine is not responsible for the anti-Aβ oligomerization bioactivity of the wine. Interestingly, evidence from our gel-electrophoresis study suggests that the generation of Aβ1-42 trimers might be relatively resistant to interference by muscadine wine. Ongoing studies are exploring whether bioactive components from the wine might preferentially interfere with specific Aβ aggregation activities.

Using an independent *in vitro* PICUP assay [23,24], we continue to explore whether muscadine treatment might interferes with initial protein-protein interactions that are necessary for aggregations of Aβ peptides into HMW oligomeric species. Consistent with our observation from gel-electrophoresis studies, we found that the muscadine wine (but not ethanol) interferes with initial aggregation of synthetic Aβ1-40 peptides into small multimeric species in the PICUP assay (Fig. 6B).

We continue to explore whether polyphenol contents in muscadine might be responsible for the anti-Aβ oligomerization activity of the red wine. We prepared an alcohol-free polyphenolic extract from the muscadine wine by solid phase extraction and tested for *in vitro* anti-Aβ oligomerization activity using the PICUP assay. We found that similar to the intact muscadine wine, the polyphenolic isolate from the wine almost completely prevented aggregations of Aβ1-40 peptides in the PICUP assay (Fig. 6B).

Thus, based on observations from independent *in vitro* assays, we demonstrated that muscadine treatment might, in part, attenuated accumulations of HMW Aβ oligomers in the brain of Tg2576 mice by mechanistically inhibiting aggregations of Aβ peptides into higher-ordered oligomers. More importantly, our evidence implicated polyphenol compounds in the muscadine wine as the bioactive components that exert beneficial AD-disease modifying activity by inhibiting Aβ aggregations.

## DISCUSSION

Genetic factors are important risk factors for AD, especially for early-onset AD cases. Despite a relatively lower penetrance of genetic factors among late-onset sporadic AD cases, genetic factors remained to be highly relevant for the vast majority of late-onset sporadic AD cases, which is the most common form of AD [4]. Non-genetic factors, including modifiable life-style dietary regimens such as moderate consumption of alcoholic beverages, are receiving increasing attention in AD research, especially in light of the recent epidemiological studies indicating that moderate wine consumption may influence the relative risk for AD clinical dementia [5,25]. Our evidence that moderate consumption of a red Cabernet Sauvignon wine [6] and a red muscadine wine (in this study) significantly attenuated the accumulation of AD-type Aβ neuropathology and the development of cognitive deterioration in the Tg2576 AD mouse model supports the hypothesis that red wine consumption might provide preventive and/or therapeutic value in AD. However, despite our encouraging evidence, a major concern for translating this observation into clinical application is the health risks associated with chronic consumption of alcoholic beverages [26]. Based on this, ongoing studies are focused on identifying the bioactive component(s) in red wines and clarifying the mechanisms of action to provide the necessary basis for developing red wine mimetic(s) for applications in AD.

The potential health benefits of wine consumption are generally ascribed to the polyphenol compounds that are present in high abundance, particularly in red wines [27,28]. Since many of the wine-derived polyphenols are strong antioxidants, it is thought that red wine polyphenols may benefit AD (and other neurodegenerative disorders) by reducing the contents of reactive oxygen species in the brain [29].

Aside from potential anti-oxidant activities, our accumulating pre-clinical evidence suggests that red wine polyphenols may also benefit AD by directly modulating Aβ-related mechanisms in the brain. In particular, we recently demonstrated that polyphenolic components from a red Cabernet Sauvignon wine may protects against the onset of AD-type Aβ-neuro-pathology and cognitive deterioration in the Tg2576 AD mouse model by promoting non-amyloidogenic α-secretase activity in the brain that reduces Aβ neuropathology [6]. Similar to our observation with Cabernet Sauvignon, results from our present studies showed that polyphenolic compounds from a red muscadine wine may also exert beneficial AD-disease modifying activity *in vivo* and inhibit the onset of Aβ neuropathology and cognitive dysfunction in Tg2576 mice. However, in contrast to Cabernet Sauvignon, our new evidence suggests that bioactive polyphenols from the muscadine wine may modulate AD phenotypes by interfering with aggregation of Aβ peptides into HMW oligomeric Aβ species in the brain that play a major role in AD-type cognitive dysfunction [1,2,3]. Recent evidence suggests that multiple polyphenols found in red wines may block oligomerization of synthetic Aβ peptides *in vitro* [30-32]. Thus, it is possible that more than one polyphenolic components in the muscadine wine might have contribute to the *in vivo* efficacy of the wine to attenuated Aβ-related neuropathology and cognitive dysfunction in Tg2576 mice. Interestingly, our recent evidence demonstrated that polyphenolic components from Cabernet Sauvignon also blocked oligomerization of synthetic Aβ peptides *in vitro* (data not shown). Thus, it is possible that in our previous study [6] Cabernet Sauvignon treatment might have attenuated AD phenotypes by promoting α-secretase activity as well as inhibiting aggregation of Aβ peptides in the brain. Ongoing studies will clarify whether similar (or different) polyphenolic compounds in Cabernet Sauvignon and muscadine are responsible for anti-Aβ oligomerization activities in the brain.

In addition to providing anti-oxidant activities and/or modulating Aβ neuropathology, select dietary polyphenols may benefit AD dementia through non-Aβ related mechanisms by directly promoting cellular processes relevant to cognitive functions. In particular, Joseph et al. [33] reported that dietary supplement with blueberries, which contains high contents of polyphenols, prevented the development of cognitive impairment in a 12-months old double transgenic APP/PS1 AD mouse model by promoting cellular signaling process associated with learning and memory without detectable effect on amyloid deposition. In our studies with two red wines, we found that dietary supplement with Cabernet Sauvignon [6] or muscadine (in the present study) did not promote cognitive behavior performance in WT mice. Thus in contrast to blueberries, polyphenolic from the two red wines may benefit AD phenotypes by attenuating Aβ neuropathology instead of promoting Aβ-unrelated mechanisms.

While our evidence to date with the Cabernet Sauvignon and the muscadine wine suggests that red wine polyphenolic compounds may benefit AD by modulating multiple Aβ-related mechanisms, there is no information on the specific bioactive polyphenolic compounds that are involved. Based on our observation that the two red wines are characterized by different polyphenol component compositions, we posit that distinct polyphenols from the two red wine are

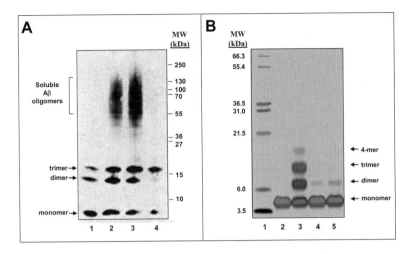

Fig. 6. Muscadine polyphenols exert anti-Aβ oligomerization bioactivity *in vitro*. (A) Muscadine wine interferes with aggregation of synthetic Aβ$_{1-42}$ peptides into HMW oligomer Aβ species, *in vitro*. Synthetic Aβ$_{1-42}$ peptides were aggregated in the absence or in the presence of muscadine wine. Aβ species were then resolved by molecular size, transblotted onto a nitrocellulose membrane, followed by immunodetection of Aβ peptides using and the 6E10 antibody. Lane 1 represents non-aggregated Aβ$_{1-42}$ peptides; Lane 2, aggregated Aβ$_{1-42}$ peptides; Lane 3, Aβ$_{1-42}$ peptides aggregated in the presence of 1.2% ethanol (the same amount of ethanol presented in the aggregation assay in the presence of muscadine in Lane 4); Lane 4, Aβ$_{1-42}$ peptides aggregated in the presence of 1 μl muscadine wine. (B) Application of an independent PICUP assay to explore the role of muscadine wine and its polyphenolic components to interfere with aggregations of synthetic Aβ$_{1-40}$ peptides. The PICUP assay is designed to explore the initial protein-protein interactions that are necessary for the formation of HMW Aβ oligomeric species. Aβ aggregates are stabilized by photo-cross-linking. Monomeric and multimeric Aβ species are separated by molecular size and visualized by silver staining. Lane 1 represents molecular weight marker; Lane 2, non-aggregated Aβ$_{1-40}$ peptide; Lane 3, aggregated Aβ$_{1-40}$ peptide; Lane 4, Aβ$_{1-40}$ peptide aggregated in the presence of muscadine wine (2 μl of the wine); Lane 5, Aβ$_{1-40}$ peptide aggregated in the presence of a polyphenolic extract from 2 μl of the muscadine wine. The position of Aβ$_{1-40}$ monomer, dimmer, trimer and 4-mer are as indicated.

responsible for anti-Aβ aggregation and the promotion of α-secretase in the brain.

The health benefits of dietary polyphenols depend not only on the type and amount consumed, and also on their bioavailability [34,35]. Accumulating *in vitro* evidence suggests that certain polyphenolic compound, include a number of polyphenolic compounds that are found in red wines and grape-by products, are capable of inhibiting the extension, and eventually destabilize the polymerization of amyloidogenic Aβ peptides [30-31]. However, with the exceptions of curcumin from turmeric spice that have been shown to attenuate AD phenotypes in an AD mouse model [36,37], it is not know whether other polyphenols could exert anti-Aβ oligomerization activity *in vivo* and modulate AD-type Aβ neuropathology and cognitive deterioration at the organism level. Based on our observation from the muscadine wine and the fact that curcumin is not found in red wines, we hypothesized there are additional dietary polyphenols from red wines (or other grape-derived products) that are capable of exerting anti-Aβ oligomerization activity *in vivo* and modulate AD phenotype. Consistent with this hypothesis, we recently found that a certain grape-seed/skin polyphenolic extract exerts anti-Aβ oligomerization activity *in vivo* and attenuated the development of AD-type Aβ-related neuropathology and cognitive deterioration in Tg2576 mice [38].

In summary, in conjunction with our previously reported pre-clinical study with a red Cabernet Sauvignon wine [6], results from our present pre-clinical study with a red muscadine wine suggest that distinct polyphenolic compounds from red wines may be bioavailable at the organism level and beneficially modulate AD phenotypes through multiple Aβ-related mechanisms. Thus, it might be possible to develop a combination of grape-derived bioactive polyphenolic compounds that modulate multiple Aβ-related mechanisms for AD prevention and/or therapy. Identification and mechanistic characterization of individual polyphenols capable of exerting bioactivity *in vivo* will promote the development of selective bioactive dietary polyphenol(s) as lead compounds for clinical testing in AD. Moreover, the information gathered will also provide the necessary basis for identifying food sources enriched in targeted bioactive polyphenols that ultimately could be incorporated as key components in the development of potential dietary guidelines for AD prevention and/or management.

## ACKNOWLEDGEMENTS

These studies were supported by 1PO1 AT004511-01 Project 1 to LH and 1PO1 AT004511-01 Project 3 to GMP, MERIT Review grant from Dept. of Veterans Affairs, and James J. Peters VA GRECC Program to GMP.

## References

[1] Cleary JP, Walsh DM, Hofmeister JJ, Shankar GM, Kuskowski MA, Selkoe DJ, Ashe KH. Natural oligomers of the amyloid-[beta] protein specifically disrupt cognitive function. *Nat Neurosci.* **8** (2005), 79-84.
[2] Lesne S, Koh MT, Kotilinek L, Kayed R, Glabe CG, Yang A, Gallagher M, Ashe KH. A specific amyloid-beta protein assembly in the brain impairs memory. *Nature* **440** (2006), 352-357.
[3] Jacobsen JD, Wu CC, Redwine JM, Comery TA, Arias R, Bowlby M, Martone R, Morrison JH, Pangalos MN, Reinhart PH, Bloom FE. Early-onset behavioral and synaptic deficits in a mouse model of Alzheimer's disease. *Proc Natl Acad Sci U S A.* **103** (2006), 5161-5166.
[4] Cummings BJ, Cotman CW. Image-Analysis of Beta-Amyloid Load in Alzheimers-Disease and Relation to Dementia Severity. *Lancet* **346** (1995), 1524-1528.
[5] Luchsinger JA, Tang MX, Siddiqui M, Shea S, Mayeux R. Alcohol Intake and Risk of Dementia. *Journal of the American Geriatrics Society* **52** (2004), 540-546.
[6] Wang J, Ho L, Zhao Z, Seror I, Humala N, Dickstein DD, Thiyagarajan M, Percival SS, Talcott ST, Pasinetti GM. Moderate consumption of Cabernet Sauvignon attenuates Abeta neuropathology in a mouse model of Alzheimer's disease. *FASEB J* **20** (2006), 2313-2320.
[7] Percival SS, Sims CA. Wine Modifies the Effects of Alcohol on Immune Cells of Mice1-3. *J. Nutr.* **130** (2000), 1091-1094
[8] B. Zoecklin, K. Fugelsang, B. Gump, F. Nury, S. Abraham, *Production Wine Analysis* Van Nostrand Reinhold New York. 1990.
[9] Talcott ST, Lee JH. Ellagic acid and flavonoid antioxidant content of muscadine wine and juice. *J. Agric. Food Chem.* **50** (2002), 3186-3192.
[10] Lee JH, Johnson JV, Talcott ST. Identification of ellagic acid congjugates and other polyphenolics in muscadine grapes by HPLC-ESI-MS. *J. Agric. Food Chem.* **53** (2005), 6003-6010.
[11] Morris R. Developments of A Water-Maze Procedure for Studying Spatial-Learning in the Rat. *Journal of Neuroscience Methods* **11** (1984), 47-60
[12] Ho L, Qin W, Pompl PN, Xiang Z, Wang J, Zhao Z, Peng Y, Cambareri G, Rocher A, Mobbs CV, Hof PR, Pasinetti GM. Diet-induced insulin resistance promotes amyloidosis in a transgenic mouse model of Alzheimer's disease. *FASEB J* **18** (2004), 902-904.
[13] Wang J, Ho L, Qin W, Rocher AB, Seror I, Humala N, Maniar K, Dolios G, Wang R, Hof PR, Pasinetti GM. Caloric restriction attenuates β-amyloid neuropathology in a mouse model of Alzheimer's disease. *FASEB J.* **19** (2005), 659-61.
[14] Vallet PG, Guntern R, Hof PR, Golaz J, Delacourte A, Robakis NK, Bouras C. A Comparative-Study of Histological and Immunohistochemical Methods for Neurofibrillary Tangles and Senile Plaques in Alzheimers-Disease. *Acta Neuropathologica* **83** (1992), 170-178
[15] McLaurin J, Kierstead ME, Brown ME, Hawkes CA, Lambermon MHL, Phinney AL, Darabie AA, Cousins JE, French JE, Lan MF, Chen F, Wing SS, Mount HT, Fraser PE, Westaway D, St George-Hyslop P. Cyclohexanehexol inhibitors of A[beta] aggregation prevent and reverse Alzheimer phenotype in a mouse model. *Nat Med.* **12** (2006), 801-808.
[16] Kin HJ, Chae SC, Lee DK, Chromy B, Lee SC, Park YC, Klein WL, Krafft GA, Hong ST. Selective neuronal degeneration induced by soluble oligomeric amyloid beta protein. *FASEB J* **17** (2003), 118-120.
[17] Wang J, Ho L, Chen L, Zhao Z, Zhao W, Qian X, Humala N, Seror I, Bartholomew S, Rosendorff C, Pasinetti GM. Valsartan lowers brain beta-amyloid protein levels and improves spatial learning in a mouse model of Alzheimer disease. *J Clin Invest.* **117** (2007), 3393-3402.
[18] Bitan G, Lomakin A, Teplow DB. Amyloid beta-protein oligomerization: prenucleation interactions revealed by photo-induced cross-linking of unmodified proteins. *J. Biol Chem.* **276** (2001), 35176-35184.
[19] United States Department of Agriculture and Health and Human Services. Dietary Guidelines for Americans 2005. http://www.health.gov/DIETARYGUIDELINES/dga2005/document/pdf/DGA2005.pdf , 2005.
[20] Hsiao K, Chapman P, Nilsen S, Eckman C, Harigaya Y, Younkin S, Yang F, Cole G. Correlative memory deficits. Aβ elevation, and amyloid plaques in transgenic mice. *Science* **274** (1996), 99-102.
[21] Zhang YW, Xu H. Molecular and cellular mechanisms for Alzheimer's disease: understanding APP metabolism. *Curr. Mol. Med.* **7** (2007), 687-696.
[22] Vassar R, Citron M. Abeta-generating enzymes: recent advances in beta- and gamma-secretase research. *Neuron* **27** (2000), 419-422.
[23] Vallers SS, Teplow DB, Bitan G. Determination of peptide oligomerization state using rapid photochemical crosslinking. *Methods Mol. Biol.* **299** (2005), 11-18.
[24] Bitan G. Structural study of metastable amyloidogenic protein oligomers by photo-induced cross-linking of unmodified proteins. *Methods Enzymol.* **413** (2006), 217-236.
[25] Luchsinger JA, Mayeux R. Dietary factors and Alzheimer's disease. *Lancet Neurol.* **3** (2004), 579-587.
[26] Reid MC, Boutros NN, O'Connor PG, Cadariu A, Concato J. The health-related effects of alcohol use in older persons: a systematic review. *Subst. Abu.s* **23** (2002), 149-164
[27] Urquiaga I, Leighton F. Plant polyphenol antioxidants and oxidative stress. *Biol. Res.* **33**(2000), 55-64.
[28] Scalbert A, Manach C, Morand C, Remesy C, Jimenez L. Dietary polyphenols and the prevention of diseases. *Crit. Rev. Food Sci. Nutr.* **45** (2005), 287-306.

[29] Ramassamy C. Emerging role of polyphenolic compounds in the treatment of neurodegenerative diseases: a review of their intracellular targets. *Eur. J. Pharmacol.* **545** (200), 51-64.

[30] Porat Y, Abramowitz A, Gazit E. Inhibition of amyloid fibril formation by polyphenols: structural similarity and aromatic interactions as a common inhibition mechanism. *Chem. Biol. Drug Des.* **67** (2006), 27-37.

[31] Ono K, Hasegawa K, Naiki H, Yamada M. Anti-amyloidogenic activity of tannic acid and its activity to destabilize Alzheimer's beta-amyloid fibrils in vitro. *Biochim. Biophys. Acta.* **1690** (2004), 193-202.

[32] Ono k, Yoshiike Y, Takashima A, Hasegawa K, Naiki H, Yamada M. Potent anti-amyloidogenic and fibril-destabilizing effects of polyphenols in vitro: implications for the prevention and therapeutics of Alzheimer's disease. *J. Neurochem,* **87** (2003), 172-181.

[33] Joseph JA, Denisova NA, Arendash G, Gorden M, Diamond D, Shukitt-Hale B, Morgan D. Blueberry supplemntation enhances signaling and prevents behaviroal deficits in an Alzheimer disease model. *Nutr. Neurosci.* **6** (2003), 153-162.

[34] Cheynier V. Polyphenols in foods are more complex than often thought. *Am. J. CLin. Nutr.* **81 (suppl)** (2005), 223S-9S.

[35] Manach C, Scalbert A, Morand C, Remesy C, Jimenez I. Polyphenols: food sources and bioavailablity. *Am. J. Clin. Nutr.* **79** (2004), 727-47.

[36] Yang F, Lim GP, Begum AN, Ubeda OJ, Simmons MR, Ambegaokar SS, Chen PP, Kayed R, Glabe CG, Frautschy SA, Cole GM. Curcumin inhibits formation of amyloid beta oligomers and fibrils, binds plaques, and reduces amyloid in vivo. *J. Biol. Chem.* **280** (2005), 5892-901.

[37] Lim GP, Chu T, Yang F, Beech W, Frautsch SA, Cole GM. The curry spice curcumin reduces oxidative damage and amyloid pathology in an Alzheimer transgenic mouse. *J. Neurosci.* **21** (2001), 8370-8377.

[38] Wang J, Ho L, Zhao W, Ono K, Rosensweig C, Chen L, Humala N, Teplow DB, Pasinetti GM. Grape-derived polypenolics prevent Abeta oligomerization and attenuate cognitive deterioration in a mouse model of Alzheimer's disease. *J. Neurosci.* **28** (2008), 6388-6392.

# Delivery of NGF to the Brain: Intranasal Versus Ocular Administration in Anti-NGF Transgenic Mice

Simona Capsoni[a,b], Sonia Covaceuszach[a*], Gabriele Ugolini[a*], Francesca Spirito[a], Domenico Vignone[a,b], Barbara Stefanini[a*], Gianluca Amato[a,b] and Antonino Cattaneo[b,c]
[a]Lay Line Genomics S.p.A., Rome, Italy
[b]European Brain Research Institute (Foundation EBRI Rita Levi Montalcini), Rome, Italy
[c]International School for Advanced Studies (SISSA), Trieste, Italy

**Abstract.** Nerve Growth Factor (NGF) has a great potential for the treatment of Alzheimer's disease. However, the therapeutic administration of NGF represents a significant challenge, due to the difficulty to deliver relevant doses to the brain, in a safe and non-invasive way.

We previously demonstrated the efficacy of a non invasive delivery of NGF to the brain in animal models, by an intranasal route. Recently, topical eye application of NGF was proposed, as an option for the delivery of NGF to the brain. Here, we compare the efficacy of the two delivery routes of hNGF-61, a recombinant traceable form of human NGF, in the mouse neurodegeneration model AD11. The intranasal administration appeared to be significantly more effective than the ocular one, in rescuing the neurodegenerative phenotypic hallmarks in AD11 mice. The ocular administration of hNGF-61 showed a more limited efficacy, even at higher doses. In addition, we used the rescue of the shrinkage of superior cervical ganglia as a parameter to indicate leakage of hNGF-61 in the peripheral blood circulation. We show that only through the intranasal route there are no effects on the peripheral target, suggesting that the ocular administration may be not sufficient to prevent the onset of peripheral side effects, such as pain. Thus, NGF nasal drops represent a viable and effective option to successfully deliver therapeutic NGF to the brain in a non-invasive manner.

Key words (max 10): delivery, NGF, ocular, intranasal, side effects, pharmacokinetic, Alzheimer, therapeutic window, dosage, non-invasive

## INTRODUCTION

Alzheimer's disease (AD) is the most common neurodegenerative disorder of the elderly, characterized by unremittingly progressive memory loss, as well as other cognitive impairments, and typical neuropathological hallmarks [1-3].

Among the multiple epigenetic factors that could contribute to the heterogeneous etiologies of sporadic AD, the potential relevance of alterations in the NGF system [4, 5] to neuronal loss and neurodegeneration occurring in AD has been suggested since the 1980s [6, 7] based on the neurotrophic actions of NGF on basal forebrain cholinergic neurons (BFCNs), whose degeneration is at the basis of the cholinergic deficit and cognitive decline in AD patients [8]. In preclinical studies, NGF was shown to prevent BFCN death, following axotomy [9-11] and to ameliorate the morphological and behavioral effects of their aging-dependent atrophy [9]. Decreased NGF availability to BFCNs in AD brains is not due to a decreased synthesis of NGF in the target cortical regions [12], but has been attributed, rather, to a defective transport [13-17], possibly due to a reduced expression of the TrkA receptor in both in BFCNs and neocortex of AD brain [18-20]. Making NGF available directly to BFCN cell bodies, bypass-

---

* Present Address: Rottapharm Biotech srl, Area Science Park edificio Q1, 34012 Basovizza - Trieste.
Corresponding author: Prof. Antonino Cattaneo, European Brain Research Institute, Via del Fosso di Fiorano 64/65, 00143 Rome, Italy. Tel.: 0039 06501703064; Fax: 0039 06501703335; Email: a.cattaneo@ebri.it

ing the transport system, is effective in rescuing BFCNs [21]. Notably, blocking the activity of mature NGF in the brain of adult transgenic mice leads to a progressive Alzheimer's neurodegeneration, with synaptic plasticity and memory behavioral defects [22-27]. A direct link between NGF deficits and the activation of the Alzheimer's amyloidogenic pathway has also been demonstrated in cultured neuronal lines [28].

The links between NGF and AD go, therefore, well beyond the long established neurotrophic actions on BFCNs and an updated version of the NGF hypothesis for AD neurodegeneration can be formulated [29], whereby NGF can be considered as an anti-amyloidogenic factor, that normally keeps the amyloidogenic pathway under control. Accordingly, whenever a normal neurotrophic supply of NGF to target neurons is shortened or interrupted, the amyloidogenic pathway is activated and a negative feedback, toxic neurodegenerative loop is activated.

Within this theoretical frame, the concept of therapeutic administration of human recombinant NGF (in its mature form) in AD patients is therefore well validated [29, 30] and any therapy aimed at re-establishing the correct balance between ligands (and receptors) of the NGF pathway appears to have a clear and strong rationale. However, a significant challenge to the clinical use of NGF is the difficulty associated with its effective delivery to the brain, in a safe and long-lasting manner. Indeed, the development of an NGF therapy is constrained by the dual conflicting need to achieve adequate concentrations in the relevant brain areas containing degenerating target neurons, while preventing unwanted adverse effects on non target regions or cells, such as pain. NGF is a hydrophilic protein which is not orally available and unable to bypass the blood-brain barrier following parenteral systemic routes of administration [31]. NGF has therefore been delivered, to rodent [9-11, 32] and primate [33, 34] brains, by intracerebroventricular (ICV) infusions, or by grafting into the brain parenchyma cells engineered to express recombinant human NGF (hNGF) [35]. Based on this, pilot clinical trials to test NGF effects in AD patients were performed. Both ICV [36] and ex-vivo gene therapy approaches [37] resulted in a reduction of the cognitive decline. In the ICV trial, patients suffered, however, from severe pain, driving from known effects of NGF in non target areas and therefore the treatment had to be interrupted, while in locally delivered NGF gene therapy trial, one patient died after neurosurgery and another suffered from epileptic seizures [38].

Although encouraging in terms of efficacy, these approaches for an NGF-based therapy in AD patients can be questioned in terms of general clinical applicability, due to the invasiveness of the neurosurgical procedure involved. Thus, attempts to find a non-invasive way of delivery are badly needed, in order to fully exploit the therapeutic potential of NGF as a long-lasting cholinergic and neuroprotective treatment for AD.

Intranasal delivery was shown to lead to an effective access of NGF into the brain [39, 40]. We previously demonstrated the feasibility and effectiveness of such a non-invasive strategy to deliver NGF to the brain, by showing that intranasal NGF rescues neurodegeneration [41] and behavioral deficits [25] in the AD11 animal model. Recently, the topical eye application of NGF was suggested as an alternative option for NGF non invasive delivery to the brain, on the basis of the evidence showing an enhanced expression of NGF receptors and ChAT immunoreactivity in rat BFCNs after ocular delivery [42, 43].

The success of an NGF-based therapy for human neurodegenerative disorders will be crucially dependent on a careful comparison and optimization of the different delivery options, in disease relevant neurodegeneration models. However, a side by side comparison of the efficacy of the nasal and the ocular delivery routes cannot be made, from the published studies, since the doses and the experimental systems were different.

In the present study, we undertook a side by side comparative study of the efficacy of intranasal versus ocular administration of NGF in reverting the neurodegeneration in the anti-NGF AD11 mouse neurodegeneration model.

## MATERIALS AND METHODS

### Animals experimentation

Six-month old AD11 anti-NGF mice were produced as described elsewhere [44]. Wild type (WT) mice were obtained by crossing non transgenic littermates of AD11 mice. Mice were kept under a 12 hours dark to light cycle, with food and water *ad libitum*. Experiments were performed according to the national and international laws for laboratory animal welfare and experimentation (EEC council directive 86/609, OJ L 358, 12 December 1987).

## Production of hNGF-61

Site directed mutagenesis of hNGF was performed according to the Stratagene protocol (Stratagene, La Jolla, CA) and verified by DNA sequencing. The mutein was then produced in E. coli according to Rattenholl et al. [45]. The detailed protocol is described in [46].

## ELISA assay

ELISA assay to determine mAb 4GA reactivity towards wild type hNGF and hNGF-61 was performed by coating a 96 well plate with the aforementioned two antigens, produced as described in Covaceuszach et al. [46], at a concentration of 10 µg/ml in sodium carbonate buffer 0.1M, pH 9.6 overnight at 4°C. After blocking with PBS containing 3% milk (MPBS) for one hour at room temperature, plates were first incubated with two serial dilutions (1:1, 1:10) of the supernatant of the monoclonal antibody (mAb) 4GA hybridoma cells, followed by peroxidase-conjugated anti-rat secondary antibody, and finally with TMB substrate (Tecna s.r.l., Trieste, Italy). After blocking the reaction with 0.1M $H_2SO_4$ the intensity of each colorimetric signal was measured at 450nm by a Spectra ELISA Reader (Tecan Group, Mannedorf, Switzerland).

## Intranasal delivery

hNGF-61 administration was performed on mice that were anaesthetized according to the protocol described in De Rosa et al. [25]. After anesthesia, mice were laid on their back, with the head in upright position, as described before [40, 41]. hNGF-61 (0.45 picomoles equivalent to a dose of 0.48 µg/kg of body weight) was dissolved in 48 µl of 40 mM phosphate-buffered saline (PBS, pH 7.4) and administered intranasally to AD11 mice according to the protocol described before [25, 41]. As control, AD11 mice were treated with PBS. Before starting the comparison between intranasal and ocular delivery, the frequency of administration for intranasal delivery was verified by randomly dividing AD11 mice in 4 groups, receiving respectively: PBS (n = 7), hNGF-61 (n = 8) once a week, hNGF-61 twice a week (n = 10) and hNGF-61 three times per week (every other day) (n = 8). WT mice treated with PBS were used as controls (n = 10). During the experiment comparing the intranasal and ocular administration, hNGF-61 was administered three times per week (n = 7). All administrations were repeated for 7 times.

## Ocular delivery

Six-month old AD11 received a single treatment 0.45 picomoles (n = 10) or 4.5 picomoles (n = 10) of hNGF-61 in 6 µl of bidistilled water. Mice received 3 µl applied to the ocular surface of each eye. As controls, WT (n = 9) and AD11 mice (n = 9) received 3 µl/eye of bidistilled water. The frequency of administration was three times per week for a total of seven administrations.

## Visual Object recognition test

The apparatus consisted of a square arena (60 cm x 60 cm x 30 cm) constructed in PVC with black walls and white floor. The objects to be discriminated were made of plastic, metal and glass and were too heavy to be displaced by the mice. The objects varied in size, the largest was approximately 6 cm x 6 cm x 10 cm and the smallest was approximately 5 cm x 5 cm x 8 cm. The box and objects were cleaned up between trials to stop the build-up of olfactory cues.

During the first day, mice received 2 sessions of 5 min duration in the empty box to help them habituate to the apparatus and test room. The second session was delayed 30 min with respect to the end of the first one. The next day, each mouse was first placed in the box and exposed to two identical sample objects (objects A1 and A2) for 10 min. This trial was called "sample phase". The experimenter measured the total time spent exploring each of the two objects. Then the mouse was returned to its cage. During the 24 hours retention interval the experimenter removed both objects and replaced one of the two by its identical copy (A3) (so to ensure that there was no carry-over of olfactory cues) and the other object (object B). After a delay of 24 hours, the mice were placed back in the box and exposed to the familiar object (A3, object identical to A1 and A2) and to a novel test object B for further

10 min. The objects were placed in the same locations as the previous ones. The experimenter measured again the total time spent exploring each of the two objects ("test period"). All trials were recorded using a videocamera connected to Any-maze software version 4.5 (Stoelting, Wood Dale, IL).

*Histological analysis*

After behavioral analysis, AD11 and WT mice were anaesthetized with an excess of 2,2,2-tribromethanol (400 mg/kg) and intracardially perfused with a 4% solution of paraformaldehyde in PBS. Superior cervical ganglia (SCGs), brains and spinal cord were collected and post-fixed in the same solution for 4 hours. SCGs were embedded in paraffin, sectioned at 5 μm and· stained with cresyl violet acetate (Sigma-Aldrich). Brains and spinal cord were transferred in 30% sucrose/PBS solution and then sectioned at a sliding freezing microtom (Leica, Wetzlar, Germany). Forty micrometers sections were collected in 0.05% sodium azide/PBS in 1.5 ml tubes and stored at 4°C until usage. To detected choline acetyltransferase (ChAT) in basal forebrain neurons, β-amyloid in the hippocampus and phopshorylated tau (ptau) in the entorhinal cortex, the following primary antibodies were used: goat anti-ChAT (1:500, Millipore, Billerica, MA); goat anti–NH2 terminus of β-amyloid (1:100; Santa Cruz Biotechnologies, Santa, CA), mouse anti-human ptau recognizing Ser199 (1: 10 clone AT8; Pierce Endogen, Rockford, IL) and mouse anti-human ptau recognizing Thr180 (1: 10 clone AT270; Pierce Endogen). Calcitonin-gene relate peptide (CGRP) was detected in the lumbar spinal cord using a polyclonal antibody raised in rabbit (1:100, Millipore). The phosphorylated form of CREB (pCREB) was analyzed only in WT mice 1 hour after the beginning of the intranasal treatment and 6 hours after the ocular administration, according to previous pharmacokinetic studies [40, 47]. The primary antibody used were: the antibody directed against Ser133 of pCREB (Cell Signaling Technology Inc., Danvers, MA) was used for pCREB detection. The primary antibody signal was detected using the appropriate biotinylated secondary antibody and the avidin horse radish peroxidase system by Vector Labtek (Burlingame, CA), with the exception of anti-β-amyloid antibodies that were detected using the avidin-alkaline phosphatase system (Vector Labtek).

*Stereology*

Morphometric analysis was performed using a Nikon microscope (Eclipse 1000, and the morphometry LUCIA program (Laboratory Imaging Ltd., Prague, Czechoslovakia). In all cases, the treatment status of the animals was unknown to the observer.

The total number of ChAT-positive neurons in the basal forebrain, i.e. medial septum (MS) plus diagonal band, and nucleus basalis of Meynert (NBM) were calculated according to a previous protocol [44] and to the optical fraction method. A similar approach was used to calculate the number of ptau-positive neurons and dystrophic neurites in the lateral entorhinal cortex. Briefly, the anatomical boundaries of lateral entorhinal cortex were defined as the entorhinal sulcus, for the dorsal aspect, a line parallel to and 600 μm distant from the previous, for the ventral aspect, laterally, the line separating layer I from layer II of the entorhinal cortex medially, the line separating layer VI of the entorhinal cortex from the white matter. Rostral and caudal boundaries were defined as a plane passing passing through bregma –2.75 and –3.80, respectively.

β-amyloid (Aβ)-positive clusters in the hippocampus were counted in 4 randomly chosen sections. The area of the hippocampus occupied by Aβ-positive clusters was calculated using the LUCIA program. The hippocampus was manually outlined and the total pixel area occupied by this brain region was determined. In the same manner, the area of each cluster was calculated. All cluster areas were then summed. The per cent of the brain region occupied by the clusters was then calculated.

The cumulative area occupied by CGRP-positive fibers in regions of the gray matter of the spinal cord was obtained using the LUCIA program. The numbers of stained pixels in each section were summed and converted into square micrometers via preprogrammed calibration standards. Each pixel represented an area of $0.24\ \mu m^2$. This degree of resolution allowed immunoreactivity to be detected in single fibers. The region of the spinal cord in which the areas of immunoreactive fibers were measured in a dorsal region including Rexed's laminae III–V, that normally contains CGRP-IR afferent fibers.

*Statistics*

The standard measure for the statistical analysis in the object recognition test (ORT) was the time spent exploring the two objects. The exploration of an object was defined as directing the nose to the object at a distance of $\approx 2$ cm and touching it with the nose. Turning around, climbing over or sitting on the object were not included. In the sample phase, if the exploration time was $\leq 3$ sec, the mice were discarded from the sample. Mice also were excluded from the sample if they spent $\leq 1$ sec exploring both new and familiar objects in the test phase. The discrimination index (DI) used to normalize for the total time spent exploring the objects was calculated as the difference between the time spent exploring new and old object divided by the total time spent exploring the objects [$(n - f)/(f + n)$, where $n$ represents new and $f$ represents familiar].

Statistical analyses were performed using the Sigmastat v. 3.5 program (Systat Software, San Jose, CA). The alpha was set at 0.05 and a normality and equal variance test were first performed. T-Test and Mann-Whitney Rank Sum test were used for the comparison between two groups. One Way Anova or Kruskal Wallis ANOVA were used for multiple comparisons, followed by Bonferroni or Holm-Sidak post hoc tests.

# RESULTS

*Outline of the delivery study: hNGF-61 and AD11 anti-NGF mice*

The aim of this study was to compare the ability of a prospective therapeutic recombinant hNGF protein, delivered via an intranasal or via an ocular route, to revert neurodegeneration in a disease-relevant model, the AD11 anti-NGF mouse model. In AD11 mice [44], the sequestration of endogenous NGF by the transgenically encoded anti-NGF antibody determines the observed progressive neurodegeneration and behavioral deficits [22-25].

We have previously demonstrated that the neurodegenerative phenotype in AD11 mice can be reverted by intranasal NGF administration [25, 41]. Therefore, the rescue of AD11 mice phenotype by NGF represents the most direct, straightforward and informative experimental read-out for determining the access and effectiveness of NGF to the central nervous system (CNS) and its biological potency.

For this study, we used hNGF-61, a recombinant mutant of hNGF that was developed towards its further application as a potential treatment of AD patients [46].

The design of hNGF-61 comes from the need, for a prospective "therapeutic" molecule of hNGF, to be traceable, against the background of endogenous NGF, in order to optimize its dosing when used in a clinical setting, while showing the same bioactivity profile and potency as unmodified hNGF. Thus, hNGF-61 is a "tagged" version of hNGF, whereby a single aminoacid of hNGF, Proline in position 61, was changed into a Serine, the corresponding residue in position 61 in mouse NGF (mNGF), the gold standard for NGF bioactivity. This is the minimal and most conservative change that one could introduce to tag a therapeutic NGF molecule with respect to the endogenous ones. Among the 5 positions that distinguish mNGF from hNGF (see Fig 1A), position 61 was chosen, because it is located on NGF loop III, a portion of the NGF molecule not directly involved in the structural interaction between NGF and its two receptors TrkA (Fig 1B) and p75 (Fig 1C) [48-50]. The monoclonal antibody mAb 4GA allows to distinguish the hNGF-61 (P61S) mutein from wild type hNGF (see Fig 1D). The bioactivity profile and potency of hNGF-61 fully recapitulates that of wild type hNGF [46].

We therefore undertook a comparison of the efficacy of recombinant hNGF-61 to revert neurodegeneration in AD11 mice, when delivered via an intranasal or an ocular route.

*Frequency of hNGF-61 administration*

Before undertaking the comparison of the two delivery routes of hNGF-61, we addressed the question of selecting the most appropriate frequency of intranasal administration, at the dose shown to be effective in the previous studies [25, 41]. In this study, AD11 mice to be treated with hNGF-61 were divided in three groups, which received 0.48 µg/kg of hNGF-61 once, twice and three times per week respectively. After 15 days of administration, mice underwent behavioral testing, by ORT. The sample phase of the ORT showed no differences among the different groups of treatment (Fig. 2A; Kruskal Wallis ANOVA P = 0.085). During the test phase,

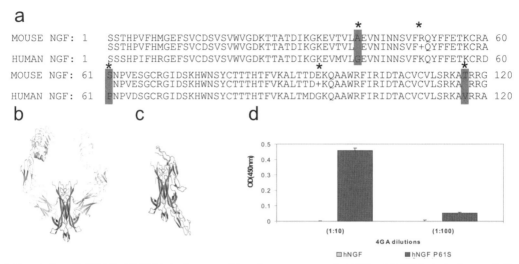

Fig. 1. Description of hNGF-61. (A) Multiple alignment of the primary structures of mNGF and hNGF (substitutions are indicated by stars and non conservative mutations are in red). (B) NGF Loop III (in red) does not interact with TrkA (in cyan) and (C) p75 (in magenta) extracellular domains. Cartoon representation produced using Pymol (www.pymol.org). (D) Comparison of binding activity of the mAb 4GA to the point mutant hNGF-61 and wild type hNGF by ELISA assay. Here shown are mean ± s.e.m.

WT mice (n = 10) explored more the new object with respect to the familiar one after a 24 hours delay (Fig 2B; paired T-test: P = 0.01), a sign of recognition memory which forms the basis for the ORT. As previously reported [25], AD11 mice treated with PBS (n = 7) did not distinguish between new and familiar objects (Fig. 2B; paired T-test P = 0.92). The same result was obtained for AD11 mice treated with hNGF-61 once (n = 8) (Fig. 2B; paired T test: P = 0.97) or twice a week ( n = 10) (Fig. 2B; paired T test: P = 0.87). On the contrary, AD11 mice treated with hNGF-61 three times a week (n = 8) explored the new object significantly more than the old one (Fig. 2B; paired T test: P = 0.006). A DI was calculated, to normalize the time spent exploring the new or the familiar object against the total time spent in exploring the two objects. The comparison of the DIs from the different groups of treatment confirmed that a memory deficit was present in AD11 mice treated with PBS ( n = 7), with respect to WT mice (n = 10) (Fig. 2C; Mann-Whitney Rank Sum Test P = 0.004) and that only the AD11 mice treated with hNGF-61 three times per week (n = 8) were able to recognize the new object from the familiar one (Fig. 2C; Kruskal-Wallis One Way ANOVA, *post hoc* Holm-Sidak test P = 0.006). In contrast, no statistical difference was observed between AD11 mice treated with hNGF-61 once or twice a week and AD11 mice treated with PBS (Fig. 2C; Kruskal-Wallis One Way ANOVA,

*post hoc* Holm-Sidak test P = 0.48 and P = 0.52, respectively).

After the behavioral analysis, mice were sacrificed and the brains were examined for the extent of neurodegeneration. In AD11 mice (n = 7) treated with PBS, a significant decrease in the number of ChAT-positive BFCNs was observed in the MS and diagonal band of Broca, with respect to WT mice (n = 10) (Mean ± S.E.M.: 3521 ± 239 vs 5192 ± 339; T-test: P = 0.002). The treatment with hNGF-61, once (n = 8) or twice (n = 10) a week, did not restore the number of ChAT-positive BFCNs in AD11 mice (Mean ± S.E.M.: 3059 ± 309 and 4518 ± 551; One Way ANOVA, *post hoc* Holm-Sidak test vs AD11 treated with PBS: P = 0.45 and P = 0.09, respectively), while the higher frequency of hNGF-61 administration (n = 8) yielded a significant rescue of the number of ChAT-positive BFCNs (Mean ± S.E.M.: 5197 ± 299; One Way ANOVA, *post hoc* Holm-Sidak test vs AD11 mice treated with PBS: P = 0.01). A similar result was observed when the number of cholinergic neurons in the nucleus basalis of Meynert (NBM) was analyzed. AD11 mice treated with PBS (n = 7) showed significant less ChAT-positive BFCNs than WT mice ( n = 10) (Fig. 2D and Fig. 3A,B; T test: P ≤ 0.001). The number of ChAT-positive BFCNs in the NBM of AD11 mice treated with hNGF-61 once (n = 8) or twice per week (n = 10) were not different from that of AD11 mice treated with PBS (Fig. 2D and Fig.

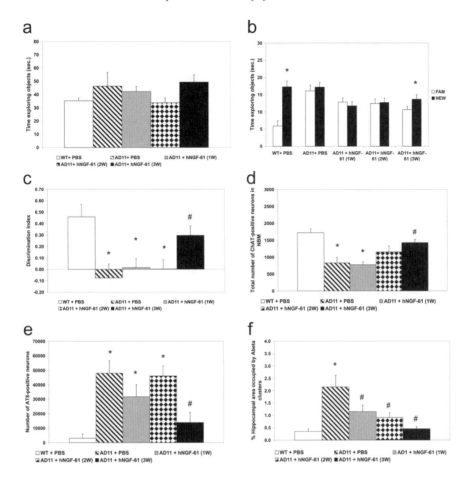

Fig. 2. Frequency of hNGF-61 intranasal administration. (A) No differences in exploration times during the sample phase of the ORT. (B) During the test phase of ORT, WT mice and AD11 mice treated 3 times per week with hNGF-61 explore more the new object than the familiar one. AD11 mice treated with PBS or 1 or 2 times per week with hNGF-61 do not discriminate between familiar and new object. (C) DI showing rescue of the memory deficit only when hNGF-61 is administered 3 times per week. (D) Quantification of cholinergic neurons in the NBM. (E) Quantification of AT8-positive neurons in the lateral entorhinal cortex. (F) Quantification of the number of Aβ clusters in the hippocampus. Here shown are mean ± s.e.m.

3C,D; One Way ANOVA, *post hoc* Holm-Sidak test vs. AD11 mice treated with PBS: P = 0.78 and 0.10). On the contrary, the administration of hNGF-61 three times per week determined the complete rescue of the number of ChAT-positive neurons in the NBM (Fig. 2D and Fig. 3E; One Way ANOVA, *post hoc* Holm-Sidak test vs AD11 mice treated with PBS: P = 0.007).

The effects of hNGF-61 treatment on ptau were quantified using antibodies against pThr181 and pSer199. At the age in which the tests were performed, the antibody against pThr181 detects an increased localization of ptau in dystrophic neurites from AD11 mice treated with PBS, with respect to WT mice. This increase in the number of ptau-positive dystrophic neurites was quantified in the lateral entorhinal cortex and was shown to be statistically significant (9440 ± 490 and 4110 ± 786, respectively in AD11 and WT mice; T-test: P ≤ 0.001). When administered once or twice a week, hNGF-61 did not decrease the number of ptau-positive dystrophic neurites, with respect to AD11 mice treated with PBS (7147 ± 1089 and 6852 ± 706, respectively for once and twice a week treatment; One Way ANOVA, *post hoc* Holm-Sidak test: P = 0.07 and P = 0.03). On the contrary, the number of ptau-positive dystrophic neurites in AD11 mice treated with hNGF-61 three times a week was significantly lower than in PBS-treated AD11 mice (5728 ± 743 vs 9440 ± 490; One Way ANOVA, *post hoc* Holm-Sidak test : P = 0.006).

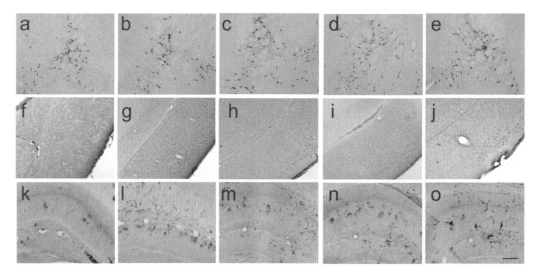

Fig. 3. Frequency of hNGF-61 administration: rescue of the neurodegenerative phenotype. ChAT-positive neurons in the NBM of (A) WT mice, (B) AD11 mice treated with PBS, (C) AD11 treated with hNGF-61 once a week, (D) AD11 treated with hNGF-61 twice a week, (E) AD11 treated with hNGF-61 three times per week. AT8-positive neurons in the lateral entorhinal cortex of (F) WT mice, (G) AD11 mice treated with PBS, (H) AD11 treated with hNGF-61 once a week, (I) AD11 treated with hNGF-61 twice a week, (J) AD11 treated with hNGF-61 three times per week. Aβ clusters in the hippocampus of (K) WT mice, (L) AD11 mice treated with PBS, (M) AD11 treated with hNGF-61 once a week, (N) AD11 treated with hNGF-61 twice a week, (O) AD11 treated with hNGF-61 three times per week. Scale bar = 200 μm.

The anti-ptau antibody against Ser199 labels, in AD11 mice, both dystrophic neurites and neuronal cell bodies. The latter were taken into consideration for the statistical analysis. AD11 mice treated with PBS show an increased localization of ptau in the cell body with respect to WT mice. This increase was quantified in the lateral entorhinal cortex and was shown to be statistically significant (Fig. 2E and Fig. 3F,G; 47988 ± 8545 and 3075 ± 3017, respectively in AD11 and WT mice; Mann Whitney Rank Sum Test: $P = 0.002$). When administered once or twice a week, hNGF-61 did not decrease the number of ptau-positive cells with respect to AD11 mice treated with PBS (Fig. 2E and Fig. 3H,I; 7147 ± 1089 and 6852 ± 706, respectively for once and twice a week treatment; One Way ANOVA, *post hoc* Holm-Sidak test: $P = 0.07$ and $P = 0.03$). On the contrary, the number of ptau-positive cells in AD11 mice treated with hNGF-61 three times a week was significantly inferior to that found in PBS-treated AD11 mice (Fig. 2E and Fig. 3J; 5728 ± 743 vs 9440 ± 490; One Way ANOVA, *post hoc* Holm-Sidak test : $P = 0.006$).

Aβ positive clusters of dystrophic neurites are found in the hippocampus of AD11 mice treated with PBS (Fig. 3K,L) [24] and their number is significantly higher than in WT mice (Fig. 2F and Fig. 3K,L; 34 ± 9 vs 11 ± 2; T-test: $P = 0.05$). Among the experimental groups of AD11 mice, treated for one, two or three times per week with hNGF-61, the number of Aβ-positive clusters of dystrophic neurites was significantly different from that calculated for PBS-treated AD11 mice only in the three times per week AD11 mice group (Fig. 2F and Fig.3M-O; 32 ± 7, 29, ± 7 and 12 ± 2; One Way ANOVA, *post hoc* Holm-Sidak test vs AD11 mice treated with PBS: $P = 0.016$, $P = 0.003$ and $P = 0.02$). The presence of Aβ clusters can also be expressed as the percentage of hippocampal area occupied by Aβ-positive dystrophic neurites (seeMethods). With this measure for Aβ a reduction after hNGF-61 treatment is also shown in AD11 mice treated one or twice per week (1.15 ± 0.24, 0.91 ± 0.18 and 0.46 ±0.07, respectively for one, two and three times per week treatments; One Way ANOVA, *post hoc* Holm-Sidak test vs AD11 mice treated with PBS: $P = 0.016$, $P = 0.003$ and $P = 0.0003$). Thus, the area of Aβ-positive clusters of dystrophic neurites in AD11 mice appears to be a parameter for Aβ more sensitive to hNGF-61 administration, as it is reduced even in the once per week treatment group.

In any case, it can be concluded that the full rescue of the behavioral and neurodegenerative phenotype was only achieved after administration of hNGF-61 three times per week. Thus, this frequency

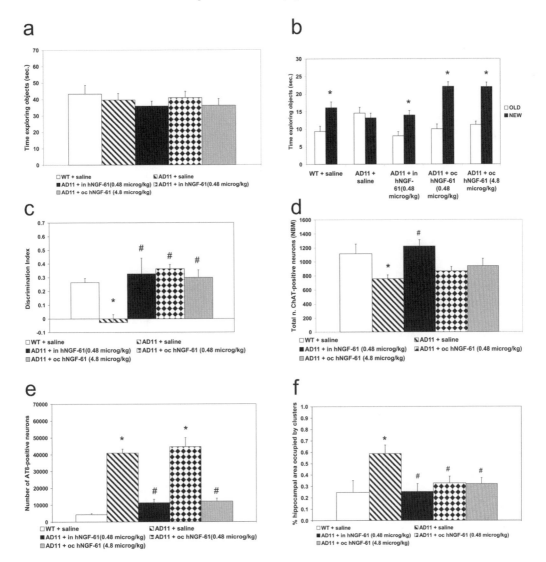

Fig. 4. Comparison of intranasal and ocular delivery. (**A**) No differences in exploration times during the sample phase of the ORT. (**B**) During the test phase of ORT, WT mice and AD11 mice treated with either intranasally or ocularly delivered hNGF-61, explore more the new object than the familiar one. AD11 mice treated with PBS do not discriminate between familiar and new object (**C**). DI showing rescue of the memory deficit in AD11 mice treated with hNGF-61. (**D**) Quantification of cholinergic neurons in the NBM. (**E**) Quantification of AT8-positive neurons in the lateral entorhinal cortex. (**F**) Quantification of the number of Aβ clusters in the hippocampus. Here shown are mean ± s.e.m.

of administration was used for the subsequent experiments.

## Comparison between intranasal and ocular administration

Having established an optimal frequency of administration, in order to compare the efficacy of different routes of hNGF-61 administration, one group of AD11 mice was treated with hNGF-61 using the intranasal route, while two other groups received hNGF-61 via the ocular route. Two doses were chosen for the ocular route: 0.45 pmol (0.48 μg/kg), the same as that used for the intranasal administration, and one ten times higher: 4.5 pmol (4.8 μg/kg).

After 15 days of hNGF-61 administration, three times per week, the possible amelioration of the recognition memory deficit was analyzed using the ORT. The sample phase of the ORT showed no differences among the different groups of treatment (Fig. 4A; One Way ANOVA: P = 0.66). During the test phase, WT mice (n = 9) explored more the new object than the familiar one, after a 24 hours delay (Fig 4B; paired T-test: P = 0.004). AD11 mice treated with PBS (n = 9) showed the expected deficit in recognition memory, with no preference between new and familiar objects (Fig. 4B; paired T-test P = 0.44). AD11 mice that received hNGF-61 through the intranasal route (n =7) explored more the new object with respect to the old one (Fig. 4B; paired T test: P = 0.005). The same result was obtained for AD11 treated with 0.48 μg/kg (n = 10) or 4.8 μg/kg ( n = 10) of hNGF-61 through the ocular route of administration (Fig. 4B; for the lower dose, Mann-Whitney Rank Sum Test: P ≤ 0.001; for the higher dose, paired T test: P = 0.002). The comparison of the DIs from the different groups of treatment confirmed that, while a memory deficit was present, as expected, in AD11 mice treated with PBS, with respect to WT mice (Fig. 4C; Mann-Whitney Rank Sum Test P = 0.001), all groups of AD11 mice treated with hNGF-61, via the intranasal or the ocular delivery route, were able to recognize the new object from the familiar one (Fig. 4C; One Way ANOVA, *post hoc* Holm-Sidak test; P ≤ 0.001 in all cases). After this behavioral test, mice were sacrificed and the analysis of the neurodegeneration was performed. In AD11 mice (n = 9) treated with PBS, a significant decrease in the number of ChAT-positive BFCNs was observed in the MS and diagonal band of Broca with respect to WT mice (n = 4) (Mean ± S.E.M.: 4113 ± 673 vs 6864 ± 704; T-test: P = 0.033). The intranasal treatment with hNGF-61 restored the number of ChAT-positive BFCNs in AD11 mice (n = 7) (Mean ± S.E.M.: 7364 ± 838; One Way ANOVA, *post hoc* Holm-Sidak test vs AD11 treated with PBS: P = 0.003). On the contrary, the ocular administration of hNGF-61 did not restore the number of cholinergic neurons in the MS and diagonal band of Broca, neither at the dose of 0.48 μg/kg (n = 10) nor at the dose of 4.8 μg/kg (n = 10) (Mean ± S.E.M.: 4444 ± 684 and 4773 ± 609, respectively; One Way ANOVA, *post hoc* Holm-Sidak test vs AD11 mice treated with PBS: P = 0.49 and P = 0.73). A similar result was observed when the NBM was analyzed. AD11 mice treated with PBS (n = 9) showed significant less ChAT-positive BFCNs in the NBM, than WT mice ( n = 5) (Fig. 4D and Fig. 5A,B; Mann-Whitney Rank Sum Test: P = 0.016). The administration of intranasal hNGF-61 three times per week determined the complete rescue of the number of ChAT-positive neurons in the NBM (Fig. 4D and Fig. 5C; One Way ANOVA, *post hoc* Holm-Sidak test vs AD11 mice treated with PBS: P = 0.0003). On the contrary, the number of ChAT-positive BFCNs in the NBM of AD11 mice treated with 0.48 μg/kg (n = 10) or 4.8 μg/kg (n = 10) hNGF-61 via the ocular route was not different from that of AD11 mice treated with PBS (Fig. 4D and Fig. 5D,E; One Way ANOVA, *post hoc* Holm-Sidak test vs. AD11 mice treated with PBS: P = 0.31 and 0.09).

As for the effects of hNGF-61 on the expression of ptau, AD11 mice treated with PBS showed an increase of ptau (pThr181) positive dystrophic neurites (Mean ± S.E.M.: 8956 ± 917 and 3084 ± 704, respectively in AD11 and WT mice; T-test: P ≤ 0.001). When administered intranasally, hNGF-61 significantly decreased the number of ptau (pThr181)-positive dystrophic neurites with respect to AD11 mice treated with PBS (Mean ± S.E.M.: 5467 ± 1258 vs 8956 ± 917; One Way ANOVA, *post hoc* Holm-Sidak test: P = 0.014). The number of ptau (pThr181)-positive dystrophic neurites in AD11 mice treated with hNGF-61, via the ocular route, at the dose of 0.48 μg/kg did not decrease the number of ptau (pThr181)- positive dystrophic neurites (Mean ± S.E.M.: 10712 ± 882; One Way ANOVA, *post hoc* Holm-Sidak test : P = 0.18), while the ocular administration of the higher dose determined a significant reduction in the number of ptau (pThr181)-positive dystrophic neurites (Mean ± S.E.M.: 5158 ± 791; One Way ANOVA, *post hoc* Holm-Sidak test : P = 0.004).

Using the anti-ptau antibody against pSer199, it was determined that AD11 mice treated with PBS showed an increased localization of ptau in the cell body, with respect to WT mice. This increase was quantified in the lateral entorhinal cortex and was statistically significant (Fig. 4E and Fig. 5F,G 40920 ± 2300 and 4307 ± 616, respectively in AD11 and WT mice; T- Test: P ≤ 0.001). When administered intranasally, hNGF-61 decreased the number of ptau (pSer199)-positive cells with respect to AD11 mice treated with PBS (Fig. 4E and Fig. 5H; 11318 ± 2063; One Way ANOVA, *post hoc* Holm-Sidak test: P = 0.07 and P = 0.03). On the contrary, after ocular administration, only the dose at 4.8 μg/kg of hNGF-61 decreased the number of ptau (pSer199)-positive neurons in AD11 mice, while the lower dose was ineffective, in this respect (Fig. 4E and Fig. 5I,J; 12154 ± 1795 and 44694 ± 5384, respectively; One Way ANOVA, *post hoc* Holm-Sidak test : P = 0.006).

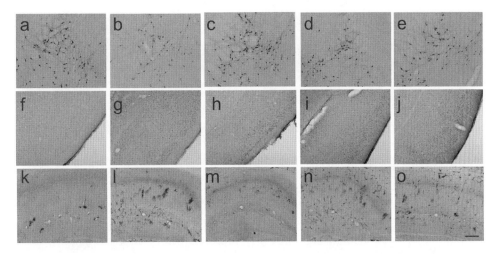

Fig. 5. Rescue of the neurodegenerative phenotype after intranasal or ocular delivery of hNGF-61. ChAT-positive neurons in the NBM of (A) WT mice, (B) AD11 mice treated with PBS, (C) AD11 treated with intranasal hNGF-61, (D) AD11 treated with ocular hNGF-61 (0.48 µg/kg), (E) AD11 treated with ocular hNGF-61 (4.8 µg/kg). AT8-positive neurons in the lateral entorhinal cortex of (F) WT mice, (G) AD11 mice treated with PBS, (H) AD11 treated with intranasal hNGF-61, (I) AD11 treated with ocular hNGF-61 (0.48 µg/kg), (J) AD11 treated with ocular hNGF-61 (4.8 µg/kg). Aβ clusters in the hippocampus of (K) WT mice, (L) AD11 mice treated with PBS, (M) AD11 treated with intranasal hNGF-61, (N) AD11 treated with ocular hNGF-61 (0.48 µg/kg), (O) AD11 treated with ocular hNGF-61 (4.8 µg/kg). Scale bar = 200 µm.

The number of Aβ-positive clusters of dystrophic neurites in PBS treated AD11 mice, was significantly higher than in WT mice (Fig. 4F and Fig. 5K,L; 16 ± 2 vs 6 ±2 T-test: P = 0.03). The number of Aβ -positive clusters of dystrophic neurites, calculated after intranasal and ocular administration of hNGF-61, was significantly different from that calculated for AD11 mice treated with PBS (Fig. 4F and Fig. 5M-O; 9.9 ± 2, 9.4 ± 2 and 8.8 ± 1, respectively for intranasal and ocular -0.48 and 4.8 µg/kg respectively- hNGF-61 administration; One Way ANOVA, *post hoc* Holm-Sidak test vs AD11 mice treated with PBS: P = 0.017, P = 0.025 and P = 0.05). The same results were obtained if the presence of Aβ clusters is expressed as percentage of hippocampal area occupied by Aβ-positive dystrophic neurites (0.25 ± 0.06, 0.32 ± 0.05 and 0.32 ± 0.04, respectively for intranasal, 0.48 and 4.8 µg/kg ocular hNGF-61 versus 0.58 ± 0.09 in AD11 mice treated with saline; One Way ANOVA, *post hoc* Holm-Sidak test vs AD11 mice treated with PBS: P = 0.017, P = 0.025 and P = 0.05).

In conclusion, the full rescue of the behavioral and neurodegenerative phenotype in AD11 mice was only achieved after intranasal administration of hNGF-61, with ocular delivery resulting only in a partial improvement, at a 10 times higher dose only. This suggests the intranasal route to be a more efficient way to deliver hNGF-61 to the brain, compared to the ocular one.

*Intranasal and ocular administration activate CREB in different brain regions*

NGF is known to induce the nuclear translocation of CREB in target cells [51]. We used CREB activation in the brain as a surrogate marker for hNGF-61 activity, after its intranasal or ocular administration. To verify which brain areas were influenced by the administration of hNGF-61, WT mice were treated with PBS, intranasally or ocularly administered hNGF-61, at the dose of 0.48 µg/kg and sacrificed 1 hour and 6 hours after the hNGF-61 administration, according to published pharmacokinetic studies for intranasal [40] and ocular [47] delivery. Intranasally administered hNGF-61 markedly increased the nuclear localization of the pCREB in the olfactory bulb (Fig. 6B), in the entorhinal cortex (Fig. 6E), in the hippocampus (Fig. 6H) and in the MS (Fig. 6K), with respect to PBS-treated WT mice (Fig. 6A,D,G,J). A very low degree of activation was found in interneurons of the neostriatum (Fig. 6M,N). Nuclear pCREB was also markedly increased in the entorhinal cortex (Fig. 6F) and in the hippocampus (Fig. 6I) after the ocular administration of hNGF-61. However, in the case of the ocular route of administration, the number of pCREB-positive cells found in the MS (Fig. 6L) was markedly inferior, with respect to the intranasal administration, and no activation of pCREB was observed

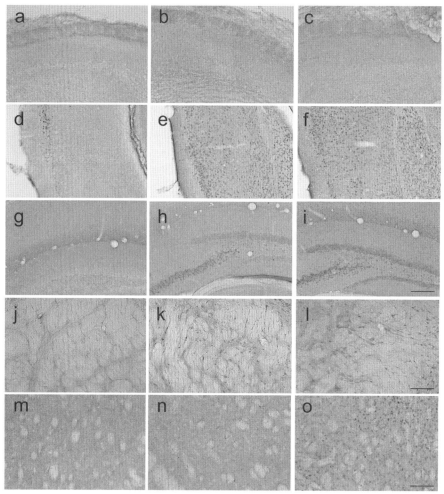

Fig. 6. pCREB localization after intranasal or ocular (0.48 µg/kg) delivery of hNGF-61 in WT mice. Olfactory bulb of (**A**) PBS, (**B**) intranasal and (**C**) ocular hNGF-61 treated mice. Entorhinal cortex (**D**) PBS, (**E**) intranasal and (**F**) ocular hNGF-61 treated mice. Hippocampus of (**G**) PBS, (**H**) intranasal and (**I**) ocular hNGF-61 treated mice. NBM of (**J**) PBS, (**K**) intranasal and (**L**) ocular hNGF-61 treated mice. Neostriatum of (**M**) PBS, (**N**) intranasal and (**O**) ocular hNGF-61 treated mice. Scale bar in **A-I**, **m-o** = 200 µm; scale bar in **J-L** = 50 µm.

in the olfactory bulb (Fig. 6C). On the other hand, a high number of neostriatal cells showed nuclear pCREB labeling after ocular delivery of hNGF-61 (Fig. 6O). Thus, ocular hNGF-61 delivery seems to induce a distinct pattern of pCREB activation in the brain, from that observed after intranasal hNGF-61 delivery, which appears to be more pronounced and more specific for those areas known to be targets of NGF action (such as the MS versus the neostriatum).

*Possible side effect issues*

In a further set of experiments, we evaluated some issues related to possible side effects, that might be of concern in choosing a way of delivery of NGF derivatives. First of all, the body weight was recorded before and during the intranasal and ocular administration. This parameter was taken into consideration since body weight reduction after NGF treatment was previously observed during the ICV clinical trial [36, 52]. The graph in Fig. 7A shows that, for all groups of treatment, there was no difference in the intergroup variability of body weight during the treatment (One Way ANOVA: WT mice: P = 0.99; PBS-treated AD11 mice: P = 0.69; intranasal hNGF-61: P = 0.79; 0.48 µg/kg ocular hNGF-61: P = 0.92; 0.48 µg/kg ocular hNGF-61: P = 0.95). The aim of a nasal or an ocular delivery of hNGF-61 is to increase the access of NGF molecules to the brain, while limiting their biodistribution to peripheral sites, where unwanted side effects, such as those related to pain, could be elicited [29, 30]. In order to evaluate, in a comparative way, possible peripheral effects after intranasal or ocular delivery of hNGF-61, SCG were analyzed, as a peripheral NGF target tissue. To this aim, the area of the medial section of the SCG was measured. As expected [44], in PBS-treated AD11 mice the SCG medial area was significantly lower than that in WT mice (Fig. 7B; Mann-Whitney Rank Sum Test: P = 0.016). Intranasally administered hNGF-61 and 0.48 µg/kg ocularly administered hNGF-61 did not increase the area of the medial section of SCG (Fig. 7B; Kruskal Wallis Anova with Bonferroni contrast vs AD11 mice treated with PBS: P = 1.00 and P = 0.17, respectively), showing that no, or little, NGF reaches this ganglion, while the treatment with 4.8 µg/kg ocularly delivered hNGF-61 determined a significant increase of SCG medial section area (Fig. 7B; Kruskal Wallis Anova with Bonferroni contrast vs AD11 mice treated with PBS: P = 0.04).

Since nasal delivery appears to be a more effective and more targeted approach than ocular delivery, potential side effects of hNGF-61, related to spinal cord nociception, were further analyzed, exclusively in WT mice intranasally treated with hNGF-61. The spinal cord is the crucial target for the possible activation of nociceptive pathways by NGF, and CGRP increased expression was taken as a nociception activation marker [53, 54]. No increased CGRP-positive fibers were found in the lumbar spinal cord, after intranasal administration of hNGF-61 (Fig. 7C; T-test: P = 0.85), confirming that little hNGF-61 reaches the spinal cord after nasal delivery.

## DISCUSSION

The rationale for NGF as a potential therapy for AD has been significantly validated by several results in anti-NGF mice [22, 25, 41] and in cell culture models [28], allowing to link deficits or abnormalities in NGF signaling, processing or transport to the activation of an amyloidogenic route of amyloidogenic Aβ precursor protein (APP) processing and full blown AD neurodegeneration in general [29].

However, its potential use for treating AD is hindered by the fact that NGF is unable to appreciably cross the blood brain barrier, making its delivery to the brain difficult. Delivery methods such as ICV and grafting of NGF-producing cell have been developed and are at the basis of clinical trials performed so far, being ICV employed in the very first clinical trials in AD patients.

The first [52] and second [36] trials were performed using mNGF, infused into the right ventricle of AD patients at a rate of 75 µg/day for 3 months. In these patients, a transient increase was observed in the scores of some episodic memory tests and there was an improvement in physiological parameters of brain function, such as nicotine binding, cortical blood flow and electroencephalographic activity [36]. No improvements were shown in the Mini-Mental Status Examination scores [36]. The trial was however discontinued after the onset of debilitating side effects due to NGF infusions. In particular, the patients lost body weight and developed severe back pain, within 2 to 14 days after the beginning of ICV infusions [36]. Thus, no further clinical trials were attempted until recently, when a new gene therapy-based approach was followed, based on the stereotaxic implantation of autologous fibroblasts, *ex vivo* modified to produce NGF [38]. After 18 months follow-up of 10 AD patients, the outcome of the trial reported a 40-50 percent reduction in the rate of cognitive decline, a significant improvement in PET scans and a significant cell growth effect due to the exposure of cholinergic neurons to the NGF [37].

Although encouraging in terms of efficacy, the results of these pilot NGF clinical trials can be questioned in terms of wide clinical applicability, due to the associated surgical risks and the high cost that might limit the application of these methods to humans [55]. Indeed, two patients suffered of cerebral hemorrhages during the surgical procedures used in the gene therapy approach [37]. Moreover, in both ICV and gene therapy approaches, the levels of NGF in the brain cannot be controlled and dosed. The issue of a correct dosing of NGF was already raised by the second ICV clinical trial, in which a very narrow window of safety was found, the side effects appearing at 200 µg/L NGF in the cerebrospinal fluid (CSF) compared to the absence of effects at 50 µg/L [36]. In addition, it is known that NGF concentration in the blood varies both from individual to individual and in a single individual at various times, and according to stress [56] and hormonal balance [57, 58]. Thus, finding a non invasive delivery to the brain and setting up a correct dosing system of the therapeutic NGF concentrations to avoid the peripheral side effects, appear to be the path to follow, to

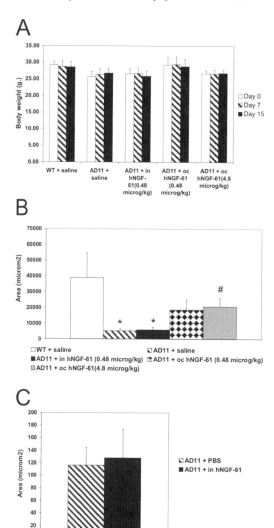

Fig. 7. Effects of intranasal and ocular delivery of hNGF-61 on the (**A**) body weight and (**B**) area of the superior cervical ganglion in AD11 mice. (C) Intranasal hNGF-61 does not induce increase in CGRP-fiber in lumbar spinal cord. Here shown are mean ± s.e.m.

fully exploit the therapeutic potential of NGF and to meet the therapeutic window for NGF clinical application [29].

In this study, we compared two non-invasive ways of delivery of NGF to the brain, which may significantly improve the prospects of using NGF clinically. We used hNGF-61, a mutein of hNGF [46], which is a "one-residue tagged" version of hNGF, that is equally potent as untagged hNGF and allows to specifically recognize the "therapeutic" NGF molecules against the background of endogenous hNGF, in human tissues. This should allow to carefully and precisely determine the therapeutic dose to be delivered and to meet the critical therapeutic window for NGF effectiveness [29]. In this paper,

we exploited the AD11 anti-NGF animal model of AD, in which a comprehensive neurodegeneration progressively builds up, as a result of NGF sequestration by a transgenically expressed anti-NGF antibody, to perform a side by side comparison of two non-invasive routes for NGF delivery to the brain, namely the nasal and the ocular routes.

Besides proof of concept studies with nasal delivery of NGF in animal models [40, 41], the delivery of peptides to the brain via the olfactory pathway is currently used in pilot clinical studies in AD patients, to deliver insulin [59] and the octapeptide ADNP [60].

The presence of anatomical connections between the brain and the nasal cavity makes the delivery of peptides and in particular of NGF feasible. Indeed, olfactory receptors cells in the olfactory epithelium are in direct in direct contact with the external environment (making them accessible for exogenous application of NGF) and the brain. Their axons project to the olfactory bulb, that makes widespread connections with the basal forebrain, the entorhinal cortex and amygdala, all regions affected by AD [61]. In addition, NGF delivered through the intranasal route may utilize a trigeminal pathway to reach the brain through the nasal epithelium [62, 63].

In the past, it was demonstrated that, through intraneuronal and extraneuronal pathways, NGF reaches the brain in a concentration sufficient to increase ChAT activity in basal forebrain cholinergic neurons [39] and to rescue neurodegeneration and behavioral deficits in anti-NGF AD11 mice [25, 41]. In this respect, it is noteworthy that by the nasal route, little build up of NGF in the blood and in the CSF is observed [39, 40], thus minimizing the possibility of activating nociceptive pathways.

Ocular delivery was first suggested by Lambiase [43] and Di Fausto [42], as a possible way of NGF delivery to the brain. In those studies, it was demonstrated that a solution of 200 µg/mL of hNGF (equivalent to a 7.69 µM solution, 0.1 nmoles of NGF per application), applied as a collyrium to the eye, was able to increase the number of ChAT-positive neurons in the basal forebrain of normal rats, or in mice in which a lesion of BFCNs was obtained by a treatment with ibotenic acid. The authors discuss a possible mechanism of this delivery approach, that could be due to: (1) an anatomical connection between the eye and the brain [64]; (2) an indirect nasal delivery, since NGF applied to the ocular conjunctiva might reach the brain via the nasal mucosa after drainage onto the nasal epithelium through the naso-lacrimal duct; (3) transport via the optic nerve [47]; or (4) NGF diffusion in the CSF surrounding the optic nerve. In any case, it is possible that both ocular and intranasal delivery routes for NGF may involve the trigeminal pathway [62, 63].

However, no data are available on the effectiveness of ocularly applied NGF in NGF deprivation models. For this reason, a side by side comparison of ocular delivery of NGF with nasal delivery of NGF, in the disease-relevant AD11 anti-NGF model, is of interest.

The data described in the present study show that, indeed, the ocular application of hNGF-61 was not able to fully rescue the neurodegeneration in AD11 mice, despite the amelioration of the behavioural deficit, revealed using the ORT, and a decrease in the accumulation of ptau in the cortex (even if only at the higher dose) and of Aβ in the hippocampus. On the other hand, the intranasal delivery of hNGF-61 confirmed to be fully effective in the AD11 model, at the lower dose of 0,48 µg/kg.

The ORT is a very sensitive test, often used to assay the effectiveness of drugs in AD mouse model, and a rescue of ORT deficit (such as that obtained by ocular hNGF-61 in this study) is not always associated to the full rescue of the neurodegeneration in the brain. Indeed, in AD11 mice, galantamine or donepezil rescue the ORT deficit but do not decrease tau phosphorylation in the cortex or Aβ accumulation in the hippocampus[65]. Similarly, in PDAPP mice, another AD mouse model [66], passive Aβ immunization reverses memory deficits, as also assessed by ORT, without however decreasing brain Aβ burden [67].

The rationale for using the same dosage and frequency of administration for the intranasal and the ocular routes was based on the need to vary only the route of administration, and to maintain the same quantity of hNGF-61 administered. The choice of limiting the dose and the frequency of administration with respect to the regimen used in previous studies was based on arguments discussed below.

Indeed, several issues might be raised, regarding not only the efficacy but also the safety of the ocular NGF delivery approach. Indeed, on one hand, the activation of pCREB in several brain areas and the increase in size of SCGs after ocular delivery provide evidence that hNGF-61 penetrates in doses sufficient to trigger biological responses, not only in the brain but also in peripheral tissues. However, on the other hand, no rescue of BFCNs, the primary NGF target in the brain, was found, even after the administration of the higher dose of 4.8 µg/kg, possibly due to a non optimal biodistribution of NGF in the different brain areas, as also confirmed by the pCREB localization experiments. It should be noted that the comparative study was performed at the NGF dose (0,48 µg/kg) and frequency (three times per week) shown to be effective, by the nasal delivery route, in rescuing neurodegeneration and memory deficits, (as also confirmed in this study). A group receiving a 10 times higher dose, via the ocular route, was included in the study, a dose still 3 to 30 times lower with respect to the studies where the ocular delivery approach was proposed [43]. However, safety issues can be raised already at these "low" dosages, for the ocular delivery, and might suggest the raising the dosage might be risky. First of all, the analysis of pCREB localization, at the dose of 0.48 μg/kg, showed that a precise and specific activation of CREB was not obtained after the ocular delivery of NGF. There was, for instance, an activation of cells in the neostriatum, which is not consistent with the number of cells expressing NGF receptors in that brain area [6], thus raising the risk of adverse effects in non targeted regions. On the other hand, after nasal delivery, the activation of CREB was found in areas that are specific for NGF signaling and/or basal forebrain projections, such as the NBM, hippocampus and cortex. Secondly, the fact that, in AD11 mice treated with ocular hNGF-61, there were peripheral effects, such as on SCG, indicates that the ocularly administered hNGF-61 might reach peripheral tissues in a more prominent way than intranasal NGF, something that is to be avoided. This is in line with a previous study, showing that NGF can reach high concentrations in the blood, after ocular application [47]. On the contrary, the intranasal delivery route, while leading to effective transport of NGF in AD relevant brain areas, determines a limited, albeit measurable, leakage into the blood stream [40]. The fact that ocular hNGF-61 can reach the blood and peripheral targets, is a concern for two reasons: first of all, we show an effect at peripheral sites using a dose (4.8 μg/kg) that is 30 times lower with respect to the first dose reported to be effective in rescuing BF cholinergic deficit (160 μg/kg [42]). Thus, we can argue that the effects at periphery with such a dose should be even more dramatic. The second concern is related to the well known nociceptive response determined by this neurotrophin, due to the stimulation of dorsal root ganglia [68]. Pain was, indeed, a side effect that prevented the completion of the first clinical trial in AD patients [36] and was observed also in clinical studies for the application of NGF to patients affected by diabetic polyneuropathy [69].

In conclusion, this comparative study demonstrates that intranasal delivery of hNGF-61 represents a more effective non-invasive strategy to increase NGF brain levels in a pharmacologically relevant context, compared to the delivery via the ocular route. The latter was shown to be less effective, even at 10 fold higher doses, and less specific in terms of potential adverse effects in non targeted regions. Thus, increasing the dosing of ocularly delivered NGF would not appear to be a viable option. Intranasal delivery was confirmed to be a good compromise solution, to meet the required therapeutic window for the effectiveness of an NGF-based therapy for neurodegenerative disorders such as AD. The hNGF-61 molecule represents the basis for the development of a class of hNGF muteins characterized by traceability and reduced nociceptive activity. Indeed, hNGF-61 has been already used to produce and characterize new muteins in which a second mutation in position R100 was introduced, in the attempt to obtain a mutein with reduced pronociceptive activity [70]. Indeed, this second mutation was inspired by the discovery of a missense point mutation in the NGFB gene (C661T, leading to the aminoacid substitution R100W) of individuals affected by a form of hereditary loss of pain perception (hereditary sensory and autonomic neuropathy type V, HSANV) [71]. A subset of hNGF-61R100 muteins have been already characterized for their reduced affinity for the p75NTR receptor and provide a basis for the development of "painless" hNGF molecules with therapeutic potential [70]. This strategy may be of particular benefit for the long-term treatment of subjects with brain degeneration disorders, since it appears that to be less invasive and less prone to systemic side effects than other currently available NGF delivery strategies.

## ACKNOWLEDGMENTS

The experimental work from the respective authors lab was funded by MIUR grants (FIRB projects n. RBIN 04H5AS and RBLA 03F, and Progetto Strategico PNR (n. RBIP063ANC_001) and by Lay Line Genomics S.p.A. The authors thank Dr. Sandro Bosco (Department of Pediatric Pathological Anatomy, University La Sapienza, Rome, Italy), for skilful assistance in some of the histological analysis presented in this manuscript.

## References

[1] Alzheimer A (1907) Uber einen eigenartige Erkrankung der Hirnrinde. *Allgemeine Zeitschrift fur Psyciatrie und PsychischGerichtliche Medizin* **64**, 146-148.

[2] Blessed G, Tomlinson BE, Roth M (1968) The association between quantitative measures of dementia and of senile change in the cerebral grey matter of elderly subjects. *Br J Psychiatry* **114**, 797-811.

[3] Selkoe DJ (2001) Alzheimer's disease: genes, proteins, and therapy. *Physiol Rev* **81**, 741-766.

[4] Levi-Montalcini R (1952) Effects of mouse tumor transplantation on the nervous system. *Ann N Y Acad Sci* **55**, 330-344.

[5] Levi-Montalcini R (1987) The nerve growth factor 35 years later. *Science* **237**, 1154-1162.
[6] Holtzman DM, Kilbridge J, Li Y, Cunningham ET, Jr., Lenn NJ, Clary DO, Reichardt LF, Mobley WC (1995) TrkA expression in the CNS: evidence for the existence of several novel NGF-responsive CNS neurons. *J Neurosci* **15**, 1567-1576.
[7] Phelps CH (1989) Trophic factors and Alzheimer's disease. *Neurobiol Aging* **10**, 584-586; discussion 588-590.
[8] Bartus RT, Dean RL, 3rd, Beer B, Lippa AS (1982) The cholinergic hypothesis of geriatric memory dysfunction. *Science* **217**, 408-414.
[9] Fischer W, Wictorin K, Bjorklund A, Williams LR, Varon S, Gage FH (1987) Amelioration of cholinergic neuron atrophy and spatial memory impairment in aged rats by nerve growth factor. *Nature* **329**, 65-68.
[10] Hefti F (1986) Nerve growth factor promotes survival of septal cholinergic neurons after fimbrial transections. *J Neurosci* **6**, 2155-2162.
[11] Kromer LF (1987) Nerve growth factor treatment after brain injury prevents neuronal death. *Science* **235**, 214-216.
[12] Goedert M, Fine A, Hunt SP, Ullrich A (1986) Nerve growth factor mRNA in peripheral and central rat tissues and in the human central nervous system: lesion effects in the rat brain and levels in Alzheimer's disease. *Brain Res* **387**, 85-92.
[13] Delcroix JD, Valletta J, Wu C, Howe CL, Lai CF, Cooper JD, Belichenko PV, Salehi A, Mobley WC (2004) Trafficking the NGF signal: implications for normal and degenerating neurons. *Prog Brain Res* **146**, 3-23.
[14] Fahnestock M, Michalski B, Xu B, Coughlin MD (2001) The precursor pro-nerve growth factor is the predominant form of nerve growth factor in brain and is increased in Alzheimer's disease. *Mol Cell Neurosci* **18**, 210-220.
[15] Mufson EJ, Conner JM, Kordower JH (1995) Nerve growth factor in Alzheimer's disease: defective retrograde transport to nucleus basalis. *Neuroreport* **6**, 1063-1066.
[16] Salehi A, Delcroix JD, Mobley WC (2003) Traffic at the intersection of neurotrophic factor signaling and neurodegeneration. *Trends Neurosci* **26**, 73-80.
[17] Scott SA, Mufson EJ, Weingartner JA, Skau KA, Crutcher KA (1995) Nerve growth factor in Alzheimer's disease: increased levels throughout the brain coupled with declines in nucleus basalis. *J Neurosci* **15**, 6213-6221.
[18] Counts SE, Nadeem M, Wuu J, Ginsberg SD, Saragovi HU, Mufson EJ (2004) Reduction of cortical TrkA but not p75(NTR) protein in early-stage Alzheimer's disease. *Ann Neurol* **56**, 520-531.
[19] Mufson EJ, Lavine N, Jaffar S, Kordower JH, Quirion R, Saragovi HU (1997) Reduction in p140-TrkA receptor protein within the nucleus basalis and cortex in Alzheimer's disease. *Exp Neurol* **146**, 91-103.
[20] Mufson EJ, Ma SY, Cochran EJ, Bennett DA, Beckett LA, Jaffar S, Saragovi HU, Kordower JH (2000) Loss of nucleus basalis neurons containing trkA immunoreactivity in individuals with mild cognitive impairment and early Alzheimer's disease. *J Comp Neurol* **427**, 19-30.
[21] Cooper JD, Salehi A, Delcroix JD, Howe CL, Belichenko PV, Chua-Couzens J, Kilbridge JF, Carlson EJ, Epstein CJ, Mobley WC (2001) Failed retrograde transport of NGF in a mouse model of Down's syndrome: reversal of cholinergic neurodegenerative phenotypes following NGF infusion. *Proc Natl Acad Sci U S A* **98**, 10439-10444.
[22] Capsoni S, Ugolini G, Comparini A, Ruberti F, Berardi N, Cattaneo A (2000) Alzheimer-like neurodegeneration in aged antinerve growth factor transgenic mice. *Proc Natl Acad Sci U S A* **97**, 6826-6831.
[23] Capsoni S, Giannotta S, Cattaneo A (2002) Early events in Alzheimer-like neurodegeneration in anti-nerve growth factor transgenic mice. *Brain Aging*, 24-43.
[24] Capsoni S, Giannotta S, Cattaneo A (2002) Beta-amyloid plaques in a model for sporadic Alzheimer's disease based on transgenic anti-nerve growth factor antibodies. *Mol Cell Neurosci* **21**, 15-28.
[25] De Rosa R, Garcia AA, Braschi C, Capsoni S, Maffei L, Berardi N, Cattaneo A (2005) Intranasal administration of nerve growth factor (NGF) rescues recognition memory deficits in AD11 anti-NGF transgenic mice. *Proc Natl Acad Sci U S A* **102**, 3811-3816.
[26] Origlia N, Capsoni S, Domenici L, Cattaneo A (2006) Time window in cholinomimetic ability to rescue long-term potentiation in neurodegenerating anti-nerve growth factor mice. *J Alzheimers Dis* **9**, 59-68.
[27] Sola E, Capsoni S, Rosato-Siri M, Cattaneo A, Cherubini E (2006) Failure of nicotine-dependent enhancement of synaptic efficacy at Schaffer-collateral CA1 synapses of AD11 anti-nerve growth factor transgenic mice. *Eur J Neurosci* **24**, 1252-1264.
[28] Matrone C, Di Luzio A, Meli G, D'Aguanno S, Severini C, Ciotti MT, Cattaneo A, Calissano P (2008) Activation of the Amyloidogenic Route by NGF Deprivation Induces Apoptotic Death in PC12 Cells. *J Alzheimers Dis* **13**, 81-96.
[29] Cattaneo A, Capsoni S, Paoletti F (2008) Towards non invasive Nerve Growth Factor therapies for Alzheimer's disease. *Journal of Alzheimer's disease* **in press**.
[30] Tuszynski MH (2007) Nerve growth factor gene therapy in Alzheimer disease. *Alzheimer Dis Assoc Disord* **21**, 179-189.
[31] Thorne RG, Frey WH, 2nd (2001) Delivery of neurotrophic factors to the central nervous system: pharmacokinetic considerations. *Clin Pharmacokinet* **40**, 907-946.
[32] Phelps CH, Gage FH, Growdon JH, Hefti F, Harbaugh R, Johnston MV, Khachaturian ZS, Mobley WC, Price DL, Raskind M, et al. (1989) Potential use of nerve growth factor to treat Alzheimer's disease. *Neurobiol Aging* **10**, 205-207.
[33] Koliatsos VE, Nauta HJ, Clatterbuck RE, Holtzman DM, Mobley WC, Price DL (1990) Mouse nerve growth factor prevents degeneration of axotomized basal forebrain cholinergic neurons in the monkey. *J Neurosci* **10**, 3801-3813.
[34] Tuszynski MH, Sang H, Yoshida K, Gage FH (1991) Recombinant human nerve growth factor infusions prevent cholinergic neuronal degeneration in the adult primate brain. *Ann Neurol* **30**, 625-636.
[35] Smith DE, Roberts J, Gage FH, Tuszynski MH (1999) Age-associated neuronal atrophy occurs in the primate brain and is reversible by growth factor gene therapy. *Proc Natl Acad Sci U S A* **96**, 10893-10898.
[36] Eriksdotter Jonhagen M, Nordberg A, Amberla K, Backman L, Ebendal T, Meyerson B, Olson L, Seiger, Shigeta M, Theodorsson E, Viitanen M, Winblad B, Wahlund LO (1998) Intracerebroventricular infusion of nerve growth factor in three patients with Alzheimer's disease. *Dement Geriatr Cogn Disord* **9**, 246-257.

[37] Tuszynski MH, Thal L, Pay M, Salmon DP, U HS, Bakay R, Patel P, Blesch A, Vahlsing HL, Ho G, Tong G, Potkin SG, Fallon J, Hansen L, Mufson EJ, Kordower JH, Gall C, Conner J (2005) A phase 1 clinical trial of nerve growth factor gene therapy for Alzheimer disease. *Nat Med* **11**, 551-555.

[38] Blesch A, Conner J, Pfeifer A, Gasmi M, Ramirez A, Britton W, Alfa R, Verma I, Tuszynski MH (2005) Regulated lentiviral NGF gene transfer controls rescue of medial septal cholinergic neurons. *Mol Ther* **11**, 916-925.

[39] Chen XQ, Fawcett JR, Rahman YE, Ala TA, Frey WH (1998) Delivery of Nerve Growth Factor to the Brain via the Olfactory Pathway. *J Alzheimers Dis* **1**, 35-44.

[40] Frey WH, Liu J, Chen XQ, Thorne RG, Fawcett JR, Ala TA, Rahman YE (1997) Delivery of $^{125}$I-NGF to the brain via the olfactory route. *Drug Delivery* **4**, 87-92.

[41] Capsoni S, Giannotta S, Cattaneo A (2002) Nerve growth factor and galantamine ameliorate early signs of neurodegeneration in anti-nerve growth factor mice. *Proc Natl Acad Sci U S A* **99**, 12432-12437.

[42] Di Fausto V, Fiore M, Tirassa P, Lambiase A, Aloe L (2007) Eye drop NGF administration promotes the recovery of chemically injured cholinergic neurons of adult mouse forebrain. *Eur J Neurosci* **26**, 2473-2480.

[43] Lambiase A, Pagani L, Di Fausto V, Sposato V, Coassin M, Bonini S, Aloe L (2007) Nerve growth factor eye drop administrated on the ocular surface of rodents affects the nucleus basalis and septum: biochemical and structural evidence. *Brain Res* **1127**, 45-51.

[44] Ruberti F, Capsoni S, Comparini A, Di Daniel E, Franzot J, Gonfloni S, Rossi G, Berardi N, Cattaneo A (2000) Phenotypic knockout of nerve growth factor in adult transgenic mice reveals severe deficits in basal forebrain cholinergic neurons, cell death in the spleen, and skeletal muscle dystrophy. *J Neurosci* **20**, 2589-2601.

[45] Kliemannel M, Rattenholl A, Golbik R, Balbach J, Lilie H, Rudolph R, Schwarz E (2004) The mature part of proNGF induces the structure of its pro-peptide. *FEBS Lett* **566**, 207-212.

[46] Covaceuszach S, Capsoni S, Ugolini G, Spirito F, Vignone D, Cattaneo A (2008) Development of a non-invasive NGF-based therapy for Alzheimer's disease. *Current Alzheimer Research* In press.

[47] Lambiase A, Tirassa P, Micera A, Aloe L, Bonini S (2005) Pharmacokinetics of conjunctivally applied nerve growth factor in the retina and optic nerve of adult rats. *Invest Ophthalmol Vis Sci* **46**, 3800-3806.

[48] He XL, Garcia KC (2004) Structure of nerve growth factor complexed with the shared neurotrophin receptor p75. *Science* **304**, 870-875.

[49] Wehrman T, He X, Raab B, Dukipatti A, Blau H, Garcia KC (2007) Structural and mechanistic insights into nerve growth factor interactions with the TrkA and p75 receptors. *Neuron* **53**, 25-38.

[50] Wiesmann C, Ultsch MH, Bass SH, de Vos AM (1999) Crystal structure of nerve growth factor in complex with the ligand-binding domain of the TrkA receptor. *Nature* **401**, 184-188.

[51] Riccio A, Pierchala BA, Ciarallo CL, Ginty DD (1997) An NGF-TrkA-mediated retrograde signal to transcription factor CREB in sympathetic neurons. *Science* **277**, 1097-1100.

[52] Olson L, Nordberg A, von Holst H, Backman L, Ebendal T, Alafuzoff I, Amberla K, Hartvig P, Herlitz A, Lilja A, et al. (1992) Nerve growth factor affects 11C-nicotine binding, blood flow, EEG, and verbal episodic memory in an Alzheimer patient (case report). *J Neural Transm Park Dis Dement Sect* **4**, 79-95.

[53] Bowles WR, Sabino M, Harding-Rose C, Hargreaves KM (2004) Nerve growth factor treatment enhances release of immunoreactive calcitonin gene-related peptide but not substance P from spinal dorsal horn slices in rats. *Neurosci Lett* **363**, 239-242.

[54] Bowles WR, Sabino M, Harding-Rose C, Hargreaves KM (2006) Chronic nerve growth factor administration increases the peripheral exocytotic activity of capsaicin-sensitive cutaneous neurons. *Neurosci Lett* **403**, 305-308.

[55] Kordower JH, Mufson EJ, Granholm AC, Hoffer B, Friden PM (1993) Delivery of trophic factors to the primate brain. *Exp Neurol* **124**, 21-30.

[56] Aloe L, Alleva E, Fiore M (2002) Stress and nerve growth factor: findings in animal models and humans. *Pharmacol Biochem Behav* **73**, 159-166.

[57] Bimonte-Nelson HA, Singleton RS, Nelson ME, Eckman CB, Barber J, Scott TY, Granholm AC (2003) Testosterone, but not nonaromatizable dihydrotestosterone, improves working memory and alters nerve growth factor levels in aged male rats. *Exp Neurol* **181**, 301-312.

[58] Sohrabji F, Miranda RC, Toran-Allerand CD (1994) Estrogen differentially regulates estrogen and nerve growth factor receptor mRNAs in adult sensory neurons. *J Neurosci* **14**, 459-471.

[59] Reger MA, Watson GS, Green PS, Wilkinson CW, Baker LD, Cholerton B, Fishel MA, Plymate SR, Breitner JC, DeGroodt W, Mehta P, Craft S (2008) Intranasal insulin improves cognition and modulates beta-amyloid in early AD. *Neurology* **70**, 440-448.

[60] Gozes I, Morimoto BH, Tiong J, Fox A, Sutherland K, Dangoor D, Holser-Cochav M, Vered K, Newton P, Aisen PS, Matsuoka Y, van Dyck CH, Thal L (2005) NAP: research and development of a peptide derived from activity-dependent neuroprotective protein (ADNP). *CNS Drug Rev* **11**, 353-368.

[61] Gaykema RP, Luiten PG, Nyakas C, Traber J (1990) Cortical projection patterns of the medial septum-diagonal band complex. *J Comp Neurol* **293**, 103-124.

[62] Thorne RG, Hanson LR, Ross TM, Tung D, Frey WH, 2nd (2008) Delivery of interferon-beta to the monkey nervous system following intranasal administration. *Neuroscience* **152**, 785-797.

[63] Thorne RG, Pronk GJ, Padmanabhan V, Frey WH, 2nd (2004) Delivery of insulin-like growth factor-I to the rat brain and spinal cord along olfactory and trigeminal pathways following intranasal administration. *Neuroscience* **127**, 481-496.

[64] Koevary SB (2003) Pharmacokinetics of topical ocular drug delivery: potential uses for the treatment of diseases of the posterior segment and beyond. *Curr Drug Metab* **4**, 213-222.

[65] Capsoni S, Giannotta S, Stebel M, Garcia AA, De Rosa R, Villetti G, Imbimbo BP, Pietra C, Cattaneo A (2004) Ganstigmine and donepezil improve neurodegeneration in AD11 antinerve growth factor transgenic mice. *Am J Alzheimers Dis Other Demen* **19**, 153-160.

[66] Masliah E, Sisk A, Mallory M, Mucke L, Schenk D, Games D (1996) Comparison of neurodegenerative pathology in transgenic mice overexpressing V717F beta-amyloid precursor protein and Alzheimer's disease. *J Neurosci* **16**, 5795-5811.

[67] Dodart JC, Bales KR, Gannon KS, Greene SJ, DeMattos RB, Mathis C, DeLong CA, Wu S, Wu X, Holtzman DM, Paul SM (2002) Immunization reverses memory deficits without reducing brain Abeta burden in Alzheimer's disease model. *Nat Neurosci* **5**, 452-457.

[68] Nicol GD, Vasko MR (2007) Unraveling the story of NGF-mediated sensitization of nociceptive sensory neurons: ON or OFF the Trks? *Mol Interv* **7**, 26-41.

[69] Petty BG, Cornblath DR, Adornato BT, Chaudhry V, Flexner C, Wachsman M, Sinicropi D, Burton LE, Peroutka SJ (1994) The effect of systemically administered recombinant human nerve growth factor in healthy human subjects. *Ann Neurol* **36**, 244-246.

[70] Covaceuszach S, Capsoni S, Marinelli S, Pavone F, Ceci M, Ugolini G, Vignone D, Amato G, Paoletti F, Lamba D, Cattaneo A (2010) In vitro receptor binding properties of a "painless" NGF mutein, linked to hereditary sensory autonomic neuropathy type V. *Biochem Biophys Res Commun* **391**, 824-829.

[71] Einarsdottir E, Carlsson A, Minde J, Toolanen G, Svensson O, Solders G, Holmgren G, Holmberg D, Holmberg M (2004) A mutation in the nerve growth factor beta gene (NGFB) causes loss of pain perception. *Hum Mol Genet* **13**, 799-805.

# Cholinomimetic Actions of Memantine on Learning and Hippocampal Plasticity

Benjamin Drever[a], William Anderson[a], Helena Johnson[a], Matthew O'Callaghan[a], Sangwan Seo[b], Deog-Young Choi[b], Gernot Riedel[a] and Bettina Platt[a,1]
[a]School of Medical Sciences University of Aberdeen, Aberdeen, Scotland, UK
[b]LG Life Sciences Ltd., Taejon, South Korea

**Abstract.** The non-competitive NMDA receptor antagonist memantine, currently prescribed for the treatment of Alzheimer's disease, is assumed to prevent the excitotoxicity implicated in neurodegenerative processes. Here, we investigated the actions of memantine on hippocampal function and signalling. In behavioural experiments using the water maze, we observed that memantine (at 2 mg/kg) reversed scopolamine-induced learning deficits in mice. When acutely applied to mouse hippocampal slices, memantine caused a significant upward shift in the population spike input-output relationship at 10 and 100 μM, and a corresponding downward shift in latency, indicative of overall enhanced synaptic transmission. This action was blocked by the muscarinic antagonist scopolamine (10 μM) but not by the NMDA antagonist MK-801 (10 μM) or the GABA antagonist bicuculline (20 μM). Further, memantine occluded potentiation induced by 50 nM carbachol (CCh), while enhancing inhibitory actions of CCh at 1 μM, suggesting additive actions. As anticipated for an NMDA antagonist, 100 μM (but not 10 μM) memantine also inhibited tetanus-induced long-term potentiation (LTP), and NMDA-induced $Ca^{2+}$ signals were blocked in cultured hippocampal neurones at 10 μM (by 88%). Overall, our data suggest actions of memantine beyond NMDA receptor antagonism, including stimulating effects on cholinergic signalling via muscarinic receptors. These interactions with the cholinergic system are likely to contribute to memantine's therapeutic potential.

Keywords. Alzheimer's disease, NMDA, neuroprotection, learning, synaptic transmission, LTP, acetylcholine, muscarinic

## INTRODUCTION

Glutamate receptors, and N-methyl-D-aspartate (NMDA) receptors in particular, have been implicated in numerous physiological processes including synaptic plasticity, learning and memory [1-3]. In addition to their physiological role, alterations in glutamatergic signalling in disease states is well documented [4,5], with excessive NMDA receptor activation causing excitotoxicity, as seen for instance in stroke and neurodegeneration [6-8]. Modulating tonic NMDA receptor activation has therefore been an attractive therapeutic target in disorders such as Alzheimer's disease (AD), albeit with little success, since NMDA receptor antagonists have been found to cause intolerable side effects [9,10]. However, one such antagonist, memantine, has recently been approved for treatment of moderate-to-severe AD (for trial data, see [11-13]).

Cellular studies indicate that memantine is a moderate affinity, uncompetitive and reversible NMDA receptor antagonist [14-17], assumed to leave physiological NMDA activation intact. Data concerning memantine's action on synaptic plasticity and learning and memory are only partly consistent with actions upon NMDA receptors: In hippocampal slices, memantine has been reported to inhibit tetanus-induced long-term potentiation (LTP) [18,19], however, it has also been shown to restore deficits in LTP [20,21,22]. Basic synaptic transmission was reported to be unaffected by memantine in some studies [19,20,23], while others found increased transmission [24] and enhanced LTP [23]. In animal models, other NMDA receptor antagonists have

---

[1]Corresponding Author: Prof. Bettina Platt, School of Medical Sciences, University of Aberdeen, Foresterhill, Aberdeen AB25 2ZD, Scotland, UK. Tel.: +44 1224 437402; Fax: +44 1224 437465; Email: b.platt@abdn.ac.uk

shown only negative effects on learning and memory [3], while memantine proved able to reverse memory deficits in transgenic AD models and enhance cognition in healthy animals [25-33]. Additionally, evidence for protection against amyloid-induced degeneration has been put forward [34], possibly due to the prevention of NMDA receptor mediated pathophysiology as shown in neurodegenerative and excitotoxicity models, while at the same time being devoid of the prominent side-effects characteristic of other NMDA blockers [9,10,18]. However, memory-enhancing effects have been questioned recently by the finding that memantine alone disrupts learning, memory and locomotor behaviour in rats and humans [35-37]. Such negative symptoms and toxicity were amplified by co-treatment with donepezil, a cholinesterase inhibitor also used in the treatment of AD [35,38]. These results contrast with previous work that revealed no interaction between memantine and clinically approved cholinesterase inhibitors [39].

Interestingly, a microdialysis study showed that both memantine and donepezil differentially affect a range of transmitter systems (e.g. dopamine, norepinephrin, serotonin and ACh) in various brain areas [40], with obvious, complex implications on learning and memory mechanisms. Moreover, memantine has been reported to directly affect nicotinic receptors (see [6] for review), and may antagonise choline evoked currents in hippocampal cultures via inhibition of α7 or α4 nACh receptors [41,42].

Thus, since memantine may have much broader effects than previously acknowledged, the present study sought to explore interactions between memantine and the cholinergic system *in vivo* and *in vitro* in greater detail since this is of relevance for therapeutic considerations. We present results that challenge the assumption that memantine's partially beneficial action in AD patients is due to its action as an NMDA receptor antagonist.

## MATERIALS AND METHODS

### Behavioural studies

#### Animals

Forty female, eight-week old NMRI mice purchased from a commercial dealer (Harlan, UK) and delivered 2–3 weeks prior to testing were used in this experiment. They were housed in groups of ten with free access to food and water. Animals were kept in a controlled holding environment with a 12 hour day-night cycle (lights on at 7 am) and fed a commercial maintenance diet. All experiments were performed in accordance with international standards on animal welfare and Home Office (UK) regulations; great care was taken to minimise any pain and discomfort.

#### Groups

Memantine hydrochloride and scopolamine hydrobromide were administered through intraperitoneal (i.p.) injection (0.1 ml per 10 g/mouse) 30 minutes prior to water maze training and testing. Animals were randomly assigned to 5 drug groups (n=8 each) and the following drug doses/combinations were given: (1) saline; (2) 0.5 mg/kg scopolamine hydrobromide; (3) 10 mg/kg memantine; (4) 10 mg/kg memantine+0.5 mg/kg scopolamine; (5) 2 mg/kg memantine+0.5 mg/kg scopolamine.

The dose of scopolamine was established previously and found to interfere with the behaviour in the memory task, yet being devoid of side effects on the visual system [43,44]. A range of doses of memantine were tested in connection with scopolamine, known to generate therapeutically relevant brain levels and effective in other models of memory impairments [16,29].

#### Apparatus

The apparatus used was the same as in previous studies in our laboratory [43,44]. The water maze was a circular white Perspex pool (150 cm diameter; 50 cm height). It was filled with water (21±1°C) and a clear Perspex, circular, rigid escape platform (10 cm diameter, 35 cm height) was submerged about 1 cm below the water surface. The pool was placed in a laboratory with plenty extra-maze cues. Swim paths of the animals in the pool were monitored with an overhead camera that was connected to an image analysing system (Ethovision, Noldus, Holland) and a video recorder for back-up.

#### Training and testing

Reference memory acquisition training was conducted over four days with six trials per day, and an inter-trial interval (ITI) of 10 min [45]. White curtains were drawn around the circumference of the pool and cue cards were suspended from the ceiling as spatial clues. These cue cards (80 × 80 cm) were placed over the pool wall at the four quadrants to generate a three-dimensional arrangement. On black background, the cards contained a white circle (NE), horizontal parallel lines (NW), a white triangle (SE) and four small squares (two black and two white – SW). The platform was positioned within the centre of one of the four pool quadrants and remained at a fixed location throughout the training phase. All possible platform locations were used and counterbalanced for each group. Both cue cards and platform location were rotated by 90° clockwise between experimental days in order to reduce the influence of any auditory or olfactory cues that are imminent in the room, and to enhance the importance of visuospatial mapping for identification of the platform location. This ensured that the platform location maintained its relation to the cue cards. Four cardinal points (N, S, E, W) were randomly used as release sites and animals were introduced into the pool facing the walls. A maximum of 60s was allotted to find the platform. If mice did not find the platform, they were placed onto it by an experimenter and remained on it for 30s. Then, the animal was removed, placed individually into a new cage and placed under a warm light. Animals were run in squads of n=5 mice with 20 animals per replication.

*Data acquisition and analysis*

The swim path of each animal was recorded in real time and converted into XY coordinates. It was replayed on the PC and the following parameters were extracted: total path length required to find and climb onto the platform (this is the most accurate parameter of spatial knowledge as it is independent of swim speed or strategy employed); swim speed as a measure of overall activity and motor coordination; thigmotaxis as a measure of anxiety calculated as percentage of time spent within a 10 cm radius of the pool wall. Data of individual swims were pooled for each day and are expressed as group means± SEM. Acquisition data were assessed by factorial analysis of variance with repeated measures (ANOVA, drug treatment as between subject and days as within-subject factor) followed by appropriate planned comparisons. Significance level was set to a probability for null hypothesis of <5%.

*Electrophysiological studies*

*Slice preparation and recordings*

For hippocampal slices, 5-week old (±1 week) C57/BL6 mice were used. Animals were maintained in acclimatised rooms (temperature: 22 °C; 12 h day/night cycle) with food and water ad libitum. All experiments were executed in compliance with local and UK Home Office Animals (Scientific Procedures) regulations.

Slice preparation and recording conditions were conducted as described previously (e.g. [46]). Briefly, animals were terminally anaesthetised with halothane and decapitated. The brains were quickly removed into ice-cold artificial cerebrospinal fluid (aCSF; composition in mM): 129.5 NaCl, 1.5 KCl, 1.3 $MgSO_4$, 2.5 $CaCl_2$, 1.5 $KH_2PO_4$, 25 $NaHCO_3$ and 10 glucose (pH 7.4, continuously gassed with 95% $O_2$/5% $CO_2$) and the hippocampus dissected. 400 μm hippocampal slices were prepared with a McIllwain tissue chopper and stored in oxygenated aCSF at 32°C for at least 1 hour before experiments commenced. Slices were placed on the nylon mesh of a recording chamber and perfused at a rate of 6–8 ml $min^{-1}$ with pre-warmed, gassed aCSF. Recordings of CA1 field population spikes were chosen as the behaviourally relevant output signal, and recorded via an aCSF-filled borosilicate glass electrode (3–7 MΩ) positioned in the pyramidal cell body layer after stimulation of the Schaffer collateral/ commissural fibres by a coated stainless steel stimulation electrode (WPI, UK, 0.5 MΩ). By increasing the stimulus intensity in a stepwise manner until saturation was reached, input/output (IO) curves of basic synaptic transmission were generated. Slices with unsuitable signals (peak amplitudes <1 mV) were excluded.

For all groups, a stimulus intensity of 50–60% of saturation was applied to run subsequent LTP experiments. A baseline response was recorded for at least 10 minutes (responses recorded every 30 seconds, response variability <10%).

To evoke tetanus-induced LTP, a theta burst tetanus was applied (5 Hz, 75 bursts of 4 stimuli (100 Hz), inter-burst interval of 200 ms for 15 seconds). Recording continued for 60 minutes post-tetanus, with LTP controls run at regular intervals to ascertain stable recording conditions.

To evoke carbachol (CCh)-induced LTP, 50 nM CCh was perfused onto the slice for 25 minutes [47]. Recording continued for 35 minutes after CCh perfusion ceased. Application of 1 μM CCh for 25 minutes was used to evoke a transient depression of POP spike amplitude.

*Data analysis*

Data acquisition and storage on a PC was conducted with the PWIN software package (Leibniz

Institute for Neurobiology, Magdeburg, Germany). IO curves of population spike amplitude and latency (artefact to peak) were generated and analysed using GraphPad Prism software (Version 4.01; GraphPad Software, San Diego, CA, USA). IO curves were compared between groups via a two-way ANOVA (treatment x stimulus).

LTP time courses were calculated relative to baseline values (=100%), with data illustrated as means+standard error of means (SEM). A two-way repeated measure ANOVA was applied to compare post-tetanus values between groups (treatment × time). Significance was rated as P<0.05=* (significant), P<0.01=** (highly significant) and P<0.001 =*** (extremely significant), in terms of effect of treatment (*) and interaction (#).

*Imaging*

*Hippocampal cultures*

Hippocampal cultures were prepared as described previously (e.g. [48,49]). Briefly, hippocampal tissue was dissected from 3-day old neonatal C57/BL6 mice and placed in ice-cold HEPES buffered solution (HBS; composition: 130 mM NaCl, 5.4 mM KCl, 1.8 mM $CaCl_2$, 1 mM $MgCl_2$, 10 mM HEPES and 25 mM D-glucose (pH 7.4). Tissue was chopped and transferred to protease solution (type XIV, at 1 mg/ml HBS; Sigma-Aldrich, Poole) for enzymatic dissociation (30 minutes, room temperature). Afterwards, tissue was twice washed in HBS and repeatedly triturated before double centrifugation. The resulting cell pellet was resuspended in 1 ml neurobasal medium supplemented with 1% foetal bovine serum, 20 μM L-glutamine (Sigma-Aldrich) and 2% B27 (Invitrogen). This solution was then plated onto poly-L-lysine coated (0.02 mg/ml, Sigma-Aldrich) coverslips within culture dishes, and incubated for 60 mins (37 °C in a humidified atmosphere, 5% $CO_2$). Subsequently, further supplemented neurobasal medium was added (final volume: 2 mls per dish). At 2 days *in vitro* (DIV), the medium was changed with the addition of 25 μM L-glutamate. Experiments were conducted using hippocampal cultures at 3–10 DIV.

*Fura-2 AM $Ca^{2+}$ imaging*

As in previous studies (e.g. [46,49]), cultures were incubated for 60 mins in the dark at room temperature with 10 μM of the cell-permeable fluorescence $Ca^{2+}$ indicator, Fura-2-AM (Cambridge Bioscience, Cambridge, UK), in HBS. Tetrodotoxin (TTX, 0.5 μM; Alomone Labs, Jerusalem, Israel), a sodium channel blocker, was added to all perfusion media to inhibit spontaneous cell firing and subsequent transmitter release. Cultures were perfused at a rate of approx. 2 ml/min. Under the microscope (Olympus, 40x objective), a suitable field of cells was selected and captured in greyscale using Openlab 5 software (Improvision, UK). Ratiometric imaging was conducted using alternating wavelengths of 340 and 380 nm, from a Xenon lamp regulated by a monochromator (Spectromaster I, Perkin-Elmer) and an emission filter (wavelength 510 nm). Background levels of fluorescence were calculated and subtracted on-line. The image acquisition rate was set to 5 seconds; ratio values were plotted against time for multiple regions of interest (ROIs, neurones and glia determined by morphological analysis and by fast neuronal responses to application of NMDA). Low $Mg^{2+}$ HBS (as HBS above but with 0.1 mM $MgCl_2$) was the standard perfusion solution used.

*Imaging data analysis*

Fluorescence ratio units were found to show highest reproducibility between experiments (as assessed for NMDA responses, variability <5%). Graphical illustration and statistical analysis of fluorescence values were conducted as % change in fluorescence relative to baseline fluorescence (% dF/F, see also [48,49]). Statistical analyses were performed using the GraphPad Prism statistics package with a Kruskal-Wallis test due to the non-parametric distribution of data, followed by a Dunn's multiple comparison post-hoc test. Percentage responder rates were calculated as responders from total neurones selected as regions of interest (ROIs).

*Compounds*

NaCl, KCl, $MgSO_4$, $CaCl_2$, $KH_2PO_4$, $NaHCO_3$, $MgCl_2$, HEPES and glucose were purchased from VWR (Poole, Dorset, UK); NMDA, CCh, memantine (as memantine hydrochloride) and MK-801 were purchased from Tocris Cookson (Tocris Cookson Ltd., Avonmouth, UK), while bicuculline (as bicuculline methobromide), and scopolamine (as scopolamine hydrobromide) were acquired from Sigma-Aldrich Inc. (Poole, Dorset, UK).

Stock solutions of the drugs were prepared in distilled water and stored at −20°C. All drugs were perfused by adding the required volume of stock solution to gassed aCSF or HBS. All drugs were perfused for the time indicated by the application bars in the figures.

## RESULTS

*Scopolamine impairs spatial learning in the water maze: reversal by memantine*

Acquisition learning of control and drug-treated mice is summarized in Fig. 1. While there was a progressive reduction in path length (Fig. 1A) in both saline and memantine-treated animals with no significant difference between the two groups, scopolamine injections significantly impaired learning over 4 days and this impairment was reversed by memantine. In an ANOVA with factorial design, we obtained an overall reliable difference between treatments ($F_{4,105}=5.8$; $P=0.0011$), with the scopolamine group significantly impaired relative to saline ($F_{1,42}=22.4$; $P=0.0003$), but all other treatment groups did not differ from controls (all $F$'s<2.7; $P$'s>0.12).

Consistent with this reversal, both memantine doses did significantly differ from the scopolamine group ($F$'s>6; $P$'s<0.02). Scopolamine treatment also caused an increase in thigmotaxis relative to controls (Fig. 1C) and this was also reversed by memantine. The overall group difference ($F_{4,105}=4.5$; $P=0.0046$) was reliant on the high thigmotaxis in the scopolamine group; taking this group out of the statistical analysis did not yield reliability ($F<1.2$). This also confirms that memantine alone did not alter levels of anxiety. Interestingly, swim speed (Fig. 1B) was not affected by scopolamine or memantine when administered alone, but when given in conjunction, the scopolamine + memantine 2 mg/kg groups swam considerably faster than all other groups ($F$'s 3.8; $P$'s<0.05). This motor effect may be responsible for the initial learning deficit in this group.

Overall, these data provide compelling evidence that memantine can, apart from its multiple and well established benefits in models of neurotoxicity and neurodegeneration, also reverse learning deficits induced by muscarinic receptor blockade. It raises a question about the pharmacological basis of this effect, since NMDA antagonism is unlikely to explain such an observation.

*Effects of memantine on basic synaptic transmission*

Initial studies with memantine (10–100 µM) in hippocampal slice experiments indicated that perfusion of the drug resulted in enhanced baseline field population spike amplitudes in the CA1 region, an effect that stabilised after ~10 minutes (Fig. 2). Interestingly, at the higher concentration of 1 mM, memantine caused a complete and irreversible inhibition of synaptic responses (data not shown).

Corresponding IO relationships established before and after perfusion of the drug confirmed that memantine at 10 µM (n=12) and 100 µM (n=31) caused a highly significant ($P$'s<0.0001) upward shift in IO curves, paralleled by a shortening of latencies (10 µM: $P<0.01$, 100 µM: $P<0.0001$). This result is indicative of overall enhanced basic synaptic transmission (Fig. 2A, B, D and E), which was found to cause a slow-onset, transient potentiation of responses to ~120% of baseline (Fig. 2C) for the application of 10 µM memantine for 25 minutes (n=15) that gradually decreased to a level just above baseline during wash (~106% at t=70, non-significant from baseline after t=54 mins).

For 100 µM memantine (Fig. 2F, n=10), we observed a fast-onset increase in amplitude to approximately 120% (saturated after 5 min), and values started to decline towards the end of memantine perfusion, suggesting desensitisation or a biphasic nature of drug action. Upon washout, amplitude values declined and stabilised 5% above baseline after 10 min of wash.

*Memantine-induced baseline effects: comparison with CCh*

The time-course and action of memantine was reminiscent of metabotropic effects described for muscarinic agonists such as CCh: A range of previous investigations indicated that low concentrations of CCh can induce an NMDA-independent, long lasting potentiation of synaptic responses in hippocampal slices, while higher concentrations lead to synaptic depression (e.g. [47,50]). Since memantine has been suggested to potentially affect the cholinergic system [34,41,51,52], as also suggested by our *in*

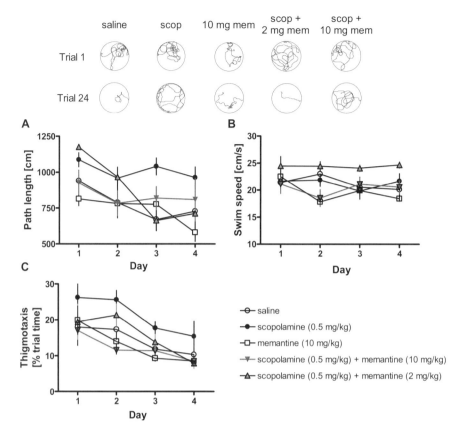

Fig. 1. Memantine reversed a learning and memory deficit induced by blockade of muscarinic receptors with scopolamine. A-C: Acquisition of a spatial reference memory paradigm in the water maze over 4 days (6 trials per day). Scopolamine treatment impaired spatial learning indicated by prolonged swimming (A) and enhanced anxiety as suggested by high levels of thigmotaxis (C). It had no effect on swim speed (B). Impairments were reversed by memantine (for details, see text). Means are shown +/−SEM. scop =scopolamine; mem =memantine.

*vivo* data, we investigated possible cholinomimetic actions of memantine. A series of experiments was conducted to I) compare the action of CCh with memantine, and II) to study interactions between these two drugs.

Initial experiments focussed on determining suitable concentrations and application times of CCh (data not shown). At concentrations of 2, 1, 0.5, 0.25 and 0.1 μM, CCh caused a transient depression of population spike amplitudes. At 50nM, CCh applied for 25 mins induced a gradually developing, long lasting increase in population spike amplitude to approximately 150% of baseline (n=10), and resulted in reliable and stable potentiation (Fig. 3A). In comparison, perfusion of 1 μM CCh to hippocampal slices led to a transient depression of population spikes in the CA1 region to approximately 75% of baseline levels (Fig. 3B, n=10).

Subsequently, the actions of memantine on the excitatory and inhibitory effects of CCh were tested. Perfusion of 10 μM memantine prevented the induction of CCh mediated LTP (Fig. 3A, n=10). Comparing the CCh time courses in the presence and absence of memantine (post-memantine values corrected for the initial baseline effect) by way of a repeated measure two-way ANOVA (treatment x time) showed a very significant difference ($P<0.01$) and an extremely significant interaction ($P<0.0001$). Further, application of 100 μM memantine (n=11) not only prevented CCh-induced LTP, but led to a transient depression of the spike amplitude in 50 nM CCh, relative to the pre-CCh baseline (Fig. 4A).

For combined applications of 1 μM CCh and memantine (Fig. 3B and 4B) it was found that 10 μM memantine (n=10) had no significant effect on

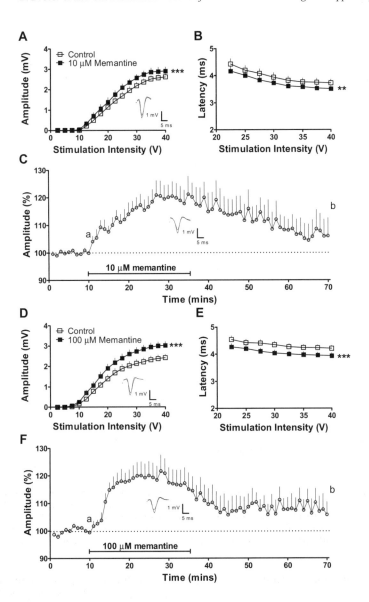

Fig. 2. Effect of memantine on basic synaptic transmission. Input-output relationships of population spike amplitude and latency in the CA1 region are shown before and after a 10-minute perfusion of memantine at 10 μM (A & B, n=12) and 100 μM (D & E, n=31). At both concentrations there was an extremely significant upward shift in the IO curve (both P's <0.001, ***) and corresponding downward shift in the latency curve (10 μM memantine: P<0.01 (**), 100 μM memantine: P<0.001). Time courses of mean population spike amplitude (as % of baseline+SEM) are shown in part C for 10 μM memantine and part F for 100 μM memantine. Superimposed on IO curves are sample traces before and after memantine application at maximum stimulus intensity, while traces prior to (a) and after (b) memantine application are superimposed on time courses. The bar depicts the duration of the memantine application.

the strength of CCh-induced depression (relative to the pre-CCh baseline, P>0.05), but 100 μM memantine (n=10) resulted in a larger transient depression when co-applied with 1 μM CCh (Fig. 4B, n=10). Comparing the amount of inhibition corrected for the initial baseline effect caused by memantine, a very significant effect of treatment on the amplitude was obtained (P<0.01) with no significant interaction (P>0.05).

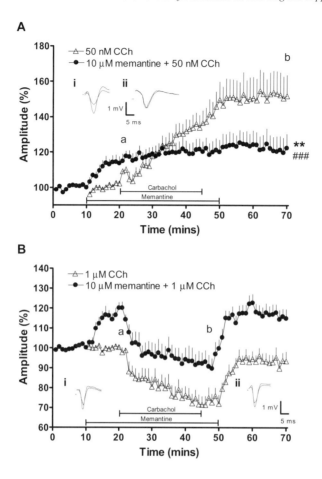

Fig. 3. Effect of memantine on CCh actions in mouse hippocampal slices. Time courses of the population spike amplitude in slices treated with 10 μM memantine in response to 50 nM CCh (A) and 1 μM CCh (B). Memantine prevented LTP induced by 50 nM CCh (treatment: **; $P<0.01$; interaction: ###; $P<0.0001$; n=10) while having no significant effect on the depressive actions of 1 μM CCh (both P's>0.05, n=11). The insets show overlapping sample traces of population spikes from timepoints a and b (as indicated on the graph) with traces i and ii corresponding to 50 nM CCh and 10 μM memantine+50 nM CCh in A and 1 μM CCh and 10 μM memantine+1 μM CCh in B.

*Pharmacology of CCh's vs. memantine's action*

Previous studies have indicated that CCh-induced LTP is an NMDA-independent phenomenon, with post-synaptic muscarinic receptor activation feeding into a point downstream of traditional NMDA-dependent tetanus induced LTP [47,50]. The inhibitory actions of higher CCh concentrations, by contrast, are mediated by the activation of presynaptic in addition to postsynaptic muscarinic receptors, with a supplementary role of the inhibitory system. Thus, the involvement of NMDA, GABA and muscarinic receptors in the effects of CCh and memantine was probed by using the respective antagonists MK-801, bicuculline and scopolamine (Figs. 5–7).

Surprisingly, treatment of slices with MK-801 prevented potentiation of population spikes by 50 nM CCh (Fig. 5A, n=12, $P<0.0001$), in contrast to previous studies where the (somewhat weaker and reversible) NMDA antagonist APV had no discernible effect on CCh induced LTP [47]. When assessing the effect of 1 μM CCh on MK-801 treated slices (n=10, data not shown), it was found that there was no significant difference ($P>0.05$) or interaction ($P>0.05$) on the amplitude time course compared to control slices.

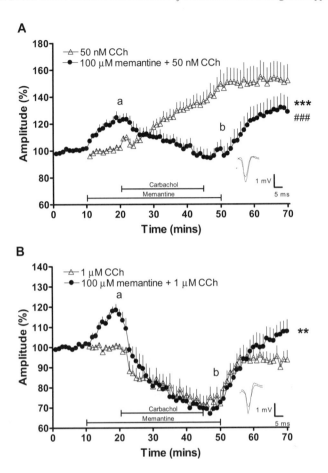

Fig. 4. Effect of memantine on CCh actions in mouse hippocampal slices. Time courses of the population spike amplitude in slices treated with 100 μM memantine in response to 50 nM CCh (A) and 1 μM CCh (B). At 100 μM, memantine also significantly affected CCh LTP (treatment: ***; $P<0.0001$; interaction: ###; $P<0.0001$; n=11) and resulted in a greater depression in response to 1 μM CCh (effect of treatment: **; $P<0.01$; n=10). The insets show overlapping sample traces of population spikes from timepoints a and b (as indicated on the graph).

In comparison, application of 10 μM MK-801 (a concentration that blocked tetanus-induced LTP, data not shown) did not affect the ability of memantine to increase excitability in slices (10 μM: n=10, Fig. 5B, 100 μM: n=12, Fig. 5C, all P's>0.05).

It was next investigated if memantine and/or CCh enhanced basic synaptic transmission through a down-regulation of inhibitory transmission (Fig. 6A–C). For this, memantine was applied to slices treated with the $GABA_A$ receptor antagonist bicuculline. Perfusion of 20 μM bicuculline caused increased population spike amplitudes and secondary spiking, characteristic of disinhibited slices (see traces). This effect reached a stable plateau after ~10 min in the drug, after which stable baseline recordings were conducted, followed by memantine or CCh perfusion in bicuculline. LTP induced by 50 nM CCh was still witnessed in bicuculline treated slices (n=13, Fig. 6A). A comparison of time courses indicated no significant effect of bicuculline

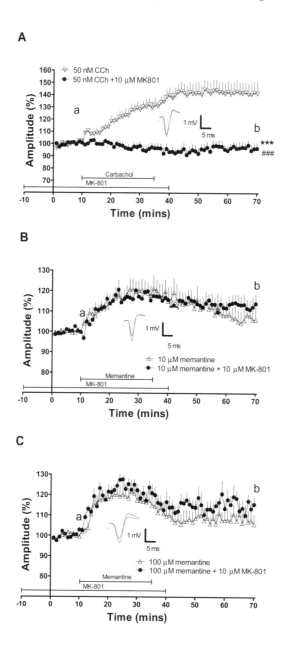

Fig. 5. The effect of MK-801 on 50 nM CCh (A), 10 μM memantine (B) and 100 μM memantine (C) on synaptic transmission in hippocampal slices. Time courses show mean population spike amplitude (% of baseline+SEM). MK-801 completely blocked potentiation caused by 50 nM CCh (interaction: ###; $P<0.0001$, treatment: ***; $P<0.0001$; n=12) and had no effect on the actions of 10 μM or 100 μM memantine (P's>0.05, n=15 and n=10, respectively). Drug application was as indicated by the bars on the time courses. The insets show overlapping traces of population spikes from the timepoints indicated on the graph (a and b).

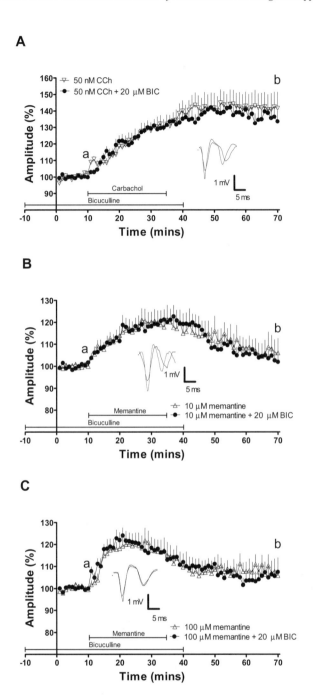

Fig. 6. The effect of bicuculline on 50 nM CCh (A), 10 μM memantine (B) and 100 μM memantine (C) on synaptic transmission in hippocampal slices. Time courses show mean population spike amplitude (% of baseline+SEM). Bicuculline had no significant effect on the actions of 50 nM CCh (n=10) or memantine at either concentration (all P's>0.05, both n=10). Drug application was as indicated by the bars on the time courses. The insets show overlapping traces of population spikes from the timepoints indicated on the graph (a and b).

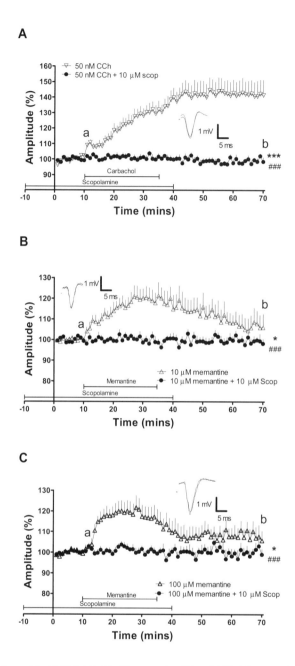

Fig. 7. The effect of scopolamine on 50 nM CCh (A), 10 μM memantine (B) and 100 μM memantine (C) on synaptic transmission in hippocampal slices. Time courses show mean population spike amplitude (% of baseline+SEM). Scopolamine completely inhibited the potentiating effects of 50 nM CCh (effect of treatment (***) and interaction (###); both $P<0.0001$; n=12), 10 μM memantine (effect of treatment (**; $P<0.05$) and interaction (###; $P<0.0001$); n=14) and 100 μM memantine (effect of treatment (**; $P<0.05$) and interaction (###; $P<0.0001$); n=11). Drug application was as indicated by the bars on the time courses. The insets show overlapping traces of population spikes from the timepoints indicated on the graph (a and b).

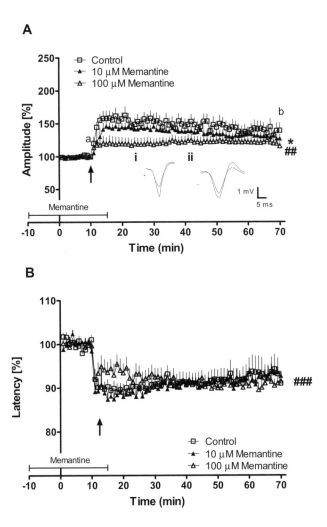

Fig. 8. Effects of memantine on tetanus-induced LTP. Time courses of mean population spike amplitude (A) and latency (B) (as % of baseline+SEM). Inset are overlapping traces of population spikes from the timepoints indicated on the graph (a and b) for (i) 10 μM memantine and (ii) 100 μM memantine-treated slices. The arrow indicates onset of tetanus, the bar depicts the memantine application. No significant difference in LTP between the control and 10 μM memantine group was observed (P>0.05), while 100 μM memantine caused a significantly reduced LTP (effect of treatment on amplitude (*; P<0.05) and interactions for both amplitude (##; P<0.01) and latency (###; P<0.0001).

on the actions of either CCh or memantine (10 μM: n=12, 100 μM: n=10) with all P's>0.05.

To confirm muscarinic receptor involvement in the drug actions of both CCh and memantine, 10 μM scopolamine was pre-applied and found to completely block the potentiating action of 50 nM CCh (Fig. 7A; n=12) and the inhibitory effects of 1 μM CCh (data not shown, both P's<0.0001). Interestingly, memantine's actions at both concentrations (Fig. 7B and C, n=11 and n=10, respectively) were also blocked, with spike amplitudes remaining at baseline levels. Time courses obtained in scopolamine differed significantly from scopolamine-free experiments (P's<0.05). These results strongly suggest that memantine acts as a stimulator of muscarinic receptors to boost basic synaptic transmission.

## NMDA antagonism by memantine

Since the pharmacological profile obtained casts doubt on the actions of memantine, we next confirmed that this compound can indeed act as an NMDA antagonist in our experimental conditions, as already suggested by previous investigations (e.g. [16,17,53]). For this, the influence of memantine on synaptic plasticity, i.e. NMDA-receptor dependent

LTP, was investigated by a 20-minute pre-application (at concentrations of 10 μM and 100 μM). At 10 μM (n=11), memantine had no discernible effect on LTP (all P's>0.05, Fig. 8), but 100 μM (Fig. 5A, n=14) caused a significant reduction in LTP of population spike amplitudes (P<0.05) relative to controls (n=14), with an additional highly significant interaction (P<0.01), indicating a time-dependent component to its action. The interaction was due to a stronger effect at early post-tetanus time points. With respect to latency, 100 μM memantine caused an extremely significant interaction (P<0.0001) compared to controls (Fig. 8B) due to similar time-dependent changes that only occurred in the early post-tetanus time window. Therefore, memantine does show the expected ability to inhibit NMDA receptors in hippocampal slices in the high micromolar range.

Further experiments set out to confirm the cellular pharmacology of memantine's actions on NMDA gated calcium channels, using $Ca^{2+}$ imaging procedures in hippocampal cultures (Fig. 9). In agreement with previous studies [54], we found the transient rise in $[Ca^{2+}]_i$ induced by NMDA (10 μM) to be reduced to 12±2% (n=31) by 10 μM memantine. At 1 μM, memantine had no significant effect on the NMDA response (n=28, Fig. 9B). In comparison, 1 μM MK-801 irreversibly reduced the NMDA response to 36±4% of controls (n=35; P<0.001). At 50 μM, MK-801 completely abolished any neuronal calcium response to 10 μM NMDA (n=41, P<0.001).

Overall, our *in vitro* data indicate that memantine does act as a weak NMDA antagonist, in addition to the observed cholinomimetic activities.

## DISCUSSION

The present study provides evidence that memantine increases cholinergic signalling and excitability in the mouse hippocampus, above and beyond its role as a non-competitive, use-dependent antagonist of NMDA receptors. It has previously been suggested that the clinical utility of memantine in the treatment of neurodegenerative disorders such as AD is related to its ability to block the pathological over-activation of NMDA receptors, while leaving their physiological activation intact. The data presented here reveal two more promising aspects to the drug's use in the treatment of such disorders. Firstly, under our conditions memantine increased basic synaptic transmission, a potentially useful effect in diseases with deficits in learning and memory. Secondly, memantine's actions were based on enhanced cholinergic (muscarinic) signalling, a transmitter system that is significantly impaired in neurodegenerative diseases such as AD. This was first revealed in behavioural experiments, where memantine was able to reverse a scopolamine-induced learning deficit, and confirmed in electrophysiological and pharmacological studies on synaptic transmission and plasticity.

We have previously shown that scopolamine selectively impairs hippocampus-dependent spatial learning in the water maze [43,44]. Since memantine reversed scopolamine-induced deficits, we explored the effect of memantine on basic synaptic transmission and LTP in the CA1 region of mouse hippocampal slices, and its pharmacology. Memantine enhanced excitability in CA1 at 10–100 μM, and reduced theta-burst-induced LTP at 100 μM. These results contrast with some previous studies. A discrepancy with the study of Frankiewicz et al. [19] is their long incubation time of 4 hrs, found to be essential for memantine to affect LTP and assumed to be due to poor penetration of the drug in the slice. Shorter application periods (e.g. 30-minute incubations of memantine) had no discernible effect on LTP (at 10–100 μM). The reason for this discrepancy remains largely unclear, but is likely to involve differences in the respective experimental conditions. For example, drug administration in our experimental set up utilised a fast solution perfusion and exchange, which would accelerate drug access and equilibration. This may also account for the differences in the effective concentration range observed for CCh.

The vast majority of studies have been carried out in conditions mimicking the pathological activation of NMDA receptors believed to cause excitotoxicity. Frankiewicz et al. [21], amongst others, used a low $Mg^{2+}$ aCSF to tonically activate NMDA receptors, which caused an enhancement of basic synaptic transmission and inhibition of LTP induction. Memantine was found to prevent both these effects (see also [20]). Moreover, memantine blocked hypoxia/hypoglycaemia induced suppression of basic synaptic transmission [55] and reversed inhibition of LTP by Aβ oligomers [25]. Together, this highlights

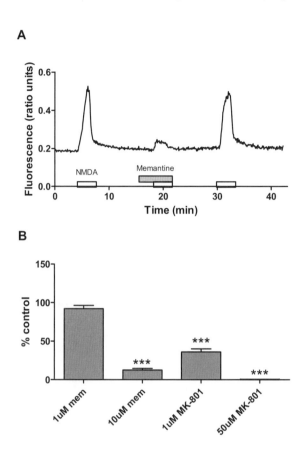

Fig. 9. Effects of memantine on NMDA-induced $Ca^{2+}$ signals. (A) Time course of NMDA (10 μM)-induced $[Ca^{2+}]_i$ responses in the absence and presence of memantine (10 μM) in hippocampal culture (mean response, n=14 ROIs). (B) Comparison between inhibitory actions of memantine and MK-801. The percentage of reduction was calculated relative to the within-experiment NMDA control (***: $P<0.001$).

the protective properties of memantine against NMDA-related excitotoxicity, but also points towards differential effects in pathological vs. physiological conditions. However, it should also be noted that memantine behaves quite differently, both in terms of the NMDA receptor subtypes with which it interacts and the voltage-dependency of its block, in conditions of physiological levels of $Mg^{2+}$ as opposed to low $Mg^{2+}$ [56].

Our results showed that memantine, at a concentration of 10 μM, had no significant effect on LTP delivered via theta-burst stimulation, while the higher concentration of 100 μM significantly reduced potentiation. The present data are consistent with memantine acting as a weak, non-competitive, voltage-dependent blocker of NMDA receptor channels (e.g. [16,17]), as was further confirmed by our imaging experiments. Moreover, this could reflect the inactivity of the drug at NR2A and NR2B containing NMDA receptors, (which are vital in LTP induction) at therapeutic concentrations, with NR2C and NR2D containing receptors thought to be preferentially inhibited [56,57]. Since therapeutic concentrations of memantine are not reaching the high micomolar range, results presented here indicate that at such concentrations memantine is devoid of detrimental effects on LTP, and thus should not affect learning and memory mechanisms. In keeping with this hypothesis, memantine at doses of 20 mg/kg*day had no effect on allocentric learning in the radial arm maze [27,58]. This reveals a positive facet to the drug's use in the treatment of AD, a disorder where cognition is retarded [59], and is in concordance with previous studies [18,19].

Inhibitory actions of memantine on NMDA receptors, on the other hand, cannot explain the excitatory effect on synaptic transmission observed here. Enhanced excitability was also previously observed *in vivo* and *in vitro* by some [23,24,60], but not in other studies [19,20]. Since this action is potentially desirable when attempting to alleviate cognitive deficits, we aimed to characterise the pharmacology and mechanisms behind this stimulatory effect on synaptic transmission. The possibility that memantine increased synaptic transmission via a bi-directional modulation of NMDA receptors or a reduction in inhibitory tone was rejected based on the negative results obtained in MK-801 and bicuculline treated slices. Instead, it was found that treatment of slices with the muscarinic antagonist scopolamine completely abolished the memantine-induced increase in synaptic transmission. Therefore, memantine behaves as a cholinomimetic drug to increase basic synaptic transmission in the CA1 region of mouse hippocampal slices by enhancing cholinergic signalling via muscarinic receptors.

A previous report has shown actions of memantine on phosphoinositide turnover, yet without a direct stimulatory action [53]. Since it was possible to block memantine's effects with scopolamine (and vice versa) it could therefore be suggested that memantine sensitises or enhances endogenous muscarinic receptor signalling. Indeed, in a recent study by Blanchard et al. that investigated the effect of memantine on agonist-induced calcium responses in HEK 293 cells, it was found that the drug decreased basal intracellular levels of calcium, increased the content of intracellular calcium stores and increased intracellular calcium release in response to CCh and subsequent store-operated calcium entry [61]. While there is limited evidence against memantine acting on AChE [39], future studies will have to establish the contribution of facilitated release or interactions with muscarinic receptors. Muscarinic stimulation is consistent with the finding that memantine reverted the cognitive deficits in mouse models of AD, but with somewhat lower efficacy than the AChE inhibitors rivastigmine or donepezil [26,29]. Disturbed glutamate functions including NMDA-receptor abnormality could conceivably account for the effects of memantine in AD mice, but this remains unexplored.

In our scopolamine model, NMDA receptor antagonism is unlikely to account for the reversal of cognitive impairment, given the acute administration and selectivity of the muscarinic antagonist; with several recent studies suggesting that scopolamine-induced amnesia is due to M1 receptor blockade [62,63]. In concordance with our data, memantine reversed scopolamine induced deficits in a one-trial taste-avoidance paradigm in one-day old chicks [64]. Nevertheless, acute systemic administration of memantine modulates a number of neurotransmitters (including dopamine, serotonin and ACh) in a heterogeneous and area-specific manner, with increased ACh levels reported in nucleus accumbens and ventral tegmental area [40]. While an overall increase in cholinergic tone stimulated by memantine could explain the reversal of scopolamine-induced learning impairments in the water maze, recent evidence suggests that NMDA receptor antagonism in the nucleus accumbens is detrimental to memory formation in the water maze in mice [65]. Moreover, the observation that other NMDA antagonists enhance hippocampal ACh levels cannot account for our *in vitro* results since such actions require intact septo-hippocampal connections and rely on GABAergic transmission [60]. Since the memantine-induced rise in excitation in hippocampal slices was blocked by scopolamine, but not MK-801 or bicuculline, we provide evidence for a muscarinic action of memantine, which suggests that the reversal of scopolamine-deficits *in vivo* may also be related to an action of memantine on the cholinergic receptors in the hippocampus.

Overall, the results presented here indicate that memantine acts as a cholinergic stimulant, and shares some of the properties of CCh. Indeed, low concentrations of both reagents enhanced synaptic transmission in the mouse hippocampus, while higher concentrations inhibited transmission. Although CCh and memantine-mediated potentiation shared similar pharmacology (i.e. blocked by scopolamine and unaltered by bicuculline), they differed in terms of long-term maintenance of potentiation (only seen for CCh) and with respect to the role of the glutamatergic system (i.e. MK-801 only blocked the actions of CCh). The explanation for these differences may lie within memantine's antagonism of NMDA receptors, which would occlude further NMDA receptor inhibition. Actions on other receptors may also contribute, for instance antagonistic effects on nicotinic ACh receptors, agonist properties at dopamine receptors or inhibitory actions on 5-HT receptor mediated events [41,52,66-68]. The issue of nicotinic receptor antagonism requires particular consideration due to the crucial role of these resceptors in learning and memory [69], with such properties also reported for other NMDA receptor antagonists including MK-801 (e.g. [42]). However, nicotinic agonists rather than antagonists are neuroprotective and improve cognition while antagonism reportedly leads to cognitive impairments [69]. In addition to this, nicotinic signalling is assumed to regulate the balance between excitation and inhibition, and thus such an interaction cannot explain our results, as shown by the results obtained in the presence of bicuculline.

With regards to the bidirectional and partly inhibitory effect of higher CCh concentrations, the present study confirmed that NMDA receptors played no role in the latter action. A possible mechanism suggested for the inhibitory effects of higher concentrations of CCh involves the activation of low affinity postsynaptic M3 and M1 receptors [50], proposed to reduce the reactivity of the hippocampus, while activation of M1 may enhance firing of pyramidal cells and interneurones. The basis for the complex pharmacology of CCh and memantine may be reliant on the local sensitivity and distribution of muscarinic receptor subtypes [31,70-72]. CCh-induced LTP, shown to be a postsynaptic phenomenon [47,50], was initially suggested to be M2 sub-type mediated, though activation of presynaptic M2 autoreceptors may restrict cholinergic transmission [73-75]. A key role of M2 receptors in the potentiating effects of CCh was seemingly confirmed in a M2 mouse knock-out model where this activity was totally abolished [76]. However, another study with M1 and M3 receptor knock-out mice has suggested a key role for the M1 receptors in the excitatory effects of the drug, with the M3 receptors playing no role [77].

To investigate further the potential cholinergic actions of memantine, its effects on CCh-induced LTP were investigated. This form of LTP was blocked completely by the muscarinic antagonist scopolamine; but also by MK-801, contrary to previous studies [47,50]. A likely reason for this discrepancy is the potency of the respective NMDA receptor antagonists used. Furthermore, numerous studies indicate a positive modulation of NMDA receptor-mediated responses by activation of muscarinic receptors [78-80]. Memantine, on the other hand, prevented CCh-induced excitation and potentiation, which suggests a possible occlusion of further muscarinic stimulation. The observations that memantine suppressed evoked responses completely at 1 mM, and facilitated the depression induced by higher concentrations of CCh, indicate activity of memantine on inhibitory mAChRs, or mAChR subtype dependent partial agonist/antagonist properties.

Collectively, it can be concluded that the well-established bi-directional modulation of synaptic transmission in CA1 by muscarinic transmission [50,81] is mimicked by memantine, as indicated by the similarities between CCh and memantine, their interactions, as well as the antagonism of scopolamine. Since the cholinergic system is affected by neurodegeneration in AD [82,83], contributing at least in part to the learning and memory deficits seen in this disorder as indicated by current therapeutic approaches boosting ACh levels [84,85], we propose that memantine's cholinergic stimulation rather than NMDA antagonism is the basis for its beneficial, albeit transient, actions in the treatment of AD. The somewhat disappointing clinical effectiveness of memantine, especially in the longer term, could potentially have been avoided by a more thorough pharmacological characterisation.

## ACKNOWLEDGEMENTS

This study was supported by a grant of the Korean Health 21 R&D Project, Ministry of Health and Welfare, Republic of Korea (A040147). The authors would like to thank Ms Gillian McKay for technical assistance.

## References

[1] C.W. Cotman, D.T. Monaghan, and A.H. Ganong, Excitatory amino acid neurotransmission: NMDA receptors and Hebb-type synaptic plasticity, *Annu. Rev. Neurosci.* **11** (1988) 61-80.

[2] G. Riedel, Function of metabotropic glutamate receptors in learning and memory, Trends Neurosci. 19 (1996) 219-224.

[3] G. Riedel, B. Platt and J. Micheau, Glutamate receptor function in learning and memory, *Behav.Brain Res.* **140** (2003) 1-47.

[4] E. Masliah, M. Alford, R. DeTeresa, M. Mallory and L. Hansen, Deficient glutamate transport is associated with neurodegeneration in Alzheimer's disease, Ann.Neurol. **40** (1996) 759-766.

[5] H.L. Scott, D.V. Pow, A.E. Tannenberg and P.R. Dodd, Aberrant expression of the glutamate transporter excitatory amino acid transporter 1 (EAAT1) in Alzheimer's disease, *J.Neurosci.* **22** (2002) 1-5.

[6] W. Danysz, C.G. Parsons, J. Kornhuber, W.J. Schmidt and G. Quack, Aminoadamantanes as NMDA receptor antagonists and antiparkinsonian agents--preclinical studies, *Neurosci. Biobehav. Rev.* **21** (1997) 455-468.

[7] M.R. Hynd, H.L. Scott and P.R. Dodd, Glutamate-mediated excitotoxicity and neurodegeneration in Alzheimer's disease, *Neurochem. Int.* **45** (2004) 583-595.

[8] L. Nowak, P. Bregestovski, P. Ascher, A. Herbet and A. Prochiantz, Magnesium gates glutamate-activated channels in mouse central neurones, *Nature* **307** (1984) 462-465.

[9] G.L. Wenk, W. Danysz and S.L. Mobley, MK-801, memantine and amantadine show neuroprotective activity in

the nucleus basalis magnocellularis, *Eur.J.Pharmacol.* **293** (1995) 267-270.

[10] G.L. Wenk, W. Danysz and S.L. Mobley, Investigations of neurotoxicity and neuroprotection within the nucleus basalis of the rat, *Brain Res.* **655** (1994) 7-11.

[11] S. Ferris, L. Schneider, M. Farmer, G. Kay, and T. Crook, A double-blind, placebo-controlled trial of memantine in age-associated memory impairment (memantine in AAMI), *Int. J. Geriatr. Psychiatry* **22** (2006) 448-455.

[12] R. McShane, A. Areosa Sastre and N. Minakaran, Memantine for dementia, *Cochrane Database Syst. Rev.* **(2)2** (2006) CD003154.

[13] F.A. Schmitt, C.H. van Dyck, C.H. Wichems and J.T. Olin, Cognitive response to memantine in moderate to severe Alzheimer disease patients already receiving donepezil: an exploratory reanalysis, *Alzheimer Dis. Assoc.Disord.* **20** (2006) 255-262.

[14] J. Bormann, Memantine is a potent blocker of N-methyl-D-aspartate (NMDA) receptor channels, *Eur. J. Pharmacol.* **166** (1989) 591-592.

[15] I. Bresink, T.A. Benke, V.J. Collett, A.J. Seal, C.G. Parsons, J.M. Henley and G.L. Collingridge,. Effects of memantine on recombinant rat NMDA receptors expressed in HEK 293 cells, *Br.J.Pharmacol.* **119** (1996) 195-204.

[16] C.G. Parsons, W. Danysz and G. Quack, Memantine is a clinically well tolerated N-methyl-D-aspartate (NMDA) receptor antagonist--a review of preclinical data, *Neuropharmacol.* **38** (1999) 735-767.

[17] C.G. Parsons, V.A. Panchenko, V.O. Pinchenko, A.Y. Tsyndrenko and O.A. Krishtal, Comparative patch-clamp studies with freshly dissociated rat hippocampal and striatal neurons on the NMDA receptor antagonistic effects of amantadine and memantine, *Eur. J. Neurosci.* **8** (1996) 446-454.

[18] H.S. Chen, Y.F. Wang, P.V. Rayudu, P. Edgecomb, J.C. Neill, M Segal, S.A. Lipton, and F.E. Jensen, Neuroprotective concentrations of the N-methyl-D-aspartate open-channel blocker memantine are effective without cytoplasmic vacuolation following post-ischemic administration and do not block maze learning or long-term potentiation, *Neurosci.* **86** (1998) 121-1132.

[19] T. Frankiewicz, B. Potier, Z.I. Bashir, G.L. Collingridge and C.G. Parsons, Effects of memantine and MK-801 on NMDA-induced currents in cultured neurones and on synaptic transmission and LTP in area CA1 of rat hippocampal slices, *Br. J. Pharmacol.* **117** (1996) 689-697.

[20] J.P Apland, and F.J Cann, Anticonvulsant effects of memantine and MK-801 in guinea pig hippocampal slices, *Brain Res. Bull.* **37** (1995) 311-316.

[21] T. Frankiewicz and C.G. Parsons, Memantine restores long term potentiation impaired by tonic N-methyl-D-aspartate (NMDA) receptor activation following reduction of Mg2+ in hippocampal slices, *Neuropharmacol.* **38** (1999) 1253-1259.

[22] W. Zajaczkowski, T. Frankiewicz, C.G. Parsons and W. Danysz, Uncompetitive NMDA receptor antagonists attenuate NMDA-induced impairment of passive avoidance learning and LTP, *Neuropharmacol.* **36** (2005) 961-971.

[23] C.A. Barnes, W. Danysz and C.G. Parsons, Effects of the uncompetitive NMDA receptor antagonist memantine on hippocampal long-term potentiation, short-term exploratory modulation and spatial memory in awake, freely moving rats, *Eur. J. Neurosci.* **8** (1996) 565-571.

[24] W. Dimpfel, Effects of memantine on synaptic transmission in the hippocampus in vitro *Arzneimittelforschung* **45** (1995 ) 1-5.

[25] H. Martinez-Coria, K.N. Green, L.M. Billings, M. Kitazawa, M. Albrecht, G. Rammes, C.G. Parsons, S. Gupta, P. Banerjee, F.M. LaFerla, Memantine improves cognition and reduces alzheimer's-like neuropathology in transgenic mice, *Am. J. Pathol.* **176** (2010) 870-880.

[26] R. Minkeviciene, P. Banerjee and H. Tanila, Memantine improves spatial learning in a transgenic mouse model of Alzheimer's disease, *J. Pharmacol. Exp. Ther.* **311** (2004) 677-682.

[27] S. Nakamura, N. Murayama, T. Noshita, R. Katsuragi and T. Ohno, Cognitive dysfunction induced by sequential injection of amyloid-beta and ibotenate into the bilateral hippocampus; protection by memantine and MK-801, *Eur. J. Pharmacol.* **548** (2006) 115-122.

[28] R. Minkeviciene, P. Banerjee, H. Tanila, Cognition-enhancing and anxiolytic effects of memantine, *Neuropharmacology* **54** (2008) 1079-1085.

[29] D. Van Dam, D. Abramowski, M. Staufenbiel and P.P. De Deyn, Symptomatic effect of donepezil, rivastigmine, galantamine and memantine on cognitive deficits in the APP23 model, *Psychopharmacol.* **180** (2005) 177-190.

[30] D. Van Dam and P.P. De Deyn, Cognitive evaluation of disease-modifying efficacy of galantamine and memantine in the APP23 model, *Eur. Neuropsychopharmacol.* **16** (2006) 59-69.

[31] M.T. Vilaro, G. Mengod and J.M. Palacios, Advances and limitations of the molecular neuroanatomy of cholinergic receptors: the example of multiple muscarinic receptors, *Prog.Brain Res.* **98** (1993) 95-101.

[32] K. Yamada, M. Takayanagi, H. Kamei, T. Nagai, M. Dohniwa, K. Kobayashi, S. Yoshida, T. Ohhara, K. Takuma and T. Nabeshima, Effects of memantine and donepezil on amyloid beta-induced memory impairment in a delayed-matching to position task in rats, *Behav.Brain Res.* **162** (2005) 191-199.

[33] P.R. Zoladz, A.M. Campbell, C.R. Park, D. Schaefer, W. Danysz and D.M. Diamond, Enhancement of long-term spatial memory in adult rats by the noncompetitive NMDA receptor antagonists, memantine and neramexane, *Pharmacol.Biochem.Behav.* **85** (2006) 298-306.

[34] J.J. Miguel-Hidalgo, X.A. Alvarez, R. Cacabelos and G. Quack, Neuroprotection by memantine against neurodegeneration induced by beta-amyloid(1-40), *Brain Res.* **958** (2002) 210-221.

[35] C.E. Creeley, D.F. Wozniak, J. Labruyere, G.T. Taylor, and J.W. Olney, Low doses of memantine disrupt memory in adult rats, *J.Neurosci.* **26** (2006) 3923-3932.

[36] T.H. Rammsayer, Effects of pharmacologically induced changes in NMDA receptor activity on human timing and sensorimotor performance, Brain Res. **1073-1074** (2006) 407-416.

[37] M.M. Schugens, R. Egerter, I. Daum, K. Schepelmann, T. Klockgether and P.A. Loschmann, The NMDA antagonist memantine impairs classical eyeblink conditioning in humans, *Neurosci. Lett.* **224** (1997) 57-60.

[38] C.E. Creeley, D.F. Wozniak, A. Nardi, N.B. Farber, and J.W. Olney, Donepezil markedly potentiates memantine neurotoxicity in the adult rat brain. *Neurobiol. Aging* **29** (2008) 153-167.

[39] G.L. Wenk, G. Quack, H.J. Moebius and W. Danysz, No interaction of memantine with acetylcholinesterase in-

hibitors approved for clinical use, *Life Sci.* **66** (2000) 1079-1083.

[40] E. Shearman, S. Rossi, B. Szasz, Z. Juranyi, S. Fallon, N. Pomara, H. Shershen and A. Lajtha, Changes in cerebral neurotransmitters and metabolites induced by acute donepezil and memantine administration: A microdialysis study, *Brain Res. Bull.* **69** (2006) 204-213.

[41] Y. Aracava, E.F. Pereira, A. Maelicke and E.X.Albuquerque, Memantine blocks alpha7* nicotinic acetylcholine receptors more potently than N-methyl-D-aspartate receptors in rat hippocampal neurons, *J. Pharmacol. Exp. Ther.* **312** (2005) 1195-1205.

[42] B. Buisson and D. Bertrand, Open-channel blockers at the human alpha4beta2 neuronal nicotinic acetylcholine receptor. *Mol. Pharm.* **53** (1998) 555-563.

[43] L. Robinson, D. Harbaran and G. Riedel, Visual acuity in the water maze: sensitivity to muscarinic receptor blockade in rats and mice, *Behav. Brain Res.* **151** (2004) 277-286.

[44] E.v.L. Roloff, D. Harbaran, J. Micheau, B. Platt and G. Riedel, Dissociation of cholinergic function in spatial and procedural learning in rats. *Neurosci.* **146** (2007) 875-889.

[45] S. Deiana, C.R. Harrington, C.M. Wischik and G. Riedel. Methylthioninium chloride reverses cognitive deficits induced by scopolamine: comparison with rivastigmine, *Phsychopharmacol* **202** (2009) 53-65.

[46] D.J. Koss, K.P Hindley, G. Riedel, and B. Platt, Modulation of hippocampal calcium signalling and plasticity by serine/threonine protein phosphatases, *J. Neurochem.* **102** (2007).1009-1023.

[47] J.M. Auerbach and M Segal, A novel cholinergic induction of long-term potentiation in rat hippocampus, *J. Neurophysiol.* **72** (1994) 2034-2040.

[48] A.J. Drysdale, D Ryan, R.G. Pertwee, and B. Platt, Cannabidiol-induced intracellular Ca2+ elevations in hippocampal cells, *Neuropharmacol.* **50** (2006) 621-631.

[49] D. Ryan, A.J. Drysdale, R.G. Pertwee and B. Platt, Interactions of cannabidiol with endocannabinoid signalling in hippocampal tissue, *Eur. J. Neurosci* **25** (2007) 2093-2102.

[50] J.M. Auerbach and M Segal, Muscarinic receptors mediating depression and long-term potentiation in rat hippocampus, *J. Physiol.* **492** (1996) 479-493.

[51] V.L. Arvanov, H.C. Chou, R.C. Chen and M.C. Tsai, Presynaptic and postsynaptic actions of memantine at cholinergic central synapse of Achatina-fulica, *Comp. Biochem. Physiol.* **107** (1994) 305-311.

[52] G. Rammes, R. Rupprecht, U. Ferrari, W. Zieglgansberger, and C.G. Parsons, The N-methyl-D-aspartate receptor channel blockers memantine, MRZ 2/579 and other amino-alkyl-cyclohexanes antagonise 5-HT(3) receptor currents in cultured HEK-293 and N1E-115 cell systems in a non-competitive manner, *Neurosci.Lett.* **306** (2001) 81-84.

[53] R. Mistry, R. Wilke, R.A. Challiss, Modulation of NMDA effects on agonist-stimulated phosphoinositide turnover by memantine in neonatal rat cerebral cortex, *Br. J. Pharmacol.* **114** (1995) 797-804.

[54] D. Ferger and J. Krieglstein, Determination of intracellular $Ca^{2+}$ concentration can be a useful tool to predict neuronal damage and neuroprotective properties of drugs, *Brain Res.* **732** (1996) 87-94.

[55] T. Frankiewicz, A. Pilc and C.G. Parsons, Differential effects of NMDA-receptor antagonists on long-term potentiation and hypoxic/hypoglycaemic excitotoxicity in hippocampal slices, *Neuropharmacol.* **39** (2000) 631-642.

[56] S.E. Kotermanski, J.W. Johnson, Mg2+ imparts NMDA receptor subtype selectivity to the alzheimer's drug memantine, *J. Neurosci.* **29** (2009) 2774-2779.

[57] T.E. Bartlett, N.J. Bannister, V.J. Collett, S.L. Dargan, P.V. Massey, Z.A. Bortolotto, S.M. Fitzjohn, Z.I. Bashir, G.L. Collingridge, D. Lodge, Differential roles of NR2A and NR2B-containing NMDA receptors in LTP and LTD in the CA1 region of two-week old rat hippocampus, *Neuropharmacology.* **52** (2007) 60-70.

[58] W. Zajaczkowski, G. Quack and W. Danysz, Infusion of (+) -MK-801 and memantine -- contrasting effects on radial maze learning in rats with entorhinal cortex lesion, *Eur.J.Pharmacol.* **296** (1996) 239-246.

[59] M.P. Mattson, Pathways towards and away from Alzheimer's disease, *Nature* **430** (2004) 631-639.

[60] M.G. Giovannini, D. Mutolo, L. Bianchi, A. Michelassi and G. Pepeu, NMDA receptor antagonists decrease GABA outflow from the septum and increase acetylcholine outflow from the hippocampus: A microdialysis study, *J Neurosci* **14** (1994) 1358-1365.

[61] A.P. Blanchard, G. Guillemette, G. Boulay, Memantine potentiates agonist-induced Ca2+ responses in HEK 293 cells, *Cell. Physiol. Biochem.* **22** (2008) 205-214.

[62] J. Espinosa-Raya, M. Espinoza-Fonseca, O. Picazo, and J. Trujillo-Ferrara, Effect of a M1 allosteric modulator on scopolamine-induced amnesia, *Med. Chem.* **3** (2007) 7-11.

[63] A. Salmon, C. Erb, E. Meshorer, D. Ginzberg, Y. Adani, I. Rabinovitz, G. Amitai and H. Soreq, Muscarinic modulations of neuronal anticholinesterase responses. *Chem.Biol.Interact.* **157-158** (2005) 105-113.

[64] T.A. Barber, M.K. Haggarty, Memantine ameliorates scopolamine-induced amnesia in chicks trained on taste-avoidance learning, *Neurobiol. Learn. Mem.* **93** (2010) 540-545.

[65] V. Ferretti, F. Sargolini, A. Oliverio, A. Mele and P. Roullet, Effects of intra-accumbens NMDA and AMPA receptor antagonists on short-term spatial learning in the Morris water maze task, *Behav. Brain Res.* **179** (2007) 43-49.

[66] P.D. Maskell, P. Speder, N.R. Newberry and I. Bermudez, Inhibition of human alpha 7 nicotinic acetylcholine receptors by open channel blockers of N-methyl-D-aspartate receptors, *Br. J. Pharmacol.* **140** (2003) 1313-1319.

[67] G. Reiser and R. Koch, Memantine inhibits serotonin-induced rise of cytosolic Ca2+ activity and of cyclic GMP level in a neuronal cell line, *Eur. J. Pharmacol.* **172** (1989) 199-203.

[68] P. Seeman, C. Caruso, M. Lasaga, Memantine agonist action at dopamine D2High receptors, *Synapse* **62** (2008) 149-153.

[69] F. Dajas-Bailador and S. Wonnacott, Nicotinic acetylcholine receptors and the regulation of neuronal signalling. *Trends Pharm. Sci.* **25** (2004) 317-324.

[70] A.I. Levey, S.M. Edmunds, V. Koliatsos, R.G. Wiley and C.J. Heilman, Expression of m1-m4 muscarinic acetylcholine receptor proteins in rat hippocampus and regulation by cholinergic innervation, *J.Neurosci.* **15** (1995) 4077-4092.

[71] P.R. Lewis and C.C. Shute, The cholinergic limbic system: projections to hippocampal formation, medial cor-

tex, nuclei of the ascending cholinergic reticular system, and the subfornical organ and supra-optic crest, *Brain* **90** (1967) 521-540.

[72] T.A. Milner, R. Loy. and D.G. Amaral, An anatomical study of the development of the septo-hippocampal projection in the rat. *Brain Res.* **284**, (1983) 343-371.

[73] M. McKinney, J.H. Miller and P.J. Aagaard, Pharmacological characterization of the rat hippocampal muscarinic autoreceptor, *J Pharmacol Exp Ther.* **264** (1993) 74-78.

[74] R. Pohorecki, R. Head and E.F. Domino, Effects of selected muscarinic cholinergic antagonists on [3H]acetylcholine release from rat hippocampal slices, *J. Pharmacol. Exp. Ther.* **244** (1988) 213-217.

[75] S.T. Rouse, S.M. Edmunds, H. Yi, M.L. Gilmor and A.I. Levey, Localization of M2 muscarinic acetylcholine receptor protein in cholinergic and non-cholinergic terminals in rat hippocampus, *Neurosci. Lett.* **284** (2000) 182-186.

[76] T. Seeger, I. Fedorova, F. Zheng, T. Miyakawa, E. Koustova, J. Gomeza, A.S. Basile, C. Alzheimer and J. Wess, M2 muscarinic acetylcholine receptor knock-out mice show deficits in behavioral flexibility, working memory, and hippocampal plasticity, *J. Neurosci.* **24** (2004) 10117-10127.

[77] T. Shinoe, M. Matsui, M.M. Taketo and T. Manabe, Modulation of synaptic plasticity by physiological activation of M1 muscarinic acetylcholine receptors in the mouse hippocampus, *J.Neurosci.* **25** (2005) 11194-11200.

[78] VB. Aramakis, A.E. Bandrowski and J.H. Ashe, Role of muscarinic receptors, G-proteins, and intracellular messengers in muscarinic modulation of NMDA receptor-mediated synaptic transmission, *Synapse* **32** (1999) 262-275.

[79] J. Harvey, R. Balasubramaniam and G.L. Collingridge, Carbachol can potentiate N-methyl-D-aspartate responses in the rat hippocampus by a staurosporine and thapsigargin-insensitive mechanism, *Neurosci.Lett.* **162** (1993) 165-168.

[80] H. Markram and M. Segal, The inositol 1,4,5-trisphosphate pathway mediates cholinergic potentiation of rat hippocampal neuronal responses to NMDA, *J. Physiol.* **447** (1992) 513-533.

[81] Y. Shimoshige, T. Maeda, S. Kaneko, A. Akaike and M. Satoh, Involvement of M2 receptor in an enhancement of long-term potentiation by carbachol in Schaffer collateral-CA1 synapses of hippocampal slices, *Neurosci.Res.* **27** (1997) 175-180.

[82] R.T. Bartus, R.L. Dean, B. Beer, and A.S. Lippa, The cholinergic hypothesis of geriatric memory dysfunction, *Science* **217** (1982) 408-414.

[83] E.K. Perry, The cholinergic hypothesis--ten years on, *Br. Med. Bull.* **42** (1986) 63-69.

[84] R.S. Doody, Current treatments for Alzheimer's disease: cholinesterase inhibitors, *J. Clin. Psychiatry* **64** (2003) 11-17.

[85] E.F. Pereira, C. Hilmas, M.D. Santos, M. Alkondon, A. Maelicke and E.X. Albuquerque, Unconventional ligands and modulators of nicotinic receptors, *J. Neurobiol.* **53** (2002) 479-500.

# Cognitive Performances of Cholinergically Depleted Rats Following Chronic Donepezil Administration

Debora Cutuli[a,c], Francesca Foti[a,c], Laura Mandolesi[a,b], Paola De Bartolo[a,c], Francesca Gelfo[a,b], Daniela Laricchiuta[a,c] and Laura Petrosini[a,c]
[a]IRCCS Santa Lucia Foundation, Rome, Italy
[b]School of Movement Sciences (DiSIST), University of Naples "Parthenope", Naples, Italy
[c]Department of Psychology, University of Rome "Sapienza", Rome, Italy

**Abstract.** Since acute and chronic administration of the acetylcholinesterase inhibitors (AChE-Is), namely of donepezil, improves cognitive functions in patients afflicted by mild to moderate dementia and reverses memory deficits in experimental models of learning and memory, it seemed interesting to assess the effects of chronic donepezil treatment on cognitive functions in adult rats with forebrain cholinergic depletion. Lesions were performed by means of intracerebroventricular injections of the immunotoxin 192 IgG-saporin. The cognitive functions of lesioned animals treated or not with donepezil were compared with those of intact animals. Cholinergic depletion affected working memory functions, weakened procedural competencies, affected the acquisition of localizing knowledge, and evoked remarkable compulsive and perseverative behaviors. In lesioned animals, chronic donepezil treatment ameliorated localizatory capabilities, performances linked to cognitive flexibility and procedural abilities. Furthermore, it attenuated compulsive deficits. The present data indicate positive effects of chronic donepezil treatment on specific cognitive performances, suggesting that an aimed use of AChE-Is, targeting some symptoms more than others, may be beneficial in the case of cholinergic hypofunction. The animal model used in the present research may provide an efficient method for analyzing cognition-enhancing drugs before clinical trials.

Keywords: Acetylcholinesterase inhibitors, 192 IgG-saporin, Forebrain cholinergic system, Cholinergic neurotransmission, Mnesic functions, Cognitive flexibility, Behavioral testing.

## INTRODUCTION

Models of memory hypothesize that learning is associated with increased acetylcholine (ACh) transmission [1,2,3]. Clinical studies report that disruption of cholinergic transmission results in impaired memory and attention functions in normal aging as well as in pathological conditions, such as Mild Cognitive Impairment (MCI) [4, 5] and Alzheimer's Disease (AD) [6, 7]. As the progressive loss of cholinergic neurons in the neocortex and in the hippocampus [8] and the resulting reduction in ACh levels correlate with the degree of cognitive decline in AD ("cholinergic hypothesis" of AD) [9, 10, 11, 12, 13, 14], pharmacological treatments directed toward cholinergic system have been suggested as potentially useful in AD [4, 15]. To date, pharmacological treatment of mild to moderate AD continues to target the cholinergic system, reporting some success in maintaining and/or enhancing cognitive functioning [16, 17, 18].

At present, donepezil, a second generation acetylcholinesterase inhibitor (AChE-I), is the first-line palliative treatment for augmenting, or at least maintaining, cognitive functioning and attenuating the behavioural disturbances that influence mnesic and attentional functions [19, 20, 21, 22, 23, 24]. At the cholinergic synapse, donepezil prevents the breakdown of ACh thus elevating and further maintaining its level to stimulate postsynaptic muscarinic and nicotinic receptors [25]. Neuroimaging studies of

---

Corresponding Author: Prof. Laura Petrosini, Dept. of Psychology, University of Rome "Sapienza", Via dei Marsi 78, 00185 Rome, Italy, Tel/Fax : +39 0649917522; E-mail: laura.petrosini@uniroma1.it

donepezil-treated subjects have reported increased fronto-parietal activation correlated with improved mnesic performances in MCI patients [22], and increased metabolism of hippocampal [26], prefrontal [26] and parietal posterior regions [27, 28] in AD patients.

There are also many reports that acute treatments with donepezil reverse memory deficits in experimental models of learning and memory [29, 30, 31, 32, 33, 34]; namely, it improves performances in the Radial Arm Maze [34, 35], Morris Water Maze [36], spatial and visual recognition tasks [37, 38, 39] and serial reaction time tasks [37]. Experimental studies have reported the beneficial effects of chronic donepezil treatment on the behavioral deficits elicited by reduced activity or by the loss of cholinergic neurons that occurs in aging [40, 41] or in models of dementia [42, 43, 44].

The influence of donepezil treatment was also demonstrated in intact rodents [34, 45]. In a recent study [46], the specific profile of donepezil action on cognitive functions in the presence of unaltered cholinergic system was analyzed. In particular, we demonstrated that chronic treatment with donepezil ameliorated memory functions and explorative strategies, augmented responsiveness to the context and reduced anxiety levels. On the contrary, it did not affect spatial span, motivational levels or associative learning.

On such a basis, it seemed worthwhile to analyze the behavioral effects of chronic donepezil treatment in the presence of cholinergic depletion.

Different animal models have been developed to study cholinergic dysfunctions. In particular, attempts to model Alzheimer-like deficits in animals have focused on producing cholinergic hypofunction by manipulating basal forebrain nuclei [47, 48]. Specific damage in cholinergic projections to frontal, temporal and parietal neocortical areas and hippocampal regions is induced by injecting 192 IgG-saporin in the lateral ventricles [48, 49]. This immunotoxin inhibits ribosomal protein synthesis when it is taken up into cells expressing the low affinity neurotrophin receptor p75, thereby killing the cholinergic cell [50]. Cognitive effects of variable severity are reported following intracerebroventricular (i.c.v.) injections of immunotoxin [49, 51, 52, 53, 54, 55, 56, 57]. Possible factors contributing to this variability might be lesion procedures, lesion extent, behavioral tests employed as well as concomitant cerebellar Purkinje cell loss. In fact, some cerebellar Purkinje cells express p75, and a portion of these cells are killed after i.c.v. injections of 192 IgG-saporin [51]. Even though some researches suggested that cerebellar damage was minimal [54], and that Purkinje cell loss is insufficient to explain all deficits following cholinergic depletion [58], a control for Purkinje cell loss and for the presence of symptomatology of cerebellar origin is needed in studies that use i.c.v. 192 IgG-saporin. In spite of these pitfalls, the cholinergic depletion of the neocortical and hippocampal regions represents an attractive experimental model for analyzing the impact of cholinergic deafferentation on learning and memory functions and the efficacy of pro-cholinergic drugs in antagonizing the cognitive impairment elicited by the immuno-toxic lesion. Since donepezil may confer neuroprotection and prevent apoptotic neuronal death, up-regulating nicotinic receptors and activating protective cascades [59, 60, 61, 62, 63], the donepezil treatment we used started one week before the lesion and then lasted up to the end of behavioral testing [64]. Experimental subjects were mainly tested in spatial tests. In fact, spatial competencies are of fundamental importance to mobile organisms whose spatial memories organized in distinct but coordinated frames allow well structured adaptive capacities. In particular, by means of Morris Water Maze (MWM) we analyzed competence in building spatial cognitive mapping and navigational strategies; by means of Radial Arm Maze (RAM) we analyzed spatial reference and working memory and procedural competencies; by means of Open Field with objects (OF) we evaluated the spatial and discriminative competence; by means of Serial Learning Task (SLT), we analyzed the cognitive flexibility linked to the capacity to switch efficiently between spatially changing response rules. Given the heavy involvement of associative parietal areas in many spatial aspects, to compare behavioral outcomes and structural features of neocortical neurons, the synaptic connectivity of the parietal pyramidal neurons was also analyzed as described in the companion paper [65].

## MATERIALS AND METHODS

### Experimental design

Forty-one male Wistar rats (300-350 g) kept in standard laboratory conditions (08.00-20.00 light, food and water ad libitum) were used in the present experiments. The animals were maintained according to the guidelines for ethical conduct developed by the European Communities Council Directive of 24 November 1986 (86/609/EEC). Rats were randomly assigned to three experimental groups. The first group comprised rats treated with donepezil that received bilateral injections of 192 IgG-saporin into lateral ventricles (Donepezil-treated Lesioned, DL group). The second group comprised rats that received bilateral injections of 192 IgG-saporin into lateral ventricles to obtain a cholinergic depletion of neocortical and hippocampal regions and received saline without drug (Non-treated Lesioned, NL group). Finally, the third group comprised sham-lesioned rats that received saline without drug (Non-treated Control, NC group).

*Drug*

Following the drug regimen previously used to analyze donepezil effects on cognitive performances in intact rats [46], donepezil (Eisai Inc.) was administered i.p. daily at a dosage of 0.5 mg/kg for 3 weeks; then, for the next 7 weeks it was administered daily at a dosage of 0.2 mg/kg. The drug was dissolved in 0.5 ml of 0.9% NaCl solution. The same volume of saline, but without the drug, was administered i.p. daily to the control animals. The initial dose of 0.5 mg/kg was chosen to habituate the rats to a high drug dosage that would be reduced during the behavioral testing phase. The successively administrated 0.2 mg/kg dose of donepezil was demonstrated to be sufficient to influence repeated behavioral testing but not to elicit heavy side effects [46]. In DL group, drug treatment began one week before surgery and continued with the same schedule just described (Figure 1). All injections were administered at the end of the testing sessions to avoid any acute drug effects.

*Surgery*

Rats were anesthetized with Zoletil 100 (Tiletamine and Zolazepam: 50 mg/kg i.p.- Virbac s.r.l., Milan, Italy) and Rompun (Xylazine: 10 mg/Kg i.p.- Bayer s.p.a., Milan, Italy). In the animals to be lesioned, immunotoxin 192 IgG-saporin (Chemicon International Inc., Harrow, UK) was bilaterally injected into lateral ventricles through a 10-µL Hamilton syringe at the following coordinates: AP: - 0.8 mm (from the bregma); ML: ± 1.4 mm (from the midline); DV: - 3.5 mm (from the dura). 4 µL of the immunotoxin 192 IgG-saporin diluted in Phosphate Buffered Saline (PBS) (1 µg/1µL) was injected at a rate of 1 µL/min for 4 minutes. At the end of administration, the needle was left in situ for 5 min. In NC rats, the same volume of PBS, but containing no saporin, was injected into lateral ventricles. The animals were placed in separate cages and allowed to recover from anaesthesia and surgical stress for 24 hours. Then, they were pair housed in a standard cage (40 x 26 x 18 cm) containing wood shavings.

*Subjects*

Out of the 31 rats in the DL and NL groups, 3 animals died during surgery or behavioral testing. Data obtained from 8 other animals that finished the behavioral testing were not included in the statistical analyses because the histological analyses indicated the incorrectness of the lesion. Thus, the DL and NL groups comprised 10 lesioned animals each. To equalize the number of animals of the three groups, the NC group comprised 10 sham-lesioned rats that received daily i.p. saline injections.

*Behavioral testing*

As shown in Figure 1, the animals were tested in four tests administered in a counterbalanced order: Morris Water Maze (MWM), Radial Arm Maze (RAM), Serial Learning Task (SLT), and Open Field with objects (OF).

*Morris Water Maze (MWM)*

The rats were placed in a circular white pool (diameter 140 cm) filled with 24°C water (60 cm deep), made opaque by the addition of 2 l of milk. An escape platform (diameter 10 cm) submerged 2 cm below or elevated 2 cm above the water level was placed in the middle of one cardinal quadrant, 30 cm from the pool walls. The rat was released into the water from randomly varied starting points and was allowed to swim around to find the platform for 120 sec. Each rat underwent two sessions of four trials per day, with a 4-hour inter-session interval.

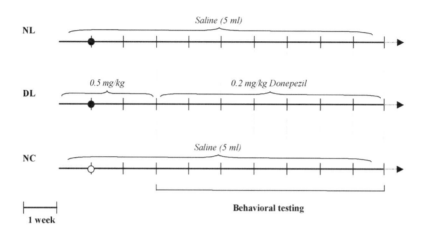

Fig. 1. Flow diagram of the experimental study. After one week of saline (5 ml; NL group) or donepezil injections (DL group), immunotoxin was administered in NL and DL animals (192 IgG-saporin i.c.v. injections = ●). Two weeks after surgery, both lesioned groups began the behavioural testing consisting of four tests: Morris Water Maze, Radial Arm Maze, Serial Learning Task and Open Field. After one week of saline i.p. injections (5 ml), NC animals received sham lesion (PBS i.c.v. injections= ○). Two weeks after sham surgery, NC animals underwent the behavioral testing, as described for lesioned animals.

When the rat reached the platform, it was allowed to remain there for 30 s. In the first four sessions, the platform was hidden in the northwest quadrant (Place I); in the next two sessions, the platform was kept visible in the northeast quadrant (Cue phase); in the final four sessions, the platform was hidden in the northeast quadrant (Place II) [66, 67, 68, 69].

The rats' trajectories in the pool were monitored by a video camera mounted on the ceiling. The resulting video signal was relayed to a monitor and to an image analyser (Ethovision, Noldus, Wageningen, The Netherlands).

The following behavioural parameters were considered in analyzing MWM performances: latencies to find the platform, total distance swum in the entire pool, distance travelled in a 20-cm peripheral annulus, heading angles (the angle formed by the actual direction of the rat's head when leaving the pool edge and a straight line from the starting location to the platform), and mean swimming velocity. Furthermore, navigational strategies put into action in reaching the platform were classified in four main categories: Circling (C), swimming in the peripheral pool areas, with inversion of swimming direction and counterclockwise and clockwise turnings in the peripheral sectors of the pool; Extended Searching (ES), swimming around the pool in all quadrants, visiting the same areas more than once; Restricted Searching (RS), swimming in some pool quadrants, not visiting some pool areas at all; direct Finding (F), swimming towards the platform without any foraging around the pool. The categorization of swimming trajectories drawn by the image analyzer was made by two different researchers who were unaware of the individual specimen's group assignment. They attributed the dominant behavior in each trial to a given category. Categorization was considered reliable only when their judgments were consistent.

*Radial Arm Maze (RAM)*

The apparatus consisted of a central platform (diameter 30 cm) from which eight arms (12.5 cm wide x 60 cm long) radiated like the spokes of a wheel. A food well (5 cm) was located at the end of each arm [70]. Prior to the habituation phase, rats were food-restricted to decrease their weight by 20%.

Full-baited maze procedure. All maze arms were baited with a piece of Purina chow at each session. The rat was placed on the central platform and allowed to make eight correct visits or sixteen (correct or incorrect) visits or to explore the maze for 15 min. The animals were submitted to two sessions a day for 5 consecutive days. The inter-session interval was 4 hours.

The following parameters were considered: total errors (number of re-visited arms divided by the total number of visits, either correct and incorrect, x 100); final spatial span (the longest sequence of correctly visited arms during the last testing session); perseverations (number of consecutive entries in an arm or in a fixed sequence of arms in up to 3 arms); percentage of 45° (or 360°) angles on the total number of angles.

Forced-choice procedure. All animals were submitted to the forced-choice paradigm 48 hours after the preceding protocol ended. In the first phase, only four arms (for example, arms 1,3,4,7) were opened and baited; the others arms remained closed. The baited arms were separated by different angles to prevent the animal from reaching the solution by adopting a stereotyped pattern. The rat was allowed to explore the open arms. Then, it spent 60 sec in its cage before being returned to the maze. In the second phase, the rat was allowed free access to all eight arms, but only the four previously closed arms were baited. This task was repeated for 5 consecutive days with a different configuration of arms closed each day to avoid any fixed search pattern.

The parameter considered was working memory errors, considered as re-entries into already visited arms (divided by the total number of visits x 100). In the second phase, this parameter was further broken down into two error subtypes: across-phase errors (APE), defined as entries into an arm entered during the first phase; within-phase errors (WPE), defined as re-entries into an arm visited earlier in the same session.

*Serial Learning Task (SLT)*

The apparatus consisted of a straight white wooden alley (cm 150 x 40 x 40) subdivided into five compartments (30 cm long) by four grey panels with two unidirectional doors (height 10 cm, width 8 cm). Each door could be locked using a pivot so that if the animal pushed the door it opened about 2 cm. The small split allowed the rat to introduce its muzzle but prevented it from going through the door. This trick permitted us to obtain sure proof that the animals were attempting to open the "wrong" door. The entire apparatus was closed with a transparent Plexiglas cover. The final rewarded compartment was darkened using a black cover.

The rats continued to be food-restricted. After 3 days of pre-training, they underwent one testing session a day for ten consecutive days. All daily sessions included 12 trials. In each trial, the goal was to reach the fifth compartment and collect the reward, by going through the open doors and not attempting to force open the closed ones. Each animal was given a sequence of open doors that remained stable for all 12 trials of a session and changed every session; thus, each animal was tested in 10 different sequences.

In each of the 12 trials of a session, the following parameters were analyzed: errors, defined as the attempt to force open any closed door (in each trial this parameter ranged from 4 to 0); correct choices, defined as the longest sequence of correct choices (in each trial this parameter ranged from 0 to 4); perseverative errors, defined as the number of consecutive errors made at the same door or in a fixed sequence of doors in the twelve trials of a session; compulsive errors, defined as the repeated attempts to force open the same closed door in the same trial; position errors, defined as the number of errors made at each of the four doors during the ten sessions. As the aim of the first session was to accustom the animals to the task requirements, the results of the first session have not been pooled with the other results. The results of the remaining nine sessions were averaged in three groups of three sessions each to analyze the behavioral results displayed in the initial (2nd - 4th sessions), middle (5th - 7th sessions) and final (8th - 10th sessions) phases of the task.

*Open Field with objects (OF)*

The apparatus consisted of a circular box (diameter 140 cm) delimited by a wall 30 cm high. Five objects were present simultaneously in the open field: A) a metal bar with a conical base; B) a plunger; C) a long steel rod; D) a yellow rubber plug; E) a black cylinder with a plastic glass turned upside down on top.

During session 1 (S1), each rat was allowed to move freely in the empty open field, and its baseline level of activity was measured. From S2 to S4 (habituation phase), four objects were placed in a square arrangement in the middle annulus of the arena; the fifth one was placed in the central area. For S5 (spatial change), the spatial configuration was changed by moving objects B and E so that the initial square arrangement was changed to a polygon-shaped configuration, without any central object. This spatial arrangement was mantained in S6. During S7 (novelty), the configuration was modified by substituting object D with a green plastic half-moon object. Sessions lasted 6 min, and inter-session intervals 3 min.

All testing was recorded by a video camera; the signal was relayed to a monitor and to the previously described image analyzer. The following emotional and motor parameters were analyzed: number of defecation boluses; motionless time; total distance and distance travelled in the peripheral arena sectors (in m) during S1; number of rearings. When the objects were present in the arena (S2-S7), the time spent contacting objects was evaluated (contact was considered to take place when the rat's snout actually touched an object, or when it sniffed the object for at least 1 sec).

*Neurochemical analyses*

No difference was found in the body weight recorded the last day of testing in the four experimental groups (mean values: NC: 384 g ± 9.4; NL: 383 g ± 8.2; DL: 385 g ± 9.2). At the end of behavioral testing, the animals were deeply anesthetized and perfused with saline followed by 4% paraformaldehyde and 0.1% glutaraldehyde in PBS (4°C, pH 7.5). Brains were cryoprotected in 30% buffered sucrose solution and processed for counterstaining with thionine (to visualize the cytoarchitectonic features of the cortex and forebrain), ChAT immunohistochemistry (to visualize the cholinergic neurons), Parvalbumin (PV) immunohistochemistry (to verify selectivity of the cholinergic depletion in the basal forebrain) and Calbindin (CB) immunohistochemistry (to visualize the Purkinje cells of the cerebellum). Methodological procedures are detailed in the companion paper [65].

Lesions were assessed by counting ChAT-IR neurons in the Ch1-Ch2 and Ch4 sectors [47], the main projection regions of the basal forebrain system. All cases showing evidence of gross unspecific neuronal damage, such as cavitation, or marked ventricle dilatation provoking thinness of the cortical layers were discarded from statistical analyses. Further details on histological analyses are provided in the companion paper [65].

*Statistical analysis*

Metric unit results of animals belonging to the different experimental groups were first tested for homoscedasticity of variance and then compared by using one- or two-way analyses of variance (ANOVAs), followed by *post-hoc* multiple comparisons using Duncan's test. Differences were considered significant at the $p<0.05$ level.

## RESULTS

*Histological analysis of 192-IgG saporin lesion*

In Ch1-2 region, the mean number of ChAT-IR neurons was 778.5 ± 45.7 in un-lesioned animals and 254.5 ± 44.7 in lesioned animals, with a proportional decrease of lesioned groups versus un-lesioned group of about 67%. In the Ch4 sector, the mean number of ChAT-IR neurons was 1698.5 ± 62.2 in un-lesioned animals and 442.1 ± 90.4 in lesioned animals. Thus, the proportional decrease of lesioned groups versus un-lesioned group was about 74%.

*Neurological examination*

At the time of testing, postural and locomotor behavior of all animals was analyzed. Lesioned rats displayed normal postural reflexes and locomotor activity. When standing, lesioned animals exhibited symmetrical posture with limbs balanced under the trunk. When stepping, no ataxic symptoms, staggering or dragging hind-paws were noted. Lesioned animals were able to walk and run without any directional bias. When grooming, they were able to rear on their hind-limbs and maintain an upright posture. When swimming, all animals exhibited good coordination and kept their noses out of the water; they displayed characteristic swimming movements, with forelimb inhibition and alternate hind-limb thrusting. Their ability to turn in water was also preserved.

## Morris Water Maze

As the sessions went by, all animals displayed a progressive reduction of latencies in reaching the platform (Figure 2a). A two-way ANOVA (group x phase) on latency values revealed a significant phase effect ($F_{2,54}=37.32$; $p<0.0001$), while group effect ($F_{2,27}=0.62$; $p$ n.s.) and interaction ($F_{4,54}=0.57$; $p$ n.s.) were not significant.

Heading angles were calculated to obtain information about localizatory knowledge regarding platform position gained by the animals as the sessions went by. A two-way ANOVA (group x phase) on heading angles revealed significant group ($F_{2,27}=7.15$; $p<0.003$) and phase ($F_{2,54}=16.13$; $p<0.00001$) effects. The interaction was not significant ($F_{4,54}=0.06$; $p$ n.s.). In fact, NC and DL groups showed heading angles significantly narrower than untreated lesioned animals (NC vs NL: $p=0.01$; DL vs NL: $p=0.001$; NC vs DL: $p$ n.s.) (Figure 2b). When the total distance swum in the pool during the entire task was taken into account, a one-way ANOVA revealed a significant group effect ($F_{2,27}=3.20$; $p<0.05$). Post-hoc comparisons indicated a significant difference between DL and NC groups ($p=0.02$), given treated lesioned animals travelled longer distances than controls (Figure 2c). The percentage of peripheral swimming may provide some indication of the navigational strategies. One-way ANOVA revealed a not significant group effect ($F_{2,27}=2.09$; $p$ n.s.) indicating that all animals spent similar percentages of their swimming in the peripheral pool annulus. A one-way ANOVA on mean swimming velocity revealed a significant group effect ($F_{2,27}=7.27$; $p=0.003$) given both lesioned animals exhibited mean velocities significantly ($p<0.01$) higher than NC group (NL: $\overline{X}=26.71$ cm/s ± 0.59; DL: $\overline{X}=25.7$ cm/s ± 0.80; NC: $\overline{X}=22.6$ cm/s ± 0.53).

No significant group effect was observed when strategies were analyzed (two-way ANOVA (group x strategy): group ($F_{2,27}=1.89$; $p$ n.s.); strategy: ($F_{3,81}=45.46$; $p<0.0001$). The interaction was not significant ($F_{6,81}=1.12$; $p$ n.s.). In particular, all animals exhibited maximal percentages of Finding, intermediate of Extended and Restricted searching, and minimal of Circling (Figure 2d).

## Radial Arm Maze

### Full-baited maze procedure

As the sessions went by, all animals decreased their errors in visiting the baited arms ($F_{4,108}=4.62$; $p<0.001$); group effect was also significant ($F_{2,27}=7.84$; $p<0.002$) given that both lesioned groups made significantly more errors than the controls (DL vs NC: $p=0.04$; NL vs NC: $p=0.001$) (Figure 3a). The interaction was not significant ($F_{8,108}=1.06$; $p$ n.s.). A one-way ANOVA performed on final spatial span failed to reveal any significant group effect ($F_{2,27}=2.55$; $p$ n.s.) (Mean values: DL: $\overline{X}=5.4 ± 0.5$; NL: $\overline{X}=4.8 ± 0.5$; NC: $\overline{X}=6.3 ± 0.4$).

Both lesioned groups made a significantly higher number of perseverations than control animals (one-way ANOVA: $F_{2,27}=5.90$; $p=0.007$), as revealed by post-hoc comparisons between groups (NC vs NL: $p=0.006$; NC vs DL: $p=0.009$) (Figure 3b).

To analyze the procedural strategies put into action in visiting the eight arms the angles between the arms entered consecutively were considered. The percentages of re-entries in the same arm through 360° angles were near to zero in all experimental groups so that ANOVA application was not feasible. Conversely, the percentage of 45° angles (that is the angles performed in the most efficient strategy of consecutively visiting adjacent arms) revealed significant group effect (one-way ANOVA: $F_{2,27}=3.83$; $p=0.03$). Post-hoc comparisons revealed significant differences between NL and the remaining groups (NC vs NL: $p=0.04$; NL vs DL: $p=0.01$) while DL and NC groups were not significantly different (Figure 3c).

### Forced-choice procedure

During the first phase, no differences were found among groups regarding the number of errors made ($F_{2,27}=2.36$; $p$ n.s.), because the easier task (only four arms open) favored lesioned animals.

A one-way ANOVA on working memory errors performed in the second phase revealed a significant group effect ($F_{2,27}=8.03$; $p=0.001$) given that both lesioned groups made significantly more errors than NC animals (DL vs NC: $p=0.01$; NL vs NC: $p=0.0007$).

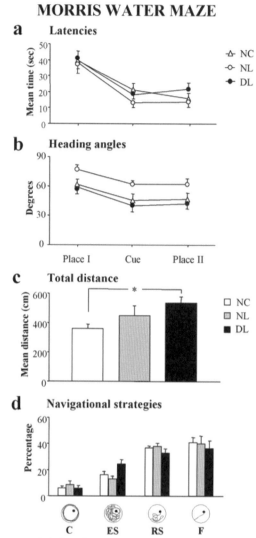

Fig. 2. Effects of chronic donepezil treatment and cholinergic depletion on the MWM performance. Mean escape latencies to reach the platform (**a**), heading angles (**b**), total distance swum in the pool (**c**), navigational strategies (**d**) displayed by the three experimental groups are depicted. In (**d**) the circular figurines illustrate the typical explorative patterns of the four main navigational strategies. The black filled circles indicate the position of the hidden platform. C: Circling strategy at pool periphery; ES: Extended Searching for the hidden platform around the pool; RS: Restricted Searching for the hidden platform in some pool quadrants, not visiting some pool areas at all; F: Finding the platform directly without any exploration around the pool. In this and the following figures, **NC**: Non treated Controls; **NL**: Non treated Lesioned animals; **DL**: Donepezil-treated Lesioned animals. Asterisks inside the graphs indicate the *post-hoc* comparisons between groups: * $p<0.05$. Vertical bars indicate SEM.

An additional error analysis revealed a significant group (One-way ANOVA: $F_{2,27}=6.21$; $p=0.006$) effect on within-phase errors (WPE). *Post-hoc* comparisons revealed a significant difference between NL and NC group ($p=0.002$), while DL and NC groups were not significantly different. When across-phase errors (APE) were considered, a significant group effect was observed ($F_{2,27}=8.03$; $p=0.001$). *Post-hoc* comparisons revealed significant differences between NL and NC group ($p=0.0007$) and DL and NC ($p=0.01$) (Figure 4).

*Serial Learning Task*

In SLT, DL and NC animals performed a similar number of errors, while NL animals made a significantly higher number of errors in comparison to the other groups, as indicated by a two-way ANOVA (group x phase) on total errors that revealed significant group ($F_{2,27}=8.97$; $p=0.001$) and phase ($F_{2,54}=15.65$; $p=0.0001$) effects. Interaction was not significant ($F_{4,54}=1.66$; $p$ n.s.). To verify if errors diminished significantly in all groups during phases one-way ANOVAs were performed. In NC and DL groups a significant learning effect resulted (NC: $F_{2,18}=15.34$; $p=0.0001$; DL: $F_{2,18}=8.21$; $p=0.003$), whereas in NL group it was not significant (NL: $F_{2,18}=1.93$; $p$ n.s.) (Figure 5a). A one-way ANOVA on the longest sequence of correct choices showed a significant group effect ($F_{2,27}=3.92$; $p=0.03$). Post-hoc comparisons revealed a significant difference between NL and NC groups ($p=0.001$) and between NL and DL groups ($p=0.05$), while no significant difference was found between DL and NC groups (NL: $\overline{X}=1.70 \pm 0.09$; DL: $\overline{X}=1.83 \pm 0.07$; NC: $\overline{X}=1.92 \pm 0.07$). A two-way ANOVA (group x phase) on perseverative errors revealed a significant group ($F_{2,27}=6.53$; $p=0.005$) and phase ($F_{2,54}=4.8$; $p=0.0001$) effects. Interaction was not significant ($F_{4,54}=1.39$; $p$ n.s.). Post-hoc comparisons indicated that both lesioned groups showed a higher number of perseverative errors than NC group (NC vs NL: $p=0.002$; NC vs DL: $p=0.04$) (Figure 5b). When compulsive errors were taken into account, a two-way ANOVA (group x phase) revealed significant group ($F_{2,27}=6.90$; $p=0.004$) and phase ($F_{2,54}=119.57$; $p=0.0001$) effects. Interaction was not significant ($F_{4,54}=1.02$; $p$ n.s.). Post-hoc comparisons indicated that NL animals' performances were significantly different from NC and DL groups' ones ($p=0.004$). DL and NC groups' performances were not different (Figure 5c). A two-way ANOVA (group x panel) on position errors showed significant group ($F_{2,27}=12.70$; $p=0.0001$) and panel ($F_{3,81}=24.97$; $p=0.00001$) effects. Also interaction was significant ($F_{6,81}=6.75$; $p=0.00001$). Post-hoc comparisons demonstrated that errors in each panel were not significantly different in DL and NC animals, while the errors made by NL animals were significantly higher in comparison to NC animals in the first ($p<0.00001$), second ($p=0.005$) and fourth ($p=0.02$) panels (Figure 5d). These data indicate that a forebrain lesion destroyed the primacy and recency patterns of serial learning and that donepezil tended to restore them in lesioned animals.

*Open Field with objects*

The absence of significant differences in the number of defecation boluses ($F_{2,27}=0.03$; $p$ n.s.) indicated that all animals exhibited comparable levels of anxiety. Also motionless time during the task was not affected by the group the animals belonging to ($F_{2,27}=2.71$; $p$ n.s.). No significant differences were observed in the number of rearings ($F_{2,27}=1.05$; $p$ n.s.). A one-way ANOVA on total distance traveled in S1 failed to reveal any significant group effect ($F_{2,27}=0.72$; $p$ n.s.). The same pattern was found when the distance traveled in the peripheral annulus was analyzed ($F_{2,27}=0.19$; $p$ n.s.). Thus, the emotional and motor parameters indicated no difference among groups. When objects were present, all animals showed habituation as revealed by a two-way ANOVA (group: $F_{2,27}=2.61$; $p$ n.s.; session: $F_{2,54}=33.46$; $p<0.00001$; interaction: $F_{4,54}=0.59$; $p$ n.s.) (Figure 6a). Contact times with displaced (DO) and non-displaced (NDO) objects during spatial change (S5) are reported in Figure 6b. While NC animals displayed significantly different contact times between DO and NDO ($F_{1,9}=22.03$; $p<0.001$), both lesioned groups did not display any difference in contact times (NL: $F_{1,9}=1.65$; $p$ n.s.; DL: $F_{1,9}=0.84$; $p$ n.s.). These data demonstrated that neither lesioned group was able to recognize the occurrence of spatial change. Novel object (NO) vs familiar objects (FO) contact times were analysed during novelty (S7), thas is, when a familiar object was substituted with a novel one. The animals belonging to all experimental groups displayed significantly different contact times between NO and FO (NC: $F_{1,9}=22.39$; $p<0.001$; NL: $F_{1,9}=11.75$; $p=0.007$; DL: $F_{1,9}=12.35$; $p=0.006$) (Figure 6c). These data demonstrated that both groups of lesioned rats displayed object novelty reactions comparable to those of control rats.

## RADIAL ARM TASK
## Full-baited procedure

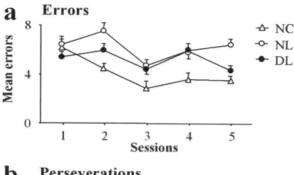

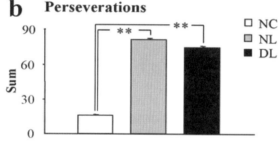

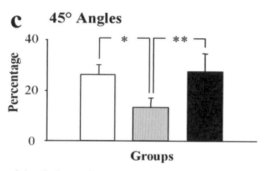

Fig. 3. Effects of chronic donepezil treatment and cholinergic depletion on the RAM performance. Total errors (**a**), perseverations (**b**) and 45° angles (**c**) displayed by the three experimental groups in the full-baited procedure are depicted. Asterisks inside the graphs indicate the *post-hoc* comparisons between groups: * $p<0.05$, ** $p<0.01$. Vertical bars indicate SEM.

*Cerebellar symptomatology*

The eventual presence of symptoms of cerebellar origin was carefully analyzed to verify whether the i.c.v. injections of 192 IgG-saporin had provoked cerebellar damage affecting behavioral performances of cholinergically depleted animals. Neither group of lesioned animals exhibited persistent circling in the MWM task (Fig. 2d) (unproductive navigational strategy displayed by HCbed animals in MWM [67]); neither did they exhibit prolonged peripheral traveling in either MWM or OF tasks (typical symptom of procedural impairment displayed by HCbed animals [67, 71]); neither did they revisit the same arm by making 360° angles (uneconomical strategy for solving the RAM task exhibited in the presence of cerebellar lesion [70]). These behavioral observations fit completely with the slight cerebellar damage detected in these lesioned animals. The loss of Purkinje cells was assessed by counting CB-immunoreactive neurons. The number of CB-IR neurons was 5090.50 ± 135.05 in unlesioned animals

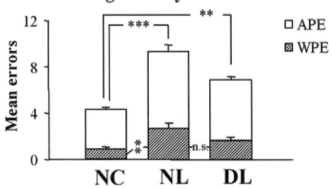

Fig. 4. Errors displayed by the three experimental groups in the forced-choice procedure. APE: across-phase errors; WPE: within-phase errors. Asterisks inside the graphs indicate the *post-hoc* comparisons between groups: ** $p<0.01$, *** $p<0.001$. Vertical bars indicate SEM.

and 4556.25 ± 193.65 in lesioned animals, with a proportional decrease of about 11% in the mean number of neurons. Details on these analyses are reported in the companion paper [65].

## DISCUSSION

Several experimental studies have addressed the cognitive effects of immuno-toxic forebrain lesions resulting in cholinergic depletion [72, 73, 74, 75, 76] and many clinical studies have addressed the effects of AChE-Is in the treatment of diseases related to cholinergic hypofunction [9, 77, 78, 79]. And yet, to date no study has analyzed in cholinergically depleted rats the effects of chronic donepezil treatment paralleling the pharmacological treatment usually prescribed to AD patients. In lesioned animals, chronic cholinergic treatment potentiated the cognitive flexibility functions, as indicated by the reduction of SLT errors and of compulsive behaviors; it facilitated the acquisition of procedural competencies, as indicated by efficient RAM explorative strategies; it ameliorated (declarative) localizatory functions, as indicated by the reduced MWM heading angles. Conversely, donepezil treatment did not succeed in restoring other functions impaired by the cholinergic lesion. In fact, it did not re-establish the abilities to build up and to use the spatial map impaired by the cholinergic lesion, as indicated by the lack of detection of spatial change in the OF; it did not reduce the lesion-induced perseverative behaviors, as indicated by perseverations in RAM and SLT. Finally, cholinergic treatment did not modify few functions left unaffected by the forebrain lesion, such as the short-term spatial memory storage, as indicated by the unmodified RAM spatial span; it did not alter motivational levels, as indicated by emotional parameters of OF and by the peripheral distances traveled in MWM and OF; it did not influence the recognition of novelty during OF and the escape latencies in MWM. The observation of unaffected MWM latencies in i.c.v. 192 IgG-saporin lesioned animals fits with previous findings on cholinergic depletion [49, 53, 80, 81, 82, 83, 84]. Both lesioned groups (DL and NL) showed increased swimming velocities. This finding was probably linked to the motor activation induced by both donepezil treatment [85] and the 192 IgG-saporin lesions [51, 86]. It cannot be discounted that the differences in the MWM performances observed in the two lesioned groups in comparison to NC animals could be linked to their profile of motor activation.

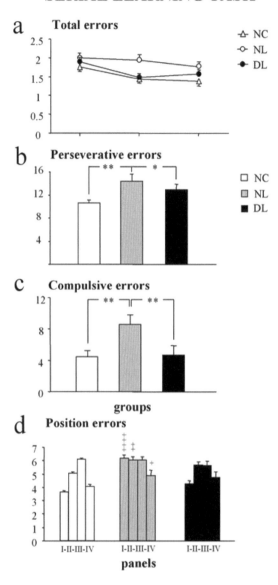

Fig. 5. Effects of chronic donepezil treatment and cholinergic depletion on the SLT performance. Total errors (a), perseverative errors (b), compulsive errors (c), and position errors (d) displayed by the three experimental groups are depicted. Asterisks inside the graphs indicate the *post-hoc* comparisons between groups: * $p<0.05$, ** $p<0.01$. In (d) the asterisks inside the graph indicate the *post-hoc* comparisons between NC *vs* NL groups ++++ $p<0.0001$, ++ $p<0.01$, + $p<0.05$. Vertical bars indicate SEM.

The enhancement of cholinergic neurotransmission allowed acquiring more pointed (although not necessarily more efficient) navigational strategies, as indicated by the narrow heading angles displayed by DL animals. Fine tuning of heading angles related to achievement of localizatory competencies was reported previously in donepezil-treated intact animals [46]; thus, it appears to be a definite action of donepezil in the absence as well as in the presence of cholinergic hypoactivity. Given that donepezil increases the extracellular ACh levels in the hippocampus [87, 88, 89], the main structure involved in building localizatory knowledge, the tuning of heading angles might be mediated by a modulating

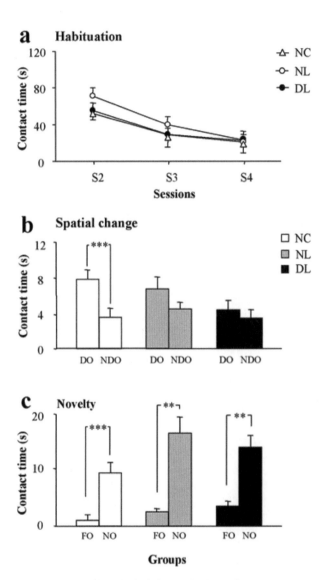

Fig. 6. Effects of chronic donepezil treatment and cholinergic depletion on the OF performance. Mean contact times with objects during habituation phase (S2-S4) (a), spatial change (b) and novelty (c) of the four experimental groups are depicted. Asterisks inside the graphs indicate the *post-hoc* comparisons between groups: ** $p<0.01$, *** $p<0.001$. Vertical bars indicate SEM. In (b), *DO* displaced objects; *NDO* non-displaced objects. In (c), *FO* familiar objects; *NO* novel object.

donepezil action on cholinergic hippocampal neurons. Some authors claim that hippocampal ACh release is more resistant to the effects of the immunotoxic lesions than that of neocortical regions [90]. Thus, it is possible to advance that AChE inhibition induced by donepezil may act on a relatively spared cholinergic substrate resulting in a more effective drug action. The present data are also in agreement with a recent report [63] indicating that donepezil preserves neurons in the CA1 hippocampal region and attenuates MWM impairments after mild traumatic brain injury.

In the full-baited RAM procedure, donepezil treatment was not able to reduce the total number of errors; however, it significantly enhanced the percentage of 45° angles in DL in comparison to NL animals. Successful performance on the RAM task involves working memory and mapping abilities as well as efficient explorative competencies. In fact, to avoid repeated entries in already visited arms the subjects have to remember the arm just visited and avoid perseverative re-visits, represent the maze in a cognitive spatial map, and/or use efficient procedural strategies. A possible explanation for the increased number of errors displayed by DL animals in spite of their higher percentage of 45° angles may take into account the presence of lesion-induced perseverative behaviors.

To distinguish procedural from working memory requirements the forced-choice procedure was used. In this procedure, DL animals significantly reduced their within-phase errors but not across-phase errors. Such a finding indicated that donepezil treatment facilitated the working memory abilities to remember the just visited arms, but failed to influence longer-term memories linked to the arms visited during the first phase of the protocol. It is noteworthy that this donepezil effect could be due not only to enhanced mnesic capabilities, but also to ameliorated attentive processes. Positive effects on RAM parameters of donepezil were demonstrated in other models of cholinergic hypofunction. Improved working memory performances were displayed by aged mice chronically treated with donepezil or with a α7-nicotinic partial agonist [64]. Beneficial effects of donepezil were reported in aged mice also in a delayed non matching to place RAM task [25]. In addition, donepezil significantly reversed scopolamine-induced errors on the RAM task [35, 91].

In the OF task, during spatial change, neither lesioned group (DL and NL) was able to recognize the presence of displaced objects, conversely during novelty all groups easily appreciated the presence of the new object. These findings on lesioned rats are consistent with previous studies reporting impaired abilities in recognizing spatial change and intact abilities in novelty recognition following 192 IgG-saporin forebrain lesions [92, 93, 94, 95]. The lesion-induced absence of spatial change detection was apparently inconsistent with tuned MWM performances displayed by the lesioned animals. It is possible that the MWM task is more affected by stress levels associated to the forced swimming in the search of an escape platform. It is reported that tasks involving high levels of stress may lead to stress-related increases in ACh release in the hippocampus [96, 97, 98] in a manner that could compensate more fully for the loss of the cholinergic functions in relatively stressful (MWM) versus less stressful (OF) tasks. Thus, the additional stress-related increase in ACh release in donepezil treated lesioned animals might contribute to the tuning of heading angles [99, 100]. It is noteworthy that the different impairment exhibited by lesioned animals in the various components of spatial function could be related to the different cholinergic "weight" of the various structures (hippocampus, parietal cortex, temporal cortex) affected by the lesion.

Besides deficiencies in working memory, spatial abilities and attentional functions, another main symptom of early AD is the impairment of executive functions including concept formation, problem solving and set shifting, that is, flexibility functions [101, 102, 103]. Cognitive flexibility is a cardinal property of declarative memory that can be demonstrated also in animals as the ability to compare and contrast information originating from separate sources to make a choice in a novel situation [64, 104, 105]. SLT is a suitable test of flexibility [106], and it appears to be very sensitive to donepezil treatment in intact animals [46]. Interestingly, DL animals showed improved performances in almost all SLT parameters, given the drug treatment was able to decrease errors, to reduce compulsive errors, and to lengthen the sequence of correct choices. Furthermore, it restored the primacy and recency effects usually present in serial learning but disrupted by the forebrain lesions. This behavioral pattern demonstrated that donepezil was able to induce efficient information processing and prompt acquisition of new responses. These data fit with other reports [37, 53, 64, 107] suggesting that donepezil improves the flexible use of long-term memory. Recently, it was demonstrated that forebrain lesions disrupt the ability to switch efficiently between conflicting response rules [106, 108]. Thus, it was hypothesized that a common impairment underlies the various flexibility behaviors. In fact, basal forebrain lesions impair learning set formation [109], configural association [110, 111], and tasks requiring divided or incremental attention [53, 73, 112, 113]. The present results indicate the marked ameliorative influence of donepezil on flexibility behaviors, underlining the role of AChE-Is as enhancers of cognitive functioning. Interestingly, we recently demonstrated that flexibility behaviors are positively influenced by experimental manipulations, such as enriched environments requiring sustained cognitive engagement [104]. Taken together, these findings support the hypothesis that forebrain may be sensitive to different factors (environmental or pharmacological) that enhance spatial cognition by potentiating the cholinergic neurotransmission [76, 114, 115, 116] and thus modulating the plasticity of cortical areas in general and of parietal cortex in particular [117, 118]. It is noteworthy that the companion paper [65] indicates that in cholinergically depleted animals chronic donepezil treatment was able to restore morphological features of parietal neurons toward control levels. Interestingly, cholinergic circuits modulate neuronal activity associated with requirements of demanding tasks [119, 120] in the hypothesis that the cortical cholinergic system functions as an optimizer of cognitive resources [113, 120]. However, there is some disagreement about how the cholinergic system promotes cognitive performances in general, and mnesic functions in particular. It is generally assumed that increased ACh transmission is associated with learning and mnesic functions [121, 122] and that ACh disruption results in impaired memory [123, 124]. Furthermore, it has been advanced that pro-cholinergic drugs may augment cognitive functioning through attenuation of neuropsychiatric disturbances that obstruct the cognitive domain of attention [18, 119, 125]. Finally, some authors argue that ACh affects interactions between mnesic and attentive functions [2, 13]. In particular, cholinergic activation may be a prerequisite of sustained attention [126], which, in turn, seems to be a prerequisite of information acquisition, recall, and adaptive responses to the context [3].

Recently, it was advanced that the beneficial effects of donepezil on lesion-induced deficits could be attributed not only to symptomatic effects exploiting the slowing of the hydrolysis of ACh at synaptic level [84, 85], but also to its neuroprotective properties, as indicated by *in vivo* and *in vitro* models of neurotoxicity [40, 41, 61, 127, 128, 129]. In fact, AChE-Is, as donepezil, have been demonstrated to protect cortical neurons against glutamate neurotoxicity via α4β2 and α7 nicotinic ACh receptors, at least partly by inhibiting the process of apoptosis [59, 60, 132, 133]. In the present research, this potential neuroprotective effect motivated the 1-week pre-treatment prior to forebrain lesioning in the DL group [64]. The choice of chronic drug regimen exploited thus all potential effects attributed to the donepezil and allowed evidencing beneficial effects, although scattered, on the spatial functions in cholinergically lesioned animals. In effect, chronic AChE inhibition does not develop a significant tolerance and ACh synthesis is not affected by chronic AChE-Is treatment. This leads to a long-lasting, large ACh concentration in the synaptic clefts and an enduring activation that may result in changes in the number, affinity and function of nicotinic (nAChR) and muscarinic (mAChR) receptors [129]. It appears that only the M2-M4 autoreceptors are constantly down-regulated in rats treated with AChE-Is [130], while other muscarinic receptors and the nicotinic receptors [5, 40, 41, 131, 132, 133, 134] may undergo changes depending on doses, treatment duration, differences in the pharmacological profile of the various AChE-Is, regions investigated, and the receptor ligands used. Therefore, as recently asserted by Pepeu and Giovannini [129], while it is possible to assume that the down-regulation of the M2-M4 autoreceptors may contribute in maintaining a high level of extracellular ACh [130], by attenuating an inhibitory control on the release, it is difficult to define whether or not changes in other cholinergic receptors play a role in the therapeutic efficacy of AChE-Is in general and of donepezil in particular.

In conclusion, the behavioral analysis of i.c.v. 192 IgG-saporin lesioned animals evidenced not generalized cognitive deficits, mimicking thus early forms of AD or MCI. This experimental model may represent a useful tool to understand the *in vivo* action of AChE-Is or novel compounds developed in AD treatment [129, 135]. Furthermore, provoking subtle symptoms it may be informative to promote tuned pharmachological strategies in cognitive decline, targeting some symptoms more than others.

## References

[1] M.G. Giovannini, F. Casamenti, Bartolini L, G. Pepeu,The brain cholinergic system as a target of cognition enhancers, *Behav Brain Res* **83** (1997), 1-5.

[2] M.B. Parent, M.G. Baxter, Septohippocampal acetylcholine: involved in but not necessary for learning and memory?, *Learn Mem* **11** (2004) 9-20.

[3] G. Pepeu, M.G. Giovannini, Changes in acetylcholine extracellular levels during cognitive processes, *Learn Mem* **11** (2004), 21-27.

[4] M. Sarter, J.P. Bruno, Developmental origins of the age-related decline in cortical cholinergic function and associated cognitive abilities, *Neurobiol Aging* **25** (2004), 1127-1139.

[5] A. Nordberg, Mechanisms behind the neuroprotective actions of cholinesterase inhibitors in Alzheimer disease, *Alzheimer Dis Assoc Disord* **20** (2006), 12-18.

[6] J. Poirier, Evidence that clinical effects of cholinesterase inhibitors are related to potency and targeting of action, *Int J Clin Pract Suppl* **127** (2002), 6-19.

[7] J. Birks, L. Flicker, Donepezil for mild cognitive impairment, *Cochrane Database Syst* **3** (2006), CD006104.

[8] E.K Perry., P.H. Gibson, G. Blessed, R.H. Perry, B.E. Tomlinson, Neurotransmitter enzyme abnormalities in senile dementia. Choline acetyltransferase and glutamic acid decarboxylase activities in necropsy brain tissue, *J Neurol Sci* **34** (1977), 247-265.

[9] P. Davies, A.J. Maloney, Selective loss of central cholinergic neurons in Alzheimer's disease. *Lancet* **2** (1976), 1403.

[10] R.T. Bartus, R.L. Dean, B. Beer, A.S. Lippa, The cholinergic hypothesis of geriatric memory dysfunction, *Science* **30** (1982), 408-414.

[11] R.T. Bartus, On neurodegenerative diseases, models, and treatment strategies: lessons learned and lessons forgotten a generation following the cholinergic hypothesis, *Exp Neurol* **163** (2000), 495-529.

[12] U. Freo, G. Pizzolato, M. Dam, C. Ori, L. Battistin, A short review of cognitive and functional neuroimaging studies of cholinergic drugs: implications for therapeutic potentials, *J Neural Transm* **109** (2002), 857-870.

[13] M. Sarter, J.P. Bruno, B. Givens, Attentional functions of cortical cholinergic inputs: what does it mean for learning and memory? *Neurobiol Learn Mem* **80** (2003), 245-256.

[14] J.W. Phillis, Acetylcholine release from the central nervous system: a 50-year retrospective, *Crit Rev Neurobiol* **17** (2005), 161-217.

[15] C.M. Smith, M. Swash, Possible biochemical basis of memory disorder in Alzheimer disease. *Ann Neurol* **3** (1978), 471-473.

[16] J.L. Cummings, The role of cholinergic agents in the management of behavioural disturbances in Alzheimer's disease. *Int J Neuropsychopharmacol* **3** (2000), 21-29.

[17] A. Clegg, J. Bryant, T. Nicholson, Clinical study of donepezil, rivastigmine and galantamine for Alzheimer's disease: a rapid and systemic review, *Health Technol Assess* **5** (2001), 1-137.

[18] G. Brousseau, B.P. Rourke, B. Burke, Acetylcholinesterase inhibitors, neuropsychiatric symptoms, and Alzheimer's disease subtypes: an alternate hypothesis to global cognitive enhancement, *Exp Clin Psychopharmacol* **15** (2007), 546-554.

[19] S.L. Rogers., R.S. Doody, R.C. Mohs, L.T. Friedhoff, Donepezil improves cognition and global function in Alzheimer disease: a 15-week, double-blind, placebo-controlled study. Donepezil Study Group, *Arch Intern Med* **158** (1998), 1021-1031.

[20] M. Shigeta, A. Homma, Donepezil for Alzheimer's disease: pharmacodynamic, pharmacokinetic, and clinical profiles, *CNS Drug Rev* **7** (2001), 353-368.

[21] H. Sugimoto, Donepezil hydrochloride: a treatment drug for Alzheimer's disease, *Chem Rec* **1** (2001), 63-73.

[22] A.J. Saykin, H.A. Wishart, L.A. Rabin, L.A. Flashman, T.L. McHugh, A.C. Mamourian, R.B. Santulli, Cholinergic enhancement of frontal lobe activity in mild cognitive impairment, *Brain* **127** (2004), 1574-1583.

[23] B. Seltzer, Donepezil: an update, *Expert Opin Pharmacother* **8** (2007), 1011-1023.

[24] C.J. Winstein, K.R. Bentzen, L. Boyd, L.S. Schneider, Does the cholinesterase inhibitor, donepezil, benefit both declarative and non-declarative processes in mild to moderate Alzheimer's disease?, *Curr Alzheimer Res* **4** (2007), 273-276.

[25] B. Bontempi, K.T. Whelan, V.B. Risbrough, G.K. Lloyd, F. Menzaghi, Cognitive enhancing properties and tolerability of cholinergic agents in mice: a comparative study of nicotine, donepezil, and SIB-1553A, a subtype-selective ligand for nicotinic acetylcholine receptors, *Neuropsychopharmacology* **28** (2003), 1235-1246.

[26] S.J. Teipel, A. Drzezga, P. Bartenstein, H.J. Möller, M. Schwaiger, H. Hampel, Effects of donepezil on cortical metabolic response to activation during (18)FDG-PET in Alzheimer's disease: a double-blind cross-over trial, *Psychopharmacology* **187** (2006), 86-94.

[27] L.Tune, P.J. Tiseo, J. Ieni, C. Perdomo, R.D. Pratt, J.R. Votaw, R.D. Jewart, J.M. Hoffman, Donepezil HCl (E2020) maintains functional brain activity in patients with Alzheimer disease: results of a 24-week, double-blind, placebo-controlled study, *Am J Geriatr Psychiatry* **11** (2003), 169-177.

[28] G. Rodriguez, P. Vitali, M. Canfora, P. Calvini, N. Girtler, C. De Leo, A. Piccardo, F. Nobili, Quantitative EEG and perfusional single photon emission computed tomography correlation during long-term donepezil therapy in Alzheimer's disease, *Clin Neurophysiol* **115** (2004), 39-49.

[29] D.H. Cheng, H. Ren, X.C. Tang, Huperzine a novel promising acetylcholinesterase inhibitor. *Neuroreport* **8** (1996), 97-101.

[30] N.M. Rupniak, S.J. Tye, M.J. Field, Enhanced performance of spatial and visual recognition memory tasks by the selective acetylcholinesterase inhibitor E2020 in rhesus monkeys, *Psychopharmacology* **131** (1997), 406-410.

[31] G.A. Higgins, M. Enderlin, R. Fimbel, M. Haman, A.J. Grottick, M. Soriano, J.G. Richards, J.A. Kemp, R. Gill, Donepezil reverses a mnemonic deficit produced by scopolamine but not by perforant path lesion or transient cerebral ischaemia, *Eur J Neurosci* **15** (2002), 1827-1840.

[32] K. Tokita, S. Yamazaki, M. Yamazaki, N. Matsuoka, S. Mutoh, Combination of a novel antidementia drug FK960 with donepezil synergistically improves memory deficits in rats, *Pharmacol Biochem Behav* **73** (2002), 511-519.

[33] F.J.van der Staay, P.C. Bouger, Effects of the cholinesterase inhibitors donepezil and metrifonate on scopolamine-induced impairments in the spatial cone field orientation task in rats, *Behav Brain Res* **156** (2005), 1-10.

[34] L.E. Wise, P.A. Iredale, R.J. Stokes, A.H. Lichtman, Combination of rimonabant and donepezil prolongs spatial memory duration, *Neuropsychopharmacology* **32** (2007), 1805-1812.

[35] H. Ogura, T. Kosasa, Y. Kuriya, Y. Yamanishi, Donepezil, a centrally acting acetylcholinesterase inhibitor, alleviates learning deficits in hypocholinergic models in rats, *Methods Find Exp Clin Pharmacol* **22** (2000), 89-95.

[36] L. Spowart-Manning, F.J. van der Staay, Spatial discrimination deficits by excitotoxic lesions in the Morris water escape task, *Behav Brain Res* **156** (2005), 269-276.

[37] D.L. Kirkby, D.N. Jones, J.C. Barnes, G.A. Higgins, Effects of anticholinesterase drugs tacrine and E2020, the 5-HT(3) antagonist ondansetron, and the H(3) antagonist thioperamide, in models of cognition and cholinergic function, *Behav Pharmacol* **l7** (1996), 513-525.

[38] G. Poorheidari, K.J. Stanhope, J.A. Pratt, Effects of the potassium channel blockers, apamin and 4-aminopyridine, on scopolamine-induced deficits in the delayed matching to position task in rats: a comparison

with the cholinesterase inhibitor E2020, *Psychopharmacology* **135** (1998), 242-255.

[39] J. Prickaerts, A. Sik, F.J. van der Staay, J. de Vente, A. Blokland, Dissociable effects of acetylcholinesterase inhibitors and phosphodiesterase type 5 inhibitors on object recognition memory: acquisition versus consolidation, *Psychopharmacology* **177** (2005), 381-390.

[40] C.A. Barnes, J. Meltzer, F. Houston, G. Orr, K. McGann, G.L. Wenk, Chronic treatment of old rats with donepezil or galantamine: effects on memory, hippocampal plasticity and nicotinic receptors, *Neuroscience* **99** (2000), 17-23.

[41] C.M. Hernandez, D.A. Gearhart, V. Parikh, E.J. Hohnadel, L.W. Davis, M.L. Middlemore, S.P. Warsi, J.L. Waller, A.V. Terry, Comparison of galantamine and donepezil for effects on nerve growth factor, cholinergic markers, and memory performance in aged rats, *J Pharmacol Exp Ther* **316** (2006), 679-694.

[42] H. Dong, C.A. Csernansky, M.V. Martin, A. Bertchume, D. Vallera, J.G. Csernansky, Acetylcholinesterase inhibitors ameliorate behavioral deficits in the Tg2576 mouse model of Alzheimer's disease, *Psychopharmacology* **181** (2005), 145-152.

[43] S. Sonkusare, K. Srinivasan, C. Kaul, P. Ramarao, Effect of donepezil and lercanidipine on memory impairment induced by intracerebroventricular streptozotocin in rats, *Life Sci* **77** (2005), 1-14.

[44] K. Yamada, M. Takayanagi, H. Kamei, T. Nagai, M. Dohniwa, K. Kobayashi, S. Yoshida, T. Ohhara, K. Takuma, T. Nabeshima, Effects of memantine and donepezil on amyloid beta-induced memory impairment in a delayed-matching to position task in rats, *Behav Brain Res* **162** (2005), 191-199.

[45] L. Spowart-Manning, F.J. van der Staay, The T-maze continuous alternation task for assessing the effects of putative cognition enhancers in the mouse, *Behav Brain Res* **151** (2004), 37-46.

[46] D. Cutuli, F. Foti, L. Mandolesi, P. De Bartolo, F. Gelfo, F. Federico, L. Petrosini, Cognitive performance of healthy young rats following chronic donepezil administration, *Psychopharmacology* **197** (2008), 661-673.

[47] M. Mesulam, E.J. Mufson, A.I. Levey, B.H. Wainer, Central cholinergic pathways in the rat: an overview based on alternative nomenclature (Ch1-Ch6), *Neuroscience* **10** (1983), 1185-1201.

[48] S. Heckers, T. Ohtake, R.G. Wiley, D.A. Lappi, C. Geula, M.M. Mesulam, Complete and selective cholinergic denervation of rat neocortex and hippocampus but not amygdala by an immunotoxin against the p75 NGF receptor, *J Neurosci* **14** (1994), 1271-1289.

[49] O. Lehmann, H. Jeltsch, C. Lazarus, L. Tritschler, F. Bertrand, J.C. Cassel, Combined 192 IgG-saporin and 5,7-dihydroxytryptamine lesions in the male rat brain: a neurochemical and behavioral study, *Pharmacol Biochem Behav* **72** (2002), 899-912.

[50] J.J. Waite, A.D. Chen, Differential changes in rat cholinergic parameters subsequent to immunotoxic lesion of the basal forebrain nuclei, *Brain Res* **918** (2001), 113-120.

[51] Waite JJ, Chen AD, Wardlow ML, Wiley RG, Lappi DA, Thal LJ (1995) 192 immunoglobulin G-saporin produces graded behavioral and biochemical changes accompanying the loss of cholinergic neurons of the basal forebrain and cerebellar Purkinje cells. Neuroscience 65, 463-476.

[52] T.J. Walsh, R.M. Kelly, K.D. Dougherty, R.W. Stackman, R.G. Wiley, C.L. Kutscher, Behavioral and neurobiological alterations induced by the immunotoxin 192-IgG-saporin: cholinergic and non-cholinergic effects following i.c.v. injection, *Brain Res* **702** (1995), 233-245.

[53] J.J. Waite, M.L. Wardlow, A.E. Power, Deficit in selective and divided attention associated with cholinergic basal forebrain immunotoxic lesion produced by 192-saporin; motoric/sensory deficit associated with Purkinje cell immunotoxic lesion produced by OX7-saporin, *Neurobiol Learn Mem* **71** (1999), 325-352.

[54] C.C. Wrenn, D.A. Lappi, R.G. Wiley, Threshold relationship between lesion extent of the cholinergic basal forebrain in the rat and working memory impairment in the radial maze, *Brain Res* **847** (1999), 284-298.

[55] A. Mattsson, S.O. Ogren, L. Olson, Facilitation of dopamine-mediated locomotor activity in adult rats following cholinergic denervation, *Exp Neurol* **174** (2002), 96-108.

[56] J.A. Vuckovich, M.E. Semel, M.G. Baxter, Extensive lesions of cholinergic basal forebrain neurons do not impair spatial working memory, *Learn Mem* **11** (2004), 87-94.

[57] M. Garcia-Alloza, N. Zaldua, M. Diez-Ariza, B. Marcos, B. Lasheras, F. Javier Gil-Bea, M.J. Ramirez, Effect of selective cholinergic denervation on the serotonergic system: implications for learning and memory, *J Neuropathol Exp Neurol* **65** (2006), 1074-1081.

[58] C.C. Wrenn, R.G. Wiley, Lack of effect of moderate Purkinje cell loss on working memory, *Neuroscience* **107** (2001), 433-445.

[59] Y. Takada, A. Yonezawa, T. Kume, H. Katsuki, S. Kaneko, H. Sugimoto, A. Akaike, Nicotinic acetylcholine receptor-mediated neuroprotection by donepezil against glutamate neurotoxicity in rat cortical neurons, *J Pharmacol Exp Ther* **306** (2003), 772-777.

[60] M. Fujiki, H. Kobayashi, S. Uchida, R. Inoue, K. Ishii, Neuroprotective effect of donepezil, a nicotinic acetylcholine-receptor activator, on cerebral infarction in rats, *Brain Res* **1043** (2005), 236-241.

[61] J. Meunier, J. Ieni, T. Maurice, Antiamnesic and neuroprotective effects of donepezil against learning impairments induced in mice by exposure to carbon monoxide gas, *J Pharmacol Exp Ther* **317** (2006), 1307-1319.

[62] Y. Takada-Takatori, T. Kume, M. Sugimoto, H. Katsuki, H. Sugimoto, A. Akaike, Acetylcholinesterase inhibitors used in treatment of Alzheimer's disease prevent glutamate neurotoxicity via nicotinic acetylcholine receptors and phosphatidylinositol 3-kinase cascade, *Neuropharmacology* **51** (2006), 474-486.

[63] M. Fujiki, T. Kubo, T. Kamida, K. Sugita, T. Hikawa, T. Abe, K. Ishii, H. Kobayashi, Neuroprotective and antiamnesic effect of donepezil, a nicotinic acetylcholine-receptor activator, on rats with concussive mild traumatic brain injury, *J Clin Neurosci* **15** (2008), 791-796.

[64] A. Marighetto, S. Valerio, A. Desmedt, J.N. Philippin, C. Trocmé-Thibierge, P. Morain, Comparative effects of the alpha7 nicotinic partial agonist, S 24795, and the cholinesterase inhibitor, donepezil, against aging-related deficits in declarative and working memory in mice, *Psychopharmacology* **197** (2008), 499-508.

[65] P. De Bartolo, F. Gelfo, L. Mandolesi, D. Cutuli, F. Foti, L. Petrosini, Effects of chronic donepezil treatment and cholinergic deafferentation on parietal pyramidal neuron morphology, *J Alzheimers Dis* **17** (2009), 177-191.

[66] R.G. Morris, P. Garrud, J.N. Rawlins, J. O'Keefe, Place navigation impaired in rats with hippocampal lesions, *Nature* **297** (1982), 681-683.

[67] L. Petrosini, M. Molinari, M.E. Dell'Anna, Cerebellar contribution to spatial event processing: Morris water maze and T-maze, *Eur J Neurosci* **8** (1996), 1882-1896.

[68] F. Federico, M.G. Leggio, P. Neri, L. Mandolesi, L. Petrosini, NMDA receptor activity in learning spatial procedural strategies II. The influence of cerebellar lesions, *Brain Res Bull* **16** (2006), 356-367.

[69] M.G. Leggio, F. Federico, P. Neri, A. Graziano, L. Mandolesi, L. Petrosini, NMDA receptor activity in learning spatial procedural strategies. I. The influence of hippocampal lesions, *Brain Res Bull* **16** (2006), 347-355.

[70] L. Mandolesi, M.G. Leggio, A. Graziano, P. Neri, L. Petrosini, Cerebellar contribution to spatial event processing: involvement in procedural and working memory components, *Eur J Neurosci* **14** (2001), 2011-2022.

[71] L. Mandolesi, M.G. Leggio, F. Spirito, L. Petrosini, Cerebellar contribution to spatial event processing: do spatial procedures contribute to formation of spatial declarative knowledge?, *Eur J Neurosci* **18** (2003), 2618-2626.

[72] J. Berger-Sweeney, S. Heckers, M.M. Mesulam, R.G. Wiley, D.A. Lappi, M. Sharma, Differential effects on spatial navigation of immunotoxin-induced cholinergic lesions of the medial septal area and nucleus basalis magnocellularis, *J Neurosci* **14** (1994), 4507-4519.

[73] M.G. Baxter, D.J. Bucci, L.K. Gorman, R.G. Wiley, M. Gallagher, Selective immunotoxic lesions of basal forebrain cholinergic cells: effects on learning and memory in rats, *Behav Neurosci* **109** (1995), 714-722.

[74] O. Lehmann, A.J. Grottick, J.C. Cassel, G.A. Higgins, A double dissociation between serial reaction time and radial maze performance in rats subjected to 192 IgG-saporin lesions of the nucleus basalis and/or the septal region, *Eur J Neurosci* **18** (2003), 651-666.

[75] K.M. Frick, J.J. Kim, M.G. Baxter, Effects of complete immunotoxin lesions of the cholinergic basal forebrain on fear conditioning and spatial learning, *Hippocampus* **14** (2004), 244-254.

[76] L. Mandolesi, P. De Bartolo, F. Foti, F. Gelfo, F. Federico, M.G. Leggio, L. Petrosini, Environmental enrichment provides a cognitive reserve to be spent in the case of brain lesion, *J Alzheimers Dis* **26** (2008), 475-480.

[77] P.T. Francis, A.M. Palmer, M. Snape, G.K. Wilcock, The cholinergic hypothesis of Alzheimer's disease: a review of progress, *J Neurol Neurosurg Psychiatry* **66** (1999), 137-147.

[78] Frisoni G.B., Canu E., Geroldi C., Brignoli B., Anglani L., Galluzzi S., V. Zacchi, O. Zanetti, Prescription patterns and efficacy of drugs for patients with dementia: physicians' perspective in Italy. *Aging Clin Exp Res* **19** (2007), 349-355.

[79] A. Musiał, M. Bajda, B. Malawska, Recent developments in cholinesterases inhibitors for Alzheimer's disease treatment, *Curr Med Chem* **14** (2007), 2654-2679.

[80] O.G. Nilsson, R.E. Strecker, A. Daszuta, A. Björklund, Combined cholinergic and serotonergic denervation of the forebrain produces severe deficits in a spatial learning task in the rat, *Brain Res* **453** (1988), 235-246.

[81] P. Riekkinen Jr, J. Sirviö, P. Riekkinen, Interaction between raphe dorsalis and nucleus basalis magnocellularis in spatial learning, *Brain Res* **527** (1990), 342-345.

[82] P. Riekkinen Jr, M. Riekkinen, J. Sirviö, R. Miettinen, P. Riekkinen, Comparison of the effects of acute and chronic ibotenic and quisqualic acid nucleus basalis lesioning, *Brain Res Bull* **27** (1991), 199-206.

[83] P. Jäkälä, M.Mazurkiewicz, J.Sirviö, P. Riekkinen Sr, P. Riekkinen Jr, The behavioral effects of serotonin synthesis inhibition and quisqualic acid induced lesions of the nucleus basalis magnocellularis in rats, *Gen Pharmacol* **24** (1993), 1141-1148.

[84] O. Lehmann, H. Jeltsch, O. Lehnardt, L. Pain, C. Lazarus, J.C. Cassel, Combined lesions of cholinergic and serotonergic neurons in the rat brain using 192 IgG-saporin and 5,7-dihydroxytryptamine: neurochemical and behavioural characterization, *Eur J Neurosci* **12** (2000), 67-79.

[85] M.D. Lindner, J.B. Hogan, D.B. Hodges Jr, A.F. Orie, P. Chen, J.A. Corsa, J.E. Leet, K.W. Gillman, G.M. Rose, K.M. Jones, V.K. Gribkoff, Donepezil primarily attenuates scopolamine-induced deficits in psychomotor function, with moderate effects on simple conditioning and attention, and small effects on working memory and spatial mapping, *Psychopharmacology* **188** (2006), 629-640.

[86] F.J. van der Staay, P. Bouger, O. Lehmann, C. Lazarus, B. Cosquer, J. Koenig, V. Stump, J.C. Cassel, Long-term effects of immunotoxic cholinergic lesions in the septum on acquisitionof the cone-field task and noncognitive measures in rats, *Hippocampus* **16** (2006), 1061-1079.

[87] T. Kosasa, Y. Kuriya, K. Matsui, Y. Yamanishi, Effect of donepezil hydrochloride (E2020) on basal concentration of extracellular acetylcholine in the hippocampus of rats, *Eur J Pharmacol* **380** (1999a), 101-107.

[88] T. Kosasa, Y. Kuriya, Y. Yamanishi, Effect of donepezil hydrochloride (E2020) on extracellular acetylcholine concentration in the cerebral cortex of rats, *Jpn J Pharmacol* **81**(1999b), 216-222.

[89] E. Shearman, S. Rossi, B. Szasz, Z. Juranyi, S. Fallon, N. Pomara, H. Sershen, A. Lajtha, Changes in cerebral neurotransmitters and metabolites induced by acute donepezil and memantine administrations: a microdialysis study, *Brain Res Bull* **9** (2006), 204-213.

[90] Q. Chang, P.E. Gold, Impaired and spared cholinergic functions in the hippocampus after lesions of the medial septum/vertical limb of the diagonal band with 192 IgG-saporin, *Hippocampus* **14** (2004),170-179.

[91] T. Masuoka, C. Kamei, The role of nicotinic receptors in the amelioration of cholinesterase inhibitors in scopolamine-induced memory deficits, *Psychopharmacology* **206** (2009), 259-65.

[92] M.C. Buhot, P. Rage, L. Segu, Changes in exploratory behaviour of hamsters following treatment with 8-hydroxy-2-(di-n-propylamino) tetralin, *Behav Brain Res* **35** (1989), 163-179.

[93] E. Save, B. Poucet, N. Foreman, M.C. Buhot, Object exploration and reactions to spatial and nonspatial changes in hooded rats following damage to parietal cortex or hippocampal formation, *Behav Neurosci* **106** (1992), 447-456.

[94] L. Ricceri, Behavioral patterns under cholinergic control during development: lessons learned from the selective immunotoxin 192 IgG saporin, *Neurosci Biobehav Rev* **27** (2003), 377-384.

[95] P. De Bartolo, D. Cutuli, L. Ricceri, F. Gelfo, F. Foti, D. Laricchiuta, M.L. Scattoni, G. Calamandrei, L. Petrosini, Does Age Matter? Behavioral and Neuro-anatomical Effects of Neonatal and Adult Basal Forebrain Cholinergic Lesions, *J Alzheimers Dis* (2010, *in press*).

[96] G.M. Gilad, The stress-induced response of the septo-hippocampal cholinergic system: a vectorial outcome of psychoneuroendocrinological interactions, *Psychoneuroendocrinology* **12** (1987), 167-184.

[97] T. Mizuno, F. Kimura, Attenuated stress response of hippocampal acetylcholine release and adrenocortical secretion in aged rats, *Neurosci Lett* **222** (1997), 49-52.

[98] K. Mizoguchi, M. Yuzurihara, A. Ishige, H. Sasaki, T. Tabira, Effect of chronic stress on cholinergic transmission in rat hippocampus, *Brain Res* **915** (2001), 108-111.

[99] R.J. Sutherland, I.Q. Whishaw, J.C. Regehr, Cholinergic receptor blockade impairs spatial localization by use of distal cues in the rat, *J Comp Physiol Psychol* **96** (1982), 563-573.

[100] L.B. Day, T. Schallert, Anticholinergic effects on acquisition of place learning in the Morris water task: spatial mapping deficit or inability to inhibit nonplace strategies?, *Behav Neurosci* **110** (1996), 998-1005.

[101] P.J. Whitehouse, Cholinergic therapy in dementia, *Acta Neurol Scand Suppl* **149** (1993), 42-45.

[102] R.J. Perry, J.R. Hodges, Attention and executive deficits in Alzheimer's disease. A critical review, *Brain* **122** (1999), 383-404.

[103] A.D. Baddeley, H.A. Baddeley, R.S. Bucks, G.K. Wilcock, Attentional control in Alzheimer's disease. *Brain* **124** (2001), 1479-1481.

[104] N.J. Cohen, H. Eichenbaum, B.S. Deacedo, S. Corkin, Different memory systems underlying acquisition of procedural and declarative knowledge, *Ann N Y Acad Sci* **444** (1985), 54-71.

[105] V. Boulougouris, J.W. Dalley, T.W. Robbins, Effects of orbitofrontal, infralimbic and prelimbic cortical lesions on serial spatial reversal learning in the rat, *Behav Brain Res* **179** (2007), 219-228.

[106] P. De Bartolo, M.G. Leggio, L. Mandolesi, F. Foti, F. Gelfo, F. Ferlazzo, L. Petrosini, Environmental enrichment mitigates the effects of basal forebrain lesions on cognitive flexibility, *Neuroscience* **154** (2008), 444-453.

[107] W.S. Chen, F.K. Wong, P.F. Chapman, D.J. Pemberton, Effect of donepezil on reversal learning in a touch screen-based operant task, *Behav Pharmacol* (2009, in press).

[108] S.M. Cabrera, C.M. Chavez, S.R. Corley, M.R. Kitto, A.E. Butt, Selective lesions of the nucleus basalis magnocellularis impair cognitive flexibility, *Behavioral Neuroscience* **120** (2006), 298-306.

[109] A.M. Bailey, M.L. Rudisill, E.J. Hoof, M.L. Loving, 192 IgG-saporin lesions to the nucleus basalis magnocellularis (nBM) disrupt acquisition of learning set formation, *Brain Res* **969** (2003), 147-159.

[110] A.E. Butt, M.M. Noble, J.L. Rogers, T.E. Rea, Impairments in negative patterning, but not simple discrimination learning, in rats with 192 IgG-saporin lesions of the nucleus basalis magnocellularis, *Behav Neurosci* **116** (2002), 241-255.

[111] A.E. Butt, J.A. Schultz, L.L. Arnold, E.E. Garman, C.L. George, P.E. Garraghty, Lesions of the rat nucleus basalis magnocellularis disrupt appetitive-to-aversive transfer learning, *Integr Physiol Behav Sci* **38** (2003), 253-271.

[112] A.A. Chiba, D.J. Bucci, P.C. Holland, M. Gallagher, Basal forebrain cholinergic lesions disrupt increments but not decrements in conditioned stimulus processing, *J Neurosci* **15** (1995), 7315-7322.

[113] J. Turchi, M. Sarter, Cortical acetylcholine and processing capacity: effects of cortical cholinergic deafferentation on crossmodal divided attention in rats, *Cogn Brain Res* **6** (1997), 147-158.

[114] G.A. Park, B.A. Pappas, S.M. Murthaa, A. Ally, Enriched environment primes forebrain choline acetyltransferase activity to respond to learning experience, *Neurosci Lett* **143** (1992), 259-262.

[115] D.D. Rasmusson, The role of acetylcholine in cortical synaptic plasticity, *Behav Brain Res* **115** (2000), 205-218.

[116] T.W. Robbins, The 5-choice serial reaction time task: behavioural pharmacology and functional neurochemistry, *Psychopharmacology* **63** (2002), 362-380.

[117] K.A. Baskerville, J.B. Schweitzer, P. Herron, Effects of cholinergic depletion on experience-dependent plasticity in the cortex of the rat, *Neuroscience* **80** (1997), 1159-1169.

[118] M.P. Kilgard, M.M. Merzenich, Plasticity of temporal information processing in the primary auditory cortex, *Nat Neurosci* **1** (1998), 727-731.

[119] T.M. Gill, M. Sarter, B. Givens, Sustained visual attention performance-associated prefrontal neuronal activity: evidence for cholinergic modulation, *J Neurosci* **20** (2000), 4745-4757.

[120] J.W. Dalley, R.N. Cardinal, T.W. Robbins, Prefrontal executive and cognitive functions in rodents: neural and neurochemical substrates, *Neurosci Biobehav Rev* **28** (2004), 771-784.

[121] A. Blokland, J. Jolles, Spatial learning deficit and reduced hippocampal ChAT activity in rats after an ICV injection of streptozotocin, *Pharmacol Biochem Behav* **44** (1993), 491-494.

[122] P.E. Gold, Acetylcholine modulation of neural systems involved in learning and memory. Neurobiol Learn Mem **80** (2003), 194-210.

[123] J. Lemière, D. Van Gool, R. Dom, Treatment of Alzheimer's disease: an evaluation of the cholinergic approach, *Acta Neurol Belg* **99** (1999), 96-106.

[124] M.A. Taffe, M.R. Weed, L.H. Gold, Scopolamine alters rhesus monkey performance on a novel neuropsychological test battery, *Brain Res Cogn Brain Res* **8** (1999), 203-212.

[125] B.J. Everitt, T.W. Robbins, Central cholinergic systems and cognition, *Annu Rev Psychol* **48** (1997), 649-684.

[126] M. Sarter, B. Givens, J.P. Bruno, The cognitive neuroscience of sustained attention: where top-down meets bottom-up, *Brain Res Brain Res Rev* **35** (2001), 146-160.

[127] A. Akaike, Preclinical evidence of neuroprotection by cholinesterase inhibitors, *Alzheimer Dis Assoc Disord* **20** (2006), 8-11.

[128] S. Akasofu, M. Kimura, T. Kosasa, K. Sawada, H. Ogura, Study of neuroprotection of donepezil, a therapy for Alzheimer's disease, *Chem Biol Interact* **175** (2008), 222-6.

[129] G. Pepeu, M.G. Giovannini, Cholinesterase inhibitors and beyond, *Curr Alzheimer Res* **6** (2009), 86-96.

[130] C. Scali, F. Casamenti, A. Bellucci, C. Costagli, B. Schmidt, G. Pepeu, Effect of subchronic administration of metrifonate, rivastigmine and donepezil on brain acetylcholine in aged F344 rats, *J Neural Transm* **109** (2002), 1067-1080.

[131] X. Zhang, J.Y. Tian, A.L. Svensson, Z.H. Gong, B. Meyerson, A. Nordberg, Chronic treatments with tacrine and (-)-nicotine induce different changes of nicotinic and muscarinic acetylcholine receptors in the brain of aged rat, *J Neural Transm* **109** (2002), 377-392.

[132] R.T. Reid, M.N. Sabbagh, Effects of cholinesterase inhibitors on rat nicotinic receptor levels in vivo and in vitro, *J Neural Transm* **115** (2008), 1437-1444.

[133] Y. Takada-Takatori, T. Kume, Y. Ohgi, T. Fujii, T. Niidome, H. Sugimoto, A. Akaike, Mechanisms of alpha7-nicotinic receptor up-regulation and sensitization to donepezil induced by chronic donepezil treatment, *Eur J Pharmacol* **590** (2008a), 150-6.

[134] Y. Takada-Takatori, T. Kume, Y. Ohgi, Y. Izumi, T. Niidome, T. Fujii, H. Sugimoto, A. Akaike (b). Mechanism of neuroprotection by donepezil pretreatment in rat cortical neurons chronically treated with donepezil, *J Neurosci Res* **86** (2008b), 3575-83.

[135] V. Antonini, O. Prezzavento, M. Coradazzi, A. Marrazzo, S. Ronsisvalle, E. Arena, G. Leanza, Anti-amnesic properties of (+/-)-PPCC, a novel sigma receptor ligand, on cognitive dysfunction induced by selective cholinergic lesion in rats, *J Neurochem* **109** (2009), 744-54.

# Subject Index

| | |
|---|---|
| acetylcholine | 297 |
| acetylcholinesterase inhibition | 3 |
| acetylcholinesterase inhibitors | 317 |
| aged dogs | 149 |
| aging | 3, 39, 109 |
| Alzheimer | 179, 277 |
| Alzheimer disease | 3, 39, 49, 59, 89, 139, 163, 297 |
| Alzheimer's disease dementia | 263 |
| aminoguanidine | 215 |
| amygdala | 49 |
| amyloid | 129, 179, 247 |
| amyloid-beta | 39, 139, 149, 163 |
| amyloid-β 42 | 89 |
| amyloid-β protein precursor (AβPP) | 263 |
| amyloid-β oligomer species | 263 |
| animal models | 39 |
| antibody | 247 |
| antioxidants | 15 |
| apoptosis | 139 |
| astrocyte | 215 |
| attention deficit disorder | 3 |
| Aβ amyloid | 49 |
| beagle | 15 |
| behavioral enrichment | 15 |
| Behavioral testing | 317 |
| beta-amyloid | 15, 77, 201 |
| Cdk4 | 139 |
| cell cycle | 139 |
| cerebellum | 77 |
| cerebral amyloid angiopathy | 247 |
| cerebral cortex | 77 |
| cholesterol | 59, 163 |
| cholinergic neurotransmission | 317 |
| clinical trial | 247 |
| cognition | 39 |
| cognitive flexibility | 317 |
| coronary artery disease | 59 |
| delay paradigm | 77 |
| delayed matching | 3 |
| delivery | 277 |
| diabetes mellitus | 179 |
| 3D images | 149 |
| dosage | 277 |
| *Drosophila* | 89 |
| drug development | 3 |
| electron microscopy | 149 |
| estrogen | 109 |
| eyeblink classical conditioning | 77 |
| forebrain cholinergic system | 317 |
| frontotemporal dementia | 49 |
| genetic screen | 89 |
| gliosis | 39 |
| glycogen synthase kinase-3 | 201 |
| high fat diet | 179 |
| high-molecular weight | 263 |
| hippocampus | 49, 77, 109, 215 |
| 27-hydroxycholesterol | 163 |
| 192 IgG-saporin | 317 |
| immunization | 247 |
| immunocytochemistry | 149 |
| immunotherapy | 15 |
| inflammation | 247 |
| insulin | 201 |
| insulin resistance | 179 |
| intranasal | 277 |
| learning | 15, 297 |
| learning and memory | 89 |
| LTP | 109, 297 |
| memory | 15 |
| Mnesic functions | 317 |
| mouse | 109 |
| muscarinic | 297 |
| neurodegeneration | 39, 89, 179 |
| neurofibrillary tangles | 49 |
| neuron loss | 15 |
| neuronal loss | 77 |
| neuroprotection | 297 |
| NGF | 277 |
| nicotinic acetylcholine receptors | 3 |
| nitrosative stress | 215 |
| NMDA | 297 |
| non-alcoholic steatohepatitis | 179 |

| | |
|---|---|
| non-human primates | 3 |
| non-invasive | 277 |
| obesity | 179 |
| ocular | 277 |
| oxidative stress | 39 |
| Parkinson's disease | 49 |
| PCNA | 139 |
| pharmacokinetic | 277 |
| phospho-histone H3 | 139 |
| phospho-Rb | 139 |
| polyphenols | 263 |
| progesterone | 109 |
| protein misfolding | 89 |
| rabbit | 59, 163 |
| rotarod | 129 |
| S100B | 215 |
| SAMP8 | 39 |
| side effects | 277 |
| sporadic Alzheimer disease | 201 |
| streptozotocin | 201, 215 |
| synaptic plasticity | 109 |
| synaptic transmission | 297 |
| tau phosphorylation | 39 |
| tau | 49, 77, 129, 201 |
| tetracycline-inducible | 139 |
| therapeutic window | 277 |
| trace paradigm | 77 |
| transgenic 2576 mice | 201 |
| transgenic animal models | 89 |
| transgenic mice | 129, 139 |
| water maze | 129 |
| working memory | 3 |
| 3xTg AD | 109 |
| 3xTg-AD mice | 149 |
| Y-maze | 129 |

# Author Index

| | | | | | |
|---|---|---|---|---|---|
| Achaval, M. | 215 | Iijima, K. | 89 |
| Agelan, A. | 77 | Iijima-Ando, K. | 89 |
| Agostinho, P. | 139 | Ittner, L.M. | 49 |
| Alamed, J. | 129 | Janle, E. | 263 |
| Amato, G. | 277 | Jiao, P. | 179 |
| Anderson, W. | 297 | Johnson, H. | 297 |
| Baudry, M. | 109 | Kim, H.-J. | 39 |
| Biasibetti, R. | 215 | Kopitz, J. | 201 |
| Bickford, P. | 129 | Kotchabhakdi, N. | 149 |
| Blurton-Jones, M. | 139 | Laferla, F.M. | 139 |
| Brinton, R.D. | 109 | Laricchiuta, D. | 317 |
| Buccafusco, J.J. | 3 | Lawton, M. | 179 |
| Camins, A. | 39 | Lee, H.-g. | 39 |
| Capsoni, S. | 277 | Leite, M.C. | 215 |
| Casadesus, G. | 39 | Lewis, J. | 129 |
| Cattaneo, A. | 277 | Longato, L. | 179 |
| Chang, J. | 39 | Lopes, J.P. | 139 |
| Chen, L.H. | 263 | Lyn-Cook, L.E. | 179 |
| Cheng, A. | 263 | Mandolesi, L. | 317 |
| Choi, D.-Y. | 297 | Mark, P. | 179 |
| Coico, R. | 231 | Marwarha, G. | 163 |
| Colton, C.A. | 247 | McGowan, E. | 129 |
| Cotman, C.W. | 15 | Morgan, D. | 129 |
| Covaceuszach, S. | 277 | Munireddy, S. | 129 |
| Cutuli, D. | 317 | Nuntagij, P. | 149 |
| De Bartolo, P. | 317 | O'Callaghan, M. | 297 |
| de la Monte, S.M. | 179 | Ono, K. | 263 |
| DeLeon, J. | 129 | Pallas, M. | 39 |
| Diamond, D.M. | 129 | Pasinetti, G.M. | 263 |
| Dickstein, D.L. | 263 | Percival, S.S. | 263 |
| Drever, B. | 297 | Perry, G. | 39 |
| Ferruzzi, M. | 263 | Petrosini, L. | 317 |
| Foti, F. | 317 | Plaschke, K. | 201 |
| Foy, M.R. | 109 | Platt, B. | 297 |
| Gelfo, F. | 317 | Prasanthi, R.P.J. | 163 |
| Ghribi, O. | 163 | Quincozes-Santos, A. | 215 |
| Gonçalves, C.-A. | 215 | Riedel, G. | 297 |
| Gordon, M.N. | 129 | Riedererf, P. | 201 |
| Götz, J. | 49 | Rodrigues, L. | 215 |
| Head, E. | 15 | Salkovic-Petrisic, M. | 201 |
| Ho, L. | 263 | Schliebs, R. | 201 |
| Hoyer, S. | 201 | Schonrock, N. | 49 |
| Humala, N. | 263 | Seo, S. | 297 |
| Hutton, M. | 129 | Siegelin, M. | 201 |

| | | | |
|---|---|---|---|
| Silbermann, E. | 179 | Valle, L.D. | 77 |
| Smith, M.A. | 39 | Vignone, D. | 277 |
| Sparks, D.L. | 59 | Vissel, B. | 49 |
| Spirito, F. | 277 | Wands, J.R. | 179 |
| Stefanini, B. | 277 | Wang, J. | 263 |
| Swarowsky, A. | 215 | Wilcock, D.M. | 247 |
| Talcott, S.T. | 263 | Woodruff-Pak, D.S. | 77, 231 |
| Teplow, D. | 263 | Xu, H. | 179 |
| Thompson, R.F. | 109 | Yamasaki, T.R. | 139 |
| Tong, M. | 179 | Zhao, W. | 263 |
| Torp, R. | 149 | Zhu, X. | 39 |
| Ugolini, C. | 277 | | |